Mycological Studies Honoring John N. Couch

Mycological Studies Honoring John N. Couch

edited by William J. Koch

THE UNIVERSITY OF NORTH CAROLINA PRESS • CHAPEL HILL
1968

Mycological Studies Honoring John N. Couch, presented as a special publication by The University of North Carolina Press, also appears as the first issue of Volume 84 of ***The Journal of the Elisha Mitchell Scientific Society*** (Spring, 1968).

To John N. Couch

Preface

The Editorial Board of the *Journal of the Elisha Mitchell Scientific Society* is pleased to dedicate this issue to John Nathaniel Couch, Kenan Professor of Botany Emeritus, at The University of North Carolina at Chapel Hill. As Dr. Couch approached his mandatory retirement last summer, the members of his department began to consider a suitable means of recognizing his extensive and valuable contributions to mycology and to the development of botanical research and education at the University of North Carolina for more than a half-century. They decided upon a *Festschrift* contributed by his former students and close mycological colleagues throughout the world. Although other journals and independent presentation as a book were also considered, publishing the *Festschrift* as an issue of this *Journal* seemed particularly appropriate, since Professor Couch has contributed so greatly to its quality and status, both as its Editor between 1945 and 1960 and as author of many distinguished papers that have appeared on its pages. The *Journal* is happy to be the vehicle for publication of the *Festschrift* and to participate in honoring Professor Couch. We believe that this issue will be a substantial addition to the literature of mycology, and that Dr. Couch will be pleased to have it dedicated to him.

This special issue was made possible by the ready response of those invited to contribute to it and the notable effort of Dr. William J. Koch, of the University of North Carolina Botany Department, who served as guest editor. Thanks are also due to Mr. Lambert Davis, Director of The University of North Carolina Press, for his cooperation in making this issue available also as a book, thus providing wider distribution in permanent form.

The response from the mycologists invited to contribute was exceptional. Those who were unable to participate were sincerely regretful, and several contributed funds to help in the publication of the issue. We regard this as evidence of the esteem with which Dr. Couch is held by his students, colleagues, and friends.

The sequence of the articles is based on current classification theory, the purpose being to put the articles in a rough approximation of the order in which Professor Couch would have discussed these articles in his two-semester advanced mycology course.

The University of North Carolina Press, as publisher of this issue as a book, will hold a copyright. However, contributors will have no difficulty in securing permission to reproduce part or all of their papers, and no special permission is required for the usual limited quotations in other scientific papers.

Dr. Couch's many friends and colleagues will be pleased to know that, while technically retired and no longer teaching, he is in excellent health and is pursuing his investigations as diligently and effectively as ever.

Victor A. Greulach, Editor
Journal of the Elisha Mitchell Scientific Society

Spring, 1968

Contents

The Career of John Nathaniel Couch

Leland Shanor
Department of Botany, University of Florida, Gainesville, Fla.

For his researches on the fungi, for the reputation he has gained as a teacher, and for his influence on the development of the Department of Botany and the University in general, Professor John Nathaniel Couch is recognized as one of the truly outstanding men who have served on the faculty of the University of North Carolina. His modesty and sincerity have earned for him the respect and affection both of students and colleagues at the University and of his professional associates in the United States and abroad. Dr. Couch and the Department of Botany of the University have become inseparably associated in the minds of botanists throughout the world and in the fond recollections of the several generations of students who have studied botany with him.

John Couch enrolled at the University in 1917 as an undergraduate transfer student to complete his college work for the baccalaureate degree; he studied under the exacting and stimulating direction of Professor William Chambers Coker for both the Master of Arts and the Doctor of Philosophy degrees; he joined the faculty of the Department in 1922, and has served his Alma Mater in a very distinguished manner for a period of four and a half decades. Over these years his talents have found their outlet for service to the University in a variety of ways: as a teacher of botany to both undergraduate and graduate students; as an enthusiastic, stimulating research director and wise personal counselor for graduate students who have worked under his direction; and as an able chairman of his department. He has served the University well as a member of a number of important committees and boards and formulated many academic policies for the University over some critical years of its development. It is in appreciation of his years of distinguished service that this volume is dedicated.

Professor Couch is internationally recognized for research contributions that have added significantly to knowledge of the fungi. When he was elected to membership in the National Academy of Sciences in 1943 in recognition of his research accomplishments, he became the second member of the University of North Carolina faculty to receive this honor. His investigations have covered a broad range of mycological research—structure, reproduction, evolution, and culture—and have included the study of representatives of most of the major groups. All of his research is characterized by meticulous attention to detail. Dr. Couch has published almost eighty research papers and two books, *The Genus Septobasidium* and *The Gasteromycetes of the Eastern United States and Canada*, co-authored with W. C. Coker.

Professor Couch's earliest interests centered on the aquatic Phycomycetes. His studies for the doctorate on *Dictyuchus*, in which he reported the first case of heterothallism in any of the water molds, continues to be considered among his most significant contributions. His early observations on this organism have provided the basis not only for the subsequent discovery of sex hormones in aquatic fungi by one of his students, but also for the further study, by his students and others, of the significance of sex hormones and their role in sexual reproduction in the Saprolegniales. In recognition of his discovery of heterothallism in *Dictyuchus*, Dr. Couch was awarded a postdoctoral National Research Council Fellowship to work for one year under Dr. A. F. Blakeslee at Cold Spring Harbor and for a second year under Dr. B. M. Duggar at the Missouri Botanical Garden.

During the tenure of this National Research Council Fellowship, Dr. Couch had the opportunity to spend two months in Jamaica in the British West Indies with the Johns Hopkins Botanical Group. It was during this time that he became thoroughly fascinated with species of the genus *Septobasidium*, a group of fungi prevalent there on living trees and shrubs. Others who had previously observed these fungi had regarded them as essentially harmless to the host plant, or had looked upon them as possibly being of some benefit since they were thought, prior to his studies, to overgrow and destroy the numerous scale insects invariably associated with the fungal growth. Dr. Couch, however, was able to demonstrate that a complicated partnership existed between the fungus colony and the scale insects associated with it; that their survival together was actually damaging to the plants on which they occurred. The details of his observations, accumulated over several years of study and announced in 1929, immediately attracted wide interest. He was invited to report upon his findings before the International Botanical Congress in Cambridge, England, the following year and again at the next meeting of the Congress in Amsterdam in 1935. His monumental contribution on the biology and taxonomy of this group, *The Genus Septobasidium*, published in 1938, brought further recognition. He was awarded the Walker Prize of the Boston Society of Natural History in 1939, "given in recognition of the most outstanding contribution in the field of Natural History during the preceeding five years." By his own collecting and from those collections sent to him by botanical correspondents around the world, Dr. Couch has demonstrated that the genus is widely distributed in nature on a variety of hosts, not only in the tropics but also in the temperate zone.

His general and continuing interest in fungi parasitic on or associated with insects, nematodes, rotifers, and other organisms is evidenced by his contributions on nematode-catching fungi and *Coelomomyces*, which parasitizes mosquito larvae, and by other studies. This enthusiasm for ascertaining the biological significance of different associations among fungi and other organisms in nature has stimulated and encouraged not only some of his own students but also other biologists to pursue studies of entomogenous fungi.

In 1949, Professor Couch announced in this journal the discovery of a new group—later established by him as a new family, Actinoplanaceae—of organisms related to *Streptomyces* and *Micromonospora* and belonging to the ray fungi, the Actinomycetales. The last fifteen years of his research activities have been substantially, though far from exclusively, concerned with a thorough investigation of their morphology, cytology, culture, and relationships to higher fungi and to the bacteria. He reported on conditions that determine formation of sporangia in this group to the 1950 meeting of the North Carolina Academy of Science and subsequently was invited to deliver the invitation lecture to the Fall Meeting of the North Carolina State College Chapter of Sigma Xi in 1952 to report upon his discoveries. A year later, he was selected to present the Fourth Annual Lecture of the Mycological Society of America for the meeting of the Society held at Madison, Wisconsin, on September 8, 1953. He was invited to present a paper on the Symposium on the Taxonomy of the Actinomycetes at the annual meetings of the Society of American Bacteriologists held at Detroit on April 28, 1957. He has appeared also on the seminar program of biological sciences departments of a number of institutions and has published a series of journal papers based on his observations on these organisms.

In addition to these major areas of research, Professor Couch has contributed significantly to knowledge of numerous species of phycomycetous orders of fungi, particularly the Chytridiales, Lagenidiales, and Blastocladiales. He has maintained also a lively interest in the algae, for many of these simple aquatic fungi live as parasites of a variety of algal species. He considered observations on only one alga sufficiently complete to be worthy of publication, his study of gametogenesis in *Vaucheria*, which appeared in the *Botanical Gazette* in 1932.

Although Professor Couch has devoted his research efforts primarily to studies in the area of fundamental research, he has had an interest in practical aspects of mycological research as well, including the mode of transmission of leaf-spot fungi as they affect the growth of peanuts, the significant role of fungi in soil and water, and the potentialities of employing *Coelomomyces* in the biological control of mosquitoes. The U.S. Public Health Service has provided substantial support for his studies on *Coelomomyces*.

Professor Couch has always been interested in teaching and has devoted himself unselfishly to instruction, both in lecturing and in laboratory teaching, of undergraduate classes and to the careful training of graduate students. His students quickly learn to admire and respect his enthusiasm for botany, especially for mycology. His devotion to his own scholarship, his insistence upon the acquisition of a background of information for sound conclusions, and his stress upon careful and meticulous attention to detail have invariably been impressed upon all new graduate students early in their graduate study. He has always been a firm believer in the importance of laboratory study in the biological sciences, for to him there is no adequate substitute for practical experience in the training of future biologists.

Professor Couch has been most conspicuously successful with upper division undergraduates and with graduate students. He has taught by example and is at his best with a group of perhaps four to ten students. In the period of his most concentrated research efforts, as at present, many were the evenings when he was still at his microscope when the last graduate student had decided to call it a day. Time was of no consequence when he wanted to observe a critical stage in the life cycle of some lowly chytrid or of one of the more elaborate water molds. He would sit far into the night and watch continuously in order to make an observation that he regarded essential but had not been able to make during the more normal working hours. Indeed, many of his graduate students will recall finding, to their amazement, upon arriving at the laboratory early in the morning that Dr. Couch had remained at his post all night. His laboratory was a work area both for himself and for his students. He was always close at hand to check student findings and would call students to peer through his microscope to share in some exciting event which he happened to be observing at the time.

Professor Couch has enjoyed taking his small classes or individual students on field trips to now famous local collecting places such as the Meeting-of-the-Waters, Battle Park, and Morgan Creek. He has always been delightful on such excursions, pointing out in rapid sequence not only the fungi along the way but also the special collecting places for algae, liverworts, mosses, and representatives of other plant groups. His devotion to the search for new organisms and for new facts concerning those better known have served as an inspiration to his students, whether in the field or in the laboratory.

Some of the practical jokes students have been known to play are an indication of his informal, harmonious relationship with students. On one occasion a member of a class "planted" a green porcelain door knob considerably off the path through Battle Park, but in plain sight of a field-trip party, just to enjoy the Professor's reaction when some other knowing student pointed to the *"Russula virescens."* After the specimen was collected and the group had been gathered around, and following a contemplative examination, Dr. Couch expounded on the inaccuracy of the determination and explained the morphological shortcomings of the "specimen" at hand. On another occasion, probably on the the first day of April, a student placed a blinker attachment on his microscope lamp. When he turned on the light it began alternately to turn on and off. At first he was puzzled and thought that perhaps the bulb was about to burn out. He then became intrigued with the regularity of what was happening and upon close scrutiny discovered the cause. At first a little chagrined, he promptly became quite amused. He enjoyed the fun with his students, for he possesses a sense of humor that is a delight to those who know him.

His success as a teacher is best indicated by the accomplishment of many of his students, twenty-seven of whom have received the Master's degree under his supervision and fifteen the doctorate. These students have held or are currently occupying important positions at numerous research institutions and universities, including many of the most outstanding ones. In recognition of his accomplishments as a teacher, Professor Couch was selected in 1955 to receive the Meritorious Teaching Award of the Association of Southeastern Biologists.

Professor Couch's interest in teaching, however, has not been limited to college and graduate students alone. He served two years as a high school teacher, first at Chapel Hill and later at Charlotte, North Carolina (1919–21). His observations during this period are recorded in the *High School Journal* for 1922 under the title "Science in the High School. A review of science teaching in the high schools of North Carolina for 1920–21." He has encouraged students who expressed a desire to enter high school teaching to do so, if he felt that they had a talent for it. He early arrived at the conviction that many more good science teachers were needed at this level of education.

The personal background and the influences which have contributed to Professor Couch's successful career as an investigator and teacher are diverse and broad. He was born in Prince Edward County, Virginia, on October 12, 1896. He was one of seven children of John and Sally Terry Couch. Since the elder John Couch was a Baptist minister, the family moved from time to time, and son John attended seven different schools in his public school career. He was graduated in 1914 from the Durham (North Carolina), High School and entered Trinity College, later to become Duke University, in 1914. He remained there for the first three years of his college career. Both in high school and at Trinity College, he studied mostly such classical subjects as language, literature, history, and mathematics. During this period he read widely, sometimes to the seeming neglect of regularly assigned work.

What he should follow as a career was not at all clear to him at this time, but law and medicine seemed to be foremost in his considerations. It was at the beginning of his third year at Trinity that he decided in favor of medicine and therefore enrolled in both biology and chemistry, his first exposure to natural science. Botany, taught by Professor J. L. Wolfe in small classes and by the laboratory method, opened to him an entirely new vista. He was invited at the beginning of the second semester of that year to join the Biology Journal Club which met two times each month. The first indication of what was to become the area of his life's work appeared at this time when he chose as the subject of his first report to the Club, "Edible and Poisonous Fungi." He so enjoyed his work in botany during the junior year at Trinity that he requested the privilege of working in Professor Wolfe's laboratory during the following summer, a request that was generously granted.

In the fall of 1917 at the beginning of his senior year, Couch transferred to the University of North Carolina to satisfy the premedical requirements, for he was expecting to enroll in the University Medical School. It was at Chapel Hill that he came under the influence of Professor W. C. Coker and decided to become a botanist instead of pursuing a career in medicine. While he was in military service in World War I, he studied for four months in 1919 at L'Université de Nancy in France, and in the summer of 1923 he studied at the University of Wisconsin with Professor E. M. Gilbert and Professor C. E. Allen. But for these two brief periods, all of his graduate education in botany was taken at Chapel Hill with Professor Coker. In 1922, Dr. Coker appointed him to an instructorship in the department. From 1925–1927, he was away from the University on postdoctoral study at Cold Spring Harbor and at the Missouri Botanical Garden. He was named Assistant Professor in 1927 on his return to Chapel Hill, was promoted to Associate Professor in 1929, and was made Professor in 1932. In 1945, he was designated Kenan Professor of Botany in recognition of his significant contributions, both in research and in teaching. Upon the retirement of Dr. Coker in 1944, he became Chairman of the Department of Botany and served in this capacity until 1960, when he asked to be relieved of administrative duties.

During his career as a member of the faculty of the University of North Carolina, Professor Couch received many honors for his work and as tribute to his abilities. In addition to those already cited, he was appointed a Special Adviser to the Chairman, Office of Scientific Research and Development in 1944. In 1955, he was elected an Honorary Foreign Member of the National Academy of Science of India. He was named, in 1961, to serve as a consultant in India to the Review Committee of the Indian University Grants Commission. In 1964, he was named to the North Carolina Governor's Science Advisory Committee, and in 1966 he was designated Consultant, U.S. Public Health Service Communicable Disease Center in Atlanta.

Professor Couch was given the Jefferson Medal and the Poteat Award of the North Carolina Academy of Science in 1937, and was the recipient of the first Gold Medal Science Award of the State of North Carolina, in 1964. He was one of fifty botanists chosen to receive the Golden Jubilee Merit Citation by the Botanical Society of America in 1956, being cited as one "whose studies of the small, the intricate, and the odd among fungi and their relatives have come to fructification in the vivid, the significant, and the delectable." Catawba College conferred an honorary Sc.D. degree on Dr. Couch in 1946 and Duke University the D.Sc. in 1965.

Professor Couch has been elected to numerous offices in the professional organizations to which he has belonged. He was president (1943) of the Mycological Society of America; a fellow, vice-president, and chairman (1962) of Section G (Botany), American Association for the Advancement of Science; chairman, Southeastern Section (1951); vice-president (1964), Botanical Society of America; president (1946–47) of the North Carolina Academy of Sciences; and president (1937–38) of the Elisha Mitchell Scientific Society. He served for fifteen years, 1946–1961, as editor of the *Journal of the Elisha Mitchell Scientific Society* and was associate editor of *Mycologia* from 1937–39. He serves on the editorial board of *Mycopathologia et Mycologia Applicata.*

In all of his professional and personal endeavors Professor Couch has had the interest, encouragement, support, and able assistance of his wife, the former Else Dorothy Ruprecht. Mrs. Couch, a graduate of Wellesley College with a major in zoology, is very talented in foreign languages and over the years has provided valuable aid not only to Dr. Couch but also to his graduate students by her many translations of research papers and articles from French and German into English. She also possesses artistic talents and has inked a number of the illustrations for Dr. Couch's early publications.

Professor and Mrs. Couch have two children. John Philip graduated from the University of North Carolina at Chapel Hill and later obtained his doctorate from Yale University. He is now on the faculty of the Department of Romance Languages of the University of North Carolina at Greensboro, thus pursuing his mother's interests in linguistics. Sally Louise (Mrs. John M. Vilas) lives in Chapel Hill with her husband, their two daughters, and their son. Graduate students recall with pleasure visits to the home of Professor and Mrs. Couch, and several generations of them delight in the memory of the antics of their growing children.

PUBLICATIONS OF JOHN NATHANIEL COUCH

1920. (with W. C. Coker) A new species of *Achlya.* Jour. Elisha Mitchell Sci. Soc. **36**: 100–101.

1922. Science in the high school. A review of science teaching in the high schools of North Carolina for 1920-1921. The High School Journal. **5**: 211–216.

1923. (with W. C. Coker) A new species of *Thraustotheca.* Jour. Elisha Mitchell Sci. Soc. **39**: 112–115.

1924. (with W. C. Coker) Revision of the genus *Thraustotheca* with a description of a new species. Jour. Elisha Mitchell Sci. Soc. **40**: 197–202.

1924. Some observations on spore formation and discharge in *Leptolegnia, Achlya,* and *Aphanomyces.* Jour. Elisha Mitchell Sci. Soc. **40**: 27–42.

1924. A dioecious water mold (*Dictyuchus monosporus*). Jour. Elisha Mitchell Sci. Soc. **40**: 116.

1925. A new dioecious species of *Choanephora.* Jour. Elisha Mitchell Sci. Soc. **41**: 141–150.

1926. Heterothallism in *Dictyuchus,* A genus of the water molds. Annals of Botany. **40**: 849–881.

1926. Notes on the genus *Aphanomyces,* with a description of a new semi-parasitic species. Jour. Elisha Mitchell Sci. Soc. **41**: 213–227.

1927. Some new water fungi from the soil, with observations on spore formation. Jour. Elisha Mitchell Sci. Soc. **42**: 227–242.

1928. (with W. C. Coker) The Gasteromycetes of the Eastern United States and Canada. University of North Carolina Press, Chapel Hill, N.C. 201 pp.

1929. A monograph of *Septobasidium.* Part I. Jour. Elisha Mitchell Sci. Soc. **44**: 242–260.

1931. Observations on some species of water molds connecting *Achlya* and *Dictyuchus.* Jour. Elisha Mitchell Sci. Soc. **46**: 225–230.

1931. *Micromyces zygogonii* Dang., parasitic on *Spirogyra.* Jour. Elisha Mitchell Sci. Soc. **46**: 231–239.

1931. The biological relationship between *Septobasidium retiforme* (B. & C.) Pat. and *Aspidiotus osborni* New and Ckll. Quart. Jour. Micros. Sci. **74**: 383–437.

1931. The biological relationship between *Septobasidium* and scale insects. (Abst.) Proc. of the V. Intern. Bot. Congress, Cambridge, 1930. Pp. 369–370.

1932. *Rhizophidium, Phlyctochytrium,* and *Phlyctidium* in the United States. Jour. Elisha Mitchell Sci. Soc. **46**: 245–260.

1932. Gametogenesis in *Vaucheria.* Bot. Gaz. **94**: 272–296

1932. The development of the sexual organs in *Leptolegnia caudata.* Amer. Jour. Bot. **19**: 584–599.

1933. Basidia of *Septobasidium* (*Glenospora*) *Curtisii.* Jour. Elisha Mitchell Sci. Soc. **49**: 156–162.

1935. New or little known Chytridiales. Mycologia **27**: 160–175.

1935. *Septobasidium* in the United States. Jour. Elisha Mitchell Sci. Soc. **51**: 1–77.

1935. A new saprophytic species of *Lagenidium,* with notes on other forms. Mycologia **27**: 376–387.

1935. An incompletely known chytrid: *Mitochytridium ramosum.* Jour. Elisha Mitchell Sci. Soc. **51**: 293–296.

1936. Structure of *Septobasidium* in relation to the association with scale insects. (Abst.) Proc. VI. Intern. Bot. Cong., Amsterdam. **2**: 154–156.

1937. Notes on the genus *Micromyces.* Mycologia **29**: 592–596.

1937. A new fungus intermediate between the rusts and *Septobasidium.* Mycologia **29**: 665–673.

1937. The formation and operation of the traps in the nematode-catching fungus, *Dactylella bembicodes* Drechsler. Jour. Elisha Mitchell Sci. Soc. **53**: 301–309.

1938. The Genus *Septobasidium.* University of North Carolina Press, Chapel Hill. 302 pp.

1939. A new chytrid on *Nitella: Nephrochytrium stellatum.* Amer. Jour. Bot. **25**: 507–511.

1939. Observations on cilia of aquatic Phycomycetes. (Abst.) Science **88**: 476.

1939. A new species of *Chytridium* from Mountain Lake, Va. Jour. Elisha Mitchell Sci. Soc. **54**: 256–259.

1939. A new *Conidiobolus* with sexual reproduction. Amer. Jour. Bot. **26**: 119–130.

1939. Technique for collection, isolation, and culture of chytrids. Jour. Elisha Mitchell Sci. Soc. **55**: 208–214.

1939. Heterothallism in the Chytridiales. Jour. Elisha Mitchell Sci. Soc. **55**: 409–414.

1939. Further studies on infection of scale insects by *Septobasidium*. (Abst.) Third Intern. Cong. for Microbiol., Abstracts of Communications. Pp. 217–218.

1939. (with Jean Leitner and Alma Whiffen) A new genus of the Plasmodiophoraceae. Jour. Elisha Mitchell Sci. Soc. **55**: 399–408.

1940. Notes on *Septobasidium* from Mexico. (Abst.) Jour. Elisha Mitchell Sci. Soc. **56**: 223–224.

1941. A new *Uredinella* from Ceylon. Mycologia **33**: 405–410.

1941. The structure and action of the cilia in some aquatic Phycomycetes. Amer. Jour. Bot. **28**: 704–713.

1941. (with Vera K. Charles, J. G. Harrar, and J. J. McKelvey, Jr.) A fungous parasite of the mealy bug. (Abst.) Phytopathology **31**: 5.

1942. Studies of lower fungi with particular reference to sex and nutrition (Abst.) Amer. Phil. Soc. Year Book 1941: 153–154.

1942. A new fungus on crab eggs. Jour. Elisha Mitchell Sci. Soc. **58**: 158–162.

1942. (with Alma J. Whiffen) Observations on the genus *Blastocladiella*. Amer. Jour. Bot. **29**: 582–591.

1943. The rediscovery of *Nadsonia*, a yeast with heterogamic conjugation. (Abst.) Jour. Elisha Mitchell Sci. Soc. **59**: 118.

1944. The yeast *Nadsonia* in America. Jour. Elisha Mitchell Sci. Soc. **60**: 11–16.

1945. Observations on the genus *Catenaria*. Mycologia **37**: 163–193.

1945. Revision of the genus *Coelomomyces* parasitic in insect larvae. Jour. Elisha Mitchell Sci. Soc. **61**: 124–136.

1945. Review of J. S. Karling: *Simple Holocarpic Biflagellate Phycomycetes*. Mycologia **37**: 794–795.

1945. Review of J. C. Gilman: *A Manual of Soil Fungi*. Science **102**: 385.

1946. Two species of *Septobasidium* from Mexico with unusual insect houses. Jour. Elisha Mitchell Sci. Soc. **62**: 87–94.

1947. (with H. R. Dodge) Further observations on *Coelomomyces*, parasitic on mosquito larvae. Jour. Elisha Mitchell Sci. Soc. **63**: 69–79.

1948. A new genus of the Auriculariaceae. (Abst.) Jour. Elisha Mitchell Sci. Soc. **64**: 169.

1949. A new species of *Ancylistes* on a saccoderm desmid. Jour. Elisha Mitchell Sci. Soc. **65**: 131–136.

1949. A new group of organisms related to *Actinomyces*. Jour. Elisha Mitchell Sci. Soc. **65**: 315–318.

1949. The taxomony of *Septobasidium polypodii* and *S. album*. Mycologia **41**: 427–441.

1950. *Actinoplanes*, a new genus of the Actinomycetales. Jour. Elisha Mitchell Sci. Soc. **66**: 87–92.

1951. Further observations on the motile cells of *Actinoplanes*. (Abst.) Jour. Elisha Mitchell Sci. Soc. **67**: 176–177.

1951. Review of Arney Robinson Childs, ed., The Private Journal of Henry William Ravenel, *in* The Garden of the New York Botanical Garden **1**: (Nov.-Dec. 1951) 188.

1952. (with W. J. Koch) Further studies on the motile spores of *Actinoplanes*. (Abst.) Jour. Elisha Mitchell Sci. Soc. **68**: 138.

1953. The occurrence of thin-walled sporangia in *Physoderma zeaemaydis* on corn in the field. Jour. Elisha Mitchell Sci. Soc. **69**: 182-184.

1953. A new form connecting *Actinoplanes* with *Streptomyces*. (Abst.) Jour. Tenn. Acad. Sci. **28**: 178.

1953. (with E. Kathleen Goldie-Smith and W. J. Koch) A demonstration of gross cultural characters, sporangia, and swimming spores in the new genus *Actinoplanes*. (Abst.) Jour. Tenn. Acad. Sci. **28**: 178–179.

1954. The genus *Actinoplanes* and its relatives. Trans. New York Acad. Sci. **16**: 315–318.

1954. (with Velma D. Matthews) William Chambers Coker. Mycologia **46**: 372–383.

1954. (with H. R. Totten) William Chambers Coker. Jour. Elisha Mitchell Sci. Soc. **70**: 116–118.

1955. A new genus and family of the Actinomycetales, with a revision of the genus *Actinoplanes*. Jour. Elisha Mitchell Sci. Soc. **71**: 148–155.

1955. Actinosporangiaceae should be Actinoplanaceae. Jour. Elisha Mitchell Sci. Soc. **71**: 269.

1955. The family Actinoplanaceae. Bacteriological Reviews **19**: 272.

1957. Actinoplanaceae. Bergey's Manual of Determinative Bacteriology (7th ed.) Williams and Wilkins Co., Baltimore. Pp. 825–829.

1957. A new horizon in soil microbiology. Proceedings of the Nat. Acad. of Sci. (India), Allahabad, 37, Section B, Part II, 69–73.

1958. Taxonomic criteria in Actinoplanaceae. Jour. Elisha Mitchell Sci. Soc. **74**: 95.

1959. (with C. E. Miller) Lyophilization of the Actinoplanaceae. Mycologia **51**: 146–150.

1960. Some fungal parasites of mosquitoes. *In* Biological Control of Insects of Medical Importance. Amer. Inst. of Bio. Sci., Washington. Pp. 35–48.

1961. (with C. J. Umphlett) Germination of the resting sporangium of *Coelomomyces*. (Abst.) ASB Bull. **8**: 33.

1962. Validation of the family Coelomomycetaceae and certain species and varieties of *Coelomomyces*. Jour. Elisha Mitchell Sci. Soc. **78**: 135–138.

1962. (with W. J. Koch) Induction of motility in the spores of some Actinoplanaceae (Actinomycetales). (Abst.) Science **138**: 987.

1963. Some new genera and species of the Actinoplaceae. Jour. Elisha Mitchell Sci. Soc. **79**: 53–70.

1963. (with C. J. Umphlett) *Coelomomyces* infections, *in* E. A. Steinhaus, ed., *Insect Pathology, an Advanced Treatise.* Academic Press, New York. Pp. 150–187.

1964. The name *Ampullaria* Couch has been replaced by *Ampullariella.* International Bulletin of Bacterial Nomenclature and Taxonomy, **14**: 137.

1964. A proposal to replace the name *Ampullaria* Couch with *Ampullariella.* Jour. Elisha Mitchell Sci. Soc. **80**: 29.

1967. Sporangial germination in *Coelomomyces punctatus* and infection of *Anopheles quadrimaculatus.* Joint U.S.-Japan Seminar on "Microbial Control of Insect Pests."

1967. (with C. E. Bland) Observations on the chromatinic bodies of two strains of the Actinoplanaceae. (Abst.) Jour. Elisha Mitchell Sci. Soc. **83**: 171–172.

Ultrastructure of *Pilimelia anulata* (Actinoplanaceae)[1]

Charles E. Bland[2]
Department of Botany, University of North Carolina at Chapel Hill, N. C.

Sporangial structure and development of several genera of Actinoplanaceae (Actinomycetales) have been studied by Rancourt and Lechevalier (1964), Lechevalier and Holbert (1965), and Lechevalier, Lechevalier, and Holbert (1966). However, little has been reported concerning the ultrastructural aspects of either the hyphae or the sporangia of members of this family. This report describes the ultrastructural features of the hyphae and sporangia of a keratinophilic member of the Actinoplanaceae, *Pilimelia anulata.*

Materials and Methods

Stock cultures of *P. anulata* Kane (Kane, 1966), maintained in continuous culture on peptone Czapek's agar, were provided for study by Miss Wilma Kane. Colonies of the organism were removed from agar culture, inoculated into sterile distilled water containing samples of white human hair, and allowed to stand at room temperature for 1 to 2 weeks or until sporangial production was abundant. Heavily infested hairs were then fixed and embedded for electron microscopy.

Numerous small segments of infested hair about 2 mm. long were fixed for 2 hours in 3% glutaraldehyde in 0.1 M cacodylate buffer at pH 7.2. This material was post-fixed for 3 hours in Kellenberger's standard fixative (Kellenberger, Ryter, and Séchaud, 1958). Following this, the material was washed for 3 hours in Kellenberger buffer containing 0.5% uranyl acetate and placed in distilled water overnight. After dehydration with ethyl alcohol, the material was placed in propylene oxide for two changes of 15 minutes each and embedded in Araldite 6005 (R. F. Cargille Lab. Inc., Cedar Grove, N.J.) according to the method of Luft (1961). Sections were made with a Porter-Blum MT-1 microtome equipped with a diamond knife. The sections were floated on 5% ethyl alcohol and mounted on bare grids. Sections were then stained for 30 minutes with 2% aqueous uranyl acetate. Mounts were examined with a Zeiss EM-9 electron microscope at instrumental magnifications ranging from 7,000 to 40,000. Photographs were taken on Kodak Kodalith LR film.

Observations

Colonies of *P. anulata* were found to be composed of two hyphal types: the substrate or vegetative hyphae, and the palisade or sporangium-forming hyphae. This was as reported by Couch (1963) for other members of the Actinoplanaceae. The vegetative hyphae were distributed throughout the inner matrix of the infested hairs (Fig. 1). Closer observation revealed the hyphae to be highly branched and irregularly catenulate (Fig. 2).

The nuclear material, which consisted of small fibers embedded within a less dense matrix, occupied the central portion of the hyphae (Figs. 3, 4). An intracytoplasmic membranous component was frequently encountered in the substrate hyphae (Figs. 3, 4). This membranous component was usually closely associated with the nuclear material of the substrate hyphae (Fig. 4).

The substrate hyphae were observed to give rise to the palisade or sporangium-forming hyphae by simple branches through the outer scales of the hair (Fig. 5). These palisade hyphae did not branch and were terminated characteristically by a sporangium.

In cross-section (Fig. 6) the sporangia were seen as hexagonal structures containing parallel rows of round spores. The spores measured 0.31 to 0.39 μ in diameter.

In longitudinal section (Fig. 7), the sporangia contained parallel rows of elongate spores which measured about 0.85 μ x 0.32 μ. These chains of spores were seen to arise as branches from a source at the base of the sporangium (Figs. 7, 8). There were indications (Fig. 8) that the sporangial wall is a continuation of the outer layer of the wall of the palisade hyphae and that the wall of the spores is formed from the inner layer of the palisade hyphal wall. This is evidenced by the similarity in structure between the wall of the spores and the inner layer of the wall of the palisade hypha (Fig. 8). The process of sporangial formation appeared to be as described previously by

---

[1] This report constitutes a portion of a thesis to be submitted to the University of North Carolina in partial fulfillment of the requirements for the Ph.D. degree in Botany.

[2] N.D.E.A. Title IV fellow.

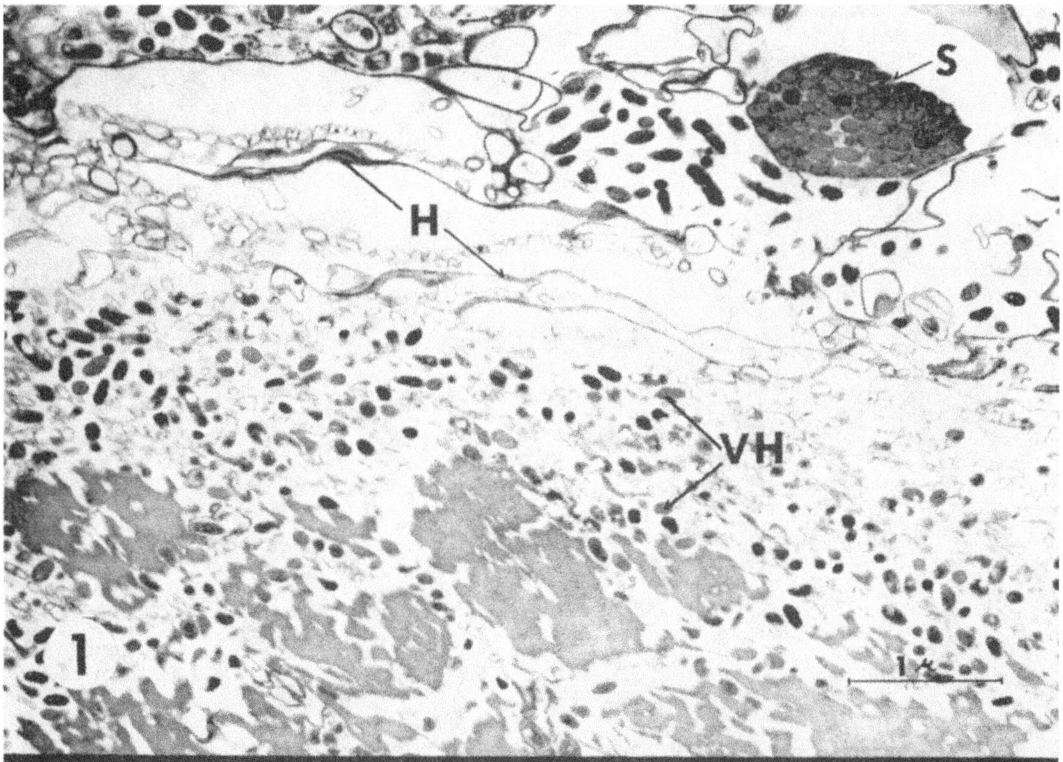

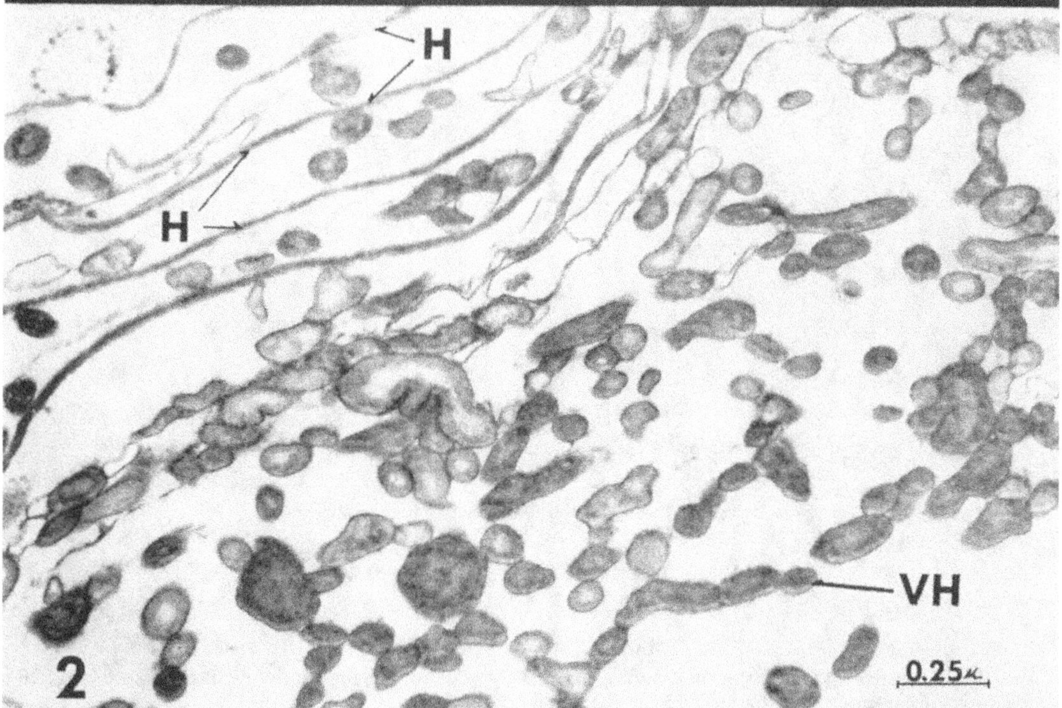

FIG. 1. Cross-section of a hair heavily infested with *P. anulata*, showing the distribution of the vegetative hyphae (VH) within the hair. Note the mature sporangium (S) lying just outside the surface scales of the hair (H).

FIG. 2. Cross-section of hair showing branching of vegetative hyphae (VH) just below the surface scales of the hair (H).

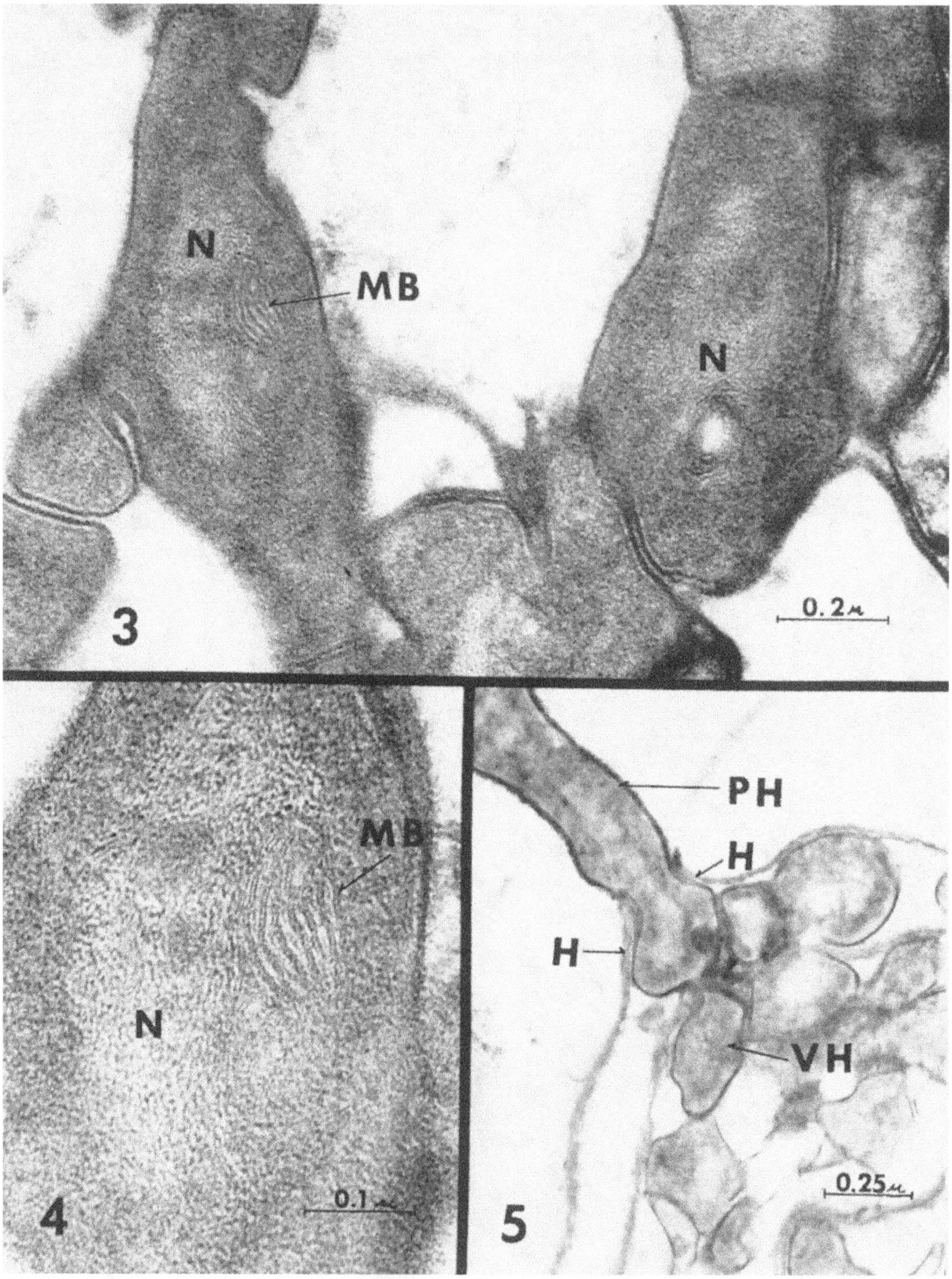

FIG. 3. Vegetative hyphae as seen growing inside hair. An intracytoplasmic membranous body (MB) can be seen closely appressed to the nuclear material (N).

FIG. 4. Enlargement of a portion of Fig. 3 to illustrate nuclear material (N) and the membranous body (MB).

FIG. 5. Vegetative hypha (VH) branching through the surface scales of the hair (H) to form a palisade hypha (PH).

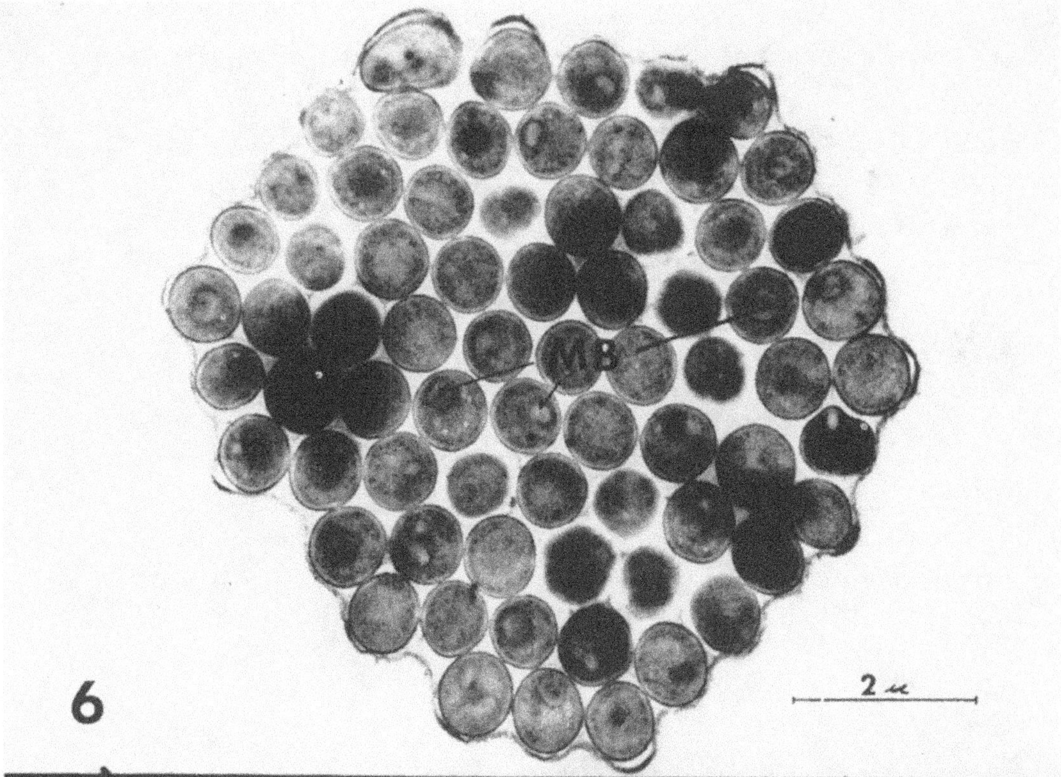

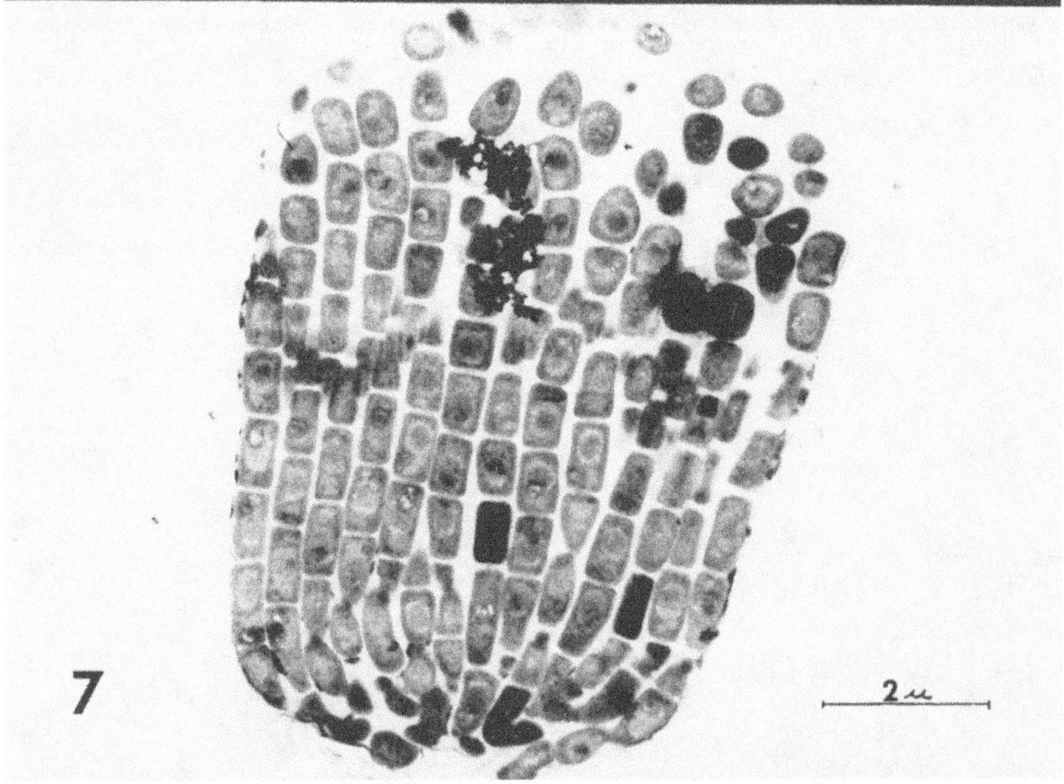

FIG. 6. Cross-section of a mature sporangium. Note the hexagonal shape of the sporangium which contains parallel rows of seemingly round spores. Membranous bodies (MB) can be seen in some of the spores.

FIG. 7. Longitudinal section through a mature sporangium. Note the branching of the chains of spores to form parallel rows within the sporangium.

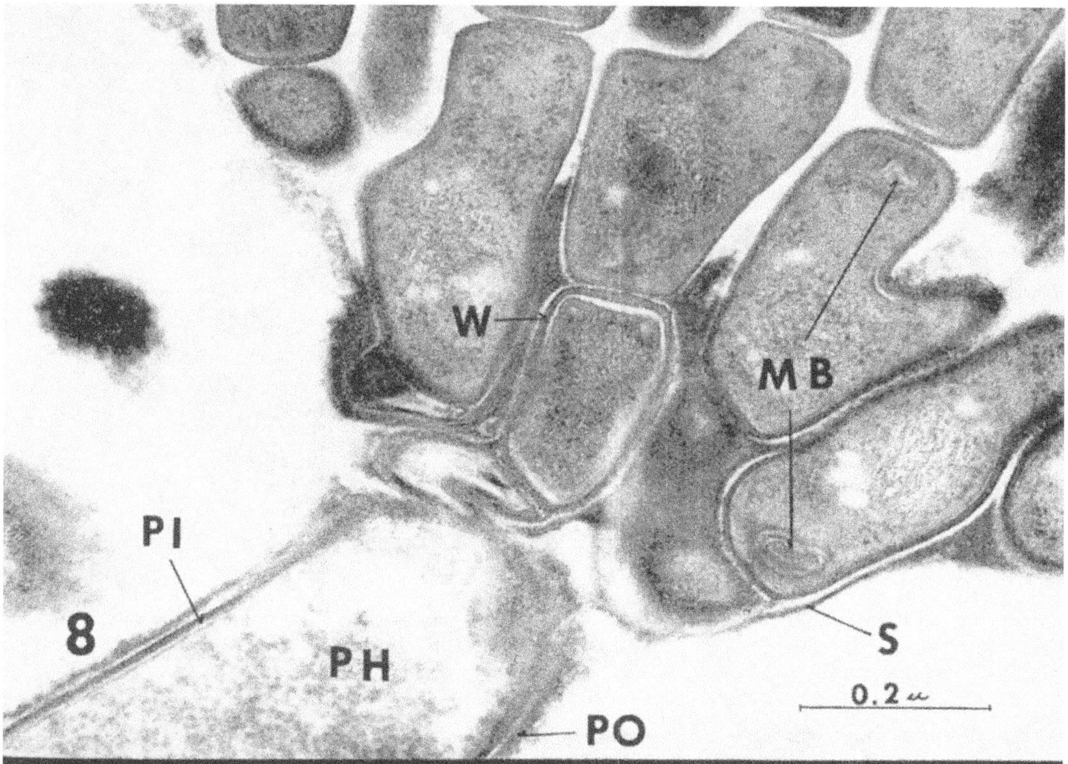

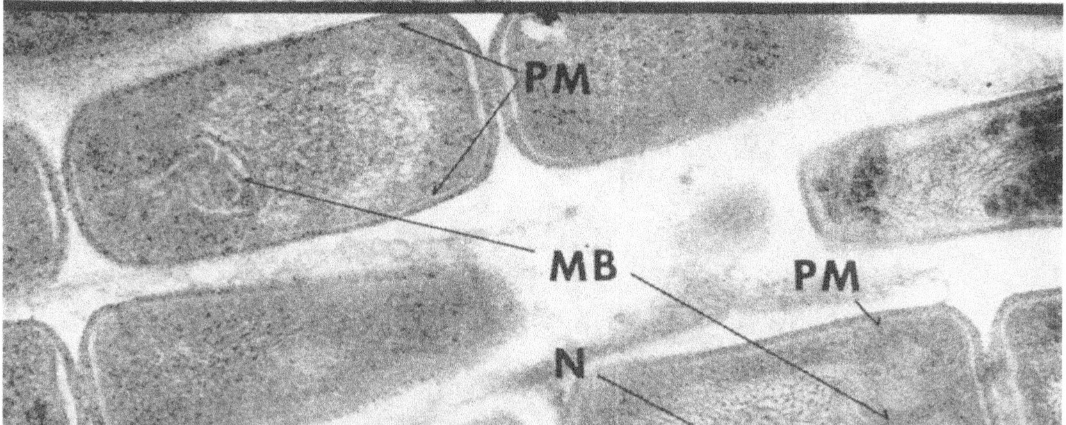

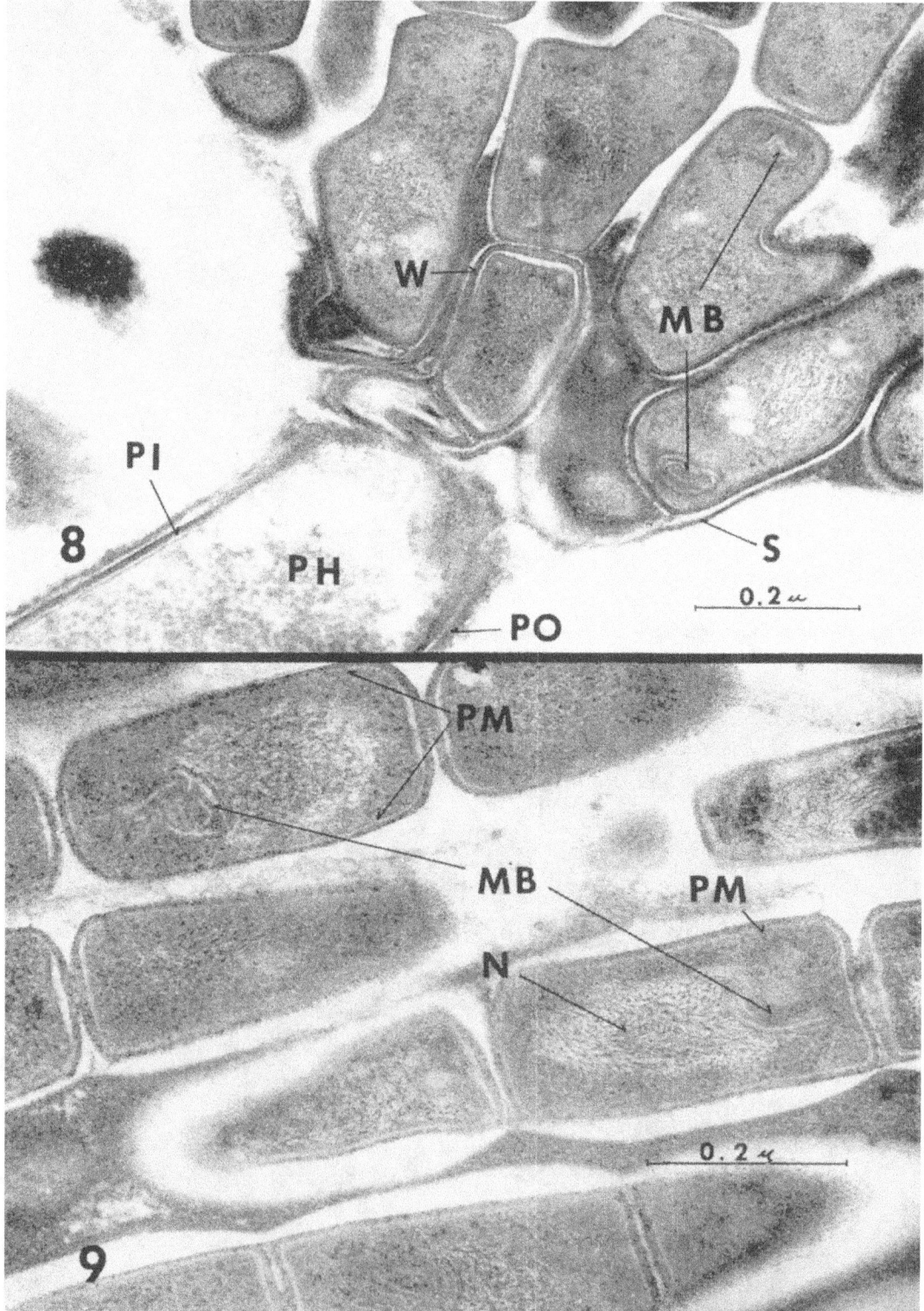

Fig. 8. Section through the base of a sporangium. The sporangial wall (S) appears to be a continuation of the outer wall (PO) of the palisade hypha (PH) with the wall of the spores (W) being formed by the inner wall (PI) of the palisade hypha. MB, membranous body.

Fig. 9. Longitudinal section through mature spores. The membranous bodies (MB) can be seen closely appressed to the finely fibrillar nuclear material (N). PM, plasma membrane.

Lechevalier et al. (1966) for other members of the Actinoplanaceae.

The nuclear material of the spores, like that of the substrate hyphae, consisted of a finely fibrous material which occupied generally a central region within the spores (Fig. 9). The nuclear region made up about 35 per cent of the internal contents of each spore.

A single intracytoplasmic membranous component was present in almost every spore sectioned (Figs. 8, 9, 10). This membranous component consisted of a coiled system of mem-

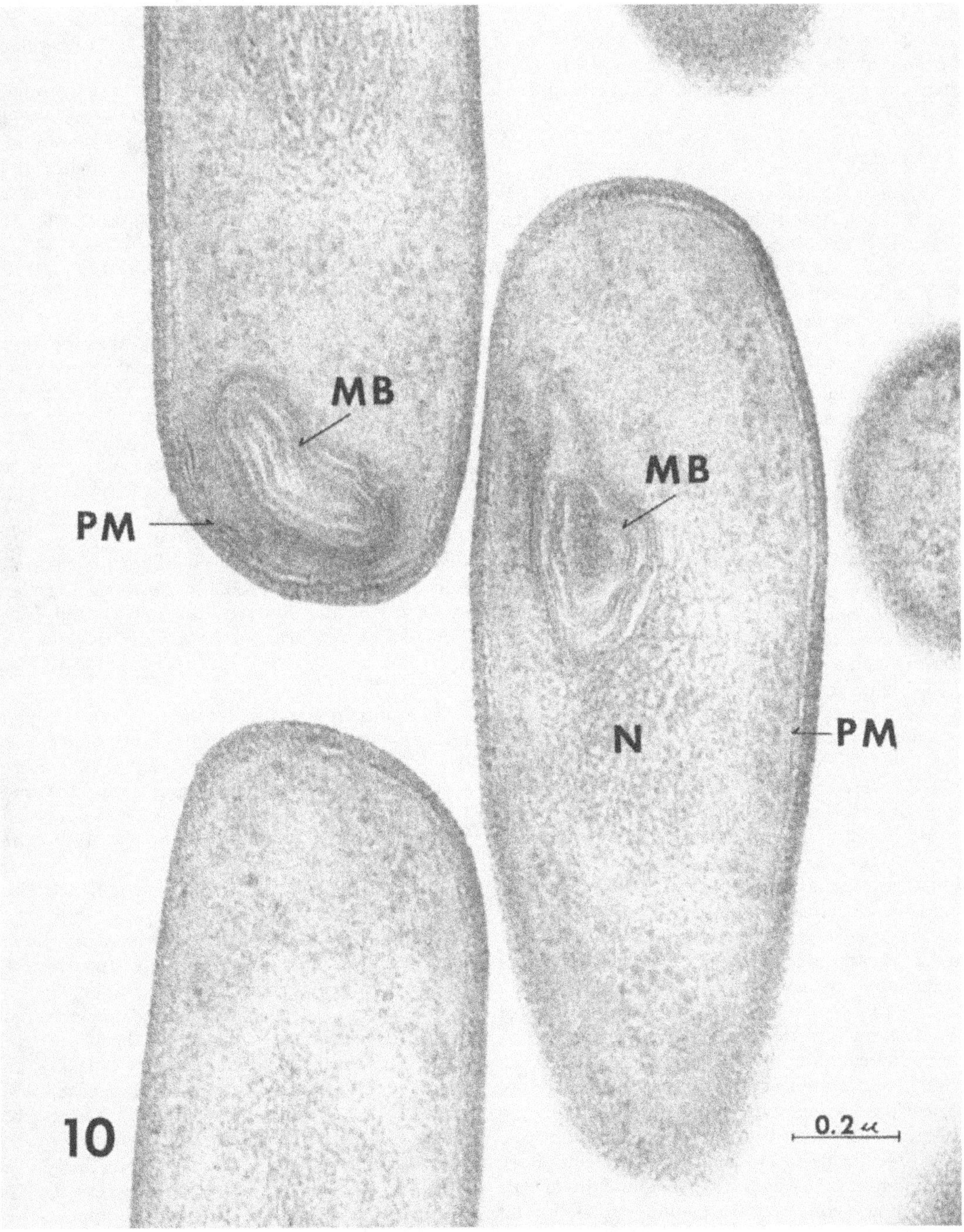

FIG. 10. Longitudinal section through mature spores illustrating apparent origin of membranous body (MB) from the plasma membrane (PM). N, nuclear material.

branes which in size and structure resembled closely the plasma membrane of the spore. Indications were that the system of membranes within the spores arose from the plasma membrane (Figs. 9, 10). Each membrane consisted of two dense layers, each of which was approximately 1 to 2 mμ thick, separated by a less dense layer of about 2 to 3 mμ. The membranous component was usually located near one end of the elongate spores and was most often closely appressed to the nuclear material (Fig. 10).

Discussion

Although Couch (1954) considered at one time that the Actinoplanaceae might represent a connecting link between the bacteria and the lower fungi, he placed the family in the bacterial order Actinomycetales (Couch, 1950). It is clearly evident through cell-wall studies (Becker, Lechevalier, and Lechevalier, 1965, Yamaguchi, 1965; and Szaniszlo, 1967), developmental studies (Lechevalier et al., 1966), nuclear staining properties (Bland and Couch, unpublished data), and now fine structural observations that these organisms are closely related to the Actinomycetales and thus to the eubacteria (Cummins and Harris, 1958). The lack of an organized nucleus with a discrete nuclear membrane, the method of flagellation, and the lack of various intracytoplasmic organelles tyical of the cells of higher organisms clearly align the Actinoplanaceae with the bacteria and not with the lower fungi.

The nuclear material of *P. anulata,* consisting of a fine fibrous network embedded in a less dense material, closely resembles that described for *Streptomyces sp.* by Hopwood and Glauert (1960). However, the nuclear material described by Stuart (1959) for *Streptomyces noursei* or by Moore and Chapman (1959) for a streptomycete showed little similarity to that of *P. anulata.* These differences in the structure of the nuclear material may be caused by differences in fixation procedures. It appears, however, that the structure of the nuclear material of *P. anulata* is very similar to the now almost standard image of bacterial nuclear material—a network of small fibers embedded within a less dense matrix.

The intracytoplasmic membrane system of *P. anulata,* which was here observed for the first time in a member of the Actinoplanaceae, is a significant component of the cytoplasm since membranous bodies were found in practically every spore sectioned and were abundant in the hyphae. Indeed, it appears that the membrane system of *P. anulata* is as extensive as that described for *Streptomyces coelicolor* (Glauert and Hopwood, 1960). The sizes given by Glauert and Hopwood for the unit membranes of *S. coelicolor* compare favorably with the size of the membranes of *P. anulata.* Because of the similarity in size and structure, it seems likely that the intracytoplasmic membranes of *P. anulata* are comparable to the mesosomes described for bacilli by Fitz-James (1960).

Several possible functions for the membranous bodies of bacteria have been suggested. Because of the proximity of the membranous bodies of *P. anulata* to the nuclear material, a function in division of the nuclear material as described for *Bacillus subtilis* by Ryter and Jacob (1964) might be considered likely. It is also possible that this body could function in genetic transformation as described by Wolstenholme, Vermeulen, and Venema (1966) for *B. subtilis.* Other investigators have proposed that the functions of membranous bodies include: carrying the respiratory enzyme system (Mitchell and Moyle, 1956), and containing enzymes for synthesis of new cell-wall material (Glauert, Brieger, and Allen, 1961). Whether the membranous bodies of *P. anulata* serve all, one, or none of these functions is unknown. However, because of its extensive membrane system, *P. anulata* along with other members of the Actinoplanaceae would probably serve as excellent subjects for the study of the nature and function of the membranous bodies of bacteria.

Summary

The ultrastructural features of the hyphae and sporangia of a keratinophilic member of the Actinoplanaceae, *Pilimelia anulata,* were studied with electron microscopy. Colonies growing on hair were found to be composed of two hyphal types: substrate or vegetative hyphae, and palisade or sporangium-forming hyphae. The substrate hyphae, which branched throughout the inner matrix of the hair, contained numerous intracytoplasmic membranous bodies which were often closely appressed to the finely fibrillar nuclear material.

The sporangia, which appeared to be formed by a process similar to that described by Lechevalier, Lechevalier, and Holbert (1966) for other members of the Actinoplanaceae, contained parallel rows of cylindrical spores that originated from a source at the base of the sporangium. A single membranous body was observed in almost every spore sectioned. The membranous bodies of the spores appeared to arise from the plasma membrane and then to become closely associated with the nuclear region. Possible functions of the membranous bodies of *P. anulata* were considered in the light of functions proposed by other investigators for similar structures.

ACKNOWLEDGMENTS

The author wishes to thank Dr. J. N. Couch for directing this work; Miss Wilma Kane for furnishing cultures of *P. anulata*; Dr. P. J. Szaniszlo, Dr. W. J. Koch, and Dr. C. J. Umphlett for reviewing the manuscript; and Dr. D. P. Costello and Dr. D. W. Misch for making the Zeiss EM-9 electron microscope available.

LITERATURE CITED

BECKER, B., M. P. LECHEVALIER, AND H. A. LECHEVALIER. 1965. Chemical composition of cell-wall preparations from strains of various form-genera of aerobic Actinomycetes. Applied Microbiol. **13**: 236–243.

COUCH, J. N. 1950. *Actinoplanes*, a new genus of the Actinomycetales. Jour. Elisha Mitchell Sci. Soc. **66**: 87–92.

———. 1954. The genus *Actinoplanes* and its relatives. Trans. New York Acad. Sci. **16**: 315–318.

———. 1963. Some new genera and species of the Actinoplanaceae. Jour. Elisha Mitchell Sci. Soc. **79**: 54–70.

CUMMINS, C. S., AND H. HARRIS. 1958. Studies on the cell-wall composition and taxonomy of Actinomycetales and related groups. J. Gen. Microbiol. **18**: 173–189.

FITZ-JAMES, P. C. 1960. Participation of the cytoplasmic membrane in the growth and spore formation of bacilli. J. Biophys. Biochem. Cytol. **8**: 507–528.

GLAUERT, A. M., AND D. A. HOPWOOD. 1960. The fine structure of *Streptomyces coelicolor*, I. The membrane system. J. Biophys. Biochem. Cytol. **7**: 479–488.

GLAUERT, A. M., AND E. M. BRIEGER, AND J. M. ALLEN. 1961. The fine structure of the vegetative cells of *Bacillus subtilis*. Exp. Cell Res. **22**: 73.

HOPWOOD, D. A., AND GLAUERT, A. M. 1960. The fine structure of *Streptomyces coelicolor*. II. The nuclear material. J. Biophys. Biochem. Cytol. **8**: 267.

KANE, W. D. 1966. A new genus of Actinoplanaceae, *Pilimelia*, with a description of two species, *Pilimelia terevasa* and *Pilimelia anulata*. Jour. Elisha Mitchell Sci. Soc. **82**: 220–230.

KELLENBERGER, E. A., A. RYTER, AND J. SÉCHAUD. 1958. Electron microscope study of DNA-containing plasms. II. Vegetative and mature phage DNA compared with normal bacterial nucleoids in different physiological states. J. Biophys. Biochem. Cytol. **4**: 671.

LECHEVALIER, H., AND P. E. HOLBERT. 1965. Electron microscopic observation of the sporangial structure of a strain of *Actinoplanes*. J. Bacteriol. **89**: 217–222.

LECHEVALIER, H., M. P. LECHEVALIER, AND P. E. HOLBERT. 1966. Elcetron microscopic observation of the sporangial structure of strains of Actinoplanaceae. J. Bacteriol. **92**: 1228–1235.

LUFT, J. H. 1961. Improvements in epoxy resin embedding methods. J. Biophys. Biochem. Cytol. **9**: 409.

MITCHELL, P., AND J. MOYLE. 1956. The cytochrome system in the plasma membrane of *Staphylococcus aureus*. Biochem. J. **64**: 19.

MOORE, R. T., AND G. B. CHAPMAN. 1959. Observations of the fine structure and modes of growth of a streptomycete. J. Bacteriol. **78**: 878–885.

RANCOURT, M. W., AND H. A. LECHEVALIER. 1964. Electron microscopic observation of the sporangial structure of an Actinomycete, *Microellobosporia flavea*. J. Gen. Microbiol. **31**: 495–498.

RYTER, A., AND F. JACOB. 1964. Étude au microscope électronique de la laison entre noyau et mésosome chez *Bacillus subtilis*. Ann. Inst. Pasteur. **107**: 384–400.

STUART, D. C. 1959. Fine structure of the nucleoid and internal membrane systems of *Streptomyces*. J. Bacteriol. **78**: 272.

SZANISZLO, P. J. 1967. Comparison of the cell-wall composition and intracellular pigmentation of some strains of Actinoplanaceae. University Microfilms, Inc.

WOLSTENHOLME, D. R., C. A. VERMEULEN, AND G. VENEMA. 1966. Evidence for the involvement of membranous bodies in the process leading to genetic transformation in *B. subtilis*. J. Bacteriol. **92**: 1111–1121.

YAMAGUCHI, T. 1965. Comparison of the cell-wall composition of morphologically distinct Actinomycetes. J. Bacteriol. **89**: 444–453.

Pigments of the Actinoplanaceae. I. Pigment Production by *Spirillospora* #1655[1]

A. Domnas

Department of Botany, University of North Carolina at Chapel Hill, N. C.

Introduction

Members of the family Actinoplanaceae were first described by Couch in 1949. Since then, many other species have been described (Couch, 1955, 1963; Cross, et al., 1963; Kane, 1966), so that today the family consists of six genera, as follows: *Actinoplanes, Ampullariella, Amorphosporangium, Streptosporangium, Microëllobosporia, Pilimelia,* and *Spirillospora.* The *Actinoplanaceae* are usually vividly colored, with hues ranging from lemon yellow, orange, red, and green to blue and black. The orange colors are usually found in close association with wall material and have been characterized as carotenoids (Szaniszlo, 1967), whereas many other pigments are soluble and can be found in the culture fluid. One such soluble substance is the complex designated Spirillomycin[1] produced by *Spirillospora* #1655 (numbering according to the Couch collection), which has the color of grape juice when found in culture filtrates. This report is a summary of data obtained concerning production and properties of Spirillomycin.

Materials and Methods

Isolate.—The isolate employed was *Spirillospora* #1655 obtained from the Actinoplanaceae collection of the Botany Department, University of North Carolina at Chapel Hill. The organism was maintained on Czapek's agar slants supplemented with 0.5% peptone and maintained at room temperature.

Media.—The media employed throughout this investigation were standard Czapek's medium (Cz), Czapek's without nitrate (Cz w/o NO_3), and Czapek's without sucrose (Cz w/o suc), purchased from Fisher Scientific Corporation. Peptone (P) was from Fisher Scientific or Difco Laboratories. Occasionally a salt medium (S) without NO_3 and sucrose was employed in this investigation. All media were sterilized by autoclaving for 15 min. at 121° C.

Chemicals.—Sugars, amino acids and vitamins employed in this study were obtained from Calbiochem. Corp. or Nutritional Biochemical Corp. and were autoclaved with the medium when possible or filter sterilized as required.

Inocula.—Several loopfuls (3–4) of *Spirillospora* #1655 were transferred from slants to 100 mls. of P (0.5%) Cz in 250-ml. Erlenmeyer flasks, which were placed on a gyrotory shaker, and grown for five days at 24.5° C. At the end of this time, 1-ml. aliquots of this culture were dispensed into fresh P (0.5%) Cz (100 mls. in 250-ml. Erlenmeyer flasks) and cultured for seven days. The mycelia were harvested, washed aseptically, and homogenized in a Waring blendor microcup under sterile conditions. One-ml. aliquots of starter culture were inoculated into 100 ml. of medium in 250-ml. Erlenmeyer flasks, or into Bellco nephelo culture flasks, as experiments dictated. When large amounts of mycelial matter or pigment were needed, 5 ml. of starter culture were inoculated into 1 liter of appropriate medium (usually P only) in a 2.5-liter low-form shake flask.

Analytical Methods

Growth of organism.—Mycelia were collected on tared filter paper, water-washed thoroughly to remove solubles, and dried at 95° C. for 24 hours. Mycelial yield was used as the index of growth of the organism.

Spectra.—Spectral analysis were performed with the aid of a Cary 14 recording spectrophotometer.

Pigment-production measurement.—Pigment production in growing cultures was measured with Bellco nephelo culture flasks fitted with side arms for the Klett-Summerson colorimeter. The flasks were allowed to stand in a rack until the mycelia settled to the bottom and the fluid was clear in the test-tube portion of the flask. The absorbency of the filtrate was measured with the Klett-Summerson colorimeter using a #54 (green) filter.

Results

Spirillomycin Production

In initial experiments, *Spirillospora* #1655 was inoculated into P (1%) Cz and maintained in the dark for several months. The standing cultures became deep blue, from the large

[1] A preliminary account of this work has been given at the North Carolina-Virginia combined meetings of the American Society for Microbiology, Bowman Gray School of Medicine, Winston-Salem, N. C., October 29, 1966.

amounts of pigment excreted into the medium, in striking contrast to light-exposed cultures, which remained colorless. As a consequence, experiments were designed to test light-dark reactions in which Bellco nephelo culture flasks containing P (1%) Cz were inoculated and either protected against or exposed to light. The organisms were grown on a gyrotory shaker at 24.5° C., and any pigment produced was measured with the Klett-Summerson colorimeter as described above. Results of such experiments are shown in Figure 1, which shows

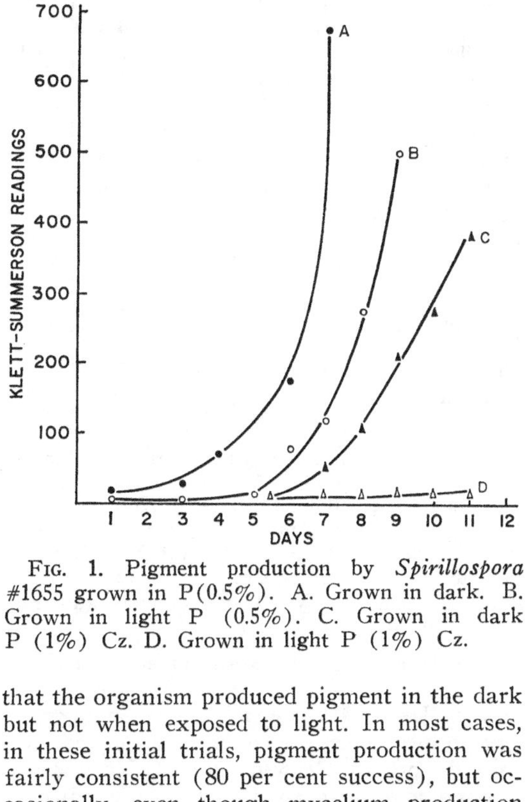

FIG. 1. Pigment production by *Spirillospora* #1655 grown in P(0.5%). A. Grown in dark. B. Grown in light P (0.5%). C. Grown in dark P (1%) Cz. D. Grown in light P (1%) Cz.

that the organism produced pigment in the dark but not when exposed to light. In most cases, in these initial trials, pigment production was fairly consistent (80 per cent success), but occasionally, even though mycelium production was abundant, no pigment was produced in dark-grown cultures.

Several reasons for this inconsistency became evident during the course of this investigation: (1) *Spirillospora* #1655 could be separated into several strains the ability of which to produce pigment varied. Some strains had pale mycelia and produced little to no pigment whereas others had mycelia of characteristic deep blue. The pale isolates are rather difficult to maintain in pure culture since they tend to revert over numerous transfers to the bluer variety. (The mechanism of reversion in this and in other isolates is the subject of another phase of investigation in this laboratory.) (2) A second reason became apparent with the discovery that certain sugars favored production of spirillomycin whereas others were inhibitory. (3) In addition, certain amino acids favored spirillomycin production whereas others did not. Also, peptone (0.5%) plus S, or simply peptone (0.5%) alone, was sufficient to produce pigment in either light or dark conditions, as Figure 1 shows. In a light-grown peptone optimum experiment, the optimum mycelium yield occurred between 3 and 4% P, but pigment was not produced at levels higher than 3 per cent (Figure 2). Optimum spirillomycin

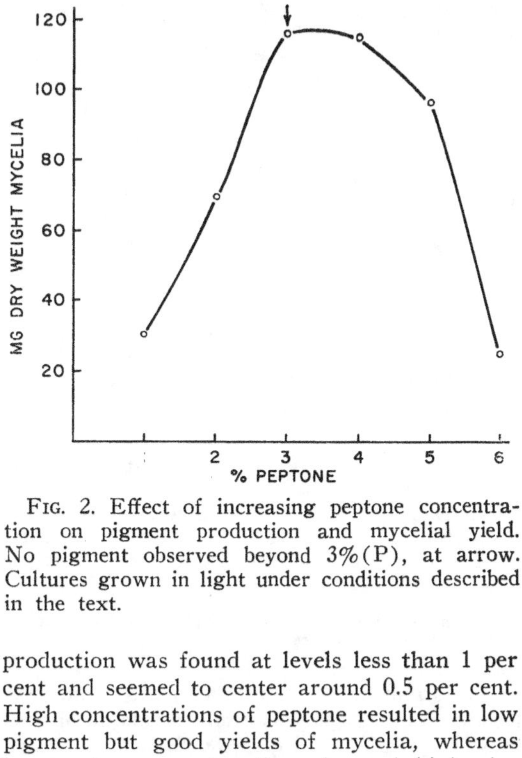

FIG. 2. Effect of increasing peptone concentration on pigment production and mycelial yield. No pigment observed beyond 3%(P), at arrow. Cultures grown in light under conditions described in the text.

production was found at levels less than 1 per cent and seemed to center around 0.5 per cent. High concentrations of peptone resulted in low pigment but good yields of mycelia, whereas low peptone concentrations favored high pigment but low mycelial yield. Peptone concentrations of 5% or greater inhibited both pigment and mycelial yield.

Effect of Carbohydrates and Nitrogen Sources on Pigment Production

To determine which substances favored pigment production, the growth of the organism on many single carbohydrate and nitrogen sources was investigated. The organism grew well on a number of sugars, but surprisingly, monosaccharides such as glucose, galactose, fructose, and mannose were not utilized despite the presence of ample peptone (Table I). Thus apparently these substances inhibit the growth

Table I

Growth and pigment production of *Spirillospora* #1655 on various carbohydrates

Carbon Source	Light: Mycelia (mg.)	Light: Pigment	Dark: Mycelia (mg.)	Dark: Pigment
L-Rhamnose	0	–	2.7	–
Glucose	4.3	–	9.5	–
Fructose	4.9	–	2.8	–
Glucose+fructose	0	–	0	–
Galactose	0	–	1.4	–
Mannose	1.2		1.9	
Mannitol	74.5	–[1]	71.4	+[2]
Inositol	48.1	–	61.0	+[3]
Sucrose	172.3	–	181.3	+[3]
Maltose	163.5	–	224.5	
Lactose	92.9	–	100.1	+[3]
Cellobiose	45.6	–	88.1	
Raffinose	104.7	–	136.4	
Amylose	185.6	–	159.0	+[2]
Amylopectin	99.9	–[1]	81.0	
Inulin	67.0	–	71.3	+[3]
Cellulose	3.5	+[2]	4.6	+[2]
Dextrin	n.g.	–	n.g.	–
Acetate	n.g.	–	n.g.	–
Propionate	n.g.	–	n.g.	–
Butyrate	n.g.	–	n.g.	–

Growth time, 15 days, at a temperature of 24.5° C. with rotary shaking, 180 oscillations/min. Inoculum 1 ml./100 ml. Cz w/o suc. + 1% P and carbohydrate sources as indicated. Monosaccharides 6, disaccharides 3, trisaccharides 2, starches, inulin 1, cellulose 0.1% (w/v).

[1]Mycelia blue at end of 15 days. [2]Blue pigment, 5 days. [3]Blue or blue-green, 15 days. n.g., no growth.

Table II

Growth and pigment production of *Spirillospora* #1655 on various nitrogen sources

Nitrogen Source	Light: Mycelia (mg.)	Light: Pigment	Dark: Mycelia (mg.)	Dark: Pigment
Glycine	1	–[3]	3.5	+[1]
Alanine	7.5	–[3]	3.1	+[2]
Aspartic acid	0	–	0	
Glutamic acid	0	–	0	
Serine	1.5	–[3]	1.3	–[3]
Ornithine	2.5	–	3.1	+[2]
Citrulline	1.0	–	2.7	+[1]
Arginine	0.5	–	4.0	+[1]
Valine	1.7	–	3.8	–
Isoleucine	10.3	–	2.7	–
Leucine	2.3	–	2.5	+[2]
Methionine	1.5	–	2.6	–
Cysteine	0	–	0	–
Histidine	0	–	0	–
Phenylalanine	2.0	–	7.0	–
Proline	2.5	–	6.1	+[2]
Peptone	73.1	+[2]	100.1	+[2]
Casein	n.g.	–	n.g.	–
Nutrient broth	22.0	–	92.8	–
Yeast extract	n.g.	–	n.g.	–

Growth time, 15 days at a temperature of 24.5° C. with rotary shaking, 180 oscillations/minutes. Inoculum, 1 ml./100 ml. Cz w/o NO_3 + amino acids as indicated. Amino acids nitrogen concentration equivalent to the amount of NO_3–N found in 100 ml. of Czapeks medium. Peptone, casein, nutrient broth, and yeast extract were added at 0.5% w/v.

[1]Blue pigment in 3 days. [2]Blue pigment in 7 days. [3]Blue mycelia.

of the organism. In contrast to the nonutilization of aldoses, polyols such as mannitol and inositol gave good growth, but best yields were obtained on disaccharides, and in particular, maltose. Light and dark were probably not significant factors in growth, since no consistent pattern in mycelial yield was observed. However, certain carbohydrates influence pigment formation under light-dark conditions as follows: n-amylose > mannitol > inulin. No pigment production was observed in the light when the organism was grown with any sugar. Other carbohydrates also favored pigment formation in the dark (Table I), but not so consistently nor in such quantity as those mentioned above. Confirmation was obtained with *Spirillospora* #1655 grown on Cz w/o sucrose in the dark, in which n-amylose, mannitol, or inulin was employed as the sole carbohydrate source. Under these conditions, blue pigment was formed, whereas cultures inoculated into Cz did not. In Cz w/o sucrose, NO_3 is the sole source of nitrogen; hence the effect of peptone is eliminated.

Table II summarizes the data obtained when *Spirillospora* was grown on individual nitrogen sources. Single amino acids in Cz w/o NO_3 gave poor yields of mycelia in both light and dark; nonetheless, pigment production with certain amino acids was quite demonstrable. Thus, an order of spirillomycin production was observed such that proline >> glycine > arginine > citrulline > ornithine > alanine = leucine. The best yields of mycelia and pigment were obtained on peptone in either light or dark. When P (1%) Cz was used, mycelial yields increased but pigment production was absent in flasks exposed to light.

Effect of Light on Spirillomycin Production by *Spirillospora*

Several experiments were performed to find means for halting pigment production in actively growing mycelia. Figure 3 shows that unless the cultures are exposed to light at a very early stage in growth, pigment yield is not retarded. Flasks uncovered before the figure of 150 on the Klett-Summerson scale was reached generally did not produce pigment over that already present. When the flask was re-covered, pigment production began again. It must be emphasized that this type of experiment was done with P (1%) Cz which contains sucrose, because cultures grown on P alone grew too rapidly and did not allow sufficient time to demonstrate the effect of light.

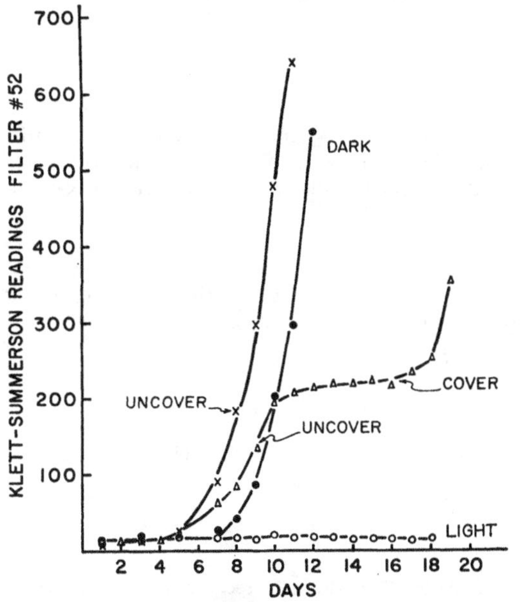

FIG. 3. Effect of light on pigment production in cultures of *Spirillospora* #1655. Flasks exposed to light and covered again as shown on the curves. Cultures grown in P(1%) Cz under conditions described in the text.

Nature of the Pigment Spirillomycin

Spirillomycin possesses some interesting properties, particularly its acid-alkaline indicator-dye characteristic, which is quite reversible. The pigment is deep marine blue in alkali but becomes red in acid, from which a fraction of the pigment will precipitate if allowed to stand for a period of time. The acid-precipitation property was used to devise a simple purification and fractionation procedure summarized in Table III. Further studies on the properties of the pigment were done on material from a fraction (P_1) which represented a water-washed precipitate. P_1 was insoluble in water (see Table IV), but soluble in alkali; acid conditions inevitably resulted in precipitation, at least with crude material and fraction P_1. When the crude filtrate was examined for absorption peaks, a curve was found with a maximum at 565 and a hump at 625–635 mμ (Fig. 4). Several fractions ($1F_1$, P_1F_2) that were not acid precipitable had shifts in their absorption maxima in response to pH changes. The major peak in strong base was ca. 645 mμ; this shifted to ca. 495 at a pH of 4 or 5. These peaks were sharp and well defined, whereas there was usually a plateau at intermediate pHs (Fig. 4).

Titration Curve

Spirillomycin is a basic substance and can be titrated with acid. From the curve, a dissociation constant of 1.2×10^{-11} was obtained.

Effect of Light on Pigment

In simple experiments, known amounts of spirillomycin were exposed to ordinary white light for given periods of time. Spirillomycin undergoes considerable initial decomposition (Fig. 5), but the rate levels off to a fixed value. The dark control was designed to measure any decomposition due to OH catalysis, since the pigment was dissolved in alkali.

Discussion

At this time, the nature of spirillomycin is difficult to discuss, since there are no data as to its structure. However, with the few facts at our disposal, speculation is permissible on the possible nature of the "indicator dye," since many such substances have been recorded in the literature. A typical class of such "indicator dyes" is the anthocyanidins found in the coloring matter of leaves and petals. Such plant substances exhibit acid-base color changes with corresponding spectral changes, have positive Erdmann reactions, and in fact have many of the characteristics shown in Table IV. It is unlikely that *Spirillospora* #1655 is producing an anthocyanidin for several reasons: (1) Spirillomycin does not have absorption maxima reported for anthocyanidins; for instance, in alkali, spirillomycin has a much higher absorption maximum than any recorded anthocyanidin. (2) The pigment does not exhibit the expected bathochromic shift when treated with $AlCl_3$ according to the method of Harborne

Table III

Fractionation of spirillomycin

Crude filtrate treated with 1 vol. glacial acetic acid/100 vols. filtrate. Allowed to stand overnight. Centrifuge precipitate formed. Supernatant (F_1) has a non-acid precipitable fraction.

↙ ↘

Filtrate 1 (F_1)
Same spectral and acid-base characteristics as crude filtrate.

Precipitate 1 (P_1)
Combined precipitates from crude filtrate washed several times with water, centrifuged, and dried *in vacuo*. Blue-black material obtained, dissolved in NaOH, and fractionated with H_2O on Sephadex-G-25.

↓

P_1F_1
Sephadex-G-25 fraction 1 precipitable by acid, collected, washed and dried, *in vacuo*. Same spectral and acid base characteristics as crude filtrate.

↓

P_1F_2
Second Sephadex-G-25 fraction. Not precipitable by acid, but has same characteristics as other fractions.

↓

P_1F_3
Third Sephadex-G-25 fraction, also not precipitable by acid. Same characteristics as others.

(1958). (3) None of the reported chromatographic methods for anthocyanidins have been successful in resolving the several components of spirillomycin.

Other substances occur in nature that have properties similar to those of spirillomycin. As early as 1908 Müller described a blue diffusible pigment found in cultures of *Streptomyces coelicolor* that had acid-base color changes. Müller stated that the pigment was obtained only in synthetic media with starch as the main carbohydrate. Beijerinck (1913) described a similar pigment from *Proactinomyces cyaneus* now identified as a *Nocardia*. This pigment was isolated and described by Gause (1946) and given the name of litmocidin, and some antibiotic properties were ascribed to it. More recently, litmocidin substances such as rhodomycetin (Shockman and Waksman, 1951) and actinorhodin (Brockman and Pini, 1947) have been found. These substances are napthazarins (Thomson, 1951) and have the general structure shown below:

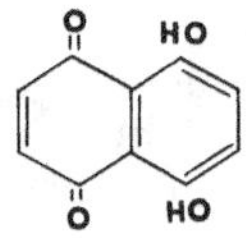

Spirillomycin has absorption maxima in acid and base much closer to actinorhodin (Bradley, 1962), and many of the characteristics reported for actinorhodin apply to spirillomycin.

However, a disturbing note is introduced upon examination of the pigment-production pattern when *Spirillospora* #1655 is grown on amino acids. As was shown, pigment production is actually stimulated by amino acids in the order proline >> glycine > arginine. This sequence coincides quite nicely with the observations of Marks and Bogorad (1960) that these amino acids in the above order were utilized preferentially for the biosynthesis of prodigiosin in *Serratia marcescens*. Indeed, blue pigments have been separated by Williams, et al. (1956) and Green, et al. (1956), from prodigiosin preparations. Thus spirillomycin may possibly be a prodigiosin or pyrrole-like pigment, or it may be that a common biosynthetic pathway for prodigiosin and spirillomycin exists.

The effect of light on pigment production is an interesting observation on pigment photodestruction. When the organism is grown in a medium containing optimal proportions of substances that will yield pigment, such as pep-

Table IV
Properties of spirillomycin

SPECTRA				
Crude material	303		575	640 mμ
In alkali, pH 11.0				645
In acid, pH 4.0		490–530		
After purification				645

Reaction with	in Acid	in Base
—	Red	Blue
Hydrosulfite	Effervescence and precipitation	Turns yellow, blue color regenerated by aeration
Zinc powder cold	—	—
Zinc powder and heat	Decolorized	—
Boracetic acid	—	—
$FeCl_3$	Turns brown-yellow	—
Pb acetate	Red precipitate	Blue-gray precipitate
Erdmann reaction with t-amyl alcohol	Becomes red, and passes into water phase. Stays fairly soluble.	Becomes blue and passes into alcohol phase

Solubilities	
Water	—
Ether	—
Ethanol	—
Acetone	—
Acid	sparingly
Pyridine	+
Phenol	+
Na_2CO_3	+
$NaHCO_3$	+
K_2CO_3	—
MP	greater than 310° C.

tone, then even though exposed to light, the rate of pigment production is greater than that due to photodecomposition. When the organism is grown in a condition that is not optimal, situations can arise whereby light will decompose the pigment, such as when the organism is tested on single amino acids.

Further interesting details were noted from the pigment-production pattern of *Spirillospora* with various carbohydrates. The best mycelia yields were obtained on disaccharides, and very poor yields obtained from aldoses. Interestingly enough, mannitol was utilized, and also favored pigment production in the dark. Mannitol is generally metabolized to fructose or fructose–6–P. Inulin was also utilized. Presumably the appropriate inulinase is present to hydrolyze the polysaccharide to constituent fructose units, yet fructose was not utilized. N-amylose, the linear portion of starch was used by *Spirillospora*, and it also favored pigment production in the dark. Presumably the hydrolysis of N-amylose proceeds with the assistance of an amylase to maltose. Maltose was well utilized, but glucose itself was not. N-amylose enhanced pigment production, but maltose did not. Clearly much remains to be clarified and many experiments may need to be redone in the light of future developments.

Summary

Spirillospora #1655, a member of the Actinoplanaceae, produces a soluble exocellular blue pigment called spirillomycin. The mode of pro-

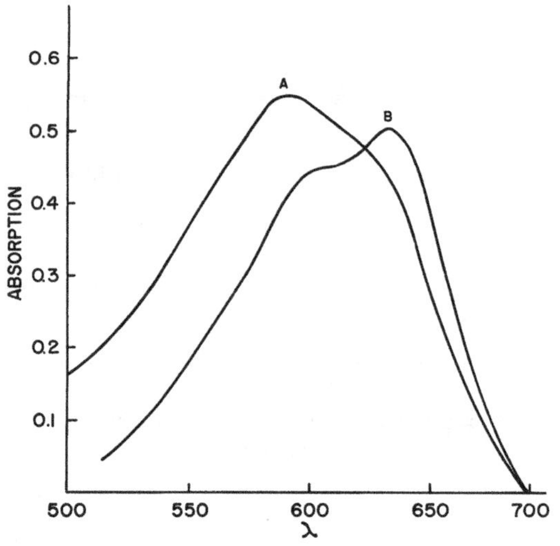

FIG. 4. Spectral characteristics of spirillomycin. A. Crude filtrate. B. Alkali-treated crude filtrate.

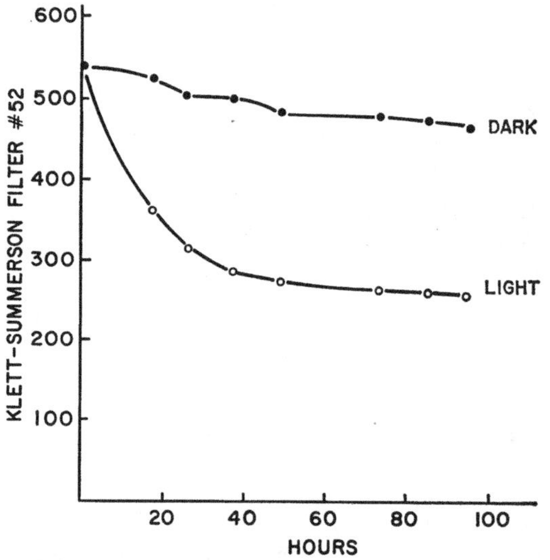

FIG. 5. Photodecomposition of spirillomycin when exposed to ordinary white light.

duction of this pigment has been studied by growing *Spirillospora* #1655 on a variety of carbohydrate and nitrogen sources under light and dark growth conditions. Starch, mannitol, and inulin favored pigment production under dark conditions and gave good mycelial yields as well. The amino acids, proline, glycine, arginine, and ornithine also induced pigment production in the dark but were not good sources for supporting growth of *Spirillospora*. However, peptone alone was found to be the best single source for pigment production in either dark or light growth conditions.

Spirillomycin was precipitated from the filtrate by acid, and a fractionation procedure was devised using Sephadex–G–25. The fractions obtained were examined for spectral characteristics and other chemical properties. Spirillomycin is an acid (red)-base (blue) indicator pigment with peaks at 645 (base) and 490 (acid) mμ. Some of the tests indicate possible relatonship to a group of substances known as naphthazarins.

ACKNOWLEDGMENT

I wish to thank the North Carolina Board of Science and Technology for its generosity in supplying funds (325–BST–11) for the purchase of equipment that made this research possible.

LITERATURE CITED

BEIJERINCK, M. W. 1913. On the composition of tyrosinase from two enzymes. Proc. Sec. Sci. Kon. Akad. Wetensch. **15**: 932–937.

BRADLEY, S. G. 1962. Biosynthesis of an actinorhodin-like pigment by *Streptomyces violaceoruber*. Developments in Industrial Microbiology **3**: 362–369.

BROCKMAN, H., AND PINI, H. 1947. Actinorhodin, a red pigment from *Streptomyces*. Naturwiss. **34**: 190.

COUCH, J. N. 1949. A new group of organisms related to the Actinomyces. Jour. Elisha Mitchell Sci. Soc. **65**: 315–318.

———. 1955. A new genus and family of the Actinomycetales, with a revision of the genus *Actinoplanes*. Jour. Elisha Mitchell Sci. Soc. **71**: 148–156.

———. 1963. Some new genera and species of the Actinoplanaceae. Jour. Elisha Mitchell Sci. Soc. **79**: 53–70.

CROSS, T., LECHEVALIER, M. P., AND LECHEVALIER, H. 1963. A new genus of the Actinomycetales: *Microëllobosporia* gen. nov. J. Gen. Microbiol. **31**: 421.

GAUSE, G. F. 1946. Litmocidin, a new antibiotic substance produced by *Proactinomyces cyaneus*. J. Bacteriol. **51**: 649–653.

GREEN, J. A., RAPPAPORT, D. A., AND WILLIAMS, R. P. 1956. Studies on pigmentation of *Serratia marcescens*. II. Characteristics of the blue and the combined red pigments of prodigiosin. J. Bacteriol. **72**: 483–487.

HARBORNE, J. B. 1958. Spectral methods of characterizing anthocyanins. Biochem. J. (London) **70**: 22–28.

KANE, W. D. 1966. A new genus of the Actinoplanaceae, *Pilimelia*, with a description of two species, *Pilimelia terevasa* and *Pilimelia anulata*. Jour. Elisha Mitchell Sci. Soc. **82**: 220–230.

MARKS, G. S., AND BOGORAD, L. 1960. Studies on the biosynthesis of prodigiosin in *Serratia marcescens*. Proc. Nat. Acad. Sci. (U.S.) **46**: 25–28.

MÜLLER, R. 1908. Eine Diphterie und eine Streptothrix mit gleichen blauen Farbstoff sowie Untersuchungen uber Streptothrizarten im Allgemeinen. Zentr. Bakteriol. Parasitenk. **1**: 46, 195–212.

SHOCKMAN, G., AND WAKSMAN, S. A. 1951. Rhodomycin—an antibiotic produced by a red pigmented mutant of *Streptomyces griseus*. Antibiotics and Chemotherapy **1**: 68–75.

SZANISZLO, P. J. 1967. Comparison of the cell-wall composition and intracellular pigmentation of some strains of Actinoplanaceae. Unpublished Ph.D. dissertation. University of North Carolina at Chapel Hill.

THOMSON, R. H. 1957. Naturally Occurring Quinones. Butterworth's Scientific Publications (London).

WILLIAM, R. P., GREEN, J. A., AND RAPPAPORT, D. A. 1956. Studies on pigmentation of *Serratia marcescens*. I. Spectral and paper chromatographic properties of prodigiosin. J. Bacteriol. **71**: 115–120.

The Nature of the Intramycelial Pigmentation of Some Actinoplanaceae[1]

PAUL J. SZANISZLO[2]

Department of Botany, University of North Carolina at Chapel Hill, N. C. 27514

Introduction

The vegetative mycelia of numerous Actinoplanaceae (Actinomycetales) frequently exhibit an orange or yellow color when grown on a variety of liquid and solid culture media. Couch (1963) reported that species of *Actinoplanes, Ampullariella,* and *Amorphosporangium* are most often characterized by orange mycelia. Kane (1966) reported that the mycelia of the species of *Pilimelia* that she studied are usually a shade of yellow. An investigation of the cell-wall composition of 48 strains of Actinoplanaceae—classified among the four genera *Actinoplanes, Ampullariella, Amorphosporangium,* and *Pilimelia* (Szaniszlo, 1967)—revealed that pigments responsible for the mycelial coloration were associated with the protoplasm and not with the cell walls. Pigment extracts were subsequently obtained for comparative purposes from all 48 strains. This report describes the nature of these pigment extracts.

Material and Methods

The name and number of each strain examined is listed in Table I. The strains of *Actinoplanes, Ampullariella,* and *Amorphosporangium* were identified by Dr. J. N. Couch (Couch, 1950, 1955, 1963), and continuous cultures of each strain are maintained in his laboratory on Czapek agar (Difco) and peptone Czapek agar (Czapek agar fortified with 5 g./1000 ml. of Difco bacto-peptone). The strains of *Pilimelia* were identified by Miss W. Kane (Kane, 1966) and are maintained by her on peptone Czapek agar.

Transfers of each strain were made from the stock cultures to peptone Czapek agar slants in 200-ml. prescription bottles. The resulting subcultures were used to inoculate 1000 ml. of Czapek dox broth (Difco) fortified with 5 g. of bacto-peptone (Difco) dispensed in 2500-ml. low-form flasks. Preparation of the inoculum was carried out by aseptically scraping the colonies from the agar slant into 50 ml. of sterile broth and blending the mixture at high speed for one minute in a pre-cooled Waring blendor microcup. The blended broth-mycelial mixture (10 ml.) was used to inoculate each culture. The cultures were incubated for five days at 24–26° C. with continuous illumination and continuous agitation (reciprocal shaking; 100 strokes/min.).

The vegetative mycelia of each strain were collected by centrifugation at 3000 x g with the Sorvall "Szent-Gyorgyi and Blum" continuous-flow system, washed repeatedly with distilled water, and disrupted to liberate the pigments, since direct extraction of the mycelia was unsuccessful. The disruption was accomplished with the Braun Homogenizer (Bronwill Scientific, Rochester, N.Y.). Packed wet mycelia were diluted (1:4, v/v) with distilled water, and 30 ml. of the resulting suspension combined with 30 g. of size-12 glass beads (Bronwill Scientific, Rochester, N.Y.) in glass-stoppered 75-ml. duran flasks. The mixtures were shaken for three minutes (4,000 oscillations/minute) with sufficient CO_2 delivered to the homogenizer's chamber to prevent excessive heating. The protoplasmic fractions containing the pigments were collected and extracted with n-hexane. The pigment extractions were conducted by forcing the pigments into the n-hexane with the slow addition of water. Care was taken to prevent emulsification of the mixture. Absorption spectra of the pigments were obtained using the Cary model 14 recording spectrophotometer.

To determine whether light affected the production of the intramycelial pigment, 1000 ml. of peptone Czapek broth in low-form flasks were inoculated with *Actinoplanes philippinensis* as previously described. Half of the flasks were wrapped in aluminum foil so that all ensuing growth occurred in the dark. All the cultures were incubated for five days under the cultural conditions previously described. The mycelia from each flask were collected and washed with distilled water by centrifugation. *In vivo* absorption spectra were obtained for turbidimetrically equal suspensions of both light- and dark-grown mycelia using the Cary model 14 recording spectrophotometer equipped

---

[1] A portion of a dissertation submitted to the Graduate School of the University of North Carolina at Chapel Hill in partial fulfillment of the requirements for the degree of Doctor of Philosophy.

[2] Present address: Laboratory of Applied Microbiology, Division of Engineering and Applied Physics, Harvard University, Cambridge, Massachusetts 02138.

Table I
Name and number of each strain examined

Name of Organism	Strain Number
Actinoplanes philippinensis	2
Actinoplanes utahensis	258, 259, 260, 261
Actinoplanes missouriensis	221, 222, 263, 267, 431, 443, 657, 825
Ampullariella regularis	28, 79, 154, 164, 168, 312, 395, 850, 915
Ampullariella digitata	33, 71, 118, 131, 137, 370, 386, 399
Ampullariella lobata	72, 74, 337
Ampullariella campanulata	65, 126, 151, 182, 640, 642, 643
Ampullariella sp.	1492, 1539
Amorphosporangium auranticolor	253, 262
Pilimelia anulata	1
Pilimelia terevasa	1
Pilimelia sp.	1777

with the Cary model 1462 scattered transmission accessory.

Results and Discussion

After the disruption of the mycelia, the pigments associated with the mycelia of each strain were easily extracted from the protoplasmic fractions. The absorption spectra obtained from the pigment extracts were relatively simple in character. In the visible region each spectrum consisted of three absorption peaks which produced a curve typically associated with the class of compounds termed carotenoids. The pigment extracts from all the strains of *Actinoplanes, Ampullariella,* and *Amorphosporangium* had identical spectral properties in n-hexane (Fig. 1). Absorption peaks were at 504, 474, and 448 mμ with an absorption maximum at 474 mμ. If the extractions were made with petroleum ether, these peaks shifted slightly to 502, 480, and 443 mμ; when made with carbon disulfide they shifted to 533, 503, and 475 mμ. Pigment extracts from the three strains of *Pilimelia* exhibited slightly different spectral properties in n-hexane (Fig. 2). Absorption peaks for the pigment extracts from these strains were shifted about 22 mμ further toward the untraviolet region of the light spectrum and were at 479, 451, and 425 mμ, with an absorption maximum at 451 mμ. While this shift in spectral properties provides an extremely useful diagnostic character for the differentiation of *Pilimelia* strains from

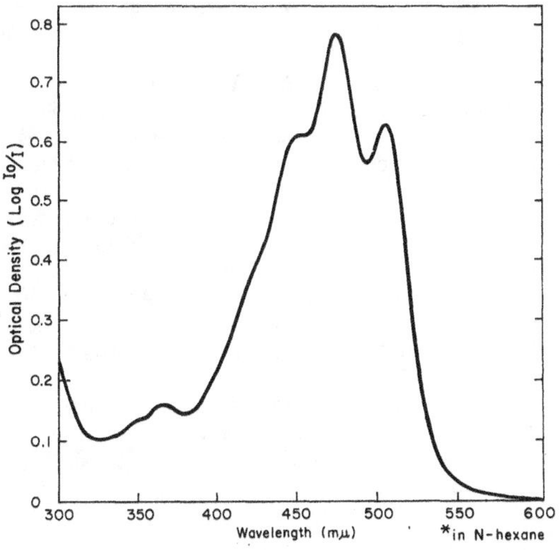

FIG. 1. Light absorption spectrum of pigment extracts obtained from all the strains of *Actinoplanes, Ampullariella,* and *Amorphosporangium.*

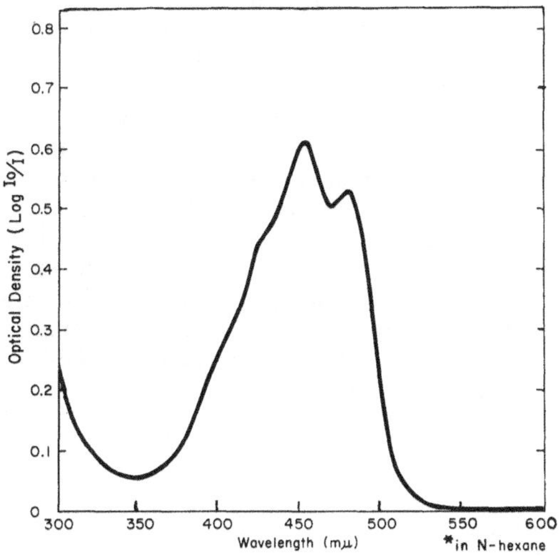

FIG. 2. Light absorption spectrum of pigment extracts obtained from all the strains of *Pilimelia.*

strains of *Actinoplanes, Ampullariella,* and *Amorphosporangium,* it probably does not represent an extreme deviation. In fact, Karrer and Jucker (1950) stated that the addition of a single conjugated double bond to a carotenoid without any other structural changes results in a displacement of the visible absorption band toward the longer wavelength by 20–22 mμ. The removal of such a conjugated double bond has the opposite effect. In view of this fact, it appears possible that a single-step mutation has caused such a conjugated double bond to be added to or subtracted from the chemical structure of the carotenoids produced by these Actinoplanaceae. This, in turn, would produce some Actinoplanaceae having intracellular yellow pigments like the strains of *Pilimelia* and some having intracellular orange pigments like the strains of *Actinoplanes, Ampullariella,* and *Amorphosporangium.* The apparent similarity of the carotenoids from all the strains, however, indicates that the relationship of these organisms may be very close.

The results for the experiment to determine whether light affected the production of the intramycelial pigment are presented in Figure 3. Almost no carotenoid synthesis occurred in the cultures grown in the dark. Light-grown cultures examined at the same time, however, exhibited significant quantities of absorbing material. Goodwin (1962) stated that there is no effect of light on carotenoid synthesis in nonphotosynthetic bacteria as in fungi. The results presented in this paper indicate that light promotes carotenoid synthesis in these Actinoplanaceae. Further research along these lines may be useful in order to confirm these results and to discover the nature of this light-promotion effect.

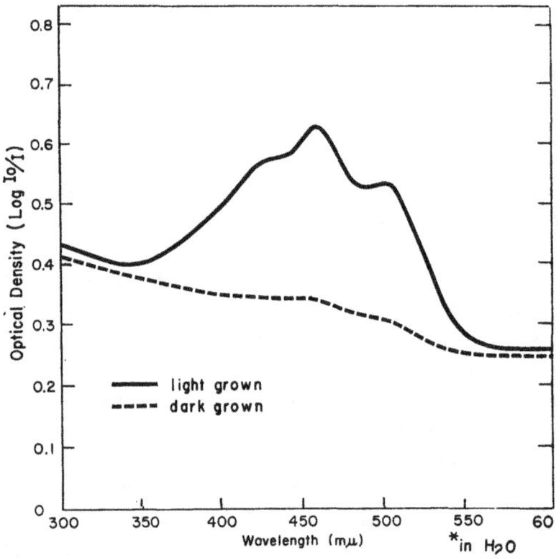

FIG. 3. *In vivo* light absorption spectra of light-grown and dark-grown vegetative mycelia of *Actinoplanes philippinensis.*

Summary

This study concerns the nature of the mycelial coloration of 48 strains of Actinoplanaceae classified among 10 species and the four genera *Actinoplanes, Ampullariella, Amorphosporangium,* and *Pilimelia.* Pigment extracts from all the strains were spectrophotometrically analyzed and were found to have spectral properties typically associated with carotenoids. The light-absorbence properties of the extracts obtained from the *Actinoplanes, Ampullariella,* and *Amorphosporangium* strains were identical and differed only slightly from the absorbence properties of the extracts obtained from the *Pilimelia* strains.

In vivo light absorption spectra of light- and dark-grown vegetative cultures of *Actinoplanes philippinensis* indicated that light may promote carotenoid synthesis in these bacteria.

LITERATURE CITED

COUCH, J. N. 1950. *Actinoplanes,* a new genus of the Actinomycetales. Jour. Elisha Mitchell Sci. Soc. **66**: 87–92.

———. 1955. A new genus and family of the Actinomycetales, with a revision of the genus *Actinoplanes.* Jour. Elisha Mitchell Sci. Soc. **71**: 148–155.

———. 1963. Some new genera and species of the Actinoplanaceae. Jour. Elisha Mitchell Sci. Soc. **79**: 53–70.

GOODWIN, T. W. 1962. Carotenoids: structure, distribution, and function. **4**: 643–675, *in* M. Florkin and M. Mason, eds., Comparative Biochemistry. Academic Press, New York and London.

KANE, W. D. 1966. A new genus of the Actinoplanaceae, *Pilimelia,* with a description of two species, *Pilimelia terevasa* and *Pilimelia anulata.* Jour. Elisha Mitchell Sci. Soc. **82**: 220–230.

KARRER, P., AND E. JUCKER. 1950. Carotenoids. Elsevier Publishing Co., Inc. Amsterdam, London, New York.

SZANISZLO, PAUL J., AND HARRY GOODER. 1967. Cell-wall composition in relation to the taxonomy of some Actinoplanaceae. J. Bacteriol. **94**: 2037–2047.

Some Nutritional Requirements of an Unidentified *Micromonospora*

A. Domnas

Department of Botany, University of North Carolina at Chapel Hill, N.C. 27514

Introduction

An organism found in the culture collection of the Mycology Laboratory, Department of Botany, University of North Carolina at Chapel Hill, was identified by Prof. Couch as a *Micromonospora.* Since the morphological and physiological properties of this species did not seem to correspond to any of those already described (Bergey, 1957), it would appear that this organism might be a new species. Indeed, one property that sets it apart from other members of the genus is the production of a soluble yellow pigment when grown on solid or in liquid cultures under appropriate conditions. This brief report summarizes the results of a nutritional study of this organism and indicates a possible procedure for enhancement of pigment production.

Materials and Methods

Isolate.—The isolation used was obtained from the culture collection of Prof. Couch, Mycology Laboratory, Department of Botany, University of North Carolina. The organism was maintained on Czapek slants supplemented by 0.5% peptone.

Media.—The media employed throughout this investigation were standard Czapek medium (Cz); Czapek without nitrate (Cz w/o NO_3), purchased from Fisher Scientific; and Czapek without sucrose (Cz w/o suc), also from Fisher Scientific. Peptone (P) was either of Difco or Fisher Scientific origin. All media were autoclaved 15 min. at 121° C.

Chemicals.—Sugars, amino acids, and vitamins employed in this study were either of Calbiochem. or Nutritional Biochemical provenience, and were autoclaved (vitamins were filter sterilized) in Cz medium unless otherwise stated.

Inocula.—Preliminary experiments indicated that the organism grew far better when Cz medium was supplemented with peptone and when the culture was swirled on a gyrotory shaker. In preliminary trials, designed to obtain approximate optimal conditions for growth, three loopfuls of organisms from P (0.5%)–Cz slants were inoculated into 125 ml. of P (0.5%)–Cz liquid medium, placed on a shaker, and allowed to grow for seven days at 24.5° C. The mycelia was harvested by sterile filtration, washed with several portions of sterile water, transferred aseptically to a sterile Waring blendor microcup, and homogenized in sterile distilled water. Aliquots (1 or 5 ml.) of the homogenate were used to inoculate flasks containing 125 ml. of P (0.5%)–Cz medium or any other medium employed. At the end of seven days the cultures were harvested by filtration on preweighed filter paper, washed thoroughly with water, dried to constant weight at 95° C. for 24 hours, and then weighed. All runs were done in duplicate, and each experimental series has been repeated at least three times.

Analytical Methods

Ammonia produced by the cultures was measured by the Conway micro-diffusion technique using Obrink (1955) vessels, followed by nesslerization according to the procedure of Koch and McMeekin (1924). Glucose was measured by the use of glucose oxidase purchased from Calbiochem. Corp. Sugars formed during growth were revealed by descending paper chromatography in two solvent systems, butanol-acetic acid and water (4:1:5) and phenol-water (4:1), and the spots were visualized with alkaline aniline phthalate.

Table I

Volume Vessel in ml.	Light & Shaking Conditions	Yield mg.
250	Dark, still.	11.2
"	Light, still.	5.0
"	Dark, shake.	119.0
"	Light, shake.	99.3
500	Dark, still.	11.8
"	Light, still.	11.3
"	Dark, shake.	200.1
"	Light, shake.	179.0
250	Light, anaerobic.	no growth

Effect of light, vessel size, and agitation on growth of *Micromonospora.* One ml. of starter culture inoculated into 125 ml. of peptone (0.5%) Czapek's contained in 250-ml. Erlenmeyer flasks, 5 ml. of starter culture inoculated into 200 ml. of same medium in 500-ml. Erlenmeyer flask. Light: F40 CW–40-watt cool-light illumination. Oscillation rate: 180/min.

Results

Effect of vessel size, agitation, and light.—Table I summarizes the data obtained on vessel size, light, and agitation, and shows that neither light nor vessel size influence the mycelial yields, whereas agitation appears to be very important, presumably in supplying oxygen. With strict anaerobiosis, no growth was observed. As a consequence of this study, all subsequent experiments were standardized on the basis peptone-Czapek medium, 125 mls. per 250 mls. Erlenmeyer flasks, and 1 ml. starting inoculum. No particular care was taken with regard to light, but the organisms were grown on a gyrotory shaker at 180 oscillations/minute at 24.5° C.

Optimal sucrose and peptone concentrations.—In order to obtain a growth curve using peptone and sucrose, it was necessary to know the optimal growth concentrations of these substances. Results of such experiments (Fig. 1)

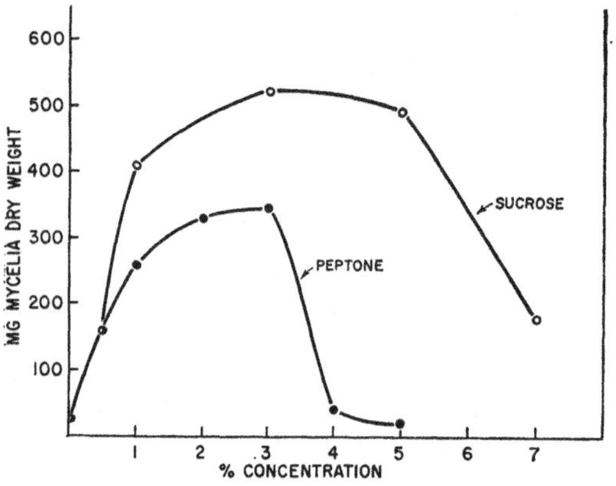

FIG. 1. Optimum peptone and sucrose concentration of *Micromonospora sp.*

showed that optimum peptone concentrations occurred between 2 and 3 per cent, and optimum sucrose between 3 and 5 per cent. It would seem logical that the mycelial yield would be of the same order of magnitude when the sucrose was held constant and peptone varied as when peptone was held constant and sucrose varied. This was not found to be so in the determination of optimal sucrose (Fig. 1, sucrose curve), and this may be due to the use of Cz w/o sucrose, which contains 0.3% sodium nitrate. Ordinarily the organism does not produce much mycelium on normal Czapek medium, but the added peptone may have resulted in utilization of nitrate as well as peptone nitrogen. Such cases are not unknown, as Cochrane (1950) noted in observing that spores of *Streptomyces griseus* would not grow on nitrate; however, mature vegetative mycelia utilized nitrate with no difficulty. In this study the *Micromonospora sp.* culture was employed as mature mycelia and may well take up nitrate as well as the amino acids of the peptone, thus accounting for the increase in mycelial mass.

Growth Curve.—A growth curve was obtained for *Micromonospora sp.* in Czapek broth supplemented with 2.5% peptone, and employing Czapek w/o NO_3 which contains 3% sucrose. Figure 2 shows that maximum yield of mycelia was obtained in approximately five days; beyond eleven days the yield of mycelia decreased, which may reflect autolytic action by the cells. During growth the organism released ammonia into the medium (Fig. 2)

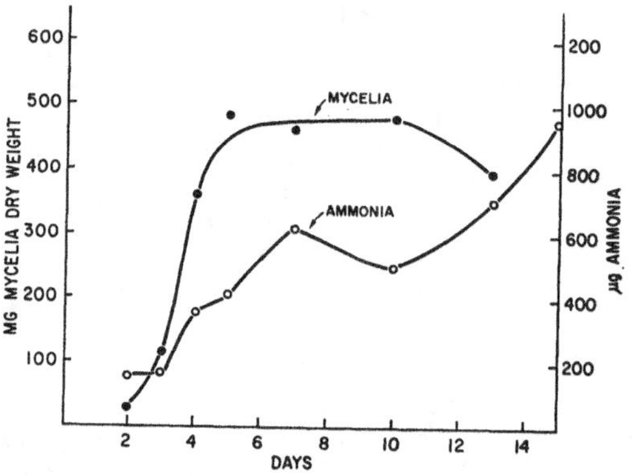

FIG. 2. Growth curve and production of ammonia in *Micromonospora sp.*

which appeared in two phases; (1) an early phase corresponding perhaps to the logarithmic growth of the organism, and (2) a later phase perhaps reflecting autolytic activity. The organism also excreted into the medium unknown substances which reacted positively to Benedicts. The original Czapek preparation contained sucrose, but little or no trace of reducing sugar was found in the medium before and after autoclaving. No reducing substances were found in cell-free autoclaved P (2.5%)-Cz media controls held at 24.5° C. on a shaker for twelve days. By use of paper chromatography several substances of carbohydrate nature were detected in seven-day-old culture filtrates. With the aid of glucose oxidase one substance could be identified as glucose; the other substances remained unidentified. The Rfs of glucose and an unknown were almost identical in both solvent systems, making it likely that fructose might be the other unknown.

Carbon and nitrogen requirements.—The organism was tested for growth on various carbon and nitrogen sources. Since 3% sucrose

Table II

Carbon source	Yield mycelia mg.	Nitrogen source	Yield mycelia mg.
Glucose	696.8	Casein	1,213.7
Maltose	675.9	Casamino acids	696.8
Galactose	604.0	Peptone	485.1
Cellobiose	592.6		
L-Arabinose	514.3		
Sucrose	511.8		
Starch	491.5		
Xylose	471.8		
Lactose	445.5		
Raffinose	404.2		
L-Rhamnose	348.2		
Carboxymethylcellulose	219.7		
Mannitol	197.3		
Fructose	no growth		
Ribose	" "		

Effect of various carbon and nitrogen sources on growth of *Micromonospora.* One ml. starter culture inoculated into 125 ml. of Czapek's w/o sucrose or Czapek's w/o nitrate as required. Sugar concentrations were made equivalent to the 3% sucrose optimum. Thus disaccharides were set at 3% (w/v), monosaccharides at 6% (w/v), etc. Carboxymethylcellulose 0.1% w/v and starch 1% w/v. Nitrogen substance concentration was equivalent to the 2.5% (w/v) peptone found to be optimum. Growth conditions were identical to those described.

was found to be optimal for *Micromonospora sp.*, this value or multiples thereof were selected for the testing of various carbohydrates with the exception of carboxymethyl-cellulose and starch, which were tested at the 0.1% and 1.0% levels respectively. The results of these carbohydrate tests are shown in Table II and are arranged in order of decreasing efficiency of utilization. The order of utilization bears a fairly close resemblance to that of *Micromonospora purpurea* as shown by Wagman and Weinstein (1966), except that *Micromonospora sp.* did not grow on fructose or ribose. Perhaps the organism possesses a sucrose transport system involving a β fructofuranosidase located in the wall or cell membrane which results in release of fructose and glucose outside the cell. The amount of glucose found at the end of a typical growth-curve study is not very high, being ca. 1.13 mg./ml. as opposed to a starting concentration of 30 mg./ml. of sucrose.

It is interesting to note that starch as well as cellulose is hydrolyzed, a property common to the *Micromonospora* and observed in *Micromonospora purpurea* (Wagman and Weinstein, (1966).

In striking contrast to the broad spectrum of the carbohydrates utilized by this organism, very few nitrogen-containing compounds were used. With the exception of some proteins, no single amino acid was utilized, nor was the organism capable of assimilating ammonia, hydroxylamine, urea, or guanidine. Slight growth was obtained on tryptophane, but only after two weeks of incubation. This does not exclude the possibility that combinations of amino acids could be used; however, this aspect was not further examined. With preformed nitrogen substances such as peptone or casein, excellent yields of mycelia were always obtained.

Effect of pH on growth.—Growth of *Micromonospora sp.* was not overly affected by pH, since the organism was capable of changing the starting pH to a final value of 7.8 in all instances, given a seven-day growth period. To obtain the data shown in Figure 3, the cultures were harvested at the end of five instead of the usual seven days, since it had been noted that for this period of time the initial pHs remained unaltered.

Vitamins.—This isolate of *Micromonospora* did not grow well on unsupplemented Czapek's medium. Growth was slow, and although final yields were fair, mycelial production was obviously greater when a preformed nitrogen source such as peptone was added. To ascertain whether increase in mycelial yield was due to peptone or to some other substances or growth factors, a vitamin study was under-

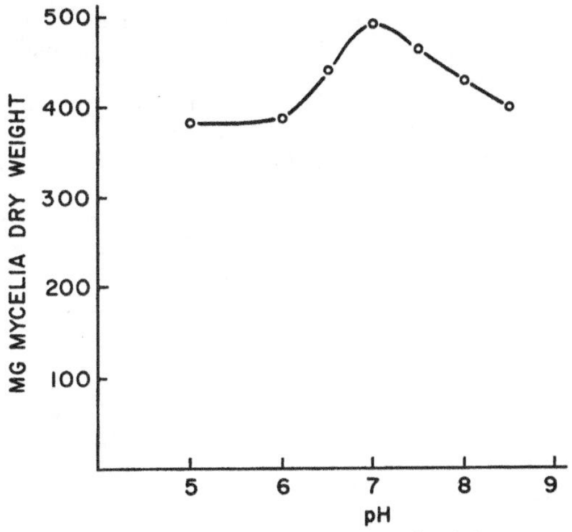

FIG. 3. Effect of pH on growth in *Micromonospora* sp.

Table III

Vitamin effect and production of yellow pigment

Vitamins added + 25 mg. Peptone	Yield Mg.	Pigment Production	Vitamins Added (no Peptone)	Yield Mg.
All	179.9	+	All	6.8
" - pyridoxine	200.7	+	" - pyridoxine	2.1
" - Biotin	199.6	+	" - Biotin	13.0
" - Nicotinic acid	228.6	+	" - Nicotinic Acid	20.7
" - Thiamine	216.6	+	" - Thiamine	5.1
" - Riboflavin	202.1	−	" - Riboflavin	n.g.

Effect of vitamins on pigment production and mycelial yield. Inoculation and incubation conditions identical to those given (see Table II). Vitamins added at the following concentrations: pyridoxine 50 μg, thiamine 7.5 μg, nicotinic acid 70 μg, riboflavin 5.0 μg, biotin 2.5 μg.

taken. Vitamins added singly to Czapek's medium had no measurable effect on increasing mycelia production. However, vitamin combinations supplemented with 25 mg. peptone per 125 ml. medium gave good yields (Table III). It was further noted that vitamin combinations and peptone (25 mg.) also enhanced pigment production, which had been up to this point an inconsistent growth property—sometimes present, more often absent. Under these conditions, however, yields of mycelia are sacrificed and, while they remain respectable, do not equal those cultured on P (2.5%)–Cz medium. (Table III).

The nature of the pigment is unknown, and the absorption spectrum has no distinguishing features (Fig. 4). No attempt was made to isolate or characterize the pigment.

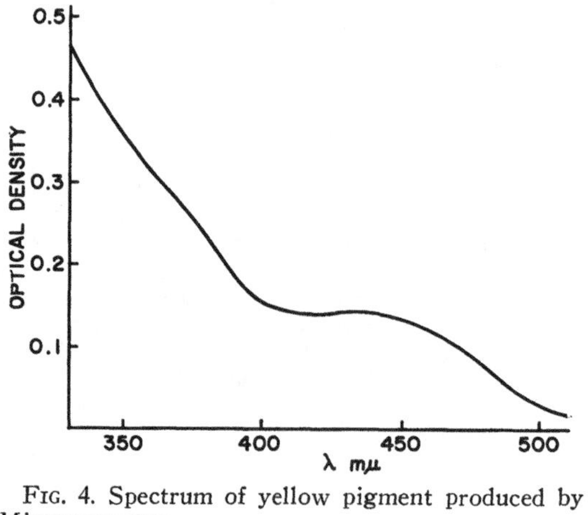

FIG. 4. Spectrum of yellow pigment produced by *Micromonospora* sp.

LITERATURE CITED

1957. Bergey's Manual of Determinative Bacteriology, 7th ed. William & Wilkins, Baltimore. Pp. 822–824.

COCHRANE, V. W. 1950. The metabolisms of a species of Streptomyces. III. The nitrate metabolism of *Streptomyces griseus.* Bull. Torrey Bot. Club. **77**: 176–180.

KOCH, F. C., AND MCMEEKIN, T. L. 1924. A new direct nesslerization microkjeldahl method and modification of the Nessler-Folin reagent. J. Am. Chem. Soc. **46**: 2066–2069.

OBRINK, K. J. 1955. A modified Conway unit for micro-diffusion analysis. Biochem. J. **59**: 134–136.

WAGMAN, G. H., AND WEINSTEIN, M. J. 1966. A chemically defined fermentation medium for the growth of *Micromonospora purpurea.* Biotechnology and Bioengineering **8**: 259–273.

Dictyostelium rosarium: A New Cellular Slime Mold with Beaded Sorocarps

KENNETH B. RAPER[a] AND JAMES C. CAVENDER[b]

[a] *Departments of Bacteriology and Botany, University of Wisconsin, Madison, Wis.*
[b] *Department of Biology, Wabash College, Crawfordsville, Ind.*

Introduction

In the course of our search for cellular slime molds from soils and other natural materials from southwestern United States and Central America (Cavender and Raper, in press), a form was twice isolated in 1962 that clearly differs from any species previously known (Raper and Cavender, 1966). An additional strain, isolated from deer dung collected in the State of Washington, was subsequently received from Professor Lindsay S. Olive and Miss Carmen Stoianovitch of Columbia University. While the three cultures are not alike in all aspects of their development, they clearly belong together and merit description as a new species within the genus *Dictyostelium.*

Mature fructifications, or sorocarps, typically present a beaded appearance due to the presence of numerous sessile spore masses, or sori, that commonly develop along the entire lengths of the supporting cellular stalks, or sorophores. The spores are globose to subglobose, and hence differ from those of all other species of *Dictyostelium* except for *D. lacteum* van Tieghem (1880), which, by comparison, is very small and delicate, and from *Acytostelium leptosomum* (Raper and Quinlan, 1958), characterized by its acellular stalks. Branching of the sorophores, which occurs quite frequently, is truly dichotomous in character and occurs in a manner quite unlike the branching in other species of *Dictyostelium,* or in *Polysphondylium.* Finally, cell aggregation usually occurs in three fairly distinct phases rather than by a continuous and progressive process leading to sorocarp formation as in most species of *Dictyostelium.*

The periodic abstriction of cell masses at the lower end of the ascending sorogen, from which the lateral sori develop, is strongly suggestive of *Polysphondylium.* But there the similarity ends, for the small cell masses, once pinched off, do not subdivide further or develop into side branches of any kind. Instead, all of the constituent myxamoebae form spores and contribute directly to the formation of the sessile sori. On the other hand, and except for the lateral sori (which vary in number depending upon the strain and the conditions of cultivation), the slime molds in question bear a definite resemblance to *Dictyostelium mucoroides* Brefeld (1869). Admittedly, inclusion of these slime molds within *Dictyostelium* requires some further extension of our present concept of this genus, which is already quite broad. Nevertheless, since they cannot be placed in *Polysphondylium* and since the creation of a new taxon intermediate between the two classic genera seems unwarranted, we believe it is appropriate to describe and assign them as a new species of *Dictyostelium.* For this the binomial *Dictyostelium rosarium* is proposed in consideration of the beaded character of the sorocarps.

A technical description of the new species follows, together with brief accounts of its cultivation, growth, and developmental history.

Technical Description

Dictyostelium rosarium sp. nov.

Sorocarpi typice erecti vel proni, interdum procidui, non ramosi vel rarius (raro duplo) dichotome ramosi, racemarii vel solitarii, magnitudine proportioneque varii, e sorophoris comparate magnos terminales soros et multos, plerumque minores, sessiles soros secundum sorophora gerentibus compositi, hinc margarita imitantes; sorophora diametro varia, plerumque 3–6 mm. in longitudinem sed ex 2 mm. usque 1 cm. varia, quin etiam longiuscula; sori globosi, lactei, terminales sori 100–250 μ in diametro, laterales sori plerumque minores, plerumque 50–150 μ in diametro, pauci usque 25 in numero, saepe in numero variabiles in eodem sorophoro; sporae globosae usque subglobosae, 4.5–6.0 μ in diametro, subtiliter granulosae cum muris comparate tenuibus.

Habitat: In superficiali solo semiaridae e *Prosopis* sylvae, Alice, Texas.

Typica cultura: CC–7, a James C. Cavender, University of Wisconsin, in anno 1962 isolata.

Sorocarps typically erect or inclined, sometimes prostrate, unbranched or less commonly branched dichotomously (rarely twice), clustered or solitary, variable in size and proportions, consisting of sorophores bearing relatively large terminal sori and numerous, usually smaller, sessile sori along the sorophores, hence presenting a beaded appearance (Fig. 1A); sorophores variable in diameter, commonly 3

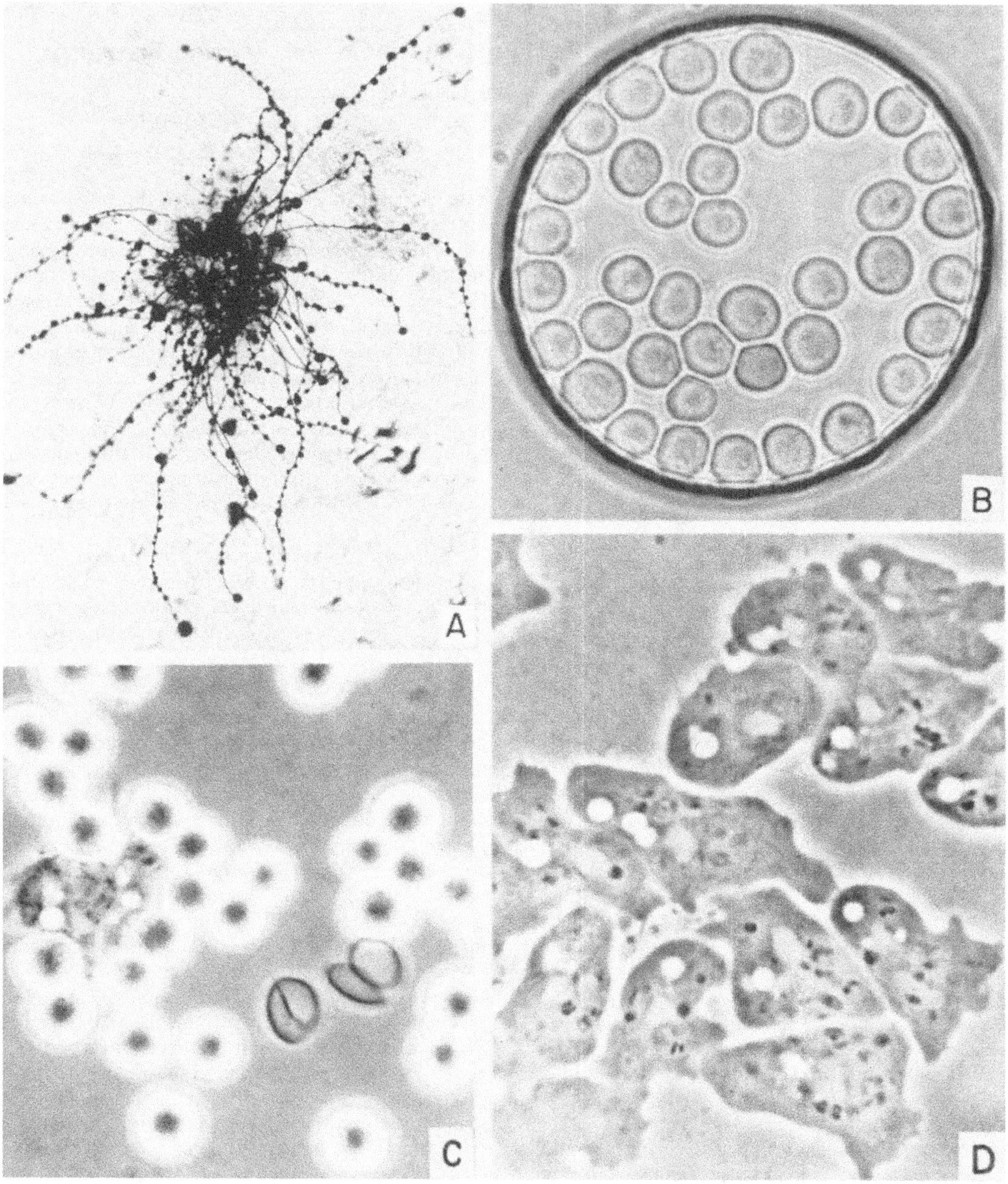

FIG. 1. A. Clustered sorocarps of *Dictyostelium rosarium* n. sp. developed in an old culture grown in association with *Escherichia coli* on 0.1% lactose-peptone agar (pH 5.5) in a glass petri plate. Incubation 20° C.; age 3 weeks. Note the many sessile sori that impart to the sorocarps their characteristic beaded appearance. X10. *B*. Mature spores in a small droplet of water as seen with normal light. X1400. *C*. Spore germination as seen with phase microscopy, showing ungerminated spores, empty spore cases, and a single myxamoeba. X1400. *D*. Vegetative myxamoebae adjacent to the margin of a bacterial colony. Note the nuclei (clear and irregular areas near the centers of the cells), the numerous food vacuoles scattered throughout the endoplasm, and the conspicuous contractile vacuoles. The myxamoebae are substantially compressed by a coverglass, hence show no filose pseudopodia. Phase microscopy. X1400.

to 6 mm. in length but ranging from 2 mm. to 1 cm. or more; sori globose, milk-white, terminal sori 100 to 250 μ in diameter, lateral sori usually smaller, mostly 50 to 150 μ in diameter, from few to 25 or more in number and often variable on the same sorophore; spores globose to subglobose (Fig. 1B), 4.5 to 6.0 μ in diameter, finely granular with walls comparatively thin.

Habitat: Surface soil from semiarid mesquite forest, Alice, Texas.

Type: Strain CC–7, isolated by James C. Cavender, University of Wisconsin, 1962.

Additional cultures investigated include: Strain CS–4, isolated by Cavender in 1962 from horse dung collected at Ciudad Serdan, Pueblo, Mexico; and strain FH–8, isolated by Carmen Stoianovitch in 1965 from deer dung collected in the State of Washington.

Of the three cultures available for study, strain CC–7 exhibits, as a rule, optimal development of the beaded sorocarps that are so characteristic of the species. Because of this, it has been investigated more exhaustively than the other isolates, and the following accounts are based upon it in large measure.

Cultivation

Dictyostelium rosarium, strain CC–7, was isolated using the technique developed in our laboratory for estimating the number and types of cellular slime molds in soil (Cavender and Raper, 1965). The sample of soil was suspended in sterile water, and small amounts of appropriate dilutions were spread over the surface of dilute hay infusion agar plates together with ca. 0.4 ml. aliquants of a dense suspension of pre-grown *Escherichia coli* cells. Cleared plaques first appeared in the bacterial film, followed by the development of wheel-like aggregations and subsequently of mature fructifications of distinctive pattern. Spores were removed from such sorocarps and transplanted to fresh plates cross-streaked with pure cultures of bacteria to obtain two-membered cultures.

A number of different bacteria were investigated as possible sources of nutrient, and of these *E. coli,* strain 281, and *Aerobacter aerogenes,* strain 900, proved to be most satisfactory. *Sarcina lutea,* strain 9341, also provided a good nutrient source if the bacteria were pre-grown, concentrated, and deposited on non-nutrient agar. Since the use of *Sarcina* afforded no obvious advantage, the two gram-negative species were selected for studying the developmental history of the new slime mold, and for comparing it with other species of *Dictyostelium* previously investigated. As the work progressed, *E. coli* was used almost exclusively as the nutrient source, and in the majority of cases a mixed suspension of bacteria and slime mold spores was used as an inoculum.

Cultures of several types have been employed, of which the following proved most useful: plates preinoculated with bacteria in two parallel streaks or spread as a broad band prior to the introduction of slime mold spores at selected sites; plates inoculated at five or six points, as a broad band, or, in most cases, as two parallel streaks with a mixed inoculum of bacteria and spores. A small loop (1 mm. diam.) was used for point inoculations, and a larger loop (3 or 5 mm. diam.) for streak cultures. If pre-grown bacteria were employed as nutrient, these were first grown in a nutrient broth for 24 to 36 hours, harvested, washed and concentrated by centrifugation, and then deposited on buffered nonnutrient agar as a dense suspension either before or concurrent with the introduction of slime mold spores.

Optimal growth and development of *Dictyostelium rosarium* was obtained with either *E. coli* or *A. aerogenes* upon substrates of relatively low nutrient content, of which lactose (0.1%)–peptone (0.1%) agar adjusted to pH 6.0 (2.05 g. KH_2PO_4 and 0.33 g. $Na_2HPO_4 \cdot 12H_2O$/1), designated "0.1L–P agar," has been used most frequently. Other media, similarly buffered, have been used successfully, including 0.1% glucose-peptone, 0.3% lactose-0.1% peptone, and 0.2% glucose-peptone agars. Good growth and development were obtained also on buffered nonnutrient agar with pre-grown cells of any of the aforementioned bacteria, but two-step cultures of this kind rarely proved to have special value. Hay infusion agar, much used in earlier studies of other cellular slime molds (Raper, 1937, 1951), did not favor optimal development, and hence was little used.

Growth of the myxamoebae can occur over a fairly wide pH range from 4.0 to $>$ 7.0; but the most useful range is much less, being approximately pH 5.0 to 6.2, with an optimum for growth and rate of development at 5.8 to 6.0. At levels near pH 7.0 myxamoeba growth is retarded, clearance of the bacterial streaks is often incomplete, and fructification is definitely suboptimal with sorocarps abnormal in form and lateral sori few in number. At pH 5.0 growth is reduced appreciably and aggregates form several hours later than at pH 6.0; the sorocarps that subsequently develop, however, are often well formed, and, perhaps significantly, limited numbers of these may arise without the substitution of clay lids (Coors unglazed porcelain) or exposure to charcoal (see p. 43).

The optimum temperature for myxamoeba growth is 23–25° C. At 30° C. growth is very slight or more often lacking, and at lower temperatures growth is progressively delayed. For example, plate cultures inoculated with parallel streaks of a mixed suspension of *E. coli* and slime mold spores may be expected to show developing aggregates within 46–48 hours when incubated at 25° C. If replicate plates are incubated at 20° C. or 15° C., this stage will not be reached until 65–70 hours and 85–90 hours, respectively. It is interesting to note, however, that the sorocarps which subsequently develop at these lower temperatures normally show an increase in the number and a decrease in the size of the lateral sori, the possible significance of which will be considered later (p. 43). Limited growth of the slime mold will occur at 10° C., and even 5°C., but for such tests pre-grown bacteria must be provided—either by outright addition or by pre-incubating the plates at a higher temperature before the spores of the slime mold are introduced. In either case, myxamoeba growth is very slow at these low temperatures, and periods of two to three weeks are required to secure fructifications, which are, as a rule, limited in number and poorly formed.

Light seems neither to stimulate nor to retard the growth of the myxamoebae, but current evidence suggests that it does accelerate fructification by several hours (Fig. 7). Incubation in one-sided light seems not to elicit any appreciable phototropic response, which is in marked contrast to the behavior of most species of *Dictyostelium.*

A saturated atmosphere is optimal for myxamoeba growth, while a slight reduction in atmospheric moisture tends to accelerate slightly but not to improve fructification. What does appear to be of primary importance for optimal sorocarp formation is some still unidentified alteration of the gaseous environment within the culture chamber at the time cell aggregation is completed. Since the effects of such alteration will later be considered at some length in relation to sorocarp development, it is sufficient to say that the postulated inimical gaseous environment can be corrected by (1) replacing the glass lid of a petri dish with an unglazed clay cover (Fig. 6), by (2) inverting the culture over a relatively thin layer of mineral oil, and (3) even more dramatically, by inverting the culture over activated charcoal (Norit).

Germination and Vegetation Growth

The onset of spore germination in *D. rosarium,* as in other species of *Dictyostelium,* is first revealed by a slight swelling of the spore body and by a lessening in the refractility of the cellulose wall, particularly as seen with phrase microscopy. This is accompanied by a limited movement of the cytoplasm within the spore, which gradually quickens as germination approaches. Meanwhile one or more contractile vacuoles appear which become increasingly prominent prior to protoplast emergence. Since the spores are globose to subglobose (Fig. 1B), they do not as a rule bulge outward at some point prior to germination, but instead show a gradual and progressive enlargement until the walls split and the nascent myxamoebae emerge. Following germination, the empty spore case typically appears as two valves, usually hinged at one side but at times separated completely (Fig. 1C).

Following germination, the myxamoebae begin almost immediately to feed by the ingestion and digestion of bacterial cells, grow to adult dimensions, and then divide. During the ensuing period of vegetative growth, the myxamoebae continue to feed and divide independently; in less than two days a very large cell population is built up and the available bacteria are almost wholly consumed. The generation time in *D. rosarium* has not been determined specifically, but it is believed to approximate the 2.6 to 2.9 hours reported for *D. discoideum* by Hohl and Raper (1963), based upon the time required to consume bacterial colonies of comparable density and to arrive at the aggregation stage.

When actively feeding inside a bacterial colony, the myxamoebae usually appear rounded or ovate in outline and are essentially immobile except for the occasional projection of lobate pseudopods to entrap additional bacteria. The feeding cells range from about 10 to 12 μ x 12 to 15 μ in diameter and, except for the conspicuous contractile vacuole(s), seemingly consist of a granular endoplasm with numerous food vacuoles containing bacteria in progressive stages of digestion, together with other vacuoles that contain globules of some amorphous waste material that is later ejected. The nucleus may or may not be apparent, and there is little evidence of an outer layer of hyaloplasm in such growing but sedentary cells.

Details of structure may be observed much more clearly in myxamoebae that are moving freely just outside the margin of the bacterial colony, as seen in Figure 1D. Such cells are inconstant in shape but commonly appear as broadly triangular or irregularly elongate and range in size from 12 to 18 μ in width by 18 to 25 μ in length. A thin, fan-like zone of hyaloplasm 2 to 4 μ wide is usually evident along the advancing front, from which comparatively broad pseudopods may be extended in rather sudden bursts. The endoplasm, comprising the bulk of the cell volume, appears finely granular and contains relatively few and generally small food vacuoles along with globules of the waste material mentioned above. A single nucleus occupies a position near the center of the myxamoeba. This appears as a clear area 2.5 to 3.0 μ in diameter that is easily deformed during cell movements and shows at its periphery two or more crescent-shaped bodies that are believed to represent nucleoli. Most conspicuous are the contractile vacuoles seen near the posterior end of the moving cells. These may occur singly and arise by the fusion of several smaller vacuoles, or they may occur in pairs (rarely more), each enlarging and discharging independently. As in other species of *Dictyostelium*, cells moving on an agar surface normally show few filose pseudopodia, and these occur most often at the posterior ends; in liquid mounts such structures are commonplace and may extend outward from almost any area of the cell surface, but most often from the anterior and posterior ends.

Aggregation

Under optimal conditions for growth—e.g., in association with *Escherichia coli* on 0.1L-P agar (pH 6.0) at 25° C.—*Dictyostelium rosarium* grows quite rapidly and will, as a rule, effectively consume the available bacteria within 40 to 44 hours following inoculation with a mixed suspension of bacteria and slime mold spores. In cultures of this age, the myxamoebae are present in a dense layer and evenly distributed over the agar surface, suggesting to the unaided eye a matte-like texture and under low magnification a delicately granular field. Soon thereafter, the process of cell aggregation begins, and in a general way this follows the same pattern observed in other species of *Dictyostelium* that produce sorocarps of comparable dimensions. There are, however, certain recognizable differences in the way this proceeds. Cell aggregation is not a continuous and uninterrupted process of myxamoeba convergence leading to sorocarp formation, but exhibits three rather distinct phases.

The first evidence of cell aggregation appears two or three hours later when myxamoebae in limited areas, up to 1.5 to 2.0 mm. in diameter and often polygonal in outline (Fig. 2A), become oriented toward ill-defined centers, producing patterns that are somewhat radial in character but lacking well-defined or persistent streams. The "streams" are, instead, thin and rather sheet-like, with the distal ends of those in one primary aggregation abutting those of the neighboring organizations. The inflowing myxamoebae are, at the same time, not conspicuously elongate or closely compacted, but tend to be fusiform in shape (Fig. $2A_1$) and semi-independent in their aggregative response. The stage just described is transitory and within an additional hour or two these small primary aggregations tend to break up, leaving the myxamoebae unevenly distributed in little hillocks and ridges of varying size and shape (Fig. 2B) until new and more commanding secondary centers of aggregation arise among them. These may arise *de novo*, or alternatively, a few of the pre-existing centers may assume dominance (Fig. 2C). In either case, these secondary aggregations may reach dimensions of 1.5 to 2.0 cm. in diameter, or even more, by the progressive extension of inflowing streams of very elongate, crowded, and uniformly oriented cells. Such aggregations, are, as a rule, quite irregular in pattern since the myxamoebae they attract are no longer uniformly distributed. As these aggregations expand, and perhaps in part because of their irregular form, the proximal areas of major streams first show a marked unevenness, then become increasingly nodular, and finally break up into masses of irregular shape and dimensions as the cells outward from these concentrations are blocked in their advance (Figs. 2D and E).

Once the large aggregations have become fragmented during this third phase of the aggregation process, the individual cell masses tend to round up, forming irregularly spaced mounds of varying size. If conditions inducive for fructification are provided, sorocarp formation will be initiated within a matter of a few hours (see pp. 37 to 43). If conditions that favor culmination are not provided, these mounds, at first characterized by their compactness and sharply defined outlines, tend to become increasingly diffuse, particularly at their margins, and over a period of two or three days will essentially disintegrate without producing any sorocarps, or at most few and generally abnormal structures. No true migration stage has been observed in *D. rosarium*, although cell masses that are so arrested in their development may show limited changes of position within their general sites of origin.

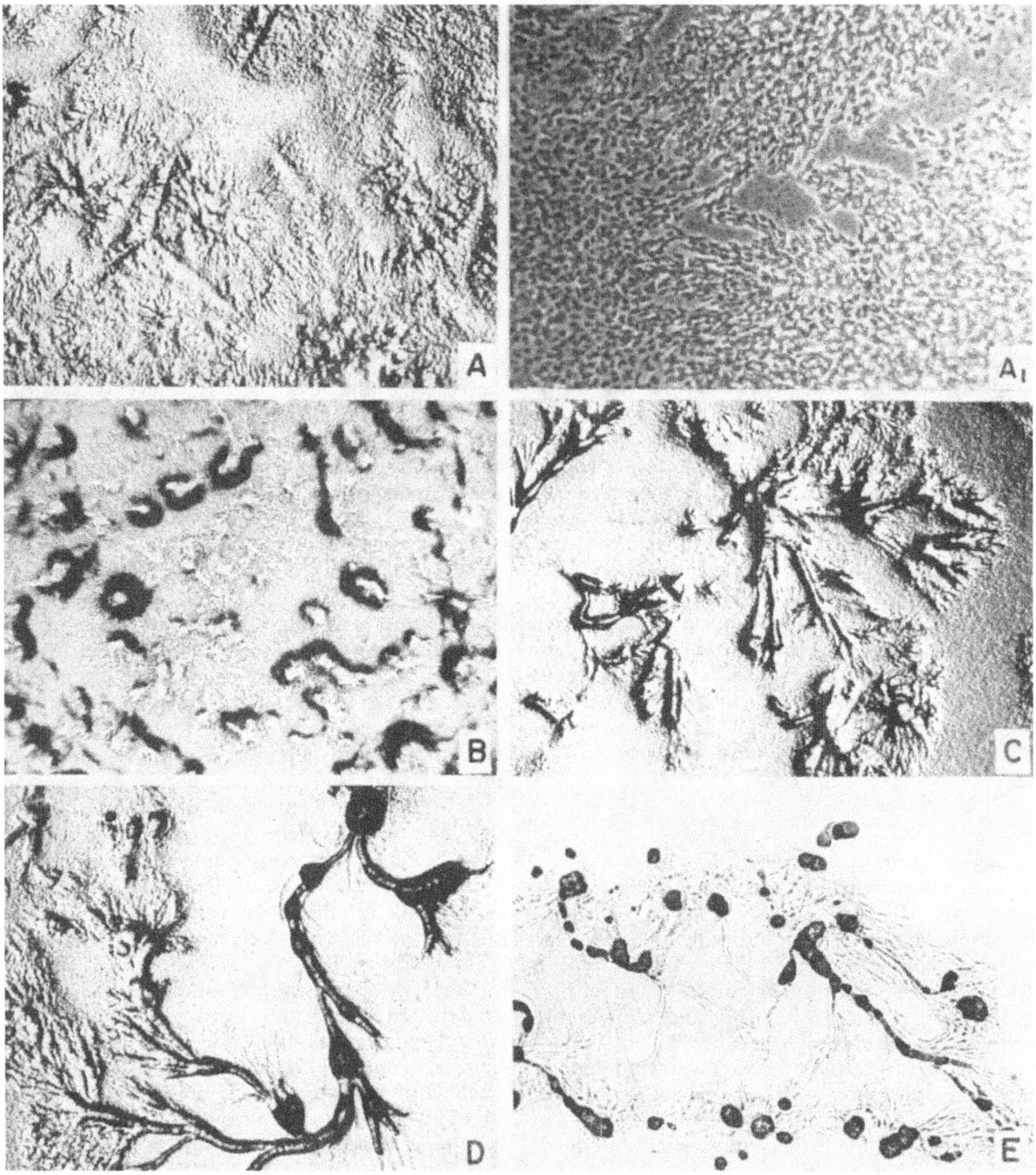

FIG. 2. Cell aggregation in *Dictyostelium rosarium* grown in association with *Escherichia coli* on 0.1L-P agar (pH 6.0) at 25° C. *A. Phase 1.* Small primary aggregations, some of which show the angular outlines that often characterize this phase. X10. *A.* Detail of the frontier between two primary aggregations. Note that the "streams" are rather diffuse, and that the converging myxamoebae tend to be fusiform in shape rather than very long and narrow. X110. *B.* A somewhat later stage during which the primary aggregations break up, leaving the myxamoebae scattered or in small, irregular clusters and ridges. X30. *C. Phase 2.* Secondary and much larger aggregations are beginning to take form by the realignment of cells that had essentially dispersed a short time before. X10. *D.* The aggregation process is further advanced and the streams nearest the center are becoming nodular due to the uneven advance of the inflowing cells. X10. *E. Phase 3.* Aggregation is nearing completion, and the few major streams that still remain are in process of becoming fragmented into irregular mounds of varying form and dimensions. This stage is normally reached at 2½ to 3 days. X5.

In any cellular slime mold that forms large aggregation, the imprint of the larger streams often persists on the agar surface, but in no other species have we observed these to be as prominent as in *D. rosarium* (Fig. 2E). This is believed to result in large part from the excessive quantity of waste material, ejected from the feeding myxamoebae as amorphous globules, that is pushed aside by and accumulates along the boundaries of the inflowing streams.

Sorocarp Formation

Under optimal conditions for fructification, to be described later, the mounds that remain after aggregation is completed soon show surface irregularities that subsequently emerge as definite papillae. Single papillae may arise from very small mounds, but more commonly these are multiple in origin and may range up to six or more for the larger cell masses. Stalk formation is then initiated in each papilla as it further develops into a low vertical column. As in other species of *Dictyostelium,* and in *Polysphondylium,* this is accomplished by the deposit of a thin, hyaline, and centrally positioned tube, the diameter of which is proportional to that of the column wherein it occurs. This tube, cellulosic in nature, is deposited in-

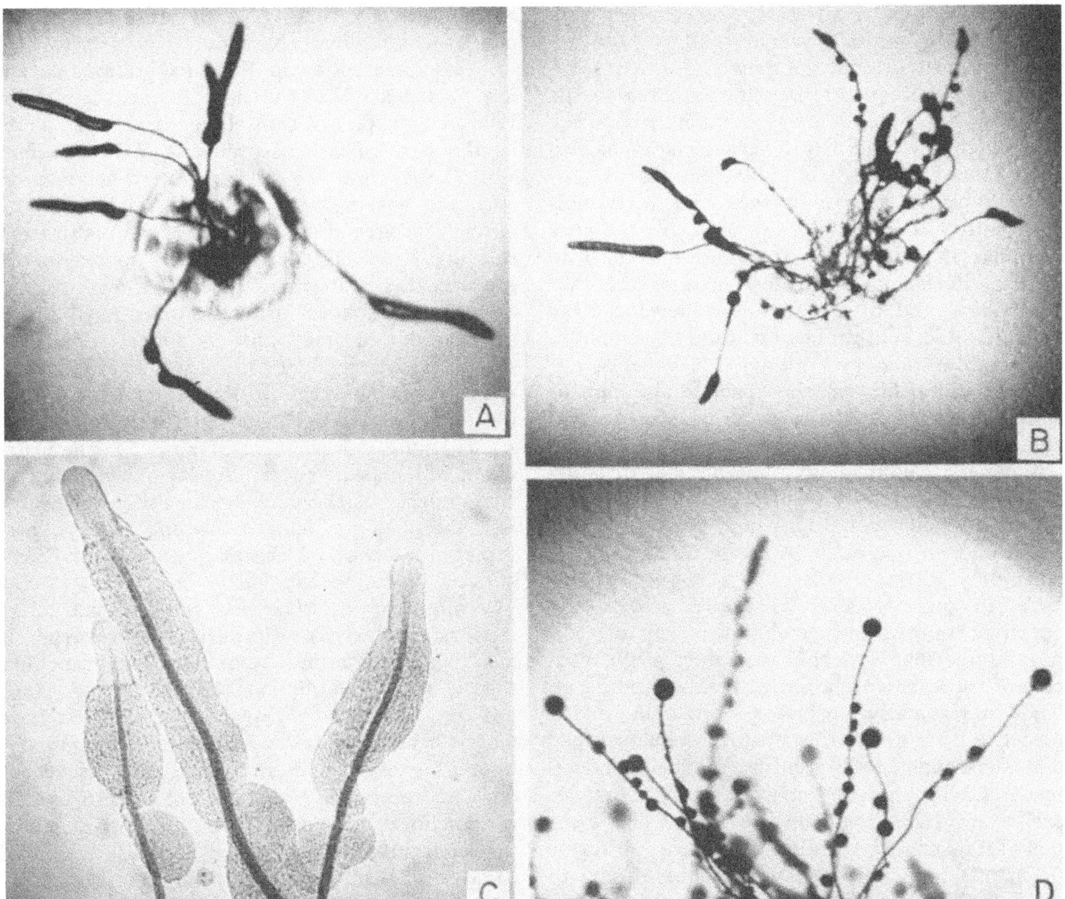

FIG. 3. Sorocarp formation in *Dictyostelium rosarium*. *A*. Seven sorogens, or masses of fructifying myxamoebae, arising from a small isolated colony of *Escherichia coli* on 0.1L-P agar. X12.5. *B*. A more advanced colony showing sorocarps in different stages of development. In some of these, small masses of cells that were left behind by the rising sorogens have already formed sessile lateral sori. X12.5. *C*. Three rising sorogens stained with chloroiodide of zinc are shown substantially enlarged. Note the stained sorophores (dark blue due to the presence of cellulose) and thin, funnel-like apices where formation of the sorophore sheath is in progress. Note also the posterior cell masses that are being abstricted and later develop into lateral sori. X75. *D*. A cluster of relatively small sorocarps viewed from the side to show their characteristic beaded appearance. An immature structure appears in the background and out of focus. X15.

tercellularly and, as it develops upward, constitutes the mold within which entrapped myxamoebae become strongly vacuolate to form the cellular elements of the stalk, or sorophore. At this early stage, the lengthening sorogen, or column of fructifying myxamoebae (Fig. 3A), is rather similar to that seen in *Dictyostelium mucoroides*, or, except for some differences in shape and the absence of a basal disk, to that described in detail by Raper and Fennell (1952) for *D. discoideum.*

Soon, however, marked differences in sorocarp formation become apparent. As the stalk continues to lengthen, and beginning soon after the sorogen has been lifted some distance above the agar surface, small masses of cells begin to be cut off from its posterior end. These remain in position on the newly formed sorophore and subsequently develop into the sessile lateral sori (Figs. 3B to D) that impart to the completed sorocarp its characteristic beaded appearance. The abstriction of groups of posteriorly placed cells may occur at fairly uniform intervals in some cases, or it may be very irregular in others; and the same may be said of the dimensions of the cell masses that are left behind and of the sori that develop from them. While substantial variation is commonplace rather than exceptional, the greatest degree of uniformity in the spacing and dimensions of the lateral sori occurs, as a general rule, when the environmental conditions are optimal for fructification. The number of lateral sori varies greatly in different sorocarps and under differing conditions and may at times reach 50 or more, while a range of 15 to 25 is commonly encountered.

As in other species of *Dictyostelium,* the apical portion of the developing sorophore appears somewhat funnel-shaped during the process of sorocarp formation (Figs. 3C and $5B_1$). The horizontally oriented pre-stalk myxamoebae first form the hyaline cellulose tube and subsequently differentiate as cellular elements within this in the same manner as described for *D. discoideum* by Raper and Fennell (1952). Meanwhile, the more posterior unoriented pre-spore cells contribute in part to the lateral sori and finally to the terminal sorus, which is, as a rule, larger than any sorus previously formed. As in *D. discoideum,* the diameter of the cellulose tube (and hence of the resulting sorophore) is roughly proportional to the diameter of the apical region of the sorogen. The length of the sorophore, on the other hand, varies greatly and is determined by the quantity of myxamoebae that have undergone a preliminary differentiation as pre-stalk cells, usually at a fairly early stage in sorocarp formation. Thus if the rising sorogen is relatively long and narrow, the resulting sorophore will be comparatively long and thin and bear numerous and small lateral sori; conversely, if the sorogen is relatively short and thick, the diameter of the sorophore will be greater, its length substantially less, and the number of lateral sori much reduced.

When cut free from the main body of the sorogen, each cell mass that remains behind is still encased by a portion of the enveloping slime sheath (Fig. 3C), which in turn persists for some time as a surface covering and serves to anchor the fractional mass securely to the sorophore (Fig. $4A_4$). With time, and as the constituent myxamoebae round up preparatory to spore differentiation, this membranous covering disappears and the fully formed spores lie free in a droplet of thin slime, as do the spores in the sori of other species of *Dictyostelium* (Figs. $4A_2$ and $4A_5$). Thus in a developing sorocarp bearing several lateral sori, one may observe progressive stages in spore maturation ranging from fully differentiated spores in the lowermost sori to myxamoebae showing little or no evidence of differentiation in the cell masses four or five positions above. This progressive change can be observed by direct microscopic observation of living material, but it is far more pronounced in preparations stained with chloroiodide of zinc, in which the older cell masses (sori) appear purple, due to the staining of the cellulose in the spore walls, and the younger ones show only the yellow-brown iodine stain characteristic of undifferentiated myxamoebae.

While *Dictyostelium rosarium* is characterized particularly by its beaded sorocarps, it also possesses a unique pattern of branching that is truly dichotomous (Figs. 5A, $5A_1$, 5B, and $5B_2$). In all other species of *Dictyostelium* where branching occurs (and in *Polysphondylium* as well), the branches develop as secondary and essentially independent structures that are anchored to the upright sorophore in much the same manner as the sorophore is anchored to the substratum. Such a branch invariably arises secondary to the main axis of the sorocarp and is discontinuous with it. Moreover the base of the branch is somewhat enlarged, usually rounded, and anchored to the supporting stalk by an apron of congealed slime left in position by the mass of cells from which it developed. Not so in *Dictyostelium rosarium.* In this slime mold the tip of the sorogen first becomes flattened and then splits into two parts, followed soon thereafter by the bifurcation of

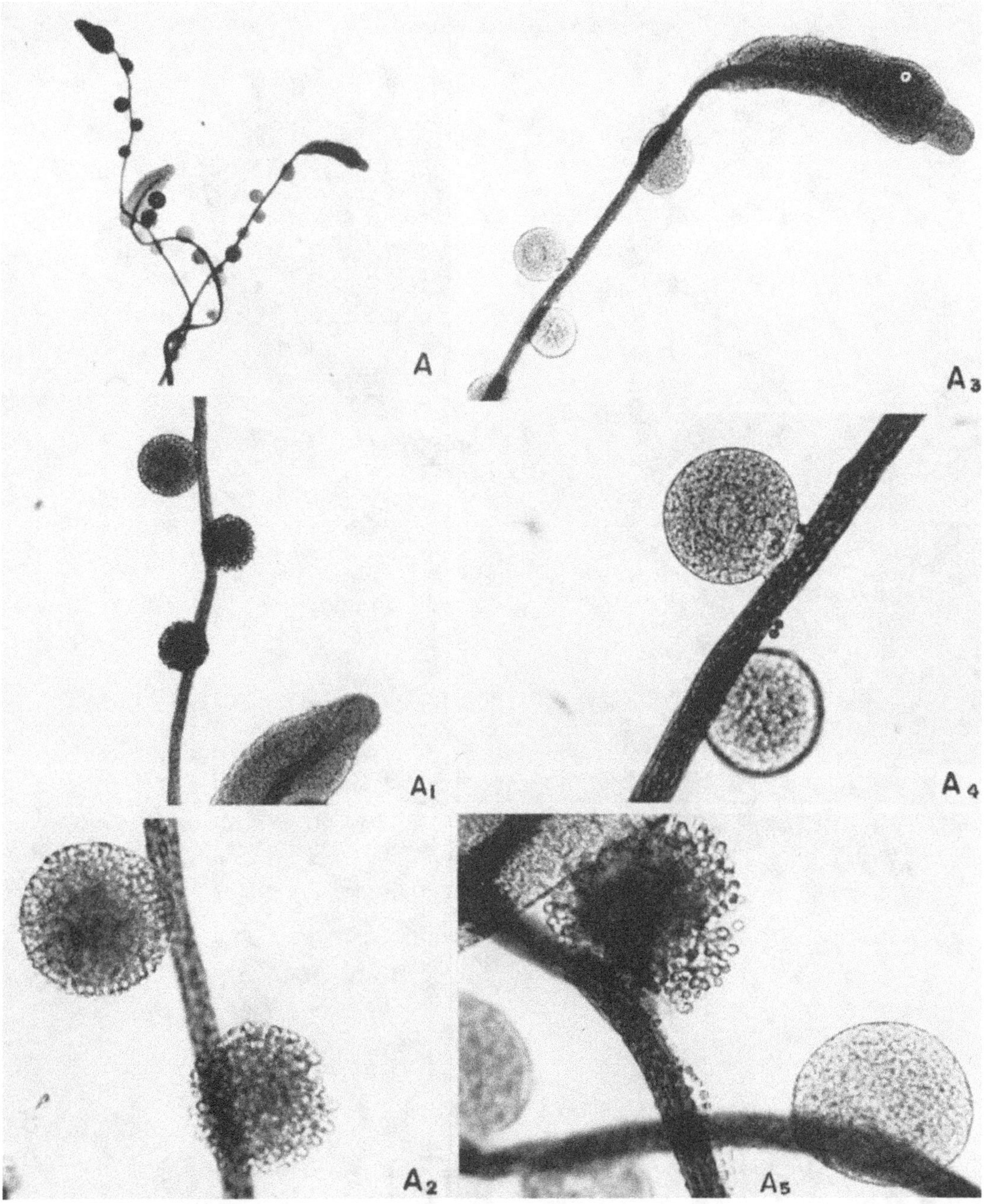

FIG. 4. Sorocarp formation in *Dictyostelium rosarium* (cont'd). *A*. Three developing sorocarps stained with chloroiodide of zinc. The one at upper left is most advanced and that at the right less so, whereas the one at left center is still relatively young. X25. *A1.* A portion of the sorocarp at left somewhat enlarged, together with a part of the sorogen of the youngest structure in which the funnel-like terminus of the sorophore may be seen. X110. A_2. Two of the three lateral sori of A_1 are further enlarged. Note that the cells in the upper (and younger) sorus have assumed a spherical shape but are still confined by a delicate membrane, or slime envelope, while those in the lower sorus appear to lie free. X270. A_3. The structure at the right in *A* is enlarged substantially and shows the characteristic apical tip of the sorogen as well as four developing lateral sori below. X110. A_4. Two of these lateral sori are shown further enlarged. Note that the uppermost, consisting of still undifferentiated cells, is clearly confined by a continuous slime covering that anchors it to the sorophore. X270. A_5. Shown here is an immature sorus (lower right) borne on the central sorocarp, and an older, fully mature sorus (upper left) formed on that at the left in *A*. Note that the spores of the latter are quite free, the slime in which they were earlier suspended having dispersed in the staining reagent. X270.

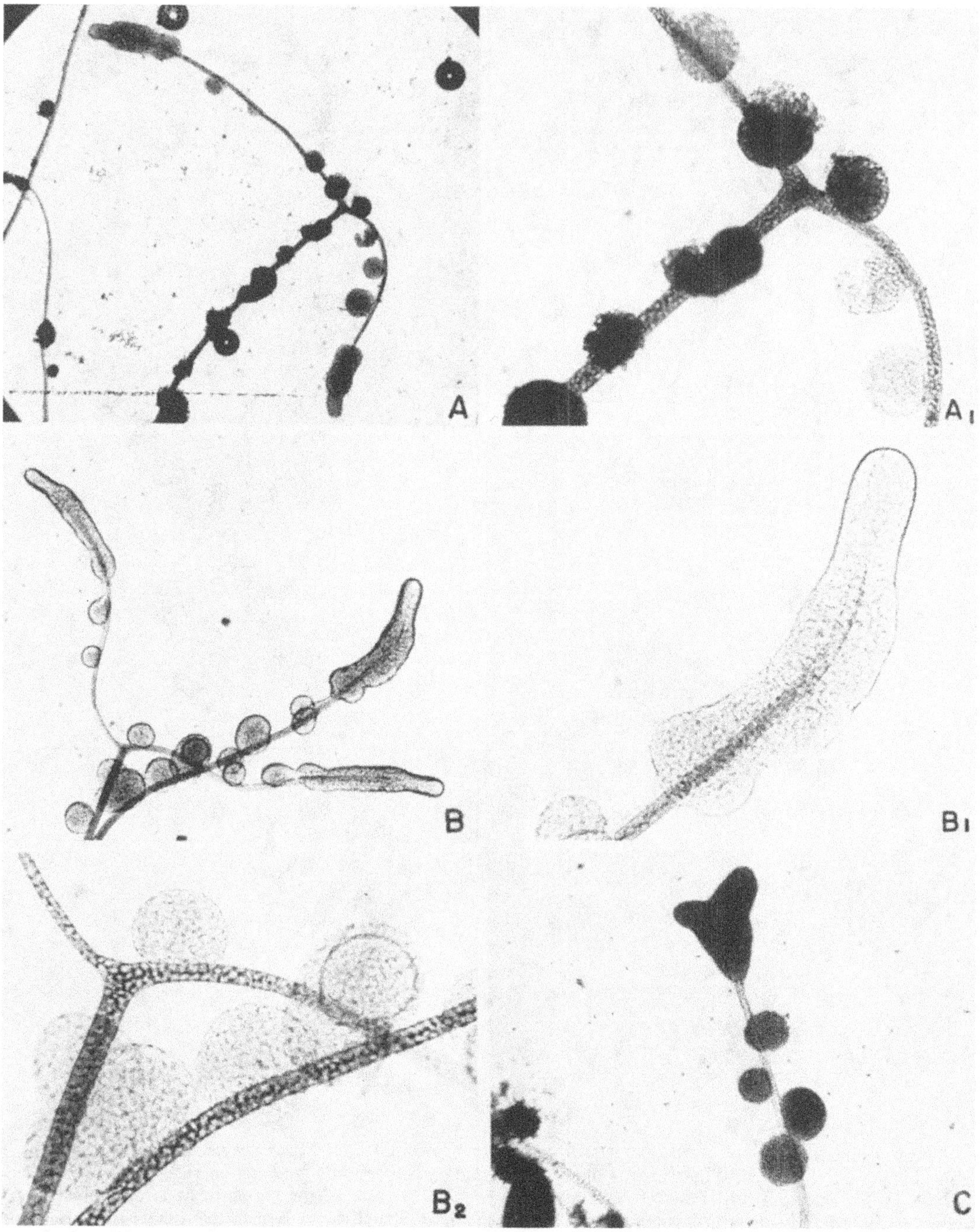

FIG. 5. Dichotomous branching of sorocarps in *Dictyostelium rosarium*. Structures were stained with chloroiodide of zinc. *A*. A comparatively large branched sorocarp bearing two sorogens within each of which stalk formation is actively taking place. X50. A_1. A portion of the same sorocarp somewhat enlarged to show the continuity of the stalk at the site of branching. Shown also are some details of both mature and immature lateral sori. X160. *B*. Branched and unbranched sorocarps are shown during the process of formation. X65. B_1. A portion of this complex is enlarged to reveal details of structure in the sorogen, shown at center right in *B*. X210. B_2. A portion of the branched sorocarp in *B* enlarged. Note the uniformity of cellular structure throughout the area of branching. X265. *C*. A sorogen in process of division. Unfortunately, the stain is too dark to reveal the newly branched sorophore within. X80.

the hyaline cellulose tube that is being formed within it. In a very real sense the phenomenon is comparable to the longitudinal splitting of a migrating pseudoplasmodium of *D. discoideum*, except in that case such division occurs when all the associated cells are still amoeboid and stalk formation has not yet begun. Even so, the situation may be more nearly comparable than seems apparent at first. Subdivision of the migrating slug in *D. discoideum* is believed to occur when the viscosity of the slime covering at the precise tip exceeds, for some reason, that of the covering at either side (Francis, 1962) and, as a result, direct forward movement of the myxamoebae is impeded while such movement still continues in a slightly subapical region. This results first in a widening of the slug apex and subsequently the emergence of two new apices at either side of the original position. The same may occur in *D. rosarium*, even though the mass of fructifying cells is completely surrounded by air rather than being only partially exposed, as would be true of a slug resting on agar. If we assume a comparable cessation of forward movement by cells at the tip of the rising sorogen of *D. rosarium*, a corresponding flattening of the frontal surface should first occur followed by the emergence of new aspices, as can actually be observed. With continued development, the circular site of cellulose deposition would soon impinge upon the barrier, and with sorophore formation continuing apace, as it obviously does, the lengthening funnel-shaped sorophore sheath might be expected to separate into parts with further extension occurring in each of the new apical tips. What this does not explain is why similar branching fails to occur in the rising sorogens of other species of *Dictyostelium*, or why only two apices have been observed to arise at any given site of arrested forward movement. While detailed histological studies of the branched sorophores have not been made, it is obvious that the transition from a single stem to a branched structure is a smooth and *continuous* one (Figs. $5A_1$ and $5B_2$). The two divergent branches are, as a rule, of approximately equal diameter, as seen in Fig. $5A_1$; but unequal divisions of the sorogen do occur, as seen in Fig. 5C, where division is in progress, and in Fig. $5B_2$, where such division had earlier occurred. While the relationship is not precise, it is of interest that the combined cross-sectional areas of the two branches in most cases approximate rather closely that of the sorophore below the point of dichotomy, suggesting neither an increase nor a reduction in the number of myxamoebae already committed to stalk formation.

Influence of the Cultural Environment

The formation of normal sorocarps in any cellular slime mold is dependent upon a favorable cultural environment, but in no other species is this so limiting or so obvious as in *Dictyostelium rosarium*. From the outset, difficulties were encountered in obtaining reasonably satisfactory fructification of this species. Cultures of a type long employed for *D. mucoroides*, *D. purpureum*, and *D. discoideum*—species that are of the same general dimensions—often failed to show any sorocarps for several days following cell aggregation, and when these did appear they were as a rule limited in number and confined to localized areas within the streaks or broad bands where the myxamoebae had grown. Growth occurred in a normal way, followed by cell aggregation that seemingly differed only in a minor way from that in the aforementioned species. But following this, development was usually arrested and few if any sorocarps were formed for a period of several days. Meanwhile the mounds of cells from which these might have developed tended to disintegrate, or at least to appear increasingly diffuse. At times, and in limited areas, these would subsequently reassemble to a limited degree, and from such later cell concentrations clusters of sorocarps might arise, oftentimes a week or more after the time of inoculation (Fig. 1A).

This pattern of behavior suggested that fructification might be favored by more osmophilic conditions, but deliberate attempts to create such an environment by increasing the rigidity of the agar base or by other means failed to elicit a more favorable response. It was also envisioned that fructification might occur as a result of reduced atmospheric moisture, as one could assume gradually occurred as the cultures aged. This possibility seemed all the more plausible in view of the responses Whittingham and Raper (1957) had previously obtained with *D. polycephalum*. Tests were made in which sterile, unglazed clay covers were substituted for glass petri plate lids at 2½ to 3 days, i.e., after aggregation had been completed and the resulting cell mounds were still intact. Satisfactory fructification was found to have occurred in these cultures when examined 15 to 18 hours later, while control plates with glass lids still showed only mounds of amoeboid cells (Fig. 6). Plate cultures at a similar stage of development were next inverted over activated charcoal (Norit) placed in the petri dish

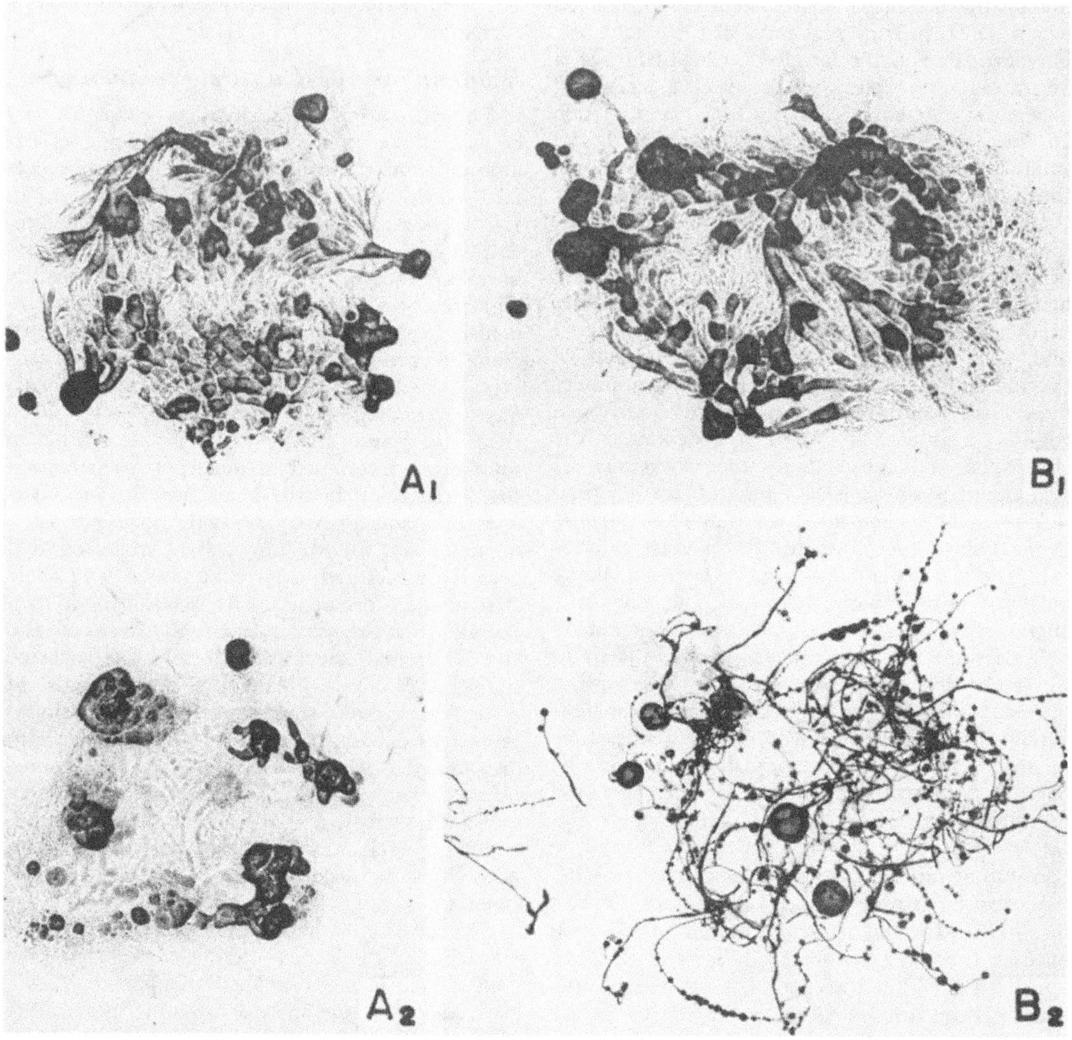

FIG. 6. Sorocarp induction in *Dictyostelium rosarium*. Cultures were grown in glass petri plates with *Escherchia coli* on 0.1L-P agar (pH 6.0) at 25° C. in fluorescent light. A_1. A small colony in which the slime mold has reached a rather advanced stage of aggregation, photographed at three days. A_2. The same colony photographed 20 hours later with incubation continued under glass. The myxamoebae have collected into irregular masses that now begin to show some evidence of disassociation. B_1. A colony similar to A_1, also photographed at three days. B_2. The same colony photographed 20 hours after an unglazed porcelain cover was substituted for the conventional glass lid. Virtually the entire cell population has now produced well-formed sorocarps. X7.

lids, and in these tests sorocarp formation was fully equal to that under the clay covers, with regard to both the time of formation and the pattern of the sorocarps produced. In each of these cases, some reduction in atmospheric moisture within the culture chambers could be assumed, either by escape through the porous clay or by absorption in the charcoal. Thus a gradual reduction in humidity seemed to afford a reasonable explanation for sorocarp induction.

At the same time, it was recalled that Bonner and Dodd (1962) had employed charcoal as one means of confirming the phenomenon of gas-induced orientation of sorocarps in *D. discoideum* and other cellular slime molds, and that Kahn (1964) had subsequently found charcoal to enhance aggregation in *Polysphondylium pallidum*, a response that he attributed to the adsorption of a gaseous suppressor of aggregation. To differentiate between the suggested beneficial effects of reduced atmospheric moisture and a possible gaseous inhibitor, cultures of several types were then prepared. Using deep Pyrex storage dishes (80 mm. deep x 100 mm. wide) over which the bottoms of petri plates containing slime mold cultures of different ages could be fitted, tests were conducted in chambers of the following types: (1) Four grams dry, sterile Norit were placed in the bottom dishes, over which cultures of the slime mold were inverted. (2) Similar amounts of Norit were deposited on 1-cm. layers of solidified nonnutrient agar in the bottom dishes, over which cultures were inverted. (3) Storage dishes were prepared as for the preceding, but with the inverted slime mold cultures sealed to the dishes with plasticine. Many tests were performed, of which the following is representative. Three-day-old streak cultures of *D. rosarium* in the post-aggregation mound stage (previously grown with *E. coli* on 0.1L-P agar (pH 6.0) at 25° C.) were inverted over the dishes containing charcoal and placed in one-sided light. Similar cultures inverted over storage dishes with and without added agar served as controls, some of these being sealed with plasticine while others remained unsealed. Incubation was continued at 25° C., and the following behavior was observed. No sorocarps developed in any of the controls during the next 36 hours. In contrast, many and well-formed sorocarps were produced within 15 to 18 hours in each of the cultures that had been inverted over charcoal. Sorocarps produced over the "dry charcoal" were, on the average, somewhat shorter than those formed over the agar and charcoal, which was already wet when the cultures were superimposed. Additionally, there was some evidence that the process of fructification had proceeded somewhat more rapidly in the former vessels than in the latter. The sorocarps present in the vessels containing both agar and charcoal, including those sealed with plasticine, were for the most part beautifully proportioned, with long, slender sorophores and many and fairly uniform lateral sori. It was thus clear that optimal fructification could proceed in a fully saturated atmosphere, albeit this might occur less rapidly than if the moisture level were slightly reduced.

In subsequent experiments, petri plates with bottoms 20 mm. deep were employed, and when the cultures grown therein had reached the mound stage, some were inverted over dry charcoal with a glass-to-glass closure, while others were inverted over charcoal placed on agar so that an airtight glass-to-agar seal was effected. Good sorocarp formation occurred within the ensuing 15- to 18-hour period in both types of culture, but it was apparent that culmination proceeded faster (three to four hours), and the resulting sorocarps were as a rule measurably shorter, in the plates with a glass-to-glass closure than in those where the glass rim of the culture-containing dish rested upon agar. In this as in the preceding tests, no sorocarps were observed in charcoal-free control plates.

In another experiment, cultures were grown in petri plates, removed when they reached the post-aggregative (mound) stage, and inverted over 1-cm. layers of sterile mineral oil contained in storage dishes of the type described above. Here, also, abundant sorocarps developed within a period of 15–18 hours, which in their general conformation approximated those produced over dry charcoal.

While we cannot at this time wholly dismiss the possible role of slightly reduced atmospheric moisture as a factor in sorocarp formation, we can assert with reasonable confidence that such reduction is not required, and that its role, if any, is one of acceleration rather than induction. Evidence accumulated to date from many types of experiments strongly suggests that sorocarp formation in this slime mold is prevented or greatly delayed by the accumulation within the culture vessel of some volatile, gaseous compound that is, apparently, self-generated during myxamoeba growth and/or cell aggregation. Sorocarp formation, it would seem, can be initiated and proceeds to completion only when this gaseous substance is reduced to a level that is no longer inhibitory. Seemingly, such reduction may be achieved in several ways, including: (1) removal, as by adsorption on charcoal or absorption in mineral oil; (2) affording a means of escape, as through porous clay lids when substituted for glass; or (3) substantial dilution, as must occur with time in cultures grown in conventional glass petri plates in which a constant, if limited, gas exchange undoubtedly occurs.

In this connection, mention should be made of certain cases in which reasonably good sorocarp formation may occur in the absence of deliberate attempts to induce or enhance this process; and these, we believe, can be explained by some modification of the "dilution effect"—namely, that the postulated gaseous inhibitor never attains or soon falls below a critical concentration. Such is thought to be the case when cultures are grown at lower temperatures at which the growth of *E. coli* and of the slime mold are both reduced and delayed. For example, in streak cultures on 0.1L-P agar (pH 6.0) at 15° C., the post-aggregative (mound) stage is not reached until five days after inoculation, but sorogens begin to appear under glass after seven or eight days, and many thin but well-formed sorocarps often develop at ten to twelve days. Here, too, culmination can be accelerated by substituting clay covers, or by exposure to charcoal at five or six days, but the significant point is that it can proceed normally in the absence of such manipulation. If the inhibitory gas is in fact self-generated by the growing and/or aggregating myxamoebae, it is only reasonable to assume that the production and concentration of this in any given culture would be lessened by (1) a reduction in the growth of the associated microorganism, and (2) the slower rate of metabolism associated with lower incubation temperatures. There is also limited evidence that fructification under glass is favored by media at pH 5.0 rather than 6.0, which may again stem from a reduction in bacterial and slime mold growth.

The nature of the postulated inhibitory gas is still unknown. From a variety of experiments we know only that it is not CO_2 and that it is essentially neutral in reaction. Investigations are in progress to determine its composition and, if possible, its mode of action. Hopefully, these questions can be resolved during the months ahead.

Light also exerts an effect by accelerating fructification, although sorocarps will develop in total darkness when given longer periods of incubation. Some measure of the light effect is shown in Figure 7. Cultures of *Dictyostelium*

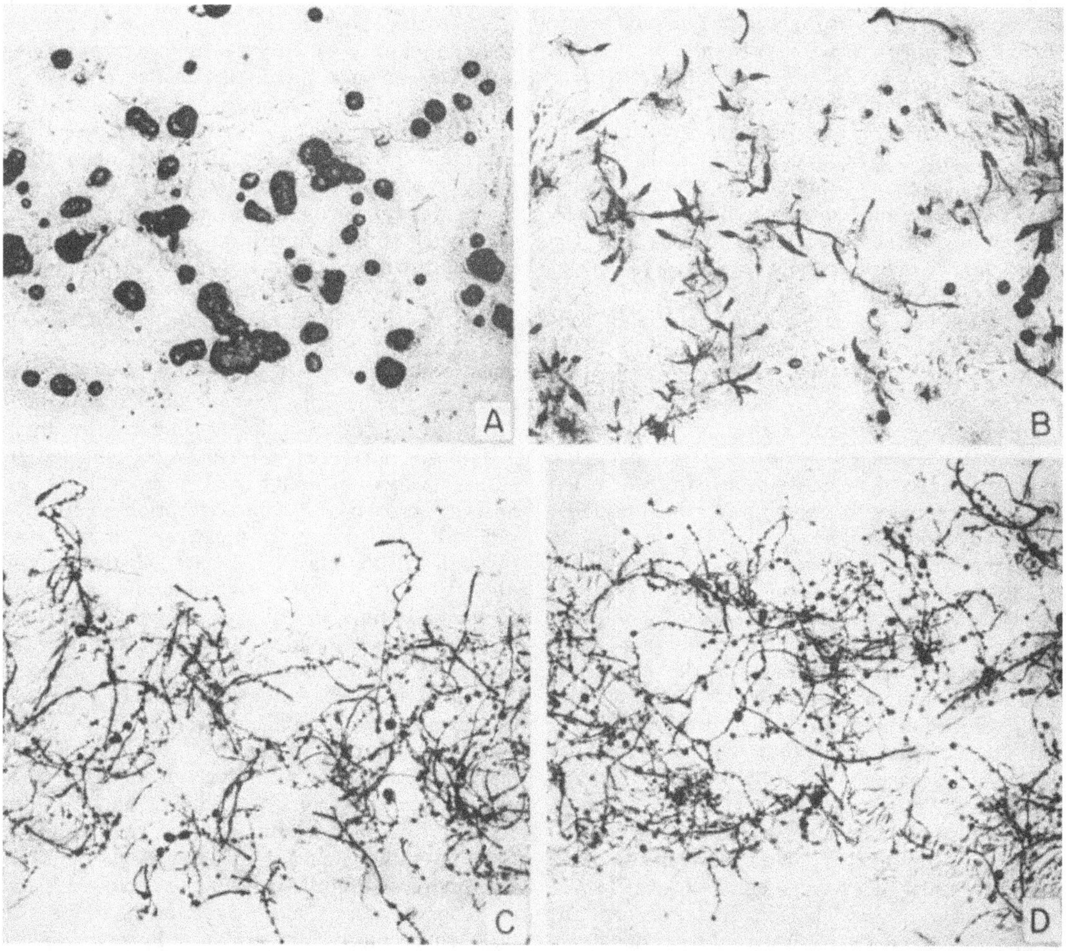

FIG. 7. Acceleration of sorocarp formation by light. Cultures of *Dictyostelium rosarium* were grown on 0.1L-P agar (pH 6.0) at 25° C. for two days in conventional petri plates, after which the glass lids were replaced by unglazed porcelain (clay) covers. All cultures were photographed at three days. *A*. Culture incubated in continuous darkness. *B*. Culture incubated two days in darkness, then in a black box with light entering through a narrow slit. *C*. Culture incubated two days in darkness, one day in light. *D*. Culture incubated in continuous light. X6.

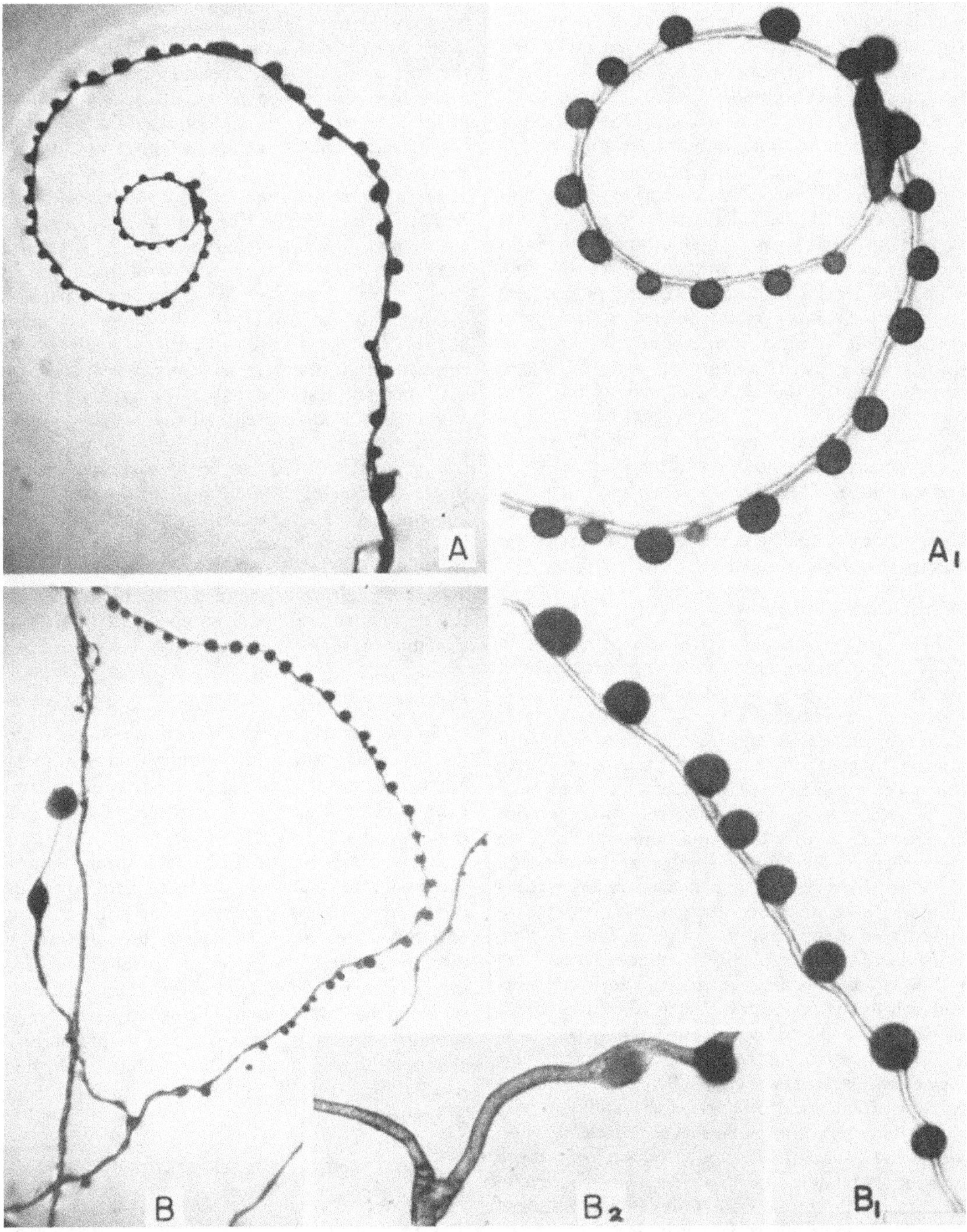

FIG. 8. Prostrate sorocarps of *Dictyostelium rosarium*, strain CS–4. Cultures grown in association with *Escherichia coli* on 0.1L-P agar and incubated at 20° C. for three days under glass, then five days under unglazed porcelain lids. *A*. Sorocarp developed in contact with the agar showing more than 50 sessile sori. X15. A_1. A portion of *A*, including the still advancing sorogen, enlarged to show greater detail. X60. *B*. A branched sorocarp in which one branch (left) produced an aerial sorus, while the sorogen of the other (right) continued to develop in contact with the agar surface. X15. B_1 and B_2. Portions of the same sorocarp enlarged. X60.

rosarium with *E. coli* were grown on 0.1L-P agar for 2 days in conventional petri dishes, after which the glass covers were removed and unglazed porcelain covers substituted for them. All cultures were photographed at three days. The culture shown in *A* was incubated in *continuous darkness.* The myxamoebae have collected into numerous rounded mounds that show no evidence of sorocarp formation. The culture shown in *B* was incubated two days in the dark, then placed in a black box with light entering only through a narrow slit (2 mm. wide) from a 15W fluorescent tube placed two feet from the box. Abundant *young* sorogens are now developing throughout the area of slime mold growth. Culture *C* was incubated two days in the dark, then placed in full light two feet from a 15W fluorescent tube. Note that many sorocarps are still in the process of development and show terminal clavate sorogens. Culture *D* was incubated in *continuous and full light* from a 15W fluorescent tube. Note that virtually all sorocarps are now mature and bear globose terminal sori.

Additional Cultures

The foregoing account is based upon strain CC–7, which represents the type and forms the primary basis for description of the new species, *Dictyostelium rosarium.* Two other isolates have been investigated, however, that possess in varying degree the characters that distinguish this species, namely: sessile lateral sori, globose to subglobose spores, and true dichotomous branching. While these obviously belong in *D. rosarium,* they differ somewhat from strain CC–7 in their general appearance in laboratory cultures. One of these, strain CS–4, isolated in 1962 from dung collected in Ciudad Serdan, Pueblo, Mexico, typically produces erect and rather stout sorocarps with large terminal sori and relatively few lateral sori if cultivated at 25° C. and exposed to charcoal or placed under a clay lid when aggregation is complete. At lower temperatures (e.g., 20° C.) sorocarps usually produce more lateral sori; and, for reasons unknown, these sorocarps often develop along the agar surface and produce very large numbers of lateral sori, as seen in Figure 8. The sorocarp shown in *8A* developed outward from the colony where it originated and, as the sorogen progressed, masses of posteriorly placed myxamoebae periodically lagged behind to form lateral sori. In A_1 the terminal portion of the same, including the sorogen, is shown enlarged four times. A branched sorocarp is shown in *B.* The branch at the left formed an aerial sorus (now collapsed); the other continued to develop along the agar surface and produced many lateral sori. Portions of the latter branch are enlarged four times to show the sori in B_1 and the area of branching in B_2, respectively. The second strain, FH–8, isolated in 1965 from deer dung collected in the State of Washington, is somewhat intermediate between strains CC–7 and CS–4 in its general appearance and development. However, it is, withal, more nearly like the latter. Whereas erect and delicately proportioned sorocarps with many lateral sori are often produced in clay covered dishes at 20° C., it is not possible at this time to anticipate their formation in either CS–4 or FH–8 with the high degree of certainty that the type strain affords.

In prostrate sorocarps of the type shown in Figure 8, the abstricted cell masses may collect on the upper surface of the sorophore, where the cells differentiate to form sori that remain intact (as shown). If situated in a more nearly horizontal position and in contact with the agar the sori will collapse when the spores mature. In other and relatively rare cases the small masses of cells may move a short distance from the sorophore and form independent sori, occasionally on very short stalks.

Epilogue

To one interested in the cellular slime molds *per se, Dictyostelium rosarium* is singularly attractive, for it exhibits many new if still unexplained developmental features. To the general biologist, and particularly to the student of morphogenesis, it should have a special appeal, for, as we have shown, it can be held for many hours in a post-aggregative but still undifferentiated state. It can also, at the pleasure of the investigator, be induced to enter its final morphogenetic phase in which the constituent myxamoebae proceed to differentiate either as strongly vacuolate stalk cells or as propagative spores. Equally important, the slime mold does these things quickly and in a reasonably orderly manner.

ACKNOWLEDGMENTS

The writers wish to express their sincere appreciation to Dr. Claude E. Vézina, Montreal, Canada, for preparing the Latin diagnosis for the new species.

This study was supported in part by research grants from the National Science Foundation (G–24953) and the National Institutes of Health (AI–04915), United States Public Health Service.

REFERENCES

BONNER, J. T., AND M. R. DODD. 1962. Evidence for gas-induced orientation in the cellular slime molds. Develop. Biol. **5**: 344–361.

BREFELD, O. 1869. *Dictyostelium mucoroides*. Ein neuer Organismus aus der Verwandschaft der Myxomyceten. Untersuchungen aus der Gesammtgebeit der Mycologie **6**: 1–34.

CAVENDER, J. C., AND K. B. RAPER. 1965. The Acrasieae in nature. I. Isolation. Am. J. Bot. **52**: 294–296.

———, AND ———. In Press. The occurrence and distribution of Acrasieae in forests of tropical and subtropical America. Am. J. Bot.

FRANCIS, D. W. 1962. The movement of pseudoplasmodia of *Dictyostelium discoideum*. Ph.D. thesis. University of Wisconsin.

HOHL, H.-R., AND K. B. RAPER. 1963. Nutrition of cellular slime molds. I. Growth on living and dead bacteria. J. Bact. **85**: 191–198.

KAHN, A. J. 1964. The influence of light on cell aggregation in *Polysphondylium pallidum*. Biol. Bull. **127**: 85–96.

RAPER, K. B. 1937. Growth and development of *Dictyostelium discoideum* with different bacterial associates. J. Agr. Res. **55**: 289–316.

———. 1951. Isolation, cultivation, and conservation of simple slime molds. Quart. Rev. Biol. **26**: 169–190.

RAPER, K. B., AND J. C. CAVENDER, 1966. Fructification in a unique cellular slime mold. Science **154**: 426 (abstract).

RAPER, K. B., AND D. I. FENNELL. 1952. Stalk formation in *Dictyostelium*. Bull. Torrey Bot. Club. **79**: 25–51.

RAPER, K. B., AND M. S. QUINLAN. 1958. *Acytostelium leptosomum*: A unique cellular slime mold with an acellular stalk. J. Gen. Microbiol. **18**: 16–32.

TIEGHEM, P. VAN. 1880. Sur quelques Myxomycètes à plasmode aggrégé. Bull. Soc. Bot. France **27**: 317–322.

WHITTINGHAM, W. F., AND K. B. RAPER. 1957. Environmental factors influencing the growth and fructification of *Dictyostelium polycephalum*. Am. J. Bot. **44**: 619–627.

Taxonomic Studies in the Myxomycetes. II. *Physarina*[1,2]

CONSTANTINE J. ALEXOPOULOS AND MEREDITH BLACKWELL
Department of Botany, University of Texas, Austin, Texas

The genus *Physarina* was erected in 1909 by von Höhnel to accommodate a new species of Myxomycetes, *P. echinocephala,* collected on decaying plant stems in Buitenzorg, Java. It differs from the closely allied genus *Diderma* in that the sporangia of *Physarina* bear numerous blunt, cylindrical, calcareous projections on the peridium. Up to 1964, *Physarina* remained a monotypic genus. In that year, Thind and Manocha (1964) described *P. echinospora* from the Mussoorie Hills of India. Both species have been known until now only from their type localities and are apparently rare.

On a trip from Austin, Texas, to Mazatlan, Mexico, in December, 1965, the first author collected plant litter from various localities for moist chamber culture. The material from each collection was placed in an unused paper envelope and sealed in the field. Upon return to Austin, the material was distributed in sterile petri dishes, each containing a sterile filter paper disc, and allowed to soak in sterile distilled water overnight. The excess water was then poured off and the dishes stored on a laboratory shelf at room temperature. Cultures were examined periodically for myxomycete fructifications.

About six weeks after the cultures were initiated, 18–20 sporangia of a physaraceous myxomycete, bearing blunt, white, more or less cylindrical, peg-like protrusions had developed on the side of a petri dish in which a dead leaf, collected on December 30, 1965, at Buenos Aires Point, Durango, Mexico, had been placed. The sporangia were subsessile and somewhat aberrant and the culture was overgrown with filamentous fungi, but the characters of *Physarina* were unmistakable. Spores obtained from one of the sporangia germinated on cornmeal agar, and small white plasmodia eventually developed. These fruited on the agar, producing typical *Physarina* sporangia. The organism has been carried in culture for three generations, but is becoming increasingly difficult to cultivate.

[1] Article I of this series was published in Mycologia **59**: 103–116, 1967.
[2] Supported by National Science Foundation Grant GB–2738.

In order to identify the Mexican isolate, a comparison with the two known species now became necessary. Prof. G. W. Martin very kindly provided a portion of the type of *P. echinospora.* The type of *P. echinocephala* is not available for examination without making a trip to London. Lister (1925), however, mentions a specimen collected by Ernst in Java which is on deposit at the Zürich herbarium. This obviously authentic specimen was borrowed by us through the courtesy of Prof. F. Markgraff.

The Mexican Isolate

The following discussion of the Mexican isolate is based on material developed on half-strength Difco corn meal agar (CM/2) prepared by dissolving 8.5 g. Difco cornmeal agar and 11.5 g. Difco Bacto agar in 1 liter distilled water.

1. *Fructifications.* The fruiting bodies of the Mexican isolate developed in agar culture are either subsessile or stipitate. The more robust and obviously better-developed fructifications are stalked (Fig. 3) and reach a total height of 0.75 mm. In well-formed fruiting bodies the stalks are 0.25–0.5 mm. long, cylindrical, somewhat fluted, brown when still moist, turning grayish-white upon drying because of the lime which encrusts them. The sporangia are depressed-globose 0.4–0.6 mm. in longest diameter. At an early stage the peridium is brown and bears peg-like protuberances (Fig. 1). When dry the entire peridium is typically covered with lime and then appears grayish-white (Fig. 2).

2. *Capillitium.* Many of the sporangia developed in culture lack capillitium. When present, however, the capillitium is distinctly didymiaceous, devoid of lime, violet-brown changing to pale at the extremities, consisting of long, branched, anastomosing threads with membranous expansions at the points of union (Fig. 4). Short, blunt branches are sometimes present on the capillitial strands.

3. *Spores.* Spores are black in mass, dark violet-brown by transmitted light, their surface unevenly covered with rather large spines (Fig. 6). In artificial culture spores are not always

well formed, as evidenced by a considerable variation in size and shape. The majority are walnut shaped, each with a prominent ridge separating it into two more or less equal halves (Fig. 5); they measure 11–12 × 12–14 μ over all. On corn meal agar, spores take a week or more to germinate. On Difco Noble agar, germination takes place sooner and the percentage of germination appears to be higher, but not enough data have been gathered to substantiate these observations.

4. *Plasmodium.* The plasmodium, as it develops on the agar surface, is a typical phaneroplasmodium, forming a fleshy anterior fan and rather prominent veins. Transparent at first, it soon changed to milky white. Because plasmodia grow very slowly and are sluggish, it is very difficult to obtain fungus-free cultures if fungal spores are present in the original inoculum. We have been unable to induce the plasmodia to grow larger than 1–2 cm. by feeding them bacteria or oat flakes. When they reach these dimensions, they either fruit or disintegrate.

Physarina echinocephala

Von Höhnel described the sporangia as chocolate brown, but Lister stated that they are pale pink or flesh-colored. Those of the Ernst collection at Zürich are whitish. It may be that the color has faded in the intervening 62 years. The sporangia of the Zürich specimen (Fig. 7) are hard, indicating they may not have been properly matured. Mounted spores have a tendency to cling together, but their general characters are easily observable (Fig. 8). We find the spores to be essentially globose, nearly smooth, very pale violet or gray rather than dark violet or brownish-violet, as described by von Höhnel and Lister respectively, and 8.5–10.5 μ diam. rather than 7–9 as given by Lister, but the larger size may be due to unfavorable conditions prevailing at the time of their formation.

Physarina echinospora

Examination of the type material reveals that the spores are not globose as originally described, but walnut-shaped, each bearing a prominent ridge along which the spore probably splits at the time of germination. The shape of the spores is clearly shown in Figure 9, and the ridge is unmistakable in Figure 10, which is a photomicrograph of a spore split under pressure applied on the coverslip. In other respects the characters of this species as we observed them from the type specimen agree with those given by Thind and Manocha.

It appears from the above that the Mexican isolate is very similar to *P. echinospora,* and we regard it as belonging to that species. The description of *P. echinospora* is hereby emended to revise the description of the spores and to characterize the plasmodium for the first time.

Figs. 1–6, 9–10.

***Physarina echinospora* Thind and Manocha**

Sporangia gregarious, stipitate, pale gray to ashen, globose to depressed-globose, erect, 0.5–0.7 mm. long diam., total height up to 1 mm.; peridium grayish white, deep cream-colored, or pale flesh-colored in dehisced sporangia, thin, tough, brittle, calcareous, covered with numerous blunt, cylindrical, limy pegs, somewhat paler than the peridium, 60–120 μ long and up to 54 μ wide, peridial surface rough and often reticulately ridged around the bases of the pegs; dehiscence irregular; stipe short, stout, calcareous, concolorous with the pegs, rough, somewhat ridged, 0.2–0.5 mm. long, continuing into the sporangium as a columella; columella large, subglobose to columnar, pale fawn, calcareous; capillitium abundant, noncalcareous, violet brown, tapering to the paler or hyaline extremities, branching and anastomosing, with membranous expansions at the points of union; spores black in mass, dark violet-brown by transmitted light, walnut-shaped, each bearing a conspicuous ridge which separates it into two more or less equal halves, the surface prominently spiny, the spines up to 1 μ long, irregularly distributed, 11–14 μ in diameter. Phaneroplasmodium milky-white.

TYPE LOCALITY: Mussoorie, India
HABITAT: Plant debris
KNOWN DISTRIBUTION: India, Mexico
ILLUSTRATIONS: Mycologia **56**: 714
MATERIAL EXAMINED: Type (IU); UTMC-1205 (TEX).

LITERATURE CITED

LISTER, A. 1925. A Monograph at the Mycetozoa, 3d ed. Revised by G. Lister. British Museum (Nat. Hist.), London. 296 pp.

THIND, K. S. AND M. S. MANOCHA. 1964. The Myxomycetes of India—XVII. Mycologia **56**: 712–717.

VON HÖHNEL. 1909. Javanische Myxomyceten. Fragmente zur Mykologie Sitz. -ber Akad. Wiss. Wien 118, Abt. **1**: 275–452.

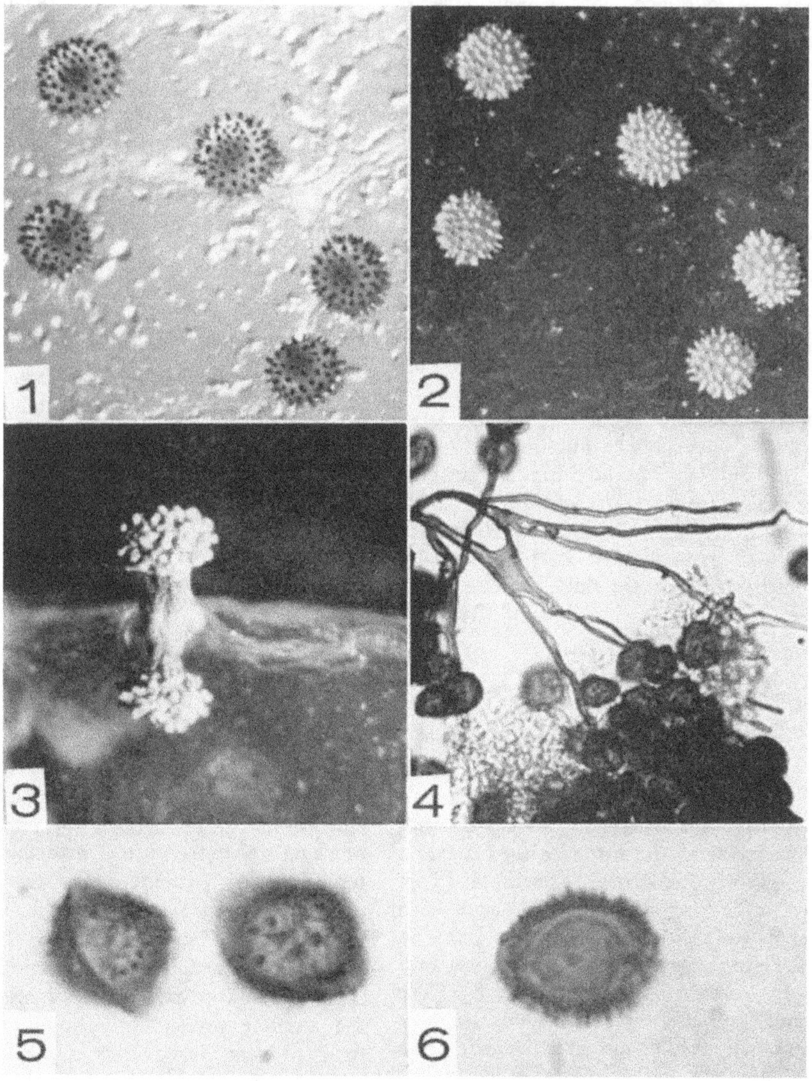

PLATE I. Figs. 1–4. *Physarina echinospora,* Mexican isolate, developed in agar culture. Fig. 1. Surface view of immature sporangia. X15. Fig. 2. Same sporangia after drying. X15. Fig. 3. Side views of two sporangia on agar. X15. Fig. 4. Capillitium and spores. X400. Figs. 5–6. Two views of spores showing ridge and spines. X1600.

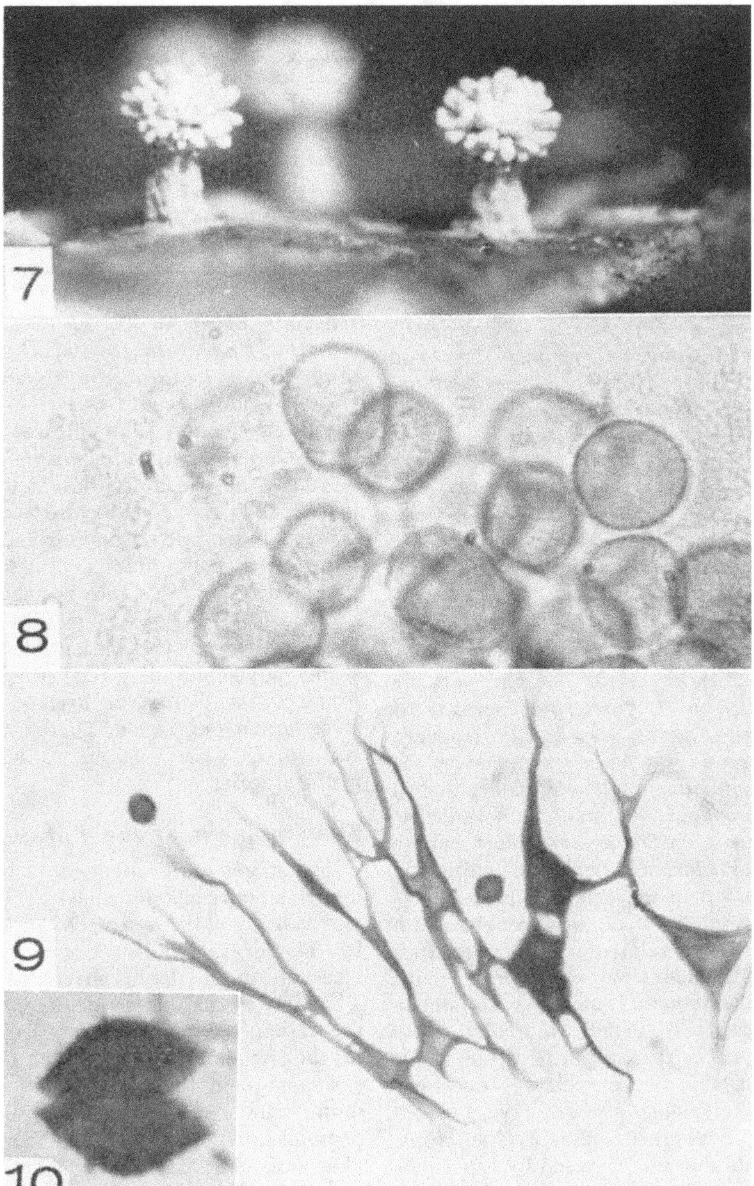

Plate II. Figs. 7–8. *Physarina echinocephala.* Fig. 7. Sporangia on leaf. X15. Fig. 8. Spores. X1600. Figs. 9–10. *P. echinospora* type. Fig. 9. Capillitium and spores. X400. Fig. 10. Spore which has split as a result of pressure applied on the coverslip to demonstrate presence of ridge. X1600.

Extension of the Host Range of *Octomyxa brevilegniae*

WILLIAM A. SHERWOOD[1]

Department of Botany, University of North Carolina at Chapel Hill, N. C. 27514

Introduction

The purpose of this article is to record the occurrence and extend the host range of *Octomyxa brevilegniae* Pendergrass as a parasite in the mycelium of *Dictyuchus pseudodictyon* Coker and Braxton, *Dictyuchus missouriensis* Couch, and *Brevilegnia unisperma* Coker and Braxton. Some observations on the development and morphology of the parasite in these species are included. These confirm the observations of Pendergrass (1950) on the development of *Octomyxa brevilegniae* in *Brevilegnia linearis* Coker.

The monotypic genus *Octomyxa* was established by Couch, Leitner, and Whiffen (1939) for an endophytic parasite which caused the hyphal tips of its host, *Achlya glomerata,* to become distorted into swollen, spherical galls. Host range studies revealed that the parasite would grow only on *A. glomerata*; because of its obligate nature, it was designated *Octomyxa achlyae.* The genus was later expanded to include *O. brevilegniae,* found by Pendergrass in a culture of *Brevilegnia linearis.* He found that it would also grow on *Geolegnia inflata,* where it did not differ morphologically or culturally from its expression in the original host.

Soils that were collected and brought from Louisiana for studies related to other research in progress in the laboratory were wetted with sterile distilled water and baited with sterile halved hempseeds in order to isolate any strains of *Dictyuchus* that might be present. Of the two strains that were isolated, one was identified as *Dictyuchus pseudodictyon* Coker and Braxton. Several of the hyphae were swollen into galls that were caused by a plasmodiophoraceous parasite determined to be *Octomyxa brevilegniae.*

Host Relationship

Experiments were carried out to determine the host range of the present species on a number of species of the water molds. Hempseed cultures of *Dictyuchus pseudodictyon* harboring the parasite were placed next to one-day-old hempseed cultures of each of the following: *Achlya caroliniana, A. colorata, A. flagellata, A. glomerata, A. hypogyna, A. orion, A. racemosa, Aphanomyces laevis, Apodachlya brachynema, Brevilegnia unisperma, Dictyuchus missouriensis, D. monosporus* (male strain), *D. monosporus* (female strain), *Geolegnia inflata, Isoachlya luxurians* (India strain), *Leptolegnia caudata, Phytophthora* sp., *Pythium debaryanum, P. undulatum, Saprolegnia diclina, S. ferax, S. litoralis, S. megasperma,* and *Thraustotheca clavata.* The cultures were examined after 48 hours, by which time infection could be easily observed in those cultures in which it occurred. *Brevilegnia unisperma, Dictyuchus missouriensis,* and *Geolegnia inflata* became infected with equal intensity as the original *D. pseudodictyon.* On the basis of overlapping host range and of its similar morphological characters in all the susceptible host species, it seems advisable to extend the host range of Pendergrass' *Octomyxa brevilegniae* to include *Brevilegnia unisperma, Dictyuchus missouriensis,* and *D. pseudodictyon* rather than to erect a new species.

Development of the Parasite

The stages in development of the parasite follow those already described by Couch, Leitner, and Whiffen (1939) for *Octomyxa achlyae* and by Pendergrass (1950) for *O. brevilegniae.* Penetration of the hyphae was not observed. The first noticeable evidence of infection was the darkened granular and reticulate protoplasm in the swollen hyphae (Fig. 1). This is an early stage in the development of the plasmodium, which seems to attract and absorb host protoplasm. Galls may contain one or more plasmodia, but rarely was more than one seen in a gall. Galls may be formed either laterally (Fig. 2) or terminally on a hypha (Fig. 3).

A plasmodium may develop into either of two types of sori, a zoosporangial sorus or a resting-spore sorus. There appears to be no difference in the plasmodial stages producing the two types of sori. The first-formed ones develop into zoosporangia while the parasite is still young (Fig. 4). Not all of the zoosporangia in a sorus mature at the same time. As the anteriorly biflagellate zoospores do mature, those nearer the center of the sorus find exit through papillae in the empty zoosporangia near the periphery of the gall.

[1] Present address: The New York Botanical Garden, Bronx, New York.

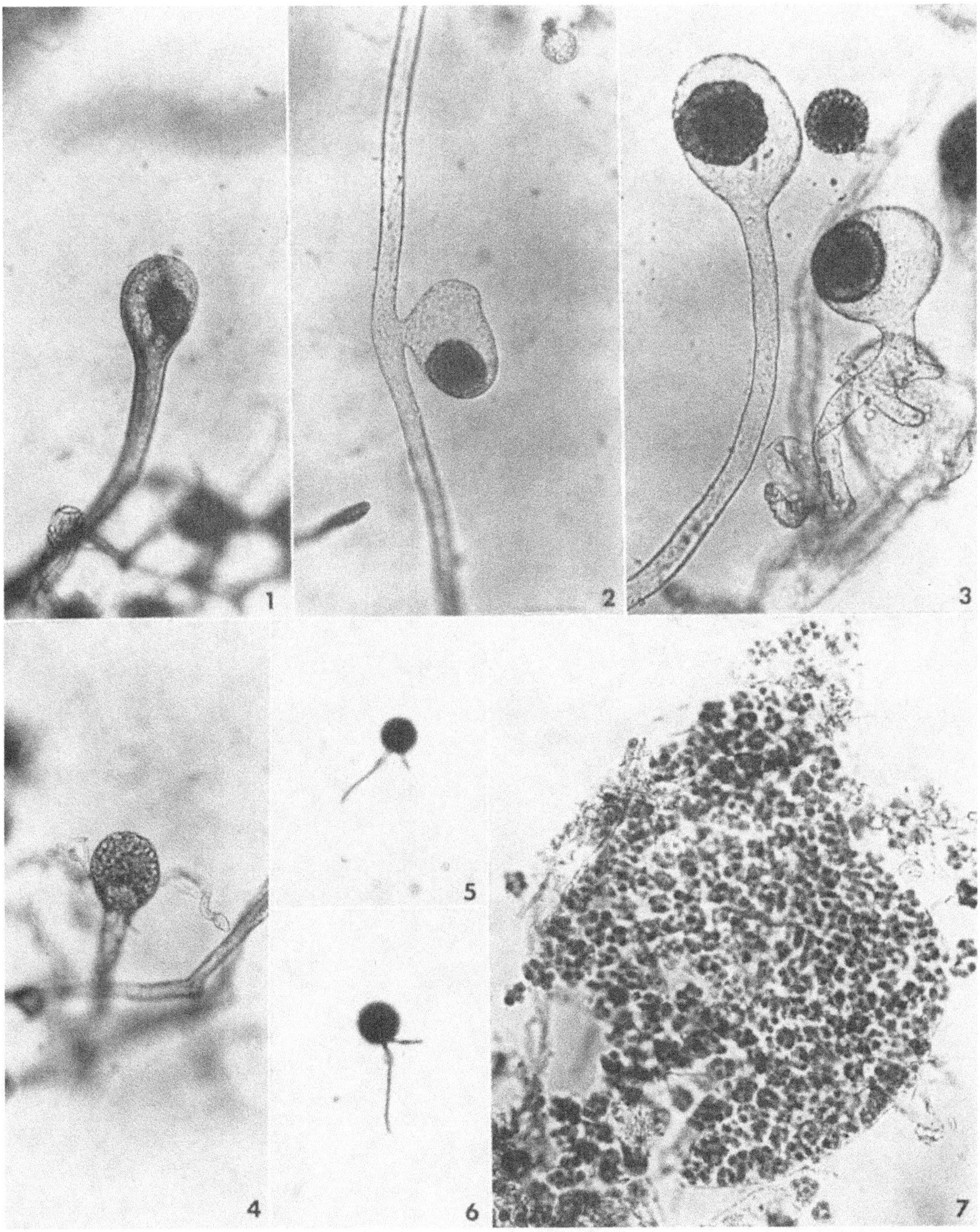

FIGS. 1–7. *Octomyxa brevilegniae*. Fig. 1. Early plasmodial stage in swollen hyphae tip of *Dictyuchus pseudodictyon*. X250. Fig. 2. Immature cystosorus formed in lateral swelling of the host hypha. X250. Fig. 3. Young cystosori formed in terminal swellings of the host hyphae. This appears to be the usual condition. X250. Fig. 4. A zoosporangial gall. X250. Figs. 5 and 6. Zoospores showing the characteristic heterocont biflagellate condition and the whiplash on the short flagellum. Stained with crystal violet. X2000. Fig. 7. Partially crushed gall showing groups of mature resting spores. X690.

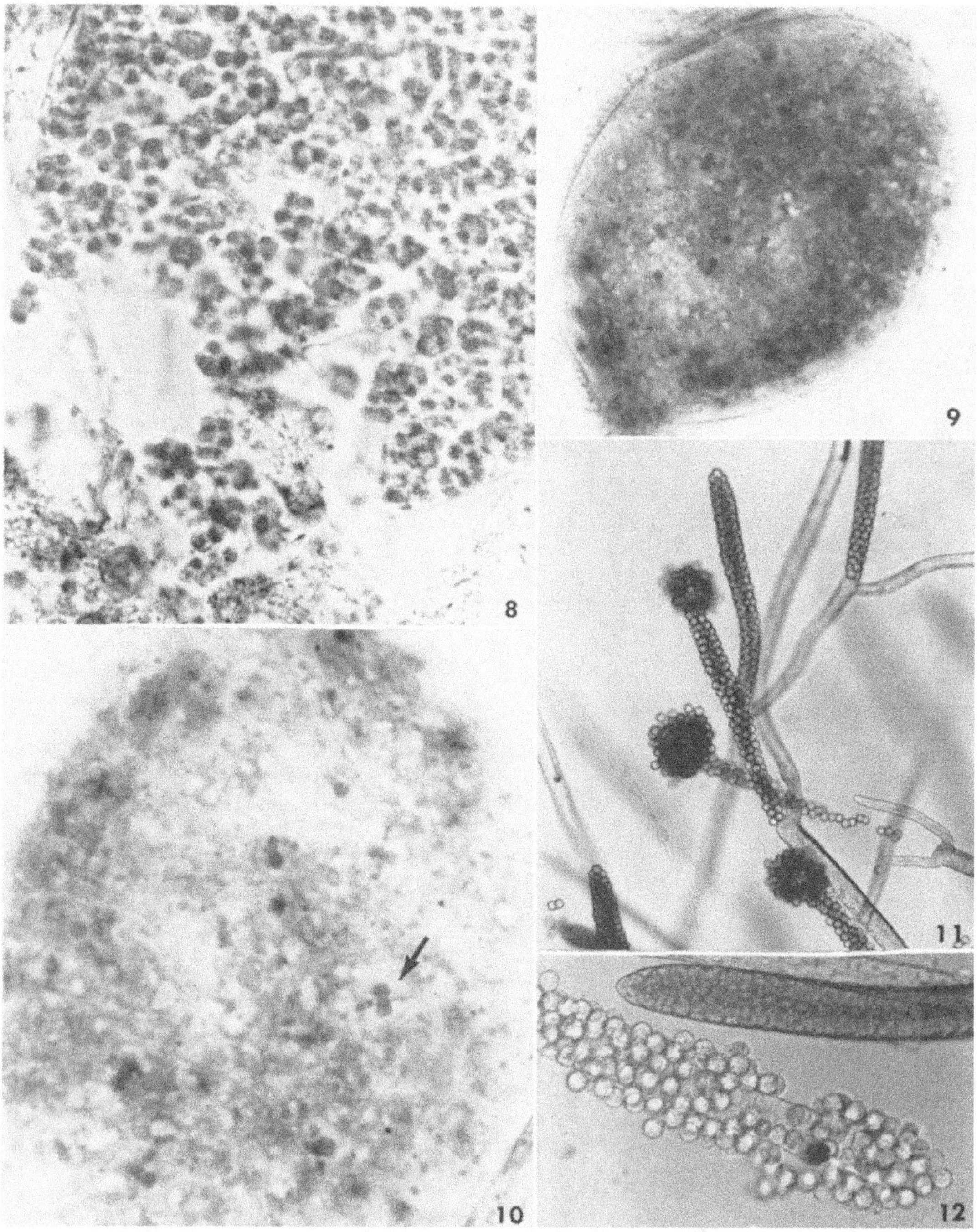

FIGS. 8–10. *Octomyxa brevilegniae.* Fig. 8. Groups of octads of spores from a partially crushed gall containing resting spores. X970. Fig. 9. Propionocarmine smear of a young plasmodium to show stages in promitotic nuclear division. X1460. Fig. 10. Same as Fig. 9. The typical dumbbell shape of the nucleolus and the ring arrangement of chromosomes is indicated by the arrow. X2680.

FIGS. 11–12. *Dictyuchus.* Fig. 11. Sporangia of the false-net type of *Dictyuchus pseudodictyon.* The typical Achlyoid type of spore discharge of the first group of spores produced is shown. X125. Fig. 12. Sporangia of the true-net type of *Dictyuchus monosporus.* The encysted spores are emerging and leave the characteristic net in the sporangium in this species. X250.

Several of the active zoospores were killed in a drop of water on a slide with fumes of 1% osmic acid for one minute. They were stained lightly by placing a drop of a 0.5% aqueous solution of crystal violet next to the cover glass and drawing it under with filter paper until the desired density was obtained. The zoospores thus observed were found to be heterocont biflagellates. The whiplash may be seen on the short flagellum in Figures 5 and 6.

The development of the resting spores occurs as the host matures. As the resting-spore sorus matures, packets of spores, usually in groups of eight, become distinguishable. The eight spores of an octad are not clearly visible in Figures 7 and 8 because of the depth of focus; however, groups showing six of the eight spores can be seen.

Several attempts were made to stain the plasmodial stage of the parasite. The most successful consisted in using Lu's (1962) propionocarmine squash technique. By this method it was possible to demonstrate the characteristic promitotic "cruciform" division of the nuclei in plasmodia undergoing vegetative growth in the Plasmodiophoraceae (Figs. 9 and 10). The chromosomes appear as distinct units around the elongated nucleolus rather than one continuous band (Fig. 10).

Summary and Conclusions

The host range of *Octomyxa brevilegniae* Pendergrass has been extended to include *Brevilegnia unisperma* Coker and Braxton, *Dictyuchus missouriensis* Couch, and *D. pseudodictyon* Coker and Braxton. Some observations on the life cycle of the parasite in these species are included. They confirm the observations of Pendergrass (1950) on the development and morphology of *O. brevilegniae* in *Brevilegnia linearis* Coker.

Couch (1931), in a key for the species of *Dictyuchus,* divided them into those having (1) sporangia of one kind, the spores emerging to leave a true net; (2) first sporangia of the *Achlya* type, later sporangia of the false-net type; (3) all sporangia normally of the false-net type. Of significant interest is the fact that the parasite affected only the two species with sporangia of the false-net type (Fig. 11), *Dictyuchus pseudodictyon* and *D. missouriensis. Dictyuchus monosporus,* with sporangia of the true-net type (Fig. 12), was not infected. This fact might be of sufficient taxonomic significance to add weight to Couch's (1931) view that species with the false-net type sporangia be included in a new genus.

Pendergrass (1950) did not observe *Dictyuchus missouriensis* to become infected in his study of host relationships. The probable explanation for this is in a difference of cultural conditions and/or possibly some variation in the strains. These conditions may also explain Couch, Leitner, and Whiffen's (1939) "obligate" parasite on the hyphae of *Achlya glomerata.* Pendergrass (1948) described *Octomyxa brevilegniae* on the basis of differences in size of the galls, zoosporangia, and resting spores and on the parasite's negative growth reaction on *Achlya glomerata.* In the present study numerous attempts were also made to induce infection of the parasite on *A. glomerata,* but all failed. These conditions may all be due to variations in strains, but they do tend to substantiate the present concept of two species of *Octomyxa. Octomyxa achlyae* appears to be an obligate parasite on *Achlya glomerata.* The host range of *O. brevilegniae* appears to be more flexible, and as a result of the present study it has been extended to include three more species of the water molds.

ACKNOWLEDGMENTS

The author wishes to thank Dr. John N. Couch, under whose direction this study was made, and Drs. W. J. Koch and C. J. Umphlett for their constructive review of the manuscript.

LITERATURE CITED

Couch, J. N. 1931. Observations on some species of water molds connecting *Achlya* and *Dictyuchus.* Jour. Elisha Mitchell Sci. Soc. **46**: 225–229.

Couch, J. N., J. Leitner, and A. Whiffen. 1939. A new genus of the Plasmodiophoraceae. Jour. Elisha Mitchell Sci. Soc. **55**: 399–408.

Lu, B. C. 1962. A new fixative and improved propionocarmine squash technique for staining fungus nuclei. Canad. Jour. Bot. **40**: 843–847.

Pendergrass, W. R. 1948. A new member of the Plasmodiophoraceae. Jour. Elisha Mitchell Sci. Soc. **64**: 132–134.

———. 1950. Studies on a Plasmodiophoraceous parasite, *Octomyxa brevilegniae.* Mycologia **42**: 279–289.

On an Unusual Fungoid Organism, *Sphaerita dinobryoni* n. sp., Living in Species of *Dinobryon*

Hilda M. Canter (Mrs. Lund)
Freshwater Biological Association, Ambleside, Westmorland, England

The presence of a hitherto undescribed organism found within the body of species of *Dinobryon* has been noted from several lakes in the English Lake District (Coll. Loughrigg Tarn 26 April 1961; Windermere north basin 3 May 1961; Grasmere and Rydal Water 2 May 1966 and Esthwaite Water 9 May 1966). Specimens occurred only in small numbers and for a short period of time. This, together with the fact that *Dinobryon* dies so quickly when removed from its natural habitat and even more rapidly when mounted on a slide, has meant that observations are incomplete. It is not unusual for cells of a *Dinobryon* population to bear simultaneously not only the above organism but also specimens of the chytrid *Rhizophidium oblongum* Canter (1954), and also an unidentified parasitic "monad"-like protozoan. Nevertheless the two stages (sporangium and resting spore) described for the organism under consideration have been encountered together so frequently that they are considered as representing parts of the same life cycle. The following description, although incomplete, will enable other workers to recognize this organism and perhaps in due course to reveal its true identity.

Life History

It is unknown what form the primary infecting body takes or how entry into the *Dinobryon* is effected. No empty spore case has ever been found attached to the algal cell wall. The incipient sporangium is first recognized as a small colorless round or oval body containing fine granular protoplasm (Fig. 1*C* and Pl. I, Fig. 1). It possesses a very thin wall and is located within the protoplast membrane at or near the posterior end of the algal cell. The sporangium enlarges, and subsequently changes in the structure of its content take place leading to the formation of a densely packed mass of elongate curved spores (Fig. 1*K* and Pl. I, Fig. 7). Exactly how these spores develop has not yet been determined. At first denser spherical areas are seen within the protoplasm, giving it a "spotted" appearance (Fig. 1*G*). Later these denser areas become more elongate and band-shaped (Fig. 1*I*). At this stage the content superficially resembles the chromosome bands found in the nucleus of an alga such as *Gymnodinium*. These curved bands represent the maturing spores. The mature sporangium is somewhat spherical (6–8.5 μ diam.) or oval to cylindrical (5 x 7 μ–8 x 16 μ) in shape. The latter specimens are found in the larger species of *Dinobryon*—e.g., *D. divergens* Imhof from Loughrigg Tarn. Never more than one infection per cell has been seen.

As noted earlier, *Dinobryon* cells are sensitive to being placed on a slide, and they soon die or become dislodged from their cases by pressure of the coverslip. Both of these conditions are usually accompanied by bursting of the algal cell. In this process the body of the alga often elongates considerably (as in Pl. I, Fig. 6) and flows toward the apex of the case which surrounds it. The cell bursts, freeing the protoplasm and chromatophores while the sporangium still remains within the *Dinobryon* envelope. A moment or so later its delicate wall also bursts and the content is liberated. No dehiscence papilla or exit tube has ever been found associated with a sporangium, and how the spores are normally liberated remains unknown. Various types of spore bodies have been seen, presumably due to the bursting of both immature and mature sporangia. What are considered to be the mature spores are curved in shape, one end possessing a slight knob while the opposite end tapers somewhat. When such spores are viewed under phase contrast, it is seen that a small portion of the spore at the tapered end contains either no protoplasm or less dense protoplasm (Fig. 1*N* and Pl. II, Fig. 1) than the rest of the spore, which appears bluish black in this light. The spore measures 3.5–4 μ long x 0.7 μ wide approx. It contains no refractive globules, and, although the protoplasm appears homogeneous, if it is mounted in acetocarmine denser areas become visible. The spore is not surrounded by any mucilaginous envelope. Other spores of this shape and size but containing protoplasm throughout their length have been seen (Fig. 1*L, M* and Pl. I, Fig. 10). They probably represent a slightly less mature stage. Even less mature spores appear to possess a thinner

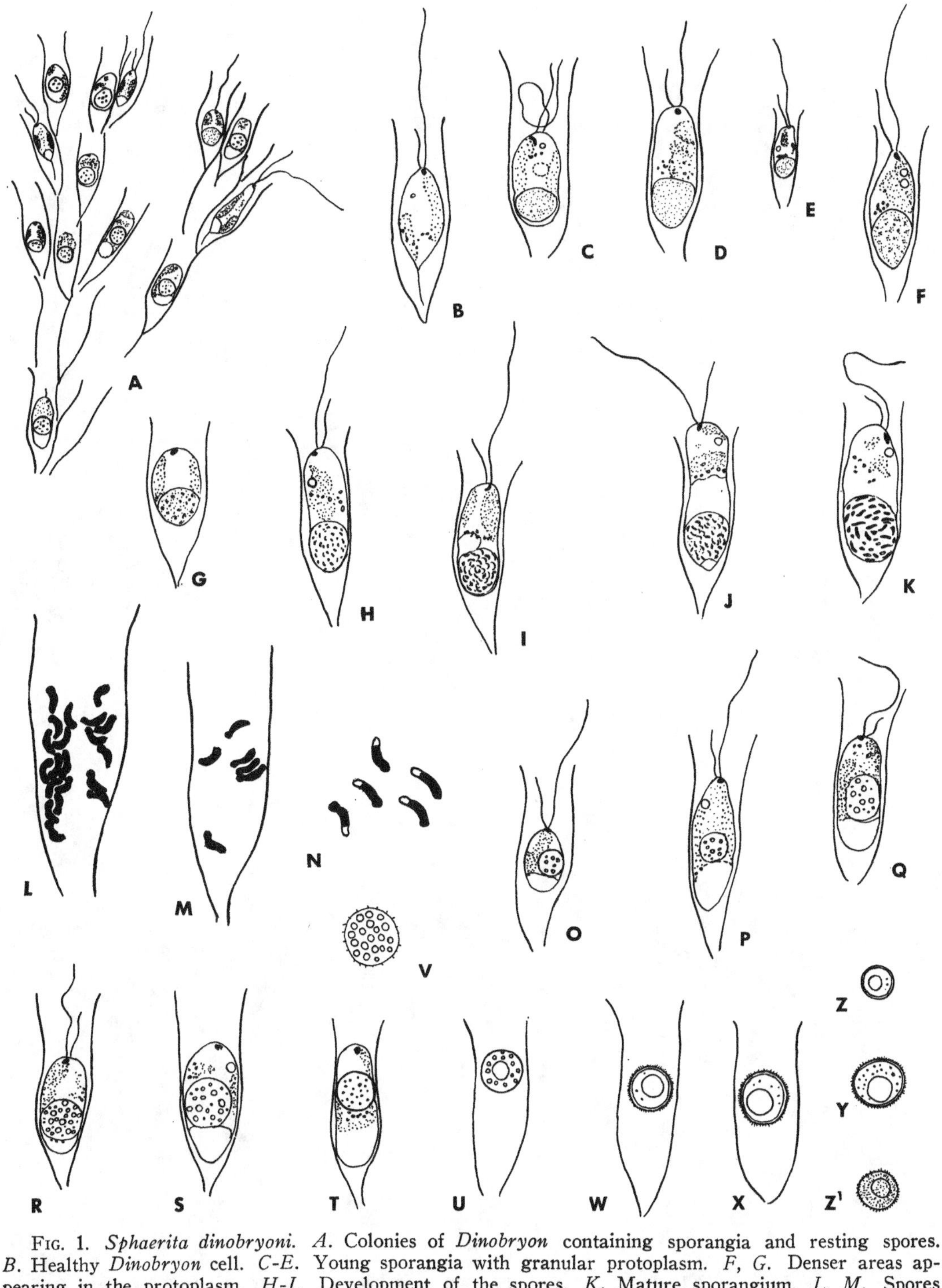

FIG. 1. *Sphaerita dinobryoni*. *A*. Colonies of *Dinobryon* containing sporangia and resting spores. *B*. Healthy *Dinobryon* cell. *C-E*. Young sporangia with granular protoplasm. *F, G*. Denser areas appearing in the protoplasm. *H-J*. Development of the spores. *K*. Mature sporangium. *L, M*. Spores liberated from the sporangium. *N*. Spores with mature content. *O-S*. Young resting spores with refractive globules. *T*. Globules becoming less obvious. *U*. Central nonrefractive sphere beginning to form. *V*. Beginning of wall ornamentation as seen in young spore. *W-Y*. Mature hirsute resting spores. *Z*. Smooth walled spore. *Z*[1]. Spore, surface view showing hirsute appearance. A, X537; L, M, X1250; N, X1500; rest, X1050. V in acetocarmine.

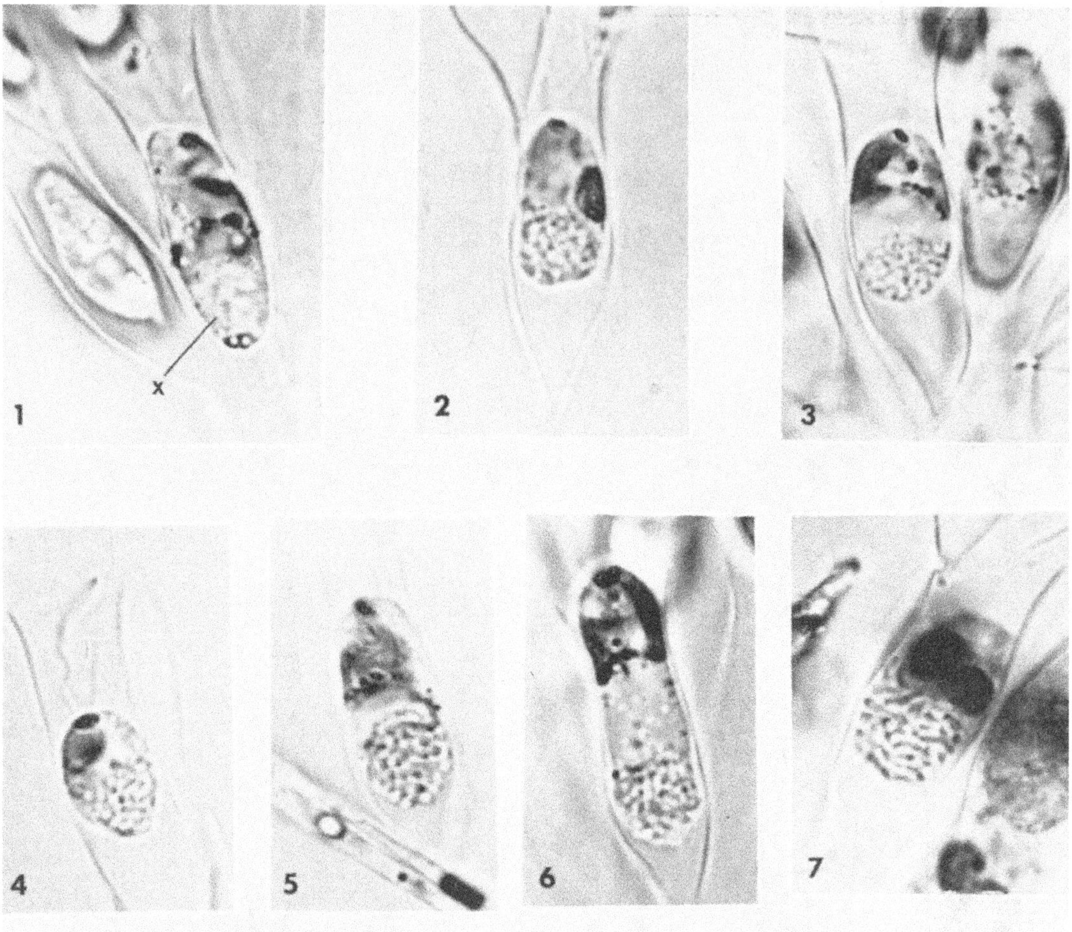

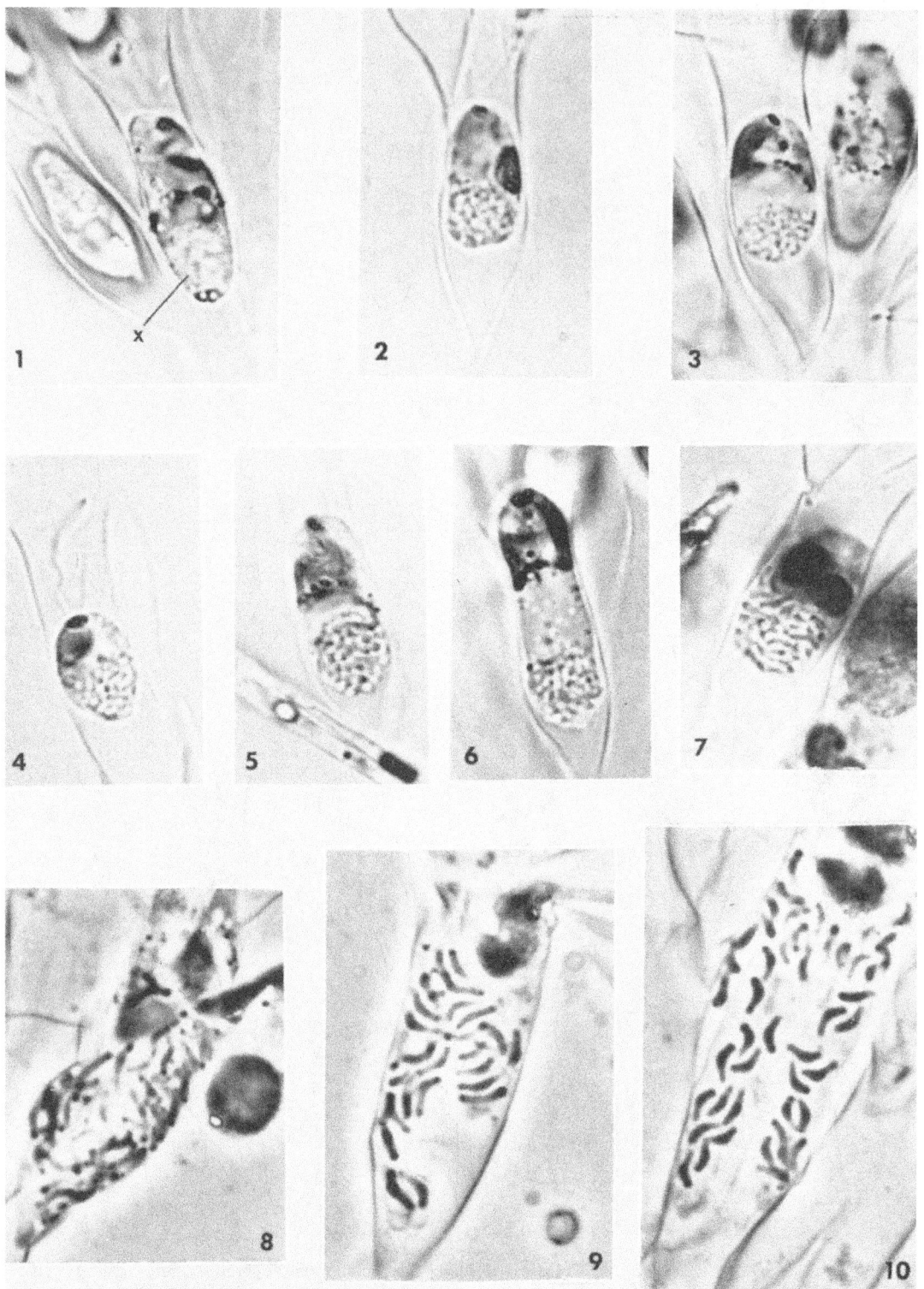

PLATE I. *Sphaerita dinobryoni.* All photographs are of living material from Grasmere (May, 1966), and were taken with a Zeiss photomicroscope using the electronic flash attachment.

Fig. 1. Young sporangium with granular content (*x*). Figs. 2–4. Early stages in spore development. Figs. 5, 6. Spores now more distinct. Fig. 7. Mature sporangium. Fig. 8. Squashed specimen showing immature spores; note thin body and conspicuous knob at one end (phase). Fig. 9. Squashed immature spores; later stage than Fig. 8 (phase). Fig. 10. Mature curved spores (phase).

Figs. 1–7 X1600; Figs. 8–10 X2000.

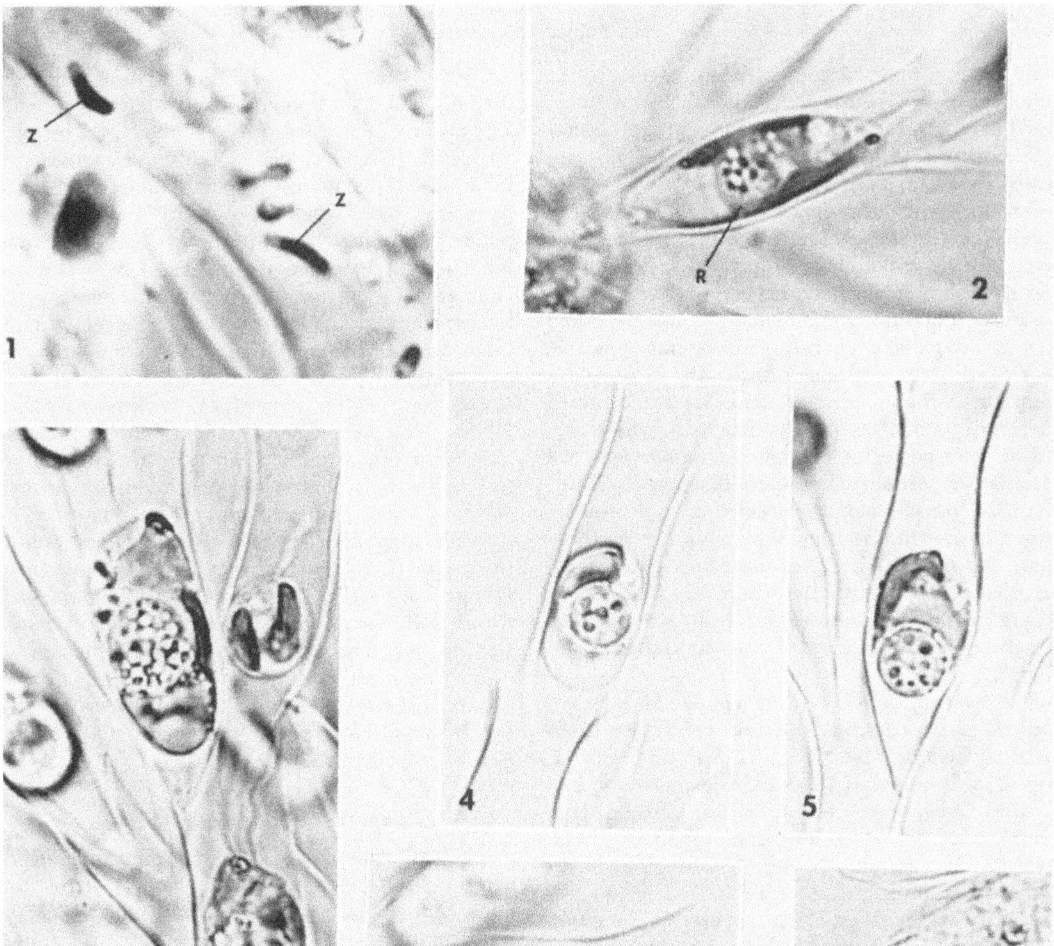

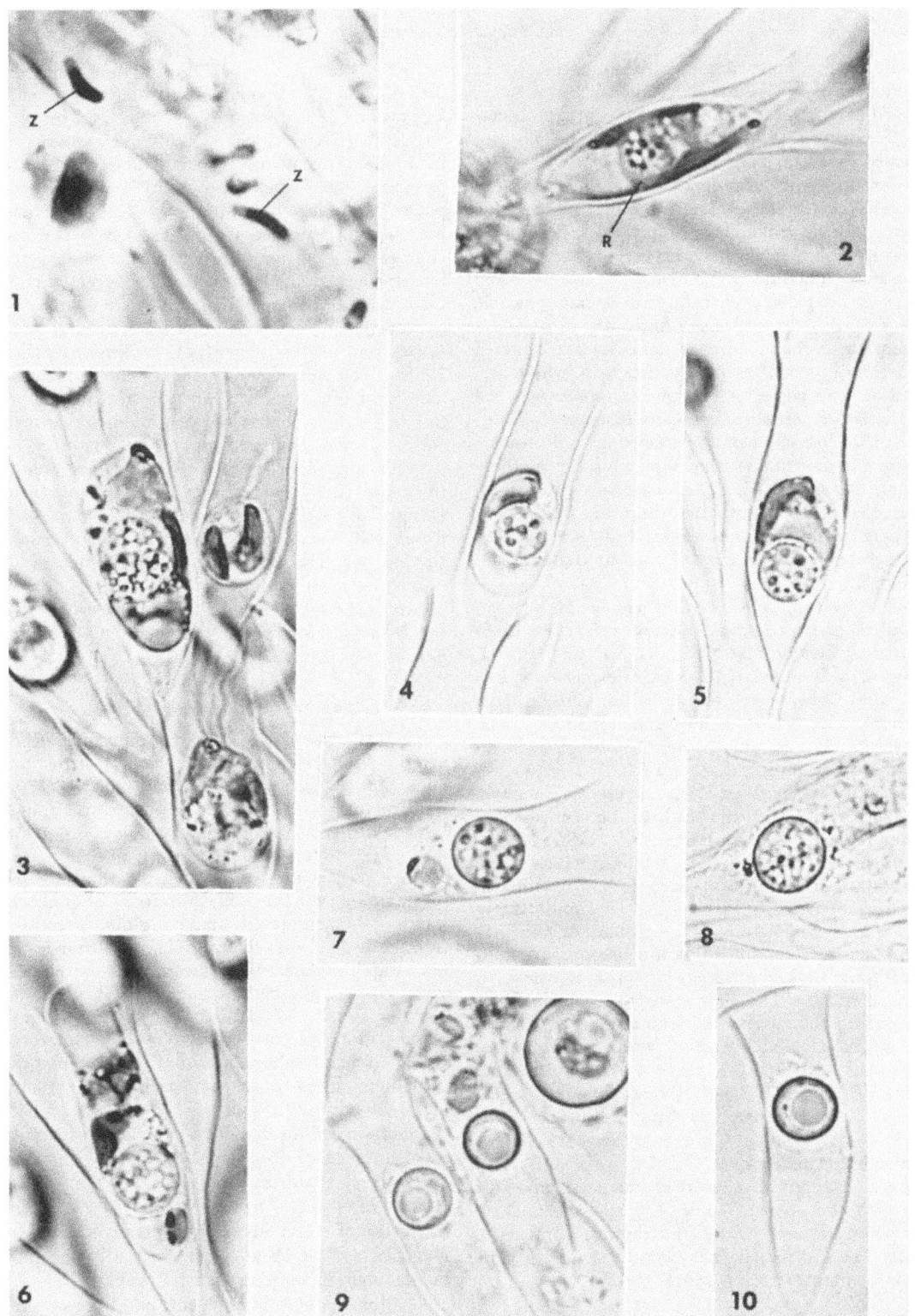

PLATE II. *Sphaerita dinobryoni.* All photographs are of living material from Grasmere (May, 1966) and were taken with a Zeiss photomicroscope using the electronic flash attachment.

Fig. 1. Two spores (*Z*) with mature content. Note less dense protoplasm at one end; uppermost in photograph (phase). Fig. 2. Very young resting spore (*R*) containing several refractive globules. Fig. 3. Young resting spore with many refractive globules, wall still delicate. Figs. 4–8. Young resting spores with a thicker wall, but content immature. Figs. 9, 10. Mature resting spores.

Fig. 1, X2560; all others, X1600.

body and a more accentuated knob-like apex. (Pl. I, Fig. 8). Further, it appears that, when freed from the sporangium, they are still embedded in a glutinous substance and do not easily separate from each other.

Exactly when the death of a *Dinobryon* cell takes place in nature due to the presence of this organism is unknown. Algal cells bearing a mature sporangium have been found which still possess functioning contractile vacuoles and waving flagella. Therefore it seems possible that death does not ensue until late in development or perhaps even at dehiscence. When a sporangium is present, the leucosin which occurs at the posterior end of the *Dinobryon* cell often becomes shrunken and displaced. Similarly, as development proceeds, the chromatophore bands in the apical part of the cell become disorganized. The posterior end of an infected cell is usually rounded instead of tapered, and it possesses no stalk-like portion attaching it to the base of the surrounding envelope.

The resting spore is first visible as a small walled sphere containing a few refractive globules of various size (Fig. 1*O, P* and Pl. II, Fig. 2). It is located toward the posterior end of the *Dinobryon* cell and lies between the leucosin and shrunken chromatophores. The sphere enlarges, and the number of globules in it increases (Fig. 1*R, S* and Pl. II, Fig. 3). If the spore is squashed at this stage or placed in acetocarmine, then the wall can be seen to bear minute, stiff, rod-like hairs (Fig. 1*V*). As development proceeds, the globules, which stain brownish in osmic acid, become less obvious in a denser matrix (Pl. II, Fig. 8) and usually disappear completely. The content of the mature spore is dominated by a dense hyaline nonrefractive sphere (which remains colorless in osmic acid). The homogeneous material surrounding the sphere may contain a few refractive granules. The outer wall of the spore thickens and finally appears as a smooth black line (Pl. II, Fig. 10), although in most specimens it is imperceptibly hirsute (Fig. 1*W-Y*, Z^1). The resting spore is spherical (4–7 μ diam.) or slightly oval (4.3 x 4.7 μ–6.5 x 8 μ) in shape. Indian ink shows no mucilage envelope around the spore. Unlike the sporangium, no *Dinobryon* cell with functioning contractile vacuoles and flagella has been observed to contain a mature resting spore.

Discussion

The curious nature of the spores produced by this organism poses difficult questions concerning its exact taxonomic position. Much of its life history is very similar to that found in the imperfectly known Chytridiaceous genus *Sphaerita*. Dangeard (1886) and Chatton and Brodsky (1909) have noted that sometimes nonmotile, presumably nonflagellate spores are produced by *S. endogena* parasitizing Rhizopoda. However these spores are spherical and not curved. Many references exist in the protozoological literature to organisms which are loosely termed "*Sphaerita*-like" or even remain unnamed. In some of these the spores are oval or club-shaped, and numerous references to them are given in the papers of Lubinsky (1955, 1955a). However none appears to be identifiable with my organism, and for the most part too few stages in the life histories are known for any real comparisons to be made. The curved, aflagellate spores are somewhat similar in shape to spores of the fungus *Protoascus* Dangeard (1903) (Maupas, 1915), some bacterial cells, and the spores of Microsporidia (Protozoa), but here the similarity ends.

In the present state of knowledge of this organism, it seems best to regard it as a species of *Sphaerita* for which the binomial *S. dinobryoni* is proposed.

***Sphaerita dinobryoni* sp. nov.**

Thallus endobioticus holocarpicusque. Sporangium sphaericum, 6–8.5 μ diam., aut ovatum aut cylindricum 5 x 7 μ–8 x 16 μ, membrana tenui, maturum sporas multas curvatas immobiles continens. Spora 3.5–4 μ long. x 0.7 μ lat. una extremitate umbone parvo praedita, extremitate opposita paululum attenuata. Protoplasma sporae ad extremitatem attenuatam minus compactum aut nullum, globulis refractivis nullis. Modus liberationis sporarum ignotus. Papilla dehiscentiae atque tubus exitus non visi. Spora quiescens sphaerica 4–7 μ diam., aut ovata 4.3 x 4.7 μ–6.5 x 8 μ. Spora iuvenis aliquot globulas refractivas continens, matura, autem, unicam sphaeram "matt" hyalinam praevalentem intus praebens; membrane crassa, levis aut minute hirsuta.

Planta in cellulis *Dinobryi sertulariae* Ehrenb. incolens (substrato typi) e lacubus Grasmere (loco typi, materia typi IMI 124294) atque Rydal water dictis lecta. Necnon in *D. divergente* Imhof e lacubus Esthwaite water, Loughrigg Tarn arque Windermere dictis lecta, omnibus lacubus in regione lacustri Anglica.

Thallus endobiotic, holocarpic. Sporangium thin walled, spherical, 6–8.5 μ diam., oval or cylindrical, 5 x 7 μ–8 x 16 μ, containing at maturity many curved nonmotile spores. Spore 3.5–4 μ long x 0.7 μ wide, one end with a small knob, the opposite end tapering slightly. Protoplasm of spore less dense or absent at tapered end, refractive globules absent. Method of liberation of spores unknown. No dehiscence papilla or exit tube seen. Resting spore spherical 4–7 μ diam., or oval 4.3 x 4.7 μ–6.5 x 8 μ. When young containing several refractive globules, but at maturity content dominated by a single hyaline matt sphere; wall thick, smooth, or minutely hirsute.

Living in cells of *Dinobryon sertularia* Ehrb. (type substratum) collected from Grasmere (type locality, type material IMI 124294) and Rydal Water. In *D. divergens* Imhof from Esthwaite Water, Loughrigg Tarn, and Windermere, all lakes in the English Lake District.

ACKNOWLEDGMENTS

Many thanks are given to Dr. Hannah Croasdale, U.S.A., for the Latin translation and to Dr. J. W. G. Lund, F. R. S., for his kind help with the manuscript.

The apparatus used in this investigation was obtained from a Royal Society Grant for Scientific Investigations and from the Central Research Fund of the University of London, to which acknowledgment is made.

REFERENCES

CANTER, H. M. 1954. Fungal parasites of the phytoplankton. III. Trans. Brit. Mycol. Soc. **37**: 111–133.

CHATTON, E., AND BRODSKY, A. 1909. Le parasitism d'une Chytridinee du genre *Sphaerita* Dangeard chez *Amoeba limax* Dujardin. Arch. Protistenk. **17**: 1–18.

DANGEARD, P. A. 1886. Recherches sur les organismes inferieurs. Annals Sci. nat. (Bot) VII **4**: 241–341.

———. 1903. Recherches sur le developpement du Perithece chez les Ascomycetes. Botaniste **9**: 1–303.

LUBINSKY, G. 1955. On some parasites of parasitic Protozoa. I. *Sphaerita hoari* sp.n.—a chytrid parasitizing *Eremoplastron bovis*. Can. J. Microbiol. **1**: 440–450.

———. 1955a. On some parasites of parasitic Protozoa. II. *Sagittospora cameroni* gen. n., sp. n.—a phycomycete parasitizing *Ophryoscolecidae*. Can. J. Microbiol. **1**: 675–684.

MAUPAS, E. 1915. Sur un champignon parasite des *Rhabditis*. Bull. Soc. Hist. nat. Afr. N. **7**: 34–49.

Physoderma hydrocotylidis and Other Interesting Phycomycetes from California[1]

FREDERICK K. SPARROW[2]

Department of Botany, University of Michigan, Ann Arbor, Mich.

The Phycomycetes of California, like everything else in that state, are interesting. This paper presents primarily observations on a species of *Physoderma* which I have been seeking, unsuccessfully, for many years throughout Southeastern United States—namely, *P. hydrocotylidis* Viégas and Teixeira, first described from Brazil. It also includes notes on various other Phycomycetes encountered in a visitor's ramblings about that fascinating state. Its wondrous array of habitats, freshwater and salt, contain innumerable varieties of these minute organisms. The methods used in trapping saprobic forms were those commonly employed (Sparrow, 1960) and involve the use of various kinds of baits, primarily pollen.

Chytridiomycetes

1. *Physoderma hydrocotylidis* Viégas and Teixeira. Parasitic on *Hydrocotyle ranunculoides*. Run-off of small freshwater pond in dunes north of Ft. Bragg, Mendocino Co., Feb. 24; March 3; April 7, 1966; pasture impoundments, James McClure Farm, Point Reys, Marin Co. April 21, 29, May 26, 1966.

Since its original description (Viégas and Teixeira, 1943) on *Hydrocotyle reniformis* in Brazil, *Physoderma hydrocotylidis* has been found in both the eastern and western parts of our country. Correspondence some years ago with W. W. Diehl of the U.S.D.A. and Prof. Lee Bonar of Berkeley yielded considerable data on its distribution. The former found it at Black Pond in Fairfax Co., Virginia, in 1923 and listed a collection made by him in Golden Gate Park, San Francisco, in 1930; also from Alameda Co., Calif., Oct. 1930, H. E. Parks, 3481. Bonar (communication and specimens in University of California Herbarium) found it in the courtyard pools of the Life Sciences Building in Berkeley in 1943, but it had disappeared by 1955. Diehl looked over material of *Hydrocotyle* in the herbarium of the Academy of Natural Sciences, Philadelphia, and U. S. National Herbarium and found *Physoderma*-infected material as follows:

Mobile, Ala., May 22, 1889—Charles Mohr 464 (Ex U. S. N. Herb. 770,075)

York Furnace, Pa., Aug. 20, 1895—Ex Herb. Jos. Crawford (Ex herb. Phila. Acad. Nat. Sci.)

St. Georges, Del., July 22, 1875—A. Commons, 563, 375 (Ex herb. Phila. Acad. Nat. Sci.)

Diehl further reports that the fungus was located in the phanerogamic herbarium of Cornell University by Dr. F. A. Guba, as follows:

Hydrocotyle sp. Cape May, N. J., Aug. 30, 1917, A. Gershoy

H. ranunculoides, Appalachicola, Fla., 1893, A. W. Chapman

H. umbellata, Riverside, Calif., Dec. 17, 1918, M. F. Barrus

A further California record was found in a 1934 collection of *Hydrocotyle* sp. from the mouth of Little River, Humboldt Co. I examined this site in 1966 but found no host plants.

Both the late Dr. Robert Johns and I searched in the field for this species at various likely localities in Texas, the Gulf states, and Florida, but failed to locate it. A thorough search in phanerogamic herbaria will undoubtedly yield sites in these areas. *Synchytrium hydrocotyles* Cook, which resembles *Physoderma* somewhat, is sporadic in central Florida. Where it occurs, however, it is abundant. These complete the records of *Physoderma hydrocotylidis* known to me.

As usual, this member of the genus does not differ significantly in its polycentric endobiotic phase from others save in one particular—namely, the wall of the resting spore. This is unusually thin—indeed, the most delicate of any congeneric form we have observed. This very faint amber-colored structure is approximately 1 μ thick. Resting spores which are 15–25 μ x 12–16 μ crowd cells just beneath

[1] Part of this work was assisted by a grant from the NSF.

[2] I am greatly indebted to Prof. Ralph Emerson for making possible an enjoyable semester of mycological teaching in Berkeley. To Drs. Melvin Fuller and Lee Bonar and the staff of the Department of Botany and Herbarium, I owe thanks for many favors. Prof. Harry Thiers of San Francisco State College was most helpful. I am also very grateful to Mr. James McClure of Point Reys for allowing me to collect in the various water impoundments on his farm.

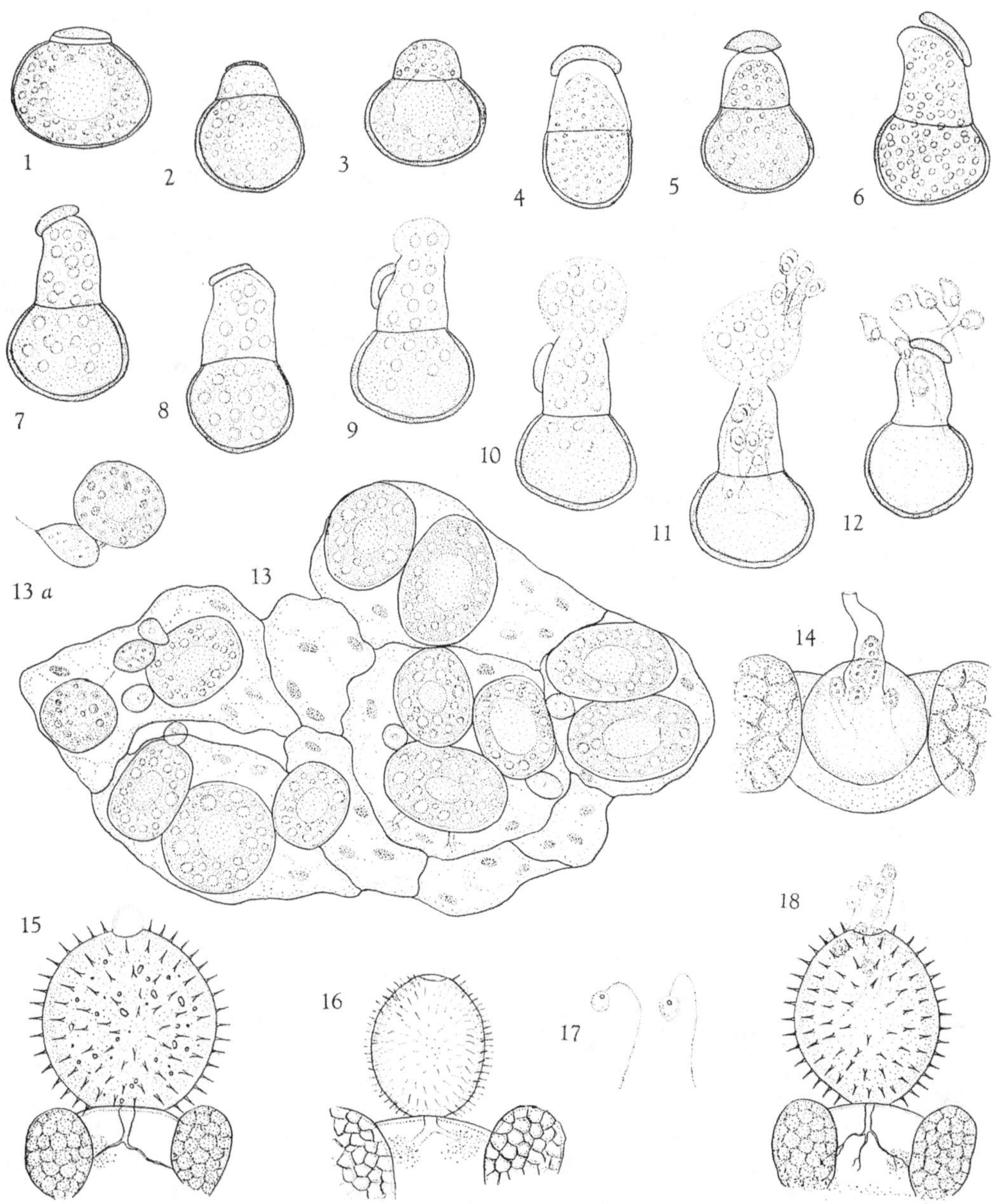

PLATE I. Figs. 1–13. *Physoderma hydrocotylidis* on *Hydrocotyle ranunculoides.* Figs. 1–3. Early stages in germination of resting spore and formation of endosporangium. Figs. 4–8. Later stages, showing organization of globules of zoospores. Figs. 9–11. Successive stages in discharge of zoospores from endosporangium. Fig. 12. Laggard zoospores emerging from an endosporangium. Fig. 13. Portion of host tissue with maturing resting spores and turbinate cells. Fig. 13a. Resting spore attached to short tube from a turbinate cell. Fig. 14. *Olpidium longicollum* in pollen grain; some of the zoospores with granular content still in sporangium; the remainder have escaped. Figs. 15–18. *Rhizophydium echinocystoides* n. sp. Fig. 15. Mature sporangium with apical discharge papilla. Fig. 16. Empty sporangium. Fig. 17. Zoospores. Fig. 18. Zoospores being discharged from sporangium. All figures X600.

the host epidermis (Fig. 13). As in other species, each arises from the tip of a short outgrowth from a 1-celled or 1-septate turbinate cell (Fig. 13a). The latter is 7–8 x 3–5 µ. There are occasional antler-like outgrowths on young spores.

Resting spores germinate soon after they reach maturity. Unlike many other species, no drying or freezing is necessary; indeed, these conditions seem distinctly inhibitory. Spores from dead and decayed leaves lying in the water germinate after as little as 19 hours when placed at 23° C. and subjected to a continuous fluorescent-light regime of 370 foot-candles.

The process of germination is no different from that described for many species (Sparrow, Griffin, Johns, 1961). It involves the circumscissile dehiscence of a cap of wall material 9–11 μ in diameter coincident with the gradual extrusion of a finger-like thin-walled "endosporangium (Figs. 1–8). The latter forms a rather distinct discharge papilla. After a sequence of cytoplasmic changes like those of other species, the papilla deliquesces and the zoospores move out in a compact, slowly enlarging mass and eventually separate, free themselves, and swim away (Figs. 9–12), precisely as earlier described for *P. menyanthis* (Sparrow, 1946). The zoospores are fusiform, 7 x 5 μ, and bear a central protruding globule and posterior flagellum. They have the same even movement with sudden change of direction seen in zoospores of other species.

As yet, there has been no evidence that an epibiotic, monocentric stage is formed, although work continues on this phase of the life history. Hundreds of resting-spore zoospores encyst on the colorless epidermal cells of the leaf, send in an unbranched rhizoid, and, in the underlying host parenchyma cells, form the primary turbinate cell.

FIG. A. Infected leaf of *Hydrocotyle* showing pustules of *Physoderma hydrocotylidis*. The left side of leaf has yellowed as result of extensive infection.

Both leaves (Fig. A) and rhizomes become infected. The former become yellow with age, and chlorosis may be accompanied by "green island" formation around the low pustules (⅓–½ mm. in diameter) of resting spores. Often, infection of the rhizome is quite extensive and may be indicated by pustules, but more often by brown, glassy, or watery areas. If infected plants are maintained in shallow pans with water just covering the surface of the rhizomes, as new leaves arise and produce pustules, one gets the distinct impression that there is systemic infection. Stunted plants are often found in the field, but apparently they are not always infected with *Physoderma*. There is very little if any hypertrophy of infected parts as compared with that produced by *Synchytrium* on the same host.

2. ***Physoderma* on *Ranunculus lobbii*.** Excellent stands of this *Ranunculus* were found 10 March, 1966, in a vernal pool[3] near Glen Ellen, Sonoma Co. On the floating leaves occasional minute brown flecks could be seen which examination proved to be composed of subepidermal clusters of resting spores of a *Physoderma*. Often these were near the tip of a lobe on the upper side and formed irregular patches 0.75 mm. in diameter around which the tissue was yellowish. There was no sure way of detecting what was *Physoderma* and what was injury or due to other fungi except by mounting a whole leaf and examining all discolored areas with a compound microscope. At that time, too, infected areas were sometimes found which were invisible macroscopically. The turbinate cells were 1–2-celled, 7–9 x 11–13 μ, the resting spores chestnut brown and 12–28 x 9–20 μ. No germination was seen. Whether this is the same species that infects other Ranunculaceae (*P. johnsii*, *P. calthae*, *P. bohemicum*, etc.) cannot be said without more material and cross-inoculation studies. The host seems to be an unrecorded one for *Physoderma*.

3. ***Physoderma heleocharidis* (Fuckel) Schroeter.** On *Eleocharis* spp.; vernal pool, Glen Ellen, Sonoma Co., March 10, 1966; shallow pool near Lower Crystal Springs Reservoir, San Mateo Co., March 23, 1966. At both sites the infection was found primarily on the lower sheath. The species was also present in two

[3] I am grateful to Mr. William Hirano for calling my attention to this site; it was exceedingly fruitful.

species of *Eleocharis* in pools in the courtyard of the Life Science Building, University of California, Berkeley, Feb. 2, 1966.

4. *Micromyces zygogonii* Dang. Parasitic on *Mougeotia* sp., sphagnum bog near Strong's Station, Willits Road (Hwy. 20), 2 mi. east of Ft. Bragg, Mendocino Co., May 14, 1966, coll. Gordon McBride.

5. *Synchytrium* on *Boisduvalia densiflora.* Vernal pool, Glen Ellen, Sonoma Co., March 10, 1966, May 7, 1966. On young plants early in season the pustules were dark red but were black in collections made in May. This material may be referable to Karling's (1964) "M 3" described on the same host in University of California Herbarium No. 502839, which, in turn, he thinks may be *S. epilobii,* or "M 2" on this host, possibly referable to *S. fulgens.* It is evident that much painstaking cross-inoculation work will be necessary to establish in *Synchytrium,* as in *Physoderma,* the limits of the species or, for that matter, just what is a species.

6. *Synchytrium papillatum* Farlow. Excellent material on *Erodium* sp. was brought into the laboratory by Mr. Hirano from Mt. Bruno, San Mateo Co., in March, 1966. The galls were brilliant red and exceedingly numerous on the lower parts of the plants. What may possibly be *Synchytrium andinum* Patouillard and Largerheim was found on *Ranunculus californicus* at the edge of a pool near Lower Crystal Springs Reservoir, San Mateo Co., March 29, 1966.

7. ? *Olpidium longicollum* Uebelmesser. On pine-pollen bait, Salton Sea, Imperial Co., Jan. 19, 1966.

This fungus is only about one-half the size of Uebelmesser's and may not belong here. It has, however, many points of resemblance other than size.

The sporangia in pollen grains (Fig. 14) were spherical, about 22 μ in diameter, and at maturity bore a single discharge tube up to 20 μ in length. The zoospores which moved through the flaring opening of the tube were 4–5 μ in diameter (one-half that of typical material) and bore a pair of bright granules in their content but no single oil globule. They were strongly amoeboid during passage through the tube, but once outside quickly assumed motility by means of their single posterior flagellum. The tube was up to 10 μ broad at its base and was, in general, stouter and shorter than that described and figured by Uebelmesser in her material from various parts of the world.

8. *Olpidium gregarium* (Nowak.) Schroeter. Parasitic in rotifer eggs, vernal pool, Glen Ellen, Sonoma Co., March 10, 1966.

Only a few eggs were found parasitized.

9. *Rhizophydium decipiens* (Br.) Fischer. On eggs of *Oedogonium* sp., vernal pool, Glen Ellen, Sonoma Co., March 10, 1966.

The broad, cushion-like sporangia rested on the surface of the parasitized egg. Within the ooplasm there was a richly branched rhizoidal system of numerous delicate elements. At maturity a single broad, nearly sessile pore formed just beneath the fertilization pore of the host oogonium. Mature sporangia contained a great number of very small zoospores, less than 2 μ in diameter, each with a minute oil droplet. In spite of the rather large exit pore, few zoospores escaped at any one time after dissolution of the papilla. Rather, they continued to swarm for several days with occasional ones darting out. Indeed, sporangia were followed within which zoospores were rapidly swarming for as many as 240 hours. This is over twice the time observed by Braun (1856), which in itself was remarkable—namely, 108 hours.

The finding in this material of a rhizoidal system of much branched, extremely delicate components satisfies this observer, at least, that this is a species of *Rhizophydium,* a suggestion earlier made by Scherffel (1926).

10. *Rhizophydium echinocystoides* n. sp. On pine-pollen bait, gross water culture of sphagnum, Big Lagoon State Park, Humboldt Co., April 6, 1966.

This striking bog chytrid has an ornamented sporangial wall. The sporangium itself is ovoid, upright, 15–25 x 10–22 μ, and its slightly amber-colored wall is beset with numerous spines of moderate length and density (Figs. 15, 16, 18). The rhizoidal system is of moderate delicacy and branched. At maturity there is a single apical, colorless, conspicuous discharge papilla. Upon its dissolution there escape numerous spherical, posteriorly uniflagellate zoospores 3–3.5 μ in diameter, each bearing a single colorless globule (Fig. 17). Epibiotic, spiny, spherical resting spores 12–15 μ in diameter with a slightly thickened wall and large central globule were found in association with the sporangia and may belong to the species, but since no germination was seen, this cannot be said with certainty.

The shape and ornamentation of the sporangium which gives this fungus its uniqueness is reflected in its name, for it recalls well the fruits of our common wild cucumber or balsam-apple.

***Rhizophydium echinocystoides* sp. nov.**

Sporangium sessile, ovoideum, pallide brunneo spinulosum, 15–25 μ alto x 10–22 μ diam. Systema rhizoidale modice ramosum, ex axe singulo. Zoosporae sphaericae, 3–3.5 μ diam., globulo singulo hyalino posteriori et flagello singulo longo postico praeditae, per porum singulam apicalem emissae. Sporae perdurantes (?) sessiles, globosae, 12–15 μ diam., membrana modice crassa, spinulosis.

In polline Pinus, in *Spagno* et detrito ad locum palustrem "Big Lagoon" dictum, Humbolt Co., California; "Bryant's Bog," Cheboygan Co., Michigan.

Sporangium sessile, ovoid, the pale-amber wall covered by fairly conspicuous spines; 15–25 μ high by 10–22 μ in diameter; rhizoidal system moderately branched, arising from the tip of a short main axis; zoospores spherical, 3–3.5 μ in diameter, with a single, basal, colorless droplet and posterior flagellum emerging through a single apical pore 3–4 μ in diameter formed upon the deliquescence of a prominent papilla; resting spores (?) epibiotic, spherical, 12–15 μ in diameter, with a moderately thick wall covered with low spines, contents with a large oil globule, germination not observed.

On pine pollen, sphagnum bog, Big Lagoon State Park, Humboldt Co., April 6, 1966; Bryant's Bog, Cheboygan Co., Mich., July 25, 1966.

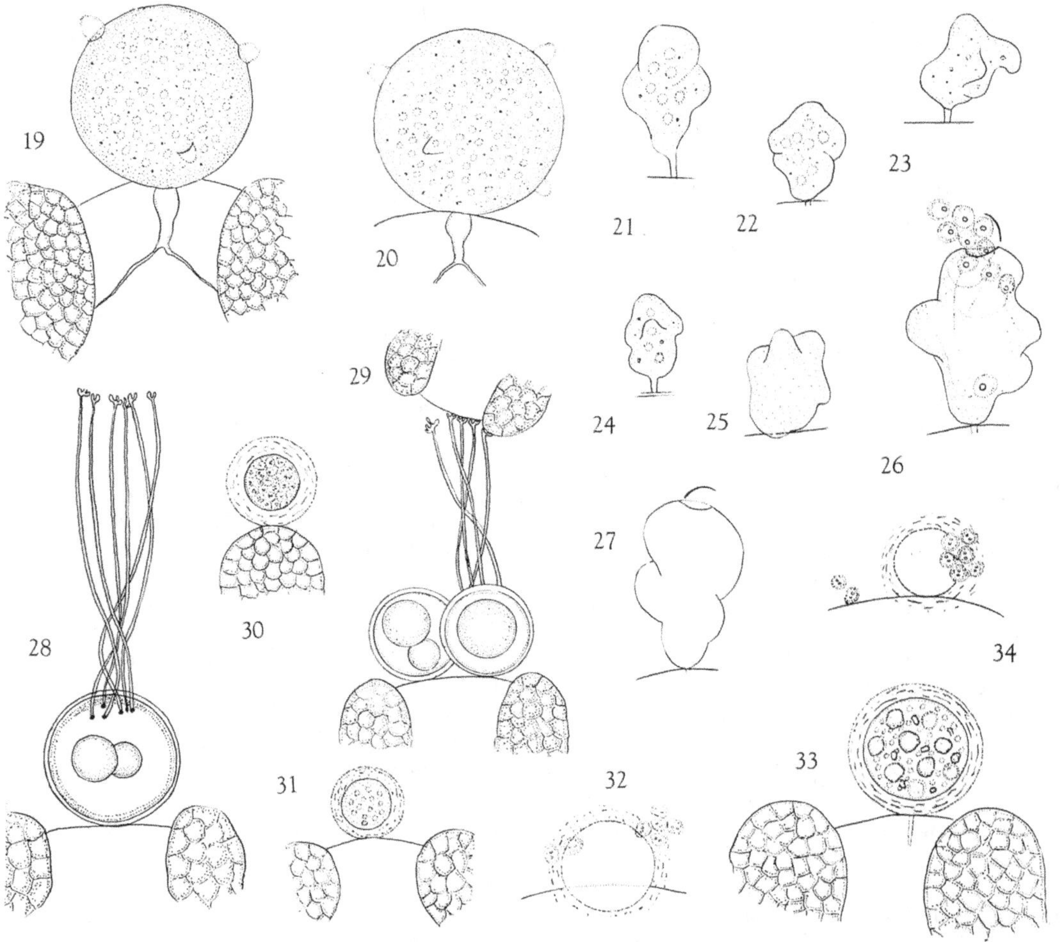

PLATE II. Figs. 19–20. *Phlyctochytrium semiglobiferum* on pollen grains. Mature sporangia with prominent discharge papillae. Figs. 21–27. – *Chytriomyces sp.* on pollen. Figs. 21–25. Various shapes of immature thalli. Fig. 26. Nearly empty operculate sporangium. Fig. 27. Empty sporangium and persistent operculum. Figs. 28, 29. *Rhizophlyctis harderi*, on pollen grains. Endobiotic part not visible. Figs. 30–34. *Thraustochytrium pachydermum* on pollen grains. All figures X600.

11. *Rhizophydium fusus* (Zopf) Fischer. On naviculoid diatoms, vernal pool, Glen Ellen, Sonoma Co., March 10, 1966.

12. *Phlyctochytrium mucronatum* Canter. On pine pollen, sphagnum bog, Big Lagoon State Park, Humboldt Co., April 6, 1966.

13. *P. furcatum* Sparrow. Pollen, sphagnum bog; same site as No. 4.

14. *Phlyctochytrium stipitatum* Kobayasi and Ookubo. On pine pollen, sphagnum bog, Big Lagoon State Park, April 6, 1966. It is probable that, as the authors of this species indicated, this fungus should be placed in a genus of its own.

15. *Phlyctochytrium dentatum* (Rosen) de Wild. Parasitic on gametangia of *Spirogyra* sp., vernal pool, Glen Ellen, Sonoma Co., March 10, 1966.

16. *P. bullatum* Sparrow. Parasitic on zygospores of *Spirogyra* sp., vernal pool, Glen Ellen, Sonoma Co., March 10, 1966.

17. *Phlyctochytrium semiglobiferum* Uebelmesser. Pine pollen bait, Salton Sea, Imperial Co., Jan. 19, 1966.

Spherical, smooth-walled sporangia up to 50 μ in diameter were formed in abundance on pollen dusted on water cultures made from wet shoreline sand. These sporangia were notable for the presence on them at maturity of 10–12 prominently protruding, somewhat conical papillae up to 5 μ long by 4 μ at the base (Figs. 19, 20). The wall was faintly amber and slightly thickened. Innumerable zoospores swarmed violently within the sporangium and, as the papillae dissolved, darted out of the pores thus formed. The zoospore body was elongate when in motion. The latter was in the nature of a series of intermittent, quick jerks. When coming to rest, the body rounded up to a diameter of 5 μ, and the minute refractive globule and long posterior flagellum was visible.

There are twice as many discharge papillae on our material as described by Uebelmesser.

18. *Podochytrium lanceolatum* Sparrow. Pine-pollen bait, pond in sand dunes, north of Ft. Bragg, Mendocino Co., Feb. 24, April 6, 1966.

It was impossible to distinguish this fungus from the one on diatoms. Whether they are actually the same, in view of the radically different substrata (the type was on *Melosira*), remains a question.

19. *Entophlyctis bulligera* (Zopf) Fischer. In vegetative cells of *Spirogyra* sp., vernal pool, Glen Ellen, Sonoma Co., March 10, 1966.

20. *Rhizophlyctis harderi* Uebelmesser. On pine pollen, wet sand, Salton Sea, Imperial Co., Calif., Jan. 19, 1966.

This is an unmistakable species by reason of the formation on the resting spores, which are spherical, epibiotic, and 18-25 μ in diameter, of a varying number of long, slender processes (Figs. 28, 29). The latter arise at the apex of the spore and extend for varying distances, up to 60 μ out into the water. Distally they form stubby, contorted branches that attach themselves to nearby pollen grains. They are easily dislodged and do not seem to penetrate the wall of the grain.

A few empty, urn-shaped, smooth sporangia that may belong to the fungus were found. No endobiotic system could be seen. If these were the sporangia, they seemed to play little part in the life history, which appeared to go rapidly from resting spore to resting spore.

Although no new pertinent information can be given, it seems that this cannot be a species of *Rhizophlyctis*. Extensive observations on its development and the endobiotic system will be required before final disposition is accomplished.

Associated with this fungus on pollen was an epibiotic form with spherical sporangia 32–48 μ in diameter with a wall covered with minute, slightly brownish incrustations. The endobiotic system arose from a thick-walled apophysis 6 μ in diameter. No other stages were seen.

21. *Chytridium confervae* (Wille) Minden. On *Tribonema* sp., vernal pool, Glen Ellen, Sonoma Co., March 10, 1966.

22. *Chytriomyces poculatus* Willoughby and Townley. Pine pollen, sphagnum, Big Lagoon State Park, Humboldt Co., April 6, 1966.

An operculate chytrid allied to this species was found in the Mendocino Co. sphagnum bog (coll. G. McCloud) mentioned in No. 4. The epibiotic sporangia were sometimes elevated on a short tube 2 μ in diameter from the surface of the substratum (Figs. 21–25, 27). The endobiotic system was not seen. The sporangial body was irregularly lobed and contorted and 9–15 μ wide by 20–23 μ high. At maturity a convex 8 μ in diameter operculum was dehisced (Fig. 26). Through the pore thus formed, a few disproportionately large posteriorly uniflagellate zoospores, 5 μ in diameter and with a large colorless globule, and an arc-like structure in the body were discharged.

Oomycetes

23. *Olpidiopsis schenkiana* Zopf. Parasitic in *Spirogyra* sp., vernal pool, Glen Ellen, Sonoma Co., March 10, 1966.

24. *Lagenidium rabenhorstii* Zopf. Parasitic in *Spirogyra* sp., same site and collection as above.

25. *Pythium dictyosporum* Raciborski. Parasitic in *Spirogyra* sp., same site as above. This is a very uncommon species of the genus.

26. *Thraustochytrium pachydermum* Scholz. Pine-pollen bait, "Badwater," Death Valley, Inyo Co., April 1, 1966.

This most unlikely site yielded considerable material of the fungus. The sporangia were perfectly spherical, 12–15 μ in diameter, and had a wall 3–5 μ thick (Figs. 30, 31). The endobiotic system could be detected only rarely and seemed to be unbranched (Fig. 33). Clumps of somewhat angular discharged zoospores clustered around many of the sporangia (Fig. 32, 34).

The species has been recovered by Scholz in European coastal soils of very high salt concentration and also from soil from the Pacific Coast near San Francisco.

LITERATURE CITED

BRAUN, A. 1856. Über *Chytridium*, eine Gattung einzelliger Schmarotzergewächse auf Algen und Infusorien. Adhandl. Berlin Akad. **1855**: 21–83, pls. 1–5.

KARLING, J. S. 1964. *Synchytrium*. xvii + 470 pp. Academic Press, New York.

SCHERFFEL, A. 1926. Einiges über neue order ungenugend bekannte Chytridineen. . . . Arch. Protistenk. **54**: 167–260, pls. 9–11.

SPARROW, F. K. 1946. Observations on chytridiaceous parasites of phanerogams. I. *Physoderma menyanthis* deBary. Amer. J. Bot. **33**: 112–118, 41 figs.

———. 1960. Aquatic Phycomycetes. xxv + 1187 pp., 91 figs. Univ. Michigan Press, Ann Arbor.

SPARROW, F. K., JOYCE E. GRIFFIN, AND R. M. JOHNS. 1961. Observations on chytridiaceous parasites of phanerogams. XI. A *Physoderma* on *Agropyron repens*. Amer. J. Bot. **48**: 850–858, 53 figs.

VIÉGAS, A. P., AND A. R. TEIXEIRA. 1943. Alguns fungos do Brazil (Phycomycetos). Bragantia **3**: 223–269, 4 figs., 22 pls.

Studies of the Motile Cells of Chytrids. IV. Planonts in the Experimental Taxonomy of Aquatic Phycomycetes

WILLIAM J. KOCH[1]

Department of Botany, University of North Carolina at Chapel Hill, N. C. 27514

Introduction

Planonts, or flagellated spores and gametes, are of outstanding use in gaining deeper insight into the taxonomic and phylogenetic meaning of Aquatic Phycomycetes. Many mycologists have recognized this, among them Scherffel (1925), Weston (1935), Sparrow (1935, 1958, 1960), Couch (1938, 1941), Bessey (1942), Martin (1955), and Waterhouse (1962). Still, little experimental work relating to the value of planonts for making immediate taxonomic decisions and resolving phyletic relationships at the species and genus levels has been done.

In the previous study of this series and at the International Botanical Congress at Montreal in a symposium dealing with the motile flagellated cells in plants, evidence was summarized for eight or more major types of motile cells in the posteriorly uniflagellate series of Phycomycetes (Koch, 1961a, 1961b). These major types were recognized mostly from light microscopic observations of the internal parts, their presence or absence, and their appearances and arrangements. Since that time electron micrographs of sectioned motile cells have given us a glimpse of the elegant future for motile cell cytology in the taxonomy of Aquatic Phycomycetes and in our search for the evolutionary meaning of these fungi. Some important papers in this area are Turian and Kellenberger (1956), Blondel and Turian (1960), Cantino, Lovett, Leak, and Lythgoe (1963), Lovett (1963), Renaud and Swift (1964), Berlin and Bowen (1964), Chambers and Willoughby (1964), Fuller and Reichle (1965), Kole (1965), Moore (1965), Fuller (1966), Reichle and Fuller (1967), Chambers, Marcus, and Willoughby (1967), and Fuller and Reichle (in press), Umphlet and Olson (in press).

Koch (1961b) has pointed out that light microscopy might reveal other features than those presently used as criteria for typing motile cells. Experimental data (Koch, 1957a) had indicated that in the zoospores of *Phlyctochytrium punctatum* the lipoidal content is especially subject to variation, and that special caution must be exercised in the taxonomic use of the lipoidal content of chytrid zoospores.

Even casual observations of the zoospores of some chytrids growing on different substrata reveal differences in certain features (Fig. 1). The experiments in the present paper will reveal more than has been abstracted (Koch, 1961a) about the size and number of lipoid globules in the motile cell body, about body shape and the degree of amoeboidness causing change in shape, about the size of the motile cell body, about flagellar length, and about the position of flagellar attachment. The results of these experiments help in further assessing the extent to which these features may be variable and with how much caution they should be used in characterizing taxa. This paper, along with the one following it by Bernstein, will tell us

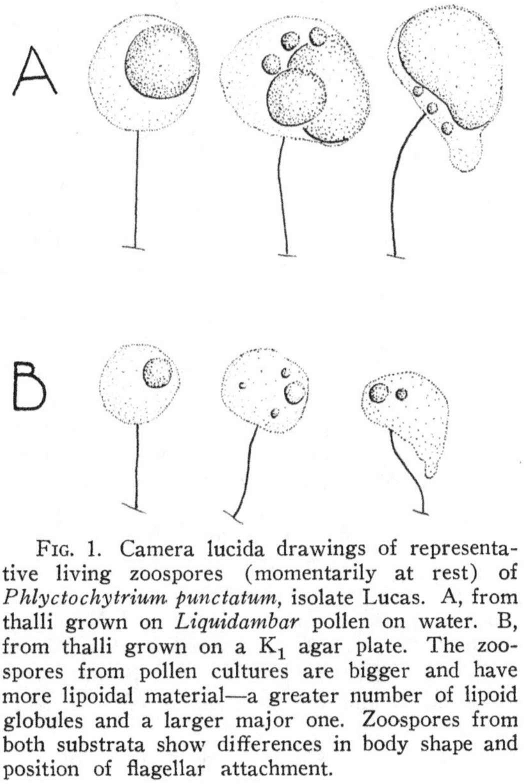

FIG. 1. Camera lucida drawings of representative living zoospores (momentarily at rest) of *Phlyctochytrium punctatum*, isolate Lucas. A, from thalli grown on *Liquidambar* pollen on water. B, from thalli grown on a K_1 agar plate. The zoospores from pollen cultures are bigger and have more lipoidal material—a greater number of lipoid globules and a larger major one. Zoospores from both substrata show differences in body shape and position of flagellar attachment.

[1] Support through a National Science Foundation Grant (GB-573) is gratefully acknowledged. John Clausz, Kate Miller, and Paige Slade were of great help in certain parts of this study.

more about inherent planont differences among isolates of a species and the importance of nutritional effects upon the response of the fungus.

Materials and Methods

The following isolates were used. They were derived either from a single spore or a single sporangium.

Chytridiales

Catenochytridium carolinianum Berdan. Collected by M. B. Huneycutt on boiled *Paspalum* leaf bait in soil from the University of North Carolina greenhouse, Coker Arboretum.

Chytriomyces hyalinus Karling. Collected on *Liquidambar* pollen bait in soil from Chapel Hill, North Carolina (CH-106) and Phenix City, Alabama (La–1). Primarily because of its flap-like operculum and unusual method of cytogamy (plasmogamy), the identity of this fungus remained uncertain until a few years ago (Koch, 1957b, 1959), and for this reason it was referred to in two earlier publications in this series as *Chytridium* sp. (Koch, 1956, 1958).

Entophlyctis sp. Collected from *Liquidambar* pollen bait in a water tank in the U.N.C. greenhouse, Coker Arboretum. This undescribed species, morphologically somewhat like *E. bulligera* (Zopf) Fischer, has been collected thirteen additional times on pollen bait of soils from the eastern United States and Canada.

Entophlyctis aurea Haskins. Collected by J. N. Couch on boiled *Paspalum* leaf bait in soil from Elberton, Georgia, and brought into pure culture by M. B. Huneycutt. The endo-operculum, zoospore size, and number of lipoid globules per zoospore make the identity of this organism questionable.

Nowakowskiella ramosa E. J. Butler. Collected on boiled *Paspalum* leaf bait in soil sent from Texas by the late Robert Johns.

Phlyctochytrium punctatum Koch. Collected on *Liquidambar* pollen bait in soil from New Orleans, Louisiana (La–6b), and on the resting spores of *Peronospora tabacina* (Lucas) in North Carolina. The other isolates were collected on pollen bait in soil and isolated by J. N. Couch (NG3 from New Guinea), C. E. Miller (MF101 from Joyce Kilmer National Forest, N. C.), T. E. Register (R17 from Cashiers, N. C.), C. J. Umphlett (CJU from California soil sample S5K498 collected by J. N. Couch), and Peggy Holland (MMH from Chapel Hill, N. C.).

Phlyctochytrium irregulare Koch. Collected on boiled *Paspalum* leaf bait in soil from the U.N.C. greenhouse, Coker Arboretum.

Rhizophydium sphaerotheca Zopf. Collected on *Liquidambar* pollen bait in soil from North Carolina.

Rhizophlyctis rosea (deBary and Woronin) Fischer. Collected on boiled *Paspalum* leaf bait in soil from Chapel Hill, N. C.

Septochytrium variabile Berdan. Collected on boiled *Paspalum* leaf bait in soil from North Carolina. The large zoospores of this isolate (6–7.5 microns in diameter) make this identification questionable.

Blastocladiales

Allomyces × *javanicus* (Kniep) Emerson & Wilson. Collected on hempseed bait in soil from Chapel Hill, N. C.

Blastocladiella sp. Collected on boiled *Paspalum* leaf bait in soil from Chapel Hill, N. C.

The agar media used were: (1) *PDP.* 150 g. diced Irish potatoes, 9 g. dextrose, 50 mg. Difco bacto-peptone, 9 g. Difco bacto-agar, and 900 ml. dist. water. Autoclave the diced potatoes in the water for 45 min. at 15 lbs. pressure. Strain the resulting potato soup through a cloth towel; restore volume; add the dextrose, peptone, and agar, and then sterilize. (2) K_1. 1.8 g. dextrose, 0.6 g. Difco bacto-peptone, 0.4 g. yeast extract, 9 g. Difco bacto-agar, and 900 ml. distilled water. (3) *F13.* 1.4 g. maltose, 50 mg. Difco bacto-peptone, 9 g. Difco bacto-agar, and 900 ml. distilled water. (4) *CMDP.* 17.1 g. Difco cornmeal agar, 8 g. dextrose, 1.8 g. Difco bacto-peptone, and 900 ml. distilled water.

The following natural substrata were used: (1) Boiled, 1-cm.-long leaf segments of *Paspalum dilatatum* Poir., paspalum grass. (2) Pollen of *Liquidambar styraciflua* L., sweet gum. (3) Halved seeds of *Cannabis sativa* L., hemp (for *Allomyces*).

The fungal isolates were grown on the natural substrata in distilled water and on agar culture media—both in petri dishes. In most experiments, and unless otherwise stated, cultures were grown on the laboratory table at room temperature. For each isolate these media or substrata were inoculated at the same time and in the same way from inoculation cultures propagated asexually.

Although some of the experiments were procedurally meticulous, for most of the experiments the degree to which they were controlled and the numbers of measurements made were not ideal. Even so, the procedures easily meet the standards currently in use, and the data should have greater accuracy than the taxo-

Table I
Flagellar length—wet

Organism	Substratum	Length of Flagellum, μ	Ratio Avg. Flagellar Length to Avg. Spore Body Length
Phlyctochytrium irregulare (GH)	Agar (K_1)	21.7 (20.6–22.5)	6.0/1
	Pollen	22.2 (20.4–23.5)	6.0/1
Entophlyctis sp. (GH–2)	Agar (PDP)	20.6 (19.4–21.5)	5.7/1
	Agar (K_1)	20.4 (18.9–21.6)	5.2/1
	Pollen	20.5	5.1/1
Phlyctochytrium punctatum (Lucas)	Agar (K_1)	19.7 (16.9–21.4)	5.3/1
	Pollen	19.5 (18.4–21.4)	3.5/1
Phlyctochytrium punctatum (La6b)	Agar (K_1)	22.0 (20.9–23.5)	5.8/1
	Pollen	22.4 (21.4–24.5)	6.1/1
Chytriomyces hyalinus (CH–106)	Agar (F13)	29.6 (25.5–33.2)	5.6/1
	Pollen	28.6 (25.5–29.1)	5.7/1
Chytriomyces hyalinus (La–1)	Agar (K_1)	29.3 (26.5–30.6)	7.0/1
	Pollen	28.1 (22.1–33.7)	6.9/1
Rhizophlyctis rosea (CH–3)	Agar (K_1)	21.7 (15.2–28.0)	4.6/1

nomists of the group have heretofore demanded.

Measurements were accurate to about 0.25 μ because the ocular micrometer unit, used with a 90X apochromatic oil-immersion objective, was 0.51 μ. Similar or even better accuracy was achieved when measuring camera lucida drawings of zoospores with a specially made ruler.

Further methods are given at the appropriate places in each of the sections.

Flagellar Length

Additional methods for taking the data in Table I follow. After zoospores had discharged from a large number of sporangia that had been placed in distilled water, a drop containing zoospores was placed on a microscope slide. The zoospores were killed with the fumes of 1% osmic acid (OsO_4) and stained darkly with crystal violet. A coverslip was added, after which the mount was blotted to maximal thinness and ringed with wax to prevent drying. Then several more slides were prepared. The slides were searched for zoospores with flagella in a single plane. Ten to twenty measurements were made, usually of camera lucida drawings. At best, flagella were not completely straight and were not completely in a single focal plane.

The data on flagellar length in Table II were taken by Kate Miller in much the same way, except for more rigorous experimental procedure: Water of triple-distilled purity was used throughout. The inocula consisted of zoospores that had been discharged for no longer than 15 minutes in each case from up to hundreds of sporangia grown on a "flood culture" on a CMDP agar plate. The cultures were incubated at 25° C. in constant fluorescent light. The zoospores to be measured were quite uniform in age, up to 15 minutes old, and prior to sampling were separated from the numerous sporangia that had produced them. Randomness was enhanced by stirring the swimming

Table II
Flagellar length—dry. *Phlyctochytrium punctatum*

Isolate	Substratum	Mean (microns)	Range (microns)	Standard Deviation	Times Exp. Performed	No. Zoospores Measured
MF101	Agar (CMDP)	14.4	9.0–22.5	2.34	3	301
	Pollen	15.0	9.0–22.5	1.79	3	298
R17	Agar (CMDP)	16.8	10.0–25.3	2.95	3	285
	Pollen	16.5	9.5–23.0	2.94	3	298

zoospores at the time of sampling. In order to get the flagella in one plane for more accurate measurements of flagellar length and in order to stain them intensely enough for clear visibility of even the whip-lash tip, the following procedure was carried out. A drop containing a random sample of swimming zoospores was placed on a clean glass microscope slide. The slide was inverted over the fumes of 1 or 2% osmic acid for 1 minute. Then a drop of freshly filtered 1% crystal violet was added. The drop was dried on a slide-warming table over a period of 5 minutes or longer. The excess stain was washed off with water and the slide was re-dried. Then the flattened, stained zoospores were covered with a permanent mounting medium and a #1 coverslip. Accurate data were obtained by measuring only camera lucida drawings, made either with ordinary or with phase contrast optics. Image magnification on the drawing board was X3720. A slide was transected with a mechanical stage as many times as necessary, and as many slides as necessary were transected in order to get drawings for measurements of 100 flagella of zoospores produced under each experimental condition. Each experiment was repeated from the beginning two times.

Unless otherwise stated, the data referred to are in Table I. In *Phlyctochytrium irregulare* the measurements of flagellar length for zoospores derived from both pollen and agar cultures are about the same. If this were the only available data, from it one would be justified in diagnosing the species as having a flagellum which is typically about 22 μ long. Here this average length is accurately expressed as a relationship with the average diameter or length of the zoospore body, a ratio of 6.0/1. Taxonomic diagnoses often indicate flagellar length in this way, stating, for example, that the flagellum is about six times as long as the spore body. That relative length may not be as accurate a measurement as flagellar length itself is indicated by the differences in the ratios for *Entophlyctis* sp., where the average length measured from the three different substrata is essentially identical. It is clearly seen in *Phlyctochytrium punctatum*: in isolate Lucas, whether grown on agar or pollen, the flagellum has essentially identical lengths, whereas the ratio of flagellar length to spore body length when agar grown agar is nearly twice the ratio when pollen grown. It is noteworthy, however, that in comparison of the two isolates of *P. punctatum*, the ratios indicate the difference between these two isolates that exists in the average lengths and the ranges of lengths of the flagellum; that is, in the isolate with the apparently longer flagellum (La6b), the flagellum is also relatively longer. In *Chytriomyces hyalinus* the difference between the ratios for isolates CH–106 and La–1 is a measure not of difference in flagellar length but rather of a difference in spore body size. In all of the species studied, there was no noticeable correlation between the length of a particular flagellum and the size of a spore body to which it was attached. Also, there may be differences between isolates and species in the length of the end portion of the flagellum, the whip-lash. This has not yet been put on a quantitative basis. This quantity is best observed with shadowed zoospores viewed with the electron microscope.

Rhizophlyctis rosea was included in Table I to show that in comparison with the other species shown, it has much greater variation in length of the flagellum—the longest flagellum is about twice as long as the shortest flagellum. The more carefully taken data in Table II question the meaning of this fact. In both of these isolates of *Phlyctochytrium punctatum* (R17 and MF101), whether grown on a natural substratum or on agar, the longest flagellum is about 2½ times as long as the shortest. One might expect a similar spread for the other isolates of *P. punctatum* (Lucas and La6b in Table I) if the more rigorous techniques under which the data shown in Table II were gathered were used. (Of course, a similarly greater spread might be found for *Rhizophlyctis rosea* as well.) Also, a comparison of the data for the isolates of *Phlyctochytrium punctatum* in Tables I and II clearly indicates a need to determine the amount of shrinkage caused by drying. The more carefully collected data (Table II) substantiate the indication (Table I) that flagellar length is a very constant quantity. In the diagnosis for *P. punctatum* (Koch, 1957), the flagellar length given is 21.7 (16.8–23.5) μ. The present data (Table II) indicate that it may be expressed more accurately as 15.7 (9.0–25.3) μ.

In conclusion, flagellar length is quite constant in the isolates of the species measured, whether they are growing on pollen or an agar culture medium. Further, as a general rule regardless of the substratum or other features of the environment in which a species might be found, length of the flagellum is one characteristic of the species which is subject to relatively little variation and would therefore be relatively reliable in helping to delimit a species, even without experimental data. Flagellar length is a constant and conservative

Table III
Zoospore body size (in microns)

Organism	Substratum	Diameter
Entophlyctis sp. (GH–2)	Agar (PDP)	3.6 (2.5–4.4)
	Agar (K_1)	3.9 (2.6–4.9)
	Pollen	4.0 (2.8–5.3)
Phlyctochytrium irregulare (GH)	Agar (K_1)	3.6 (2.0–4.9)
	Pollen	3.7 (2.6–4.6)
Phlyctochytrium punctatum (Lucas)	Agar (K_1)	3.7 (2.6–4.8)
	Pollen	5.5 (3.6–6.1)
Phlyctochytrium punctatum (La6b)	Agar (K_1)	3.8 (2.8–5.1)
	Pollen	3.7 (3.3–4.1)

Organism	Substratum	Avg.	Max.	Min.	Length Range	Width Range	Ratio Width to Length
Chytriomyces hyalinus (La-1)	Agar (K_1)	4.2×3.0	—	—	3.6–4.9	2.6–3.3	0.7
	Pollen	4.1×3.0	4.6×3.5	3.3×2.0	3.3–4.6	2.6–3.5	0.7
Chytriomyces hyalinus (CH–106)	Agar (F-13)	5.3×4.5	6.1×5.1	3.8×3.0	3.8–6.1	3.0–5.1	0.8
	Pollen	5.0×4.1	5.3×4.3	4.1×3.6	4.1–5.3	3.6–4.3	0.8

feature which is of such great value in characterizing a species that more attention should be paid to it as a matter of routine in taxonomic work. Since there may be more variation from substratum to substratum in size of the zoospore body, the ratio which expresses length of the flagellum relative to size of the zoospore body is a less reliable measure.

Body Size

Besides those given in the introduction, additional methods for taking the data in Table III follow. After zoospores had discharged from a large number of sporangia that had been placed in distilled water, a drop containing zoospores was placed on a microscope slide. The zoospores were killed with fumes of 1% osmic acid and stained lightly with crystal violet. A coverslip was added and ringed with wax to prevent drying. Additional slides were made, as necessary. Only spherical or subspherical zoospores were measured (except, of course, for *Chytriomyces*). Data for each sample were obtained by subjectively searching for and measuring the largest and the smallest zoospores and then measuring about twenty.

The data on size of zoospore body in Tables IV and V were taken in much the same way, except for more rigorous experimental procedure: Water of triple-distilled purity was used throughout. The inocula consisted of zoospores that had been discharged for no longer than 15 minutes in each case from up to hundreds of sporangia grown on a "flood culture" on a CMDP agar plate. The cultures were incubated at 25° C. in constant fluorescent light. The zoospores to be measured were quite uniform in age, up to 15 minutes old, and prior to sampling were separated from the numerous sporangia that had produced them. Randomness was enhanced by stirring the swimming zoospores at the time of sampling. A drop containing these zoospores was placed on a microscope slide. The slide was inverted over the fumes of 1 or 2% osmic acid for 1 minute. Then the killed and fixed zoospores were very darkly stained with a small amount of freshly filtered 1% crystal violet, and a #1 coverslip was added. After any excess fluid between the coverslip and the slide was blotted out from the edges of the coverslip, and assurance made that no zoospores were flattened, the mount was made air tight by applying wax to the margins of the coverslip with a bent, hot dissecting needle. Immediately, with specially made rulers, measurements were made of camera lucida images on a drawing board (X3200, Table IV; X3740, Table V). A slide was transected with a mechanical stage as may times as necessary, and as many slides as necessary were transected in order to get measurements of 100 zoospores produced under each experimental condition. Each experiment was repeated from the beginning two or three times.

The major interpretation of the data in Table III is that zoospore body size is fairly constant, but variables are indicated. Pollen-grown *Phlyctochytrium punctatum* isolate Lucas has a larger average body size than agar-grown. The reverse may be true for *Chytriomyces hyalinus* isolate CH–106. *Phlyctochytrium irregulare, P. punctatum* isolate La6b, and *Chytriomyces hyalinus* isolate La–1 show no differences in average body size from the two substrata, but *C. hyalinus* isolate CH–106 does show a difference in size range: the range in length from agar covers 2.3 μ, while from pollen it covers 1.2 μ; also, from agar the spores may be smaller or larger than they are from pollen. This is also indicated for *Phlyctochytrium punctatum* isolate La6b and *P. irregulare.* The two isolates of *P. punctatum* have different body sizes when they are grown on pollen, but from agar they are practically the same. Although in *Chytriomyces hyalinus* one isolate (CH–106) is smaller from pollen than from agar, the zoospore body of this isolate on both substrata is larger than the zoospore body of the other isolate (La–1). The length-width ratios for the two isolates of *C. hyalinus*, of course, express the difference in shape, not size, but they are included in order to show quantitatively the difference between the zoospore bodies of the two isolates in shape as well as size.

The data for the six isolates of *Phlyctochytrium punctatum* grown on CMDP agar (Table IV) and pollen (Table V) are compared with caution, even though consciously the same procedures were followed, because they were taken by different observers during different summers. John Clausz took the data in Table IV in 1965, and Kate Miller took the data in Table V in 1966. Totals of the data for all of the isolates indicate that the zoospore body of *P. punctatum* growing on CMDP agar is 3.8(2.0–6.5) μ in diameter, whereas the zoospore body from pollen cultures is 3.5(1.8–8.0) μ in diameter. In the diagnosis for *P. punctatum* (Koch, 1957) the zoospore is said to be "2.5–6.1 μ in diam. when spherical." The present data (Tables IV and V) indicate that diameter of the zoospore body of this species is more accurately stated as 3.6(1.8–8.0) μ. This slight increase in precision can hardly justify the tremendous labor involved in information gathering. However, the data do clearly indicate that these isolates of *P. punctatum,* after having been in pure culture for many

Table IV

Zoospore body diameter. *Phlyctochytrium punctatum* grown on CMDP agar

Isolate	Mean, μ	Range, μ	Standard Deviation	Times Exp. Performed	No. Zoospores Measured
MF101	3.6	2.0–5.3	0.70	3	300
R17	4.1	3.0–6.3	0.53	3	300
LUCAS	4.0	2.0–6.0	0.72	3	300
CJU	3.7	2.0–5.5	0.60	4	395
MMH	3.7	2.3–5.0	0.53	3	300
NG3	3.5	2.0–6.5	0.41	4	400

Table V

Zoospore body diameter. *Phlyctochytrium punctatum* grown on pollen

Isolate	Mean, μ	Range, μ	Standard Deviation	Times Exp. Performed	No. Zoospores Measured
MF101	3.6	2.0–5.8	0.57	3	303
R17	3.5	2.0–5.0	0.62	3	301
LUCAS	3.2	1.5–5.0	0.72	3	300
CJU	3.9	2.3–8.0	0.76	3	316
MMH	3.5	2.3–5.5	0.59	3	300
NG3	3.3	1.8–5.3	0.55	3	305

years (NG3 since 1944 and the youngest, CJU and MMH, since 1959), show but little variation in spore size from isolate to isolate and from substratum to substratum. For example, isolates Lucas and F17 are slightly smaller from pollen than from agar. A more revealing analysis of this sort (for eight isolates of *Rhizophlyctis rosea*) is presented by Bernstein in the next paper of this volume.

In conclusion, although zoospore body size is fairly constant among isolates of a species, it does have observable variances. Furthermore, size of the zoospore body may vary with the substratum on which the fungus is growing. There may be a difference in average spore size and in size range. These observations emphasize the importance, when making a taxonomic decision at the species level or below, of taking into consideration the substratum on which the fungus is growing.

Body Shape and Position of Flagellar Attachment

Shape of the zoospore body and the position of flagellar attachment to the body are two important taxonomic criteria. Precise observations on the extent of their variability within a species or genus are few. Most chytrids have zoospores with bodies that are spherical or nearly so, but other shapes occur. The posteriorly directed flagellum typically is attached at the basal end of the swimming spore; however, anteriorly or laterally attached flagella have been described for a few species.

No greater variety in body shape and position of flagellar attachment is known to occur in a species than in *Phlyctochytrium punctatum* and *Entophlyctis* sp. (Table VI), which may be described as having zoospore bodies that are spherical, subspherical, irregularly elongate, or rod-shaped, with posteriorly directed flagella that are subbasally, laterally, subapically, or apically attached. Note that the flagellum is never truly posteriorly or basally attached. If, however, I were describing these species after the casual observations of its zoospores which often attend taxonomic analyses of new species, I would probably observe and diagnose the species simply and incorrectly as having zoospores with a spherical or subspherical body and a posteriorly attached and directed flagellum. An illustration of a misconception of this sort is found in the literature for *Rhizophlyctis rosea,* which is said to have zoospores oval to spherical with posteriorly attached flagella. Here (Table VII) we see the zoospore body of *R. rosea* to be spherical, subspherical, irregularly elongate or rod-shaped, and with a posteriorly directed flagellum that is basally, subbasally, laterally, subapically, or apically attached.

The variation recorded in Tables VI and

Table VI

Phlyctochytrium punctatum, Entophlyctis sp.

	Water	Methocel	Surface of Agar Plate
Body Shape	Spherical, subspherical, irregular-elongate	Subspherical to very irregular-elongate	Some subspherical; most very irregular-elongate; often thinly rod-shaped
Flagellum Attachment	Lateral (subbasal when body spherical)	Lateral	Lateral (apical on many rod-shaped spores)
Double image	Asymmetrical	No change	No change

Table VII

Rhizophlyctis rosea, Blastocladiella sp.

	Water	Methocel	Surface of Agar Plate
Body Shape	Spherical, subspherical, slightly irregular-elongate	Same as in water except more are slightly irregular	Some subspherical; most very irregular-elongate; often rod-shaped
Flagellum Attachment	Basal or subbasal	Basal, subbasal, or lateral	Basal, subbasal, lateral, or apical (on many rod-shaped spores)
Double image	Asymmetrical	No change	No change

VII is an expression of the amoeboidness that swimming zoospores have. Gaertner (1954) has described this remarkable variety of body shapes in four closely related species of *Phlyctochytrium* when their zoospores are swimming in the thin liquid film on the surface of an agar culture plate, in gum arabic solution, or in a drop of water, not covered with a coverslip, when the drop forms a very thin layer on the slide. When unrestricted in a drop of water, the swimming spores of these species of *Phlyctochytrium* have a spherical body with a posteriorly attached flagellum; however, Gaertner found that under the other conditions the irregular or elongate zoospore swims with its flagellum inserted in front and bent back sharply along the spore body in a posterior direction. I had first noticed the very striking variety of irregular and elongate shapes in *Phlyctochytrium punctatum* while examining zoospores swimming in the thin liquid film around thalli in pure culture on various agar media (Koch, 1957, Fig. 7). Since then, I have found this to be equally true of *Entophlyctis* sp. (Table VI) and *Rhizophlyctis rosea* (Table VII) and to a lesser degree true of *Entophlyctis aurea, Rhizophidium sphaerotheca* (Table VIII), and also *Blastocladiella* sp. However, it was not found to be so with *Phlyctochytrium irregulare, Septochytrium variabile, Catenochytrium carolinianum,* and *Nowakowskiella ramosa* (Table IX), nor with *Chytriomyces hyalinus* (Table X), for in these species there was no noticeable difference in shape between the zoospores swimming in water and those on the surface of an agar culture plate.

Although body shape of swimming zoospores can be seen fairly well with phase contrast and with ordinary bright-field microscopy (Figs. 2, 3), the point at which the flagellum is attached to the body of a swimming spore can

Table VIII

Rhizophidium sphaerotheca

	Water	Methocel	Surface of Agar Plate
Body shape	Usually spherical; some subspherical.	Usually spherical; some subspherical	Usually spherical or subspherical; some irregular
Flagellum Attachment	Basal	Basal	Lateral in irregular spores; apparently basal in spherical
Double Image	Asymmetrical	No change	No change

Table IX

Phlyctochytrium irregulare, Septochytrium variabile, Catenochytrium carolinianum, Nowakowskiella ramosa

	Water	Methocel	Surface of Agar Plate
Body shape	Usually spherical; some subspherical	No change	No change
Flagellum Attachment	Basal	No change	No change
Double Image	Asymmetrical	No change	No change

Table X

Chytriomyces hyalinus

	Water	Methocel	Surface of Agar Plate
Body Shape	Ovoid, elliptical, oblong, or rarely subspherical	No change	No change
Flagellum Attachment	Basal	No change	No change
Double Image	Asymmetrical	No change	No change

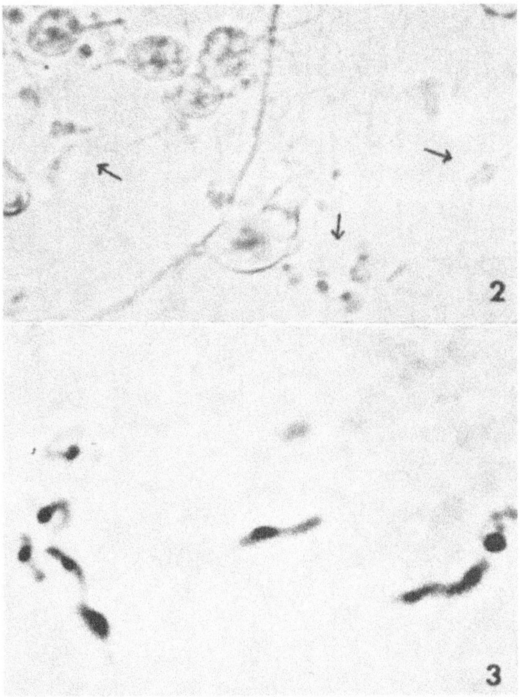

FIGS. 2, 3. Zoospores swimming on the surfaces of agar cultures as seen with the X40 dry, bright-field objective lens. Fig. 2. Zoospores of *Rhizophlyctis rosea* swimming among small thalli. X800. Fig. 3. *Phlyctochytrium punctatum* (isolate La6b). X1200.

be seen clearly only with the dark-field microscope using the X35 and X60 oil-immersion lenses. With this optical system the point of attachment of the active flagellum is seen by noting the position on the spore body of the base of the double image (e.g., Fig. 5a). Only with the X60 lens can the fine, but important, distinction between basal and subbasal attachment be ascertained; for example, on *Phlyctochytrium punctatum*, the zoospores in Fig. 5a show that the flagellum is subbasal and not truly basal or posterior in attachment, but with the X35 dark-field objective or with bright-field or phase contrast at any magnification these zoospores would have appeared to have flagella truly or completely basal.

The results of experimental observations on body shape and flagellar attachment in ten species are presented in Tables VI–XI. Comparison is made between zoospores swimming (1) in a thin film of water on the surface of an agar culture plate, (2) in ca. 1% methocel (15 CPS), and (3) in the control condition, unrestricted in water. Dark-field microscopy can be used in the latter two instances.

Tables IX–XI show that when the zoospores of *Phlyctochytrium irregulare, Septochytrium variabile, Catenochytridium carolinianum, Nowakowskiella ramosa, Chytriomyces hyalinus,* and *Allomyces* × *javanicus* are taken from water and put in methocel or on the surface of an agar plate, little or no change in body shape or position of flagellar attachment occurs. *Allomyces* is illustrated in Figure 10. In *Rhizophidium sphaerotheca* (Table VIII) no change was detected in methocel (see Fig. 10), but zoospores swimming in a thin liquid film on an agar surface show different shapes and flagellar attachments (Fig. 9). The greatest change was found in *Phlyctochytrium punctatum, Entophlyctis* sp. and *Rhizophlyctis rosea* (Tables VI, VII). This is most fully illustrated for *Phlyctochytrium punctatum* in Figures 4–7. Figures 4 and 6a show some of the great variety of shapes and positions of flagellar attachment displayed by spores swimming on agar. In methocel (Fig. 7) the zoospores are not irregular or thinly rod-shaped to the extent that they are on agar, and a greater percentage are spherical or subspherical. In water (Fig. 5) the zoospores are mostly

Table XI

Allomyces × *javanicus*

	Water	Methocel	Surface of Agar Plate
Body Shape	Usually ellipsoidal or obovoid; often slightly truncate at posterior end; rarely spherical, subspherical or oblong. Sometimes slightly irregular. (Male gamete usually elongate.)	No change	Perhaps greater percentage slightly elongate and more irregular because of amoeboidness, especially at rear.
Flagellum Attachment	Basal	No change	No change
Double image	Asymmetrical	No change	No change

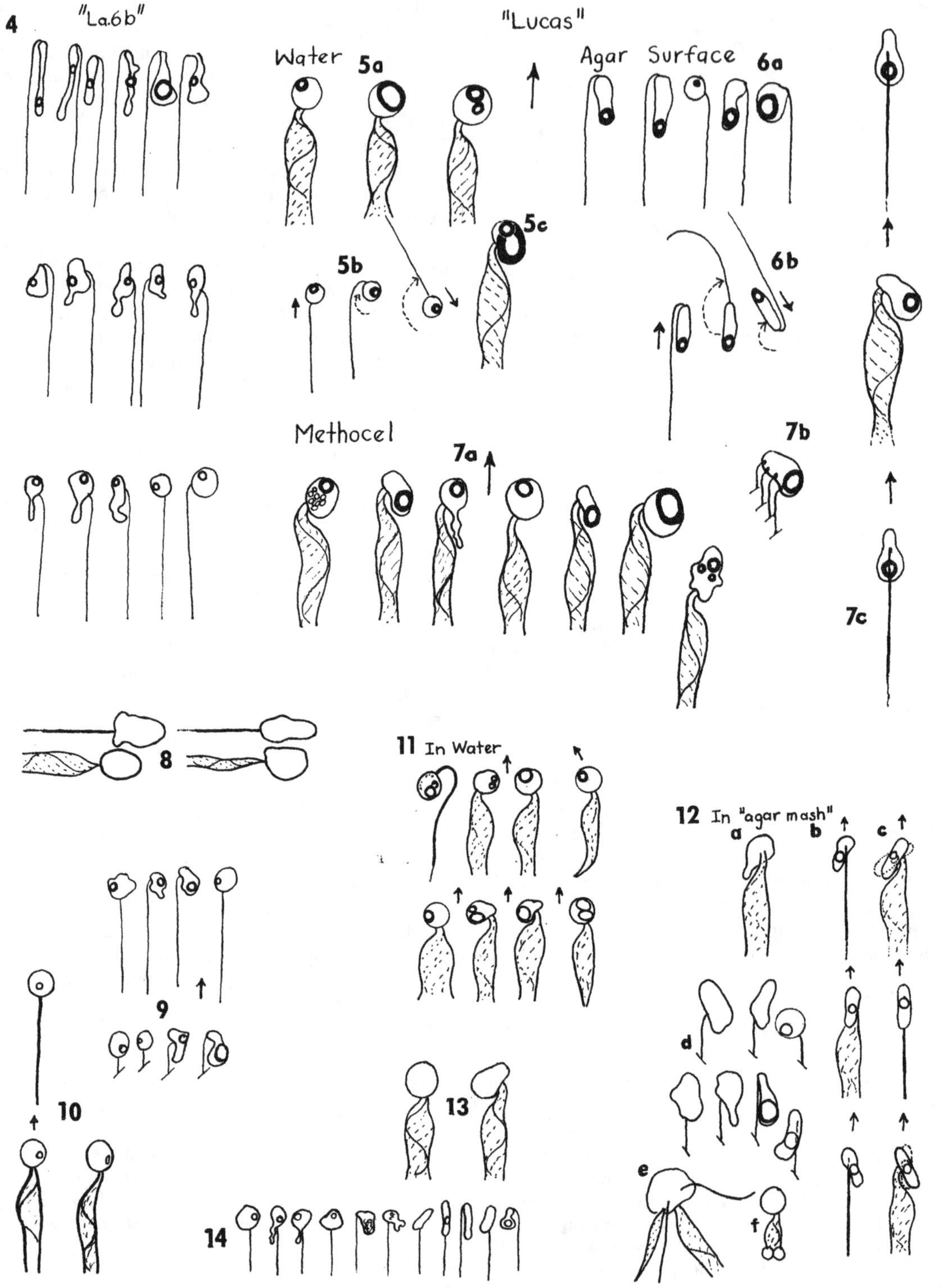

FIGS. 4–14. Not camera lucida. Figs. 4–7. *Phlyctochytrium punctatum.* Swimming zoospores, observed with bright-field, dark-field and phase-contrast microscopy, derived from pure cultures growing on agar media except for Fig. 5c. Arrows indicate direction of movement. Fig. 4. Zoospores of isolate La6b actively swimming in liquid film on surface of agar culture plate. This same variety of shapes is

spherical or subspherical, although occasionally one sees irregular and elongate spores of the sorts diagrammed for methocel (Fig. 7). Fig. 7b is a diagram included to show that in different zoospores the flagellum may be attached at different points even though the shape in all is the same. Since the differences in body shape and flagellar insertion result from amoeboid activity, it is likely that a single spore could have all of the variations shown on this plate at some time during its swimming existence. In no case is the flagellum seen to be truly basal in attachment; apparently the closest it gets to the base is shown in Fig. 5a, which is the position that I refer to as subbasal. This is also seen in *Entophlyctis* sp. (Fig. 11). *Rhizophlyctis rosea* (Figs. 13, 14) differs from *Entophlyctis* sp. and *Phlyctochytrium punctatum* in that its flagellum appears to be truly basal as well as subbasal to apical. These observations indicate that the "change" of zoospore shape and flagellar attachment in methocel and especially on an agar surface is not a complete change but is more precisely considered as a "change in degree" of irregularity of body shape and nonbasal flagellar attachment.

These features of swimming zoospores are of taxonomic value at the species level and lower. That they are meaningful below the species level is illustrated in *Phlyctochytrium punctatum,* in which there are slight but distinct differences in the very elongate zoospores of isolate La6b and isolate Lucas. The zoospores diagrammed at the upper left in Figure 4 and at the left in Figure 6a show that the rod-shaped spores are longer and relatively thinner in La6b than in Lucas and also that the lipoid globule is usually near the mid-region rather than the base. That different isolates of the same species may have zoospores with different body shapes is indicated for *Chytriomyces hyalinus* in Table III, right hand column, in which the ratio of average width to average length is given in order to express numerically the subjectively observed fact that isolate La–1 is narrower than isolate CH–106.

The limited number of species observed here do not permit assessment of the value of the features of variability and change of body shape and flagellar attachment in distinguishing taxa higher than species. It is noteworthy, however, that the three species which show this capacity most strikingly and to the greatest degree (*Phlyctochytrium punctatum, Entophlyctis* sp., and *Rhizophlyctis rosea*) have Type 5 zoospores (Koch, 1961b); that is, they have similar internal body structures and organization, and in this respect differ from the other species studied. In this connection Koch (1961b) pointed out that shape and the degree of amoeboid movement causing change in shape of the swimming cell are at present somewhat characteristic for the chytrid planont types. Typically, Types 1 and 4 are spherical; Types 2, 5, and 6 are spherical to irregular and elongate and can actively change shape; and Type 3 is ovoid and rigid.

Gaertner observed (1954) that in contrast to the *Phlyctochytrium* isolates that he studied, *Rhizophidium sphaerotheca* did not show this metabolic swimming, for in the latter the spores remained spherical and moved with a posterior flagellum. My observations of *Rhizophidium sphaerotheca* are that the zoospore has the ability to change shape and that the otherwise basal flagellum may be lateral in irregular spores. Also, the other observa-

found for the other isolates of *P. punctatum.* Fig. 5. Zoospores of isolate Lucas actively swimming in distilled water mount. Observed with X60 objective in dark field. Fig. 5b shows characteristic method of abrupt stop and change of direction. Fig. 5c shows the characteristically large lipoid globule of zoospores from thalli grown on pollen and grass leaf. Fig. 6. Same zoospores as those in Fig. 5, here swimming in liquid film on surface of agar culture. Observed with X40 dry lens. Fig. 6b shows characteristic method of abrupt stop and change of direction of elongate zoospores with apical flagella. Fig. 7. Same zoospores as those in Figs. 5 and 6, here swimming in 1% "methocel." Observed with X60 objective lens in dark field. Fig. 7b indicates the various points at which the flagellum may be inserted on a zoospore of this shape. Fig. 7c shows successive views of a zoospore which rotates as it swims forward. Fig. 8. *Allomyces* x *javanicus.* Swimming zoospores observed in dark field. Derived from hempseed cultures. Figs. 9 and 10. *Rhizophydium sphaerotheca.* Derived from pure cultures on agar. Fig. 9. Zoospores actively swimming in liquid film on surface of agar culture. Arrow indicates direction of swimming. Observed with X40 dry lens. Fig. 10. Zoospores swimming in 1% "methocel" observed in dark field with X10, X35, and X60 objective lenses. Figs. 11 and 12. *Entophlyctis* sp. Derived from pollen cultures. Fig. 11. Zoospores swimming in distilled water as seen in dark field. Arrows indicate the direction of swimming. The zoospore at the upper left is momentarily at rest. Fig. 12. Zoospores swimming in mashed culture medium as seen in dark field. Arrows indicate direction of swimming. Fig. 12e shows a monstrous zoospore with four flagella. Fig. 12f shows a zoospore swimming while at the same time withdrawing its flagellum. Figs. 13 and 14. *Rhizophlyctis rosea.* Derived from agar cultures. Fig. 13. Swimming zoospores as seen in dark field. Fig. 14. Zoospores with a variety of shapes, actively swimming on the surface of an agar culture plate. The entire flagellum is not drawn.

tions contained herein report that in contrast to *Phlyctochytrium punctatum* and in contrast to Gaertner's *Phlyctochytrium* isolates, *Phlyctochytrium irregulare* does not change. Gaertner's suggestion that these physiological features might be used as a distinguishing feature between *Phlyctochytrium* and *Rhizophidium,* when these two genera fuse into each other, therefore is presently not valid.

Lipoid Globules

Most species of chytrids have been described as having a single, conspicuous, refractive or lipoid globule. Its size, color, centricity or eccentricity, and, when eccentric, its general location in the spore body have been used to characterize some species. That more than one lipoid globule may be present also is described for quite a few species. For example, the genus *Olpidiomorpha* was established by Scherffel on the basis of one species, *O. pseudosporae,* with a zoospore distinctive in part because of its anterior ring of highly refractive globules. As another example, Karling distinguished three species of *Karlingia* in part on a similar basis: *K. hyalina,* with one large hyaline refractive globule; *K. granulata,* with numerous, minute, hyaline granules; and *K. spinosa,* with one to six golden globules.

Presently little is known about the degree of caution required in using the lipoidal content of zoospores as a taxonomic character. Much experimental work must be done in order to reveal the stability or variability of the position, size, and number of lipoid globules in the zoospores of an individual and between varieties and species, under both uniform and different conditions of growth.

Besides those given in the introduction, additional methods for taking the data in Table XII follow. After zoospores had discharged from a large number of sporangia that had been placed in distilled water, a drop containing zoospores was placed on a microscope slide. The zoospores were killed with fumes of 1% osmic acid. Staining was not needed. A coverslip was added and ringed with wax to prevent drying. Additional slides were made, as necessary. About 100 zoospores from each experimental condition were evaluated.

Table XII presents the quantative data from a set of experiments with two isolates of two species and one isolate of another species, with the substratum on which the chytrids were grown as the variable factor. The number of lipoid globules seen with the X90 apochromatic objective lens (N.A. 1.3) is expressed in one column as the average number per zoospore as well as the range of numbers that can be found in different zoospores, and in another column it is expressed as the percentage of the zoospores which have one and only one lipoid globule. Size of the single lipoid globule or of the largest lipoid globule, when one of several lipoid globules within a zoospore is conspicuously larger than the rest, is given in both absolute terms (average diameter and range of diameters) and in relative terms, as a ratio of the average diameter of the largest lipoid globule to the average diameter or width of the zoospore body. (Measurements of zoospore body size were made at the same time and are given in Table III.)

Depending on the substratum on which the fungus is growing, species or isolates may show differences in the relative amount of the spore body taken up by the large, conspicuous lipoid globule—a difference in the ratio of volume of the large lipoid globule to the volume of the spore body. This can be obvious to an observer (Fig. 1) and is expressed numerically by the ratios in the right-hand column in Table XII. In *Entophlyctis* sp. (for which the data are incomplete) one can subjectively observe that from both PDP and K_1 agars the lipoidal material does not fill more than half of the spore body, whereas from pollen the lipoidal material usually fills over half of the spore body. This same relative difference is seen even more strikingly, as indicated by the ratios, in *Phlyctochytrium irregulare, Phlyctochytrium punctatum* isolate Lucas, and *Chytriomyces hyalinus* isolate CH–106. This difference appears to come about in three ways: (1) the conspicuous lipoid globule is larger from pollen than from agar (Table XII), while the spore body has about the same diameter on both substrata (Table III), as in *Phlyctochytrium irregulare*; (2) the conspicuous lipoid globule is larger from pollen than from agar, while the spore body is also larger from pollen than from agar, as in *Phlyctochytrium punctatum* isolate Lucas; or (3) the conspicuous lipoid globule is larger from pollen than from agar, while at the same time the spore body is smaller from pollen than from agar, as in *Chytriomyces hyalinus* isolate CH–106.

Thus, it is clear that depending on the substratum on which the fungus is growing, the species or isolate may show a difference in both the average size and in the range of sizes of the single lipoid globule or of the largest of the several lipoid globules, whichever the case may be. In fact, the data for the six of the isolates, which represent four species, indicate that the conspicuous lipoid globule has a larger average diameter when the zoospores come from sporangia growing on pollen than when they come from sporangia growing on agar. The data also indicate that with the exception of isolate CH–106 of *Chytriomyces hyalinus,* all of the isolates have a greater range in lipoid globule diameter when pollen grown than when agar grown.

Table XII also shows that the species or isolate may show a difference in the average number of lipoid globules per zoospore, the range in number of lipoid bodies per zoospore, and the percentage of zoospores with just one lipoid body. Isolate Lucas of *Phlyctochytrium punctatum* shows these differences most strikingly. When pollen grown, its zoospores have a very large lipoid globule, and this is accompanied by 1–10 tiny ones, so that, unlike any of the other five isolates, there is never just one lipoid body in a zoospore; the average number is five, a big one and four tiny ones. In contrast, when agar grown, 76 per cent of its zoospores contain a single lipoid globule; some zoospores do not have any visible lipoid globules, and no spore has more than six; and the single or largest lipoid globule is small and always smaller than the big one in spores pollen grown.

It should be pointed out that just as the data on zoospore body size and shape (Table III) indicated that the two isolates of *Chytriomyces hyalinus* are different and may represent different taxa, so also do the data on lipoid globules (Table XII). These data indicate, as do subjective observations, that the biggest lipoid globule is smaller in isolate La–1 and that there is less lipoidal material in La–1 than in CH–106. In the next paper in this volume, Bernstein presents a similar but more exacting experimental study of the lipoid globules in the zoospores of eight isolates of *Rhizophlyctis rosea.* She finds interesting correlations with zoospore body size and other features of the organism.

These experiments are interesting mainly because they show clearly that the nature of the environment in which the fungus is growing may not only affect the size and shape of the zoospore body, as discussed previously, but to an even greater or more striking extent it may cause differences in the number and sizes of the lipoid globules. This re-emphasizes the importance, in identifying a collection or in formally characterizing a taxon, of taking into consideration the substratum or substrata on which the fungus grows. These may be many and varied for chytrids and other Aquatic Phycomycetes.

On the other hand, no variation in the general position of the large, conspicuous, eccentric lipoid globule (when a single one is

Table XII
Lipoid globules

Organism	Substratum	Number	% with one Lipoid Globule	Diam. Large Lipoid Globule, μ: Avg.	Range	Range Difference	Ratio Diam. Lipoid Globule to Spore Body
Entophlyctis sp. (GH–2)	Agar (PDP)	1.1 (0–5)	88%	1.7μ	0.5–2.5μ	2.6μ	0.47
	Agar (K_1)	1.3 (1–8)	85	—	—	—	—
	Pollen	3.1 (1–15)	35	—	—	—	—
Phlyctochytrium irregulare (GH)	Agar (K_1)	1.5 (0–5)	64	0.6	0.2–1.8	1.2	0.17
	Pollen	1.1 (0–3)	92	2.3	1.0–3.3	2.3	0.62
Phlyctochytrium punctatum (Lucas)	Agar (K_1)	1.6 (0–6)	76	1.0	0.3–1.7	1.4	0.27
	Pollen	5.9 (2.–11)	0	big 4.3	1.8–5.1	3.3	0.78
				tiny	0.3–1.0		
Phlyctochytrium punctatum (La6b)	Agar (K_1)	1.2 (0–5)	92	1.3	0.3–1.8	1.5	0.35 (0.17–0.44)
	Pollen	1.0 (1–2)	98	1.7	0.3–3.1	2.8	0.46 (0.14–0.78)
Chytriomyces hyalinus (La–1)	Agar (K_1)	1.5 (1–5)	73	1.0	0.3–1.6	1.3	0.33
	Pollen	1.0 (1–3)	95	1.2	0.3–2.2	1.9	0.40
Chytriomyces hyalinus (CH–106)	Agar (F13)	1.9 (1–8)	46	2.0	0.3–2.8	2.5	0.44
	Pollen	1.5 (1–3)	60	3.0	0.3–2.3	2.0	0.75

present) has been noticed in these experiments. In both isolates of *Chytriomyces hyalinus*, whether grown on pollen or agar, the large lipoid globule is in the lower part of the zoospore body. This is also true for *Catenochytridium carolinianum* and *Rhizophidium sphaerotheca*. Similarly, the large lipoid globule is located in the upper part of the zoospore body in both isolates of *Phlyctochytrium punctatum*, *P. irregulare*, *Nowakowskiella ramosa*, *Entophlyctis* sp., *Rhizophlyctis rosea*, *Entophlyctis aurea* and *Rozella allomycis*. The lipoid globule is mid-lateral in *Septochytrium variabile*. It is interesting that there are species of chytrids with Type I motile cells (Koch, 1961b) having the large lipoid globule mid-lateral (*Septochytrium variabile*), in the upper portion of the zoospore body (*Nowakowskiella ramosa*), or in the lower part (*Catenochytridium caroliniamum*).

Summary

Experiments dealing with flagellar length, body size, body shape, position of flagellar attachment, and lipoid globules of the planonts of the following Aquatic Phycomycetes are presented, interpreted, and discussed in the taxonomic context. Chytridiales: *Catenochytridium carolinianum* Berdan, *Entophlyctis* sp., *Entophlyctis aurea* Haskins (?), *Nowakowskiella ramosa* E. J. Butler, *Chytriomyces hyalinus* Karling, *Phlyctochytrium punctatum* Koch, *P. irregulare* Koch, *Rhizophydium sphaerotheca* Zopf, *Rhizophlyctis rosea* (deBary and Woronin) Fischer, *Septochytrium variabile* Berdan. Blastocladiales: *Allomyces* × *javanicus* (Kniep) Emerson and Wilson, *Blastocladiella* sp.

LITERATURE CITED

BERLIN, J. D., AND C. C. BOWEN. 1964. Centrioles in the fungus *Albugo candida*. Amer. Jour. Bot. **51**: 650–652.

BESSEY, E. A. 1942. Some problems in fungus phylogeny. Mycologia **34**: 355–379.

BLONDEL, B., AND G. TURIAN. 1960. Relation between basophilia and fine structure of cytoplasm in the fungus *Allomyces macrogynus* Em. J. Biophysic and Biochem. Cytol. **7**: 127–234.

CANTINO, E. C., J. S. LOVETT, J. S. LEAK, AND J. LYTHGOE. 1963. The single mitochondrion, fine structure, and germination of the spore of *Blastocladiella emersonii*. J. Gen. Microbiol. **31**: 393–404.

CHAMBERS, T. C., K. MARKUS, AND L. G. WILLOUGHBY. 1967. The fine structure of the mature zoosporanguim of *Nowakowskiella profuse*. J. Gen. Microbiol. **46**: 135–141.

CHAMBERS, T. C., AND L. G. WILLOUGHBY. 1964. The fine structure of *Rhizophlyctis rosea*, a soil Phycomycete. Jour. Royal Microscopical Soc. **83**: 355–364.

COUCH, J. N. 1938. Observation on cilia of aquatic Phycomycetes. (Abst.) Science **88**: 476.

———. 1941. The structure and action of the cilia in some aquatic Phycomycetes. Amer. Jour. Bot. **28**: 704–713.

FULLER, M. S. 1966. Structure of the uniflagellate zoospores of aquatic Phycomycetes. Vol. 18, pp. 67–84, of the Colston Papers. Butterworths Scientific Publications, London.

FULLER, M. S., AND R. E. REICHLE. 1965. The zoospore and early development of *Rhizidiomyces apophysatus*. Mycologia **57**: 946–961.

——— AND ———. (In press.) The fine structure of *Monoblepharella* sp. zoospores. Can. Jour. Bot.

GAERTNER, A. 1954. Beobachtungen über die Bewegusgsweise von Chytridineen zoosporen. Arkiv f. Mikrobiol. **20**: 423–426.

KOCH, W. J. 1957a. Two new chytrids in pure culture, *Phlyctochytrium punctatum* and *Phlyctochytrium irregulare*. Jour. Elisha Mitchell Sci. Soc. **73**: 108–122.

———. 1957b. Further studies in the genus *Chytridium*. (Abst.) Jour. Elisha Mitchell Sci. Soc. **73**: 239–240.

———. 1958. Studies of the motile cells of chytrids. II. Internal structure of the body observed with light microscopy. Amer. Jour. Bot. **45**: 59–72.

———. 1959. The sexual stage of *Chytriomyces*. (Abst.) Jour. Elisha Mitchell Sci. Soc. **75**: 66.

———. 1961a. Motile flagellated cells in plants: the motile cell in posteriorly uniflagellate Phycomycetes. *In* Recent Advances in Botany. Pp. 335–339. Univ. Toronto Press, Toronto.

———. 1961b. Studies of the motile cells of chytrids. III. Major types. Amer. Jour. Bot. **48**: 786–788.

KOLE, A. P. 1965. Flagella. *In* G. C. Ainsworth and A. S. Sussman, eds., The Fungi, Vol. I, 77–93. Academic Press, New York.

LOVETT, J. S. 1963. Chemical and physical characterization of "nuclear caps" isolated from *Blastocladiella* zoospores. Jour. Bact. **85**: 1235–1246.

MARTIN, G. W. 1955. Are fungi plants? Mycologia **47**: 779–792.

MOORE, R. T. 1965. The ultrastructure of fungal cells. *In* G. C. Ainsworth and A. S. Sussman, eds., The Fungi, Vol. I, 95–118. Academic Press, New York.

REICHLE, R. E. AND M. S. FULLER. 1967. The fine structure of *Blastocladiella emersonii* zoospores. Amer. Jour. Bot. **54**: 81–92.

RENAUD, F. L., AND H. SWIFT. 1964. The development of basal bodies and flagella in *Allomyces arbuscula*. J. Cell Biol. **23**: 339–354.

SCHERFFEL, A. 1925. Endophytische Phycomyceten-Parasiten der Bacillariaceen und einige neue

Monadinen. Ein Beitrag zur Phylogenie der Oömyceten (Schröter). Arch. f. Protistenk. **52**: 1–141.

SPARROW, F. K., JR. 1935. The interrelationships of the Chytridiales. *In* Proc. Zesde Int. Bot. Congress, Amsterdam. **2**: 181–183.

———. 1958. Interrelationships and phylogeny of the aquatic Phycomycetes. Mycologia **50**: 797–813.

———. 1960. Aquatic Phycomycetes, 2d. ed. Univ. Michigan Press, Ann Arbor.

TURIAN, G., AND E. KELLENBERGER. 1956. Ultrastructure du corps paranucléaire, des mitochondires et de la membrane nucleaire des gamètes *d'Allomyces macrogynus.* Exp. Cell Res. **11**: 417–422.

UMPHLETT, C. J., AND L. W. OLSON. (In press.) Cytological and morphological studies of a new species of *Phlyctochytrium.* Mycologia.

WATERHOUSE, GRACE M. 1962. Presidential address: The Zoospore. Trans. Brit. Mycol. Soc. **45**: 1–20.

WESTON, W. H. 1935. The bearing of recent investigations on the interrelationships of the aquatic *Phycomycetes.* *In* Proc. Zesde Int. Bot. Congress. Amsterdam. **1**: 266–268.

A Biosystematic Study of *Rhizophlyctis rosea* with Emphasis on Zoospore Variability[1]

Linda Beryl Bernstein[2,3]

Department of Botany, University of North Carolina at Chapel Hill, N. C.

Introduction

Koch (1961) told of an investigation of single-spore isolates of six members of the Chytridiales that revealed that these fungi can show variation in a number of zoospore features: body size, body shape, position of attachment of the flagellum, and size and number of lipoid globules. This was correlated with whether the thalli were grown on a natural substratum or on an agar medium. His findings are presented in full in the previous paper of this volume. The widespread use of these same zoospore characteristics to aid in delimiting chytrid species coupled with the frequent practice of describing species from a single culture warranted a more intensive study. The results of a study of two of these features, zoospore size and number of lipoid globules in each zoospore body, are presented in this paper. These features were examined in large enough numbers to be statistically valid.

Rhizophlyctis rosea (deBary and Woronin) Fischer was used for a number of reasons: (1) The zoospore body is usually spherical; the measurement of its diameter is an unequivocal indication of its size. (2) This species is easy to grow in both natural and artificial culture. (3) Eight isolates of *R. rosea* from several geographical locations were available in the University of North Carolina culture collection. (4) There are discrepancies in the literature about the number of lipoid globules in a zoospore. (5) *Rhizophlyctis rosea* is an interesting species phylogenetically since it shows some affinities to the Blastocladiales both on the basis of metabolic pathway information (Cantino and Hyatt, 1953) and in some features of the fine structure of its zoospore (Chambers and Willoughby, 1964).

The species has been known for a long time and has been the subject of many kinds of research. It appears to be an uncontestedly valid species with which to test the reliability of zoospore characteristics.

Materials and Methods

All cultures of the eight *Rhizophlyctis rosea* isolates were from the stocks maintaned at the University of North Carolina mycology laboratory. In all cases for which the information is on file, the cultures were isolated from either single spores or single sporangia. Throughout this paper, the isolates will be identified by their U.N.C. culture code names: 47–51, BJ–1, 128–1, 127–1, QM–517, 159–10, Vogel, and 57–5.

The two media used were Stanier's defined agar medium (Stanier, 1942), and *Paspalum dililatum* leaves in dilute salts solution (Machlis and Ossia, 1953). They will be referred to here as "agar" and "grass" respectively. Cultures were inoculated from stock agar cultures and grown at 25° C. under constant fluorescent illumination. The incubation period for agar cultures was two days and for grass, three days. Harvest at the end of this time consisted of flooding agar cultures with distilled water and pouring the water and zoospores into a beaker to mix them. Zoospores were collected directly from the grass cultures after swirling to ensure a random sample. Half-milliliter aliquots of zoospore suspension were put into equal amounts of Caulfield's fixative (Caulfield, 1957, with modifications by Koch) in test tubes and fixed for an hour. Slides of the fixed zoospores were stained with filtered, 1% crystal violet and ringed with paraffin wax.

The microscope used was a Leitz with an X90 apochromatic oil-immersion objective and an X16 ocular. The camera lucida image of the zoospore was measured directly to get the diameter information. The same zoospores that were measured were also recorded for number of lipoid globules per zoospore.

One hundred zoospores per subculture were examined. Sample size for each isolate in each medium was determined by the graph configuration method in which 100 and cumulative 100 increments to 600 of spore numbers are

[1] From a thesis submitted in partial fulfillment of the requirements for the degree of Master of Arts in Botany at the University of North Carolina at Chapel Hill.

[2] The author is indebted to Dr. William J. Koch for advice on this research.

[3] Present address: Department of Botany, University of Washington, Seattle, Wash.

graphed. In all cases, graph shape and range had become constant at a sample size of 500.

The cytology of the zoospore body was studied in living zoospores vitally stained with Janus green B and 0.01% neutral red.

A nutritional experiment compared growth of isolates 47–51 and BJ–1 on agar medium at full strength and at dilutions of the nonagar components of ½, ¼, ⅛, 1/16, and 1/32. Inoculation and growing time are as above.

Results

The first observations of several isolates of *Rhizophlyctis rosea* zoospores grown on agar indicated that there could be a great disparity in both zoospore size and globule number per zoospore among isolates. Means of the two zoospore characteristics do not give the same impression as the visual observations of two diverse isolates (such as 127–1 and BJ–1), for, as Graphs I and II show, this is a skewed average produced by large ranges and greatest number in a class appearing away from the median. An example is isolate BJ–1, in which 26 per cent of the zoospores have only one globule per zoospore; however, the mean is 3.5. The range is one through ten in the data on globules. A similar, though less pronounced, effect appears in the diameter-of-zoospore information in this isolate.

The data from the same diverse isolates grown on grass present a more nearly bell-shaped curve (Graphs I and II, dashed lines). Here, the means more nearly indicate the greatest number of zoospores with the respective characteristics.

Comparing the diameters of zoospores from the two substrata (Graph I) makes clear that the eight isolates vary much more from one another when grown on the agar medium than when grown on grass. For example, compare the positions of the medians from grass-grown zoospores in isolates 127–1, 159–10, and BJ–1

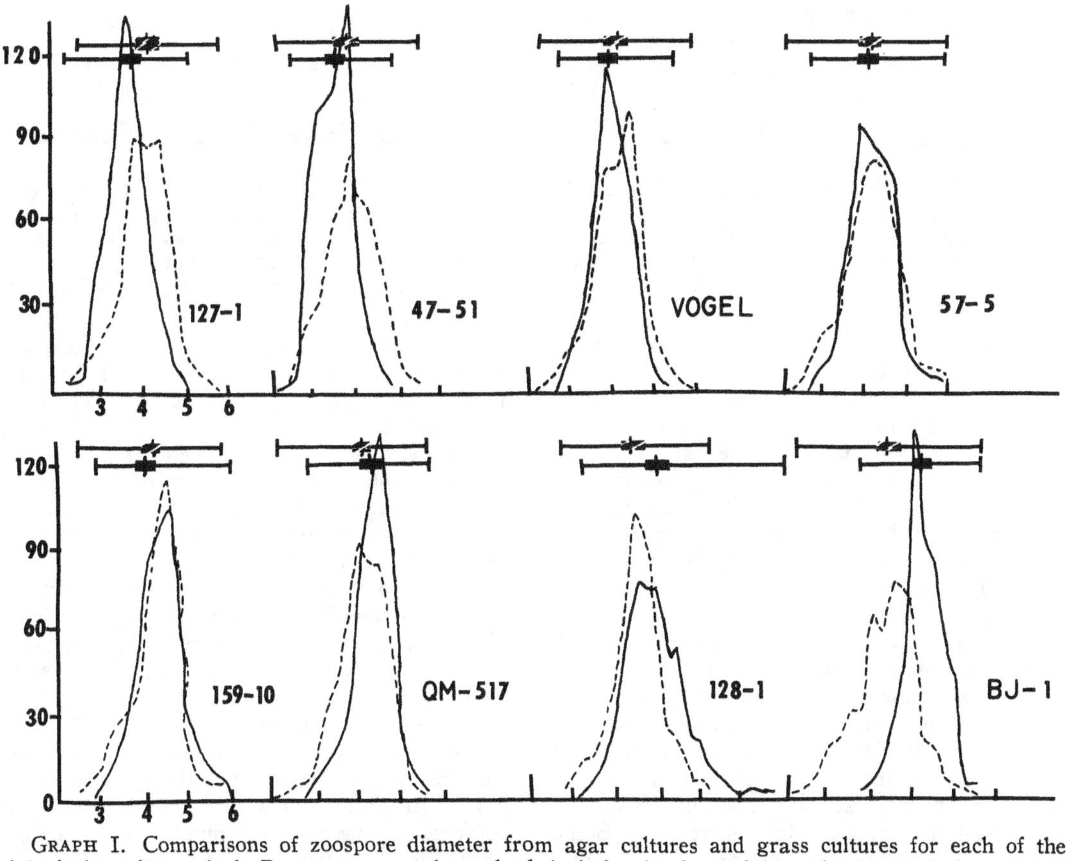

Graph I. Comparisons of zoospore diameter from agar cultures and grass cultures for each of the eight isolates is graphed. Range, mean, and standard deviation is above the graph of each isolate. From agar, the graph is a solid line and the standard deviation is shaded; from grass, broken line and a cross-hatched standard deviation. Diameter in microns is plotted on the horizontal axis, and number of zoospores in the size classes is on the vertical axis.

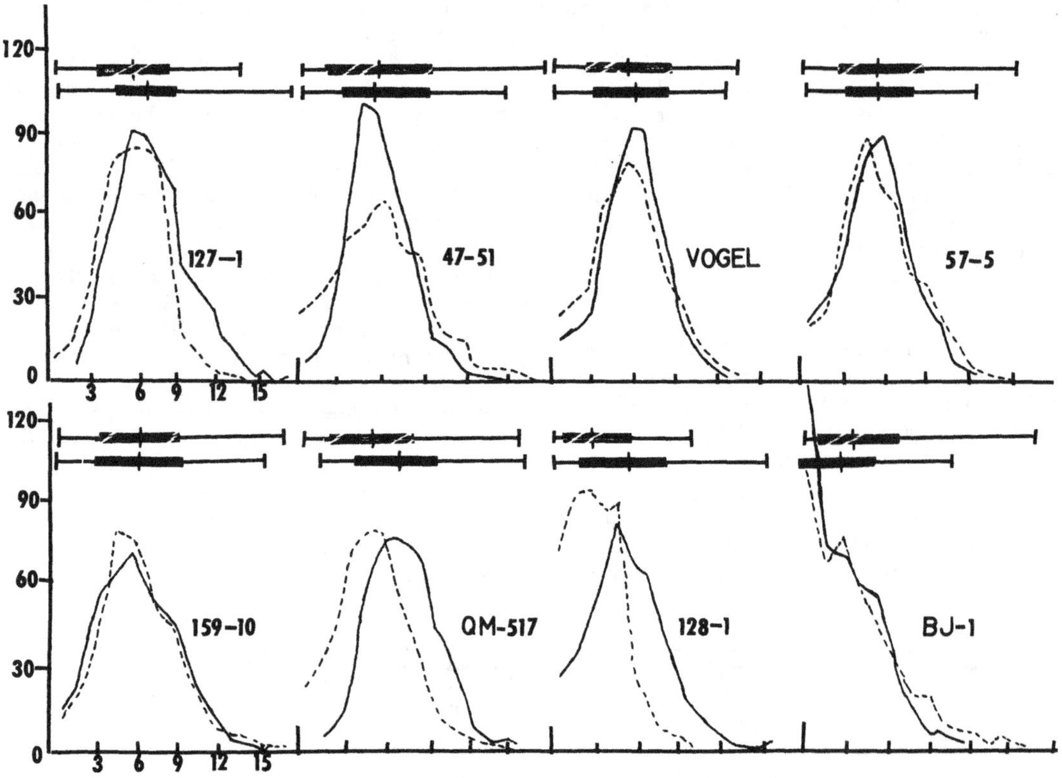

GRAPH II. Comparison of number of lipoid globules per zoospore from agar cultures and grass cultures for each of the eight isolates is graphed. Range, mean, and standard deviation is above the graph of each isolate. From agar, the graph is a solid line and the standard deviation is shaded; from grass, the graph is a broken line and the standard deviation is cross-hatched. Number of lipoid globules per zoospore is plotted on the horizontal axis, and number of zoospores in the classes is on the vertical axis.

with the median position for the same isolates from the agar cultures.

Agreement of means of the diameter data within a single isolate when grown on the two different substrata is sometimes close (as in isolates 57–5, Vogel, and 159–10); however, it is divergent in BJ–1 and 128–1. Length of range of an isolate tends either to be equal or to show a greater range on grass, the more natural culture medium.

Except for isolate BJ–1, the curves for globule number per zoospore do not show any striking differences among the isolates grown on the two substrata (Graph II). Range is greater in the agar data; it follows that the means are slightly farther apart in the data from these cultures. No generalizations can be made about closeness of means within a single isolate contrasted by substratum. Some isolates produce more zoospores with many more globules per zoospore on agar than on grass; in others, the reverse obtains.

The volume of the zoospore body that is occupied by lipoidal material is approximately 2.5 to 3 per cent; however, the small size of the lipoid globules makes it necessary to estimate their diameters; this figure can therefore be only an estimate.

Both zoospore size and number of lipoid globules in a zoospore are shown by this study to be more variable than previously reported in the literature on *Rhizophlyctis rosea*. Variation of means for these two quantities among the eight isolates falls within the range of these values given in most descriptions; however, the possible variation (range) in this species is much greater.

There is indication of a correlation of means between the number of lipoid globules per zoospore size (Graphs III and IV), especially on agar medium. This tendency is for large zoospores to have a low number of globules, and to some extent the converse. (This is striking when isolates BJ–1 and 127–1 are compared in Graph III). Variation of each hundred (which is the unit measured from a single

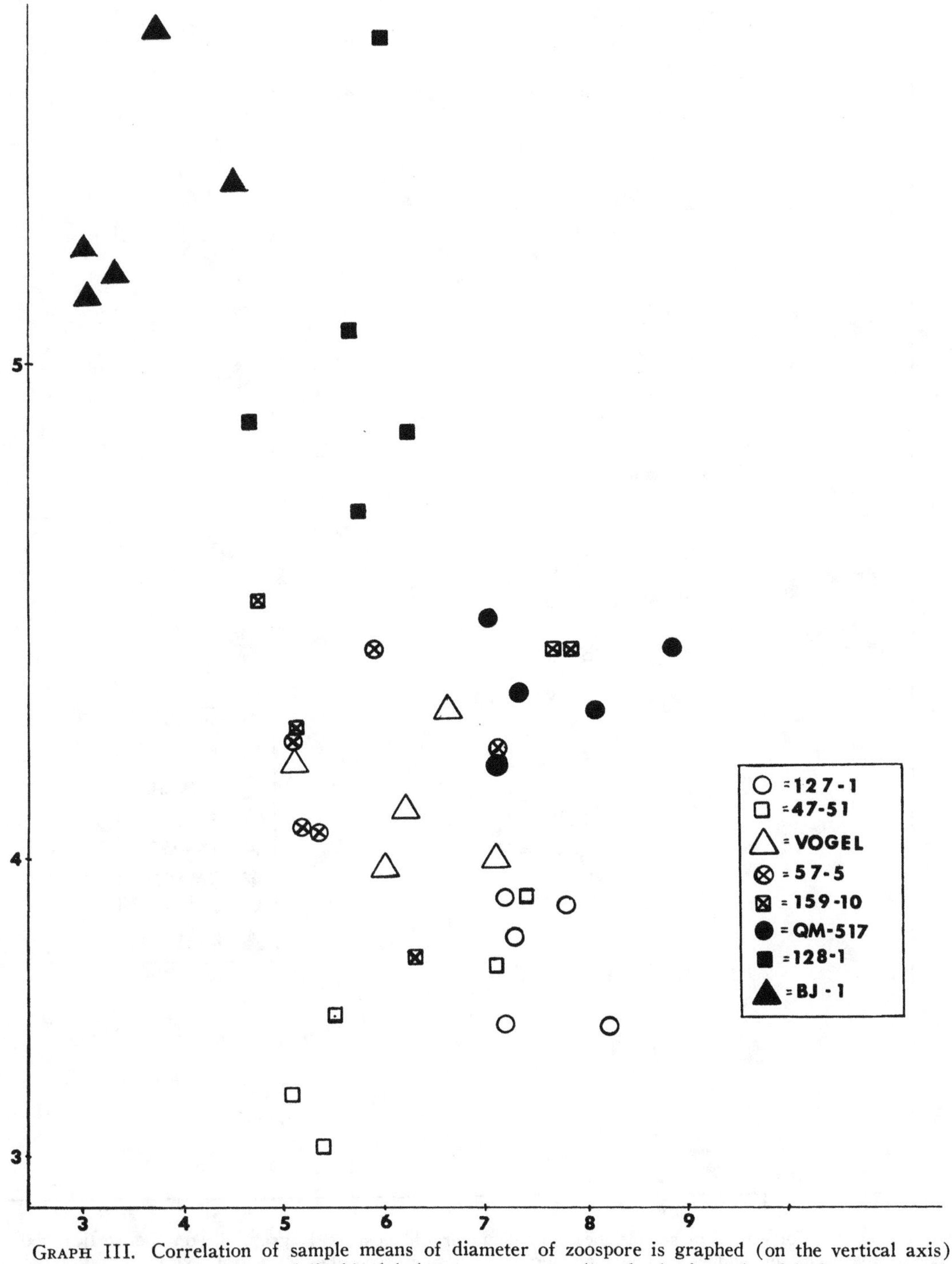

GRAPH III. Correlation of sample means of diameter of zoospore is graphed (on the vertical axis) against the means of number of lipoid globules per zoospore (on the horizontal axis) from agar cultures for each isolate.

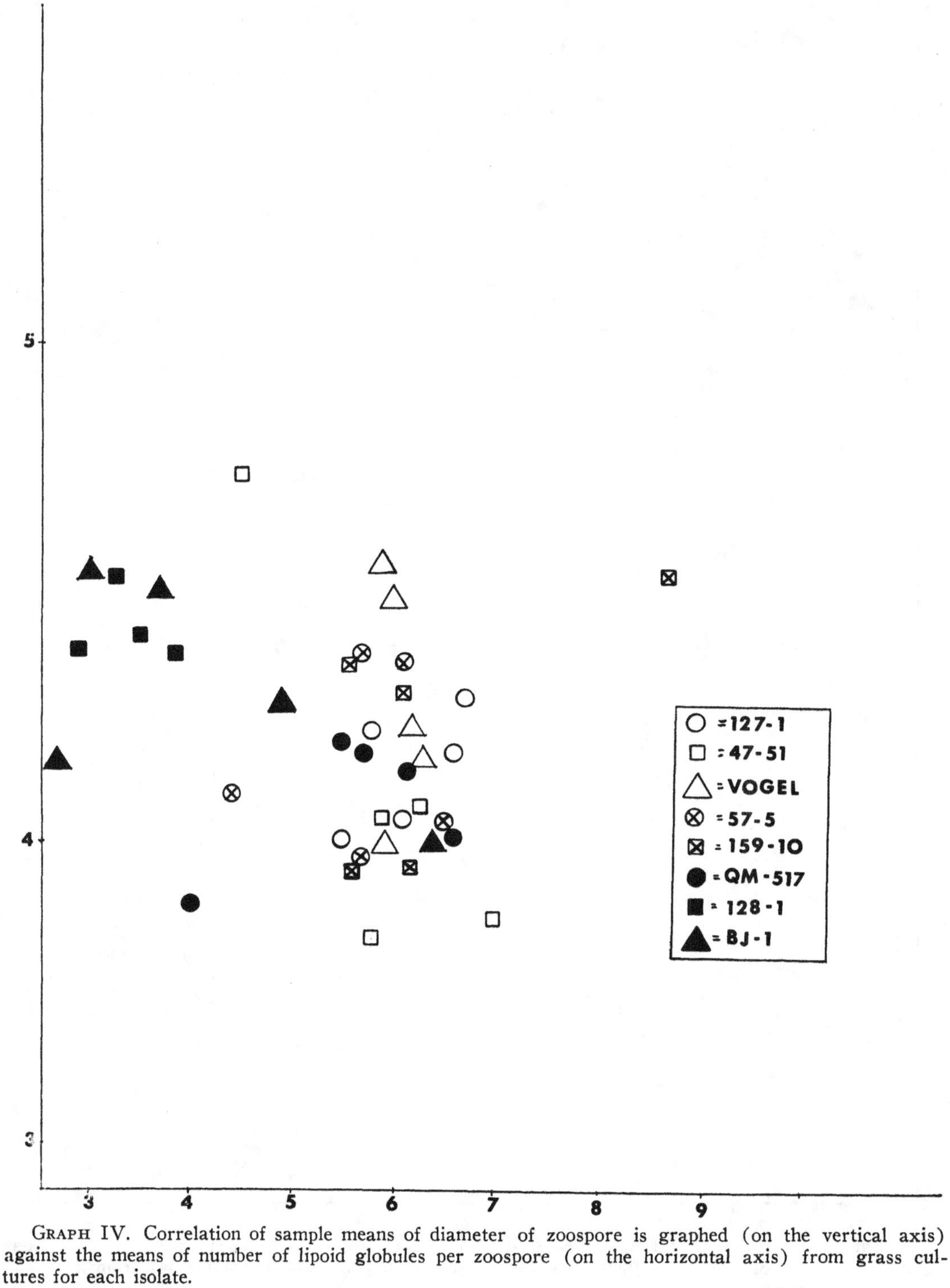

GRAPH IV. Correlation of sample means of diameter of zoospore is graphed (on the vertical axis) against the means of number of lipoid globules per zoospore (on the horizontal axis) from grass cultures for each isolate.

subculture of an isolate) is sometimes very great, as seen in isolate 159–10 on both grass and agar. There is a noticeable overlap most apparent from grass cultures (Graph IV), when the two features are correlated among the isolates. Only one of the isolates is distinct from all of the others when the two measurements are compared; BJ–1 produces zoospores that are very large and with very few globules when grown on agar. However, when this isolate is grown on grass, it is much closer to being the extreme end of a continuum than a discrete entity, as it appears in data from agar cultures.

Only two hundred zoospores (one hundred each from agar cultures of 47–51 and BJ–1) were examined to record actual correlation of these two features. It appears to agree with the other data; however, the sample size is not large enough to provide strong evidence.

Some qualitative observations were made on the arrangement of structures in the zoospore body (Plate II). The lipoid globule or globules are arranged peripherally around the nucleus. The nucleus fills about one-third of the zoospore body. It is invariably near the posterior of the cell and appears to be connected to an internal continuation of the flagellum. The nucleus contains a nucleolus and has a thin (a fraction of a micron in width), shiny rim the interpretation of which is not certain (see discussion). Frequently in vitally stained material an anterior vacuole is seen.

Sporangium size disparities were obvious and fairly consistent within an isolate; this was especially apparent from agar cultures. The sporangia in BJ–1 and 128–1 are very much larger (Fig. 1, Pl. I) than those in the other isolates which all have sporangial dimensions similar to one another (Fig. 2, Pl. I; Fig. 1, Pl. II). The figures are of typical and not extreme examples of the isolates they represent. One hundred sporangia each from agar cultures were measures for three isolates. Two "small" isolates, 127–1 and 47–51, had means of 62.1 and 88.2 microns respectively. The mean of BJ–1 was 131.3 microns. Range in BJ–1 was 90–200 microns; in 47–51, 60–145 microns; and in 127–1, 47–100 microns. In addition to the striking size differences, the large sporangia of isolates BJ–1 and 128–1 have very thick and twisted rhizoids and short exit papillae; in some cases it is impossible to distinguish a tube extension above the surface of the sporangium (Fig. 1, 5B, Pl. I). In all of the other isolates, (Figs. 2, 5A, Pl. I), the exit papillae are approximately one-seventh as long as the sporangium is wide, and the rhizoids are fine and predominantly straight.

Sporangia of all the isolates grown on grass are extremely variable and did not present a particular set of characteristics distinguishing

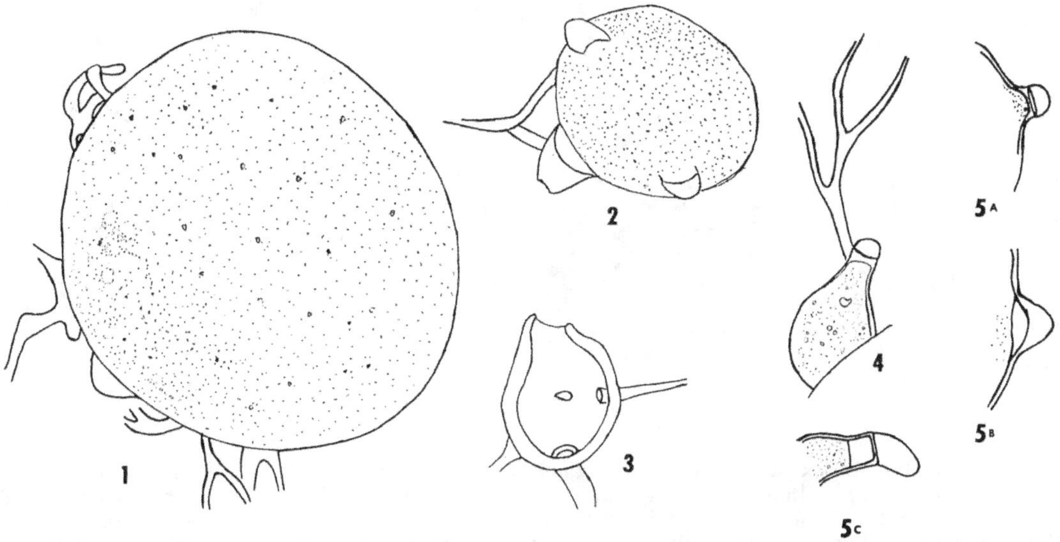

PLATE I. All drawings were made with a camera lucida from wet mounts of live material, at a microscope magnification of X640. Fig. 1. Sporangium of isolate 128–1 from a defined agar culture. Fig. 2. Sporangium of isolate 159–10, grown on agar medium. Fig. 3. Empty sporangium of isolate 159–10 from a grass culture; attachment of rhizoids to the inside of the sporangium can be seen, also interbiotic habit. Fig. 4. Young sporangium of isolate 128–1 from a grass culture. Fig. 5. Comparison of exit papilla shape and form: A. from an agar culture of isolate QM–517; B. from an agar culture of isolate 128–1; C. from a grass culture of isolate 128–1.

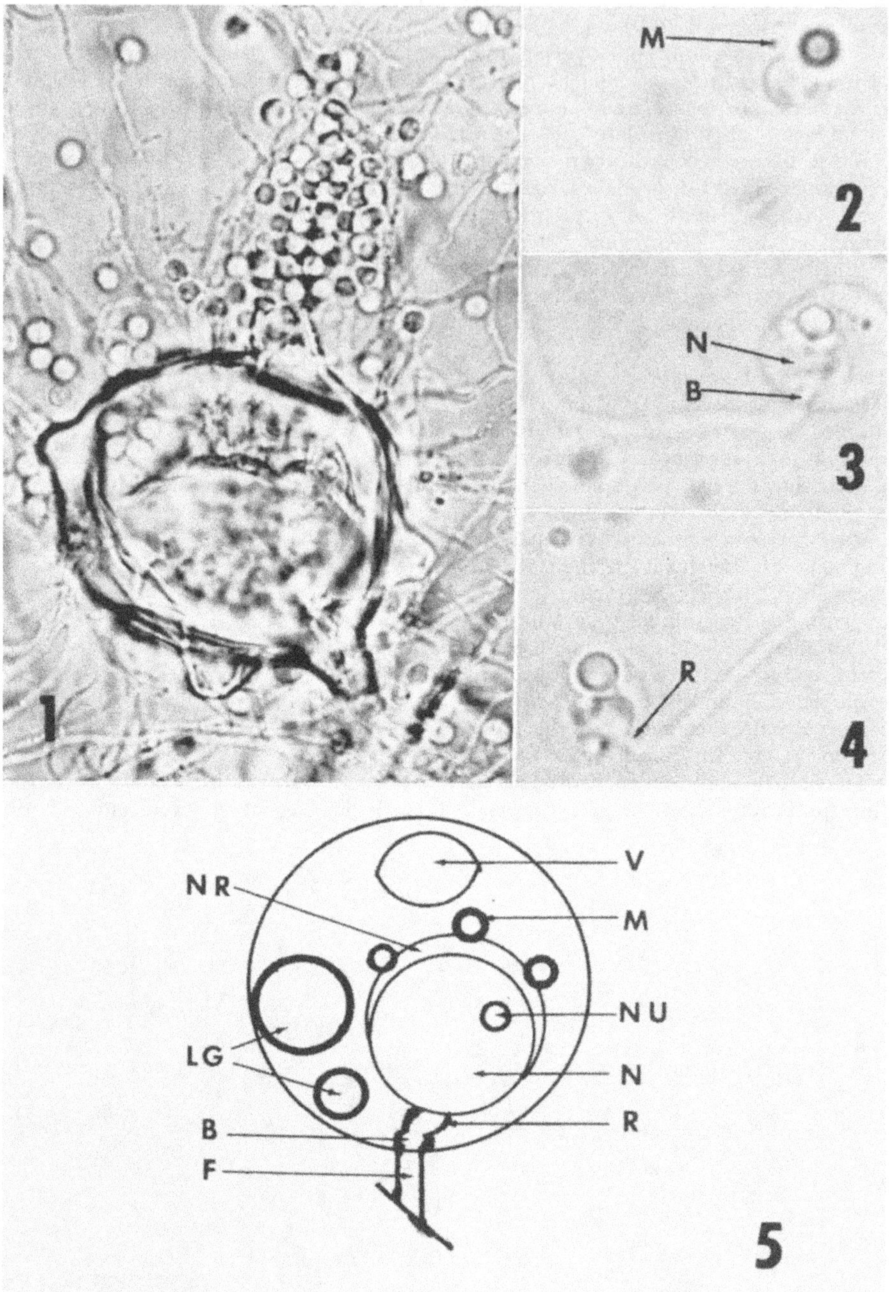

PLATE II. Fig. 1. Recently emptied sporangium of isolate 47–51 showing typical size and long exit papillae; from an agar culture. X640. Figs. 2, 3, and 4. Zoospores from isolate BJ–1 grown on agar. All are vitally stained with Janus green B for mitochondria and 0.01% neutral red for vacuoles. X3500 (approximately). Fig. 5. Diagramatic representation of the organelles of the *Rhizophlyctis rosea* zoospore. B, blepharoplast; F, flagellum; LG, lipoid globule; M, mitochondrion; N, nucleus; NR, shiny nuclear ring; NU, nucleolus; R, rhizoplast; V, vacuole.

any one isolate from any other (Figs. 3, 4, 5C, Pl. I). The other extreme of exit tube length from its norm on agar is figured for isolate 128–1 from a grass culture.

When the total data of this research are reviewed, it becomes apparent that the cultures grown on the agar medium show a consistent constellation of features accompanying the two quantitative characters of the zoospore under more intensive study. In six of the isolates (127–1, 47–51, Vogel, 57–5, 159–10, and QM–517), the growth of relatively small zoospores with a relatively large number of lipoid globules in a zoospore is associated with small sporangia having long exit papillae and narrow, highly branched rhizoids. These isolates show a tendency toward an evenly spaced growth of sporangia and will release zoospores upon rewetting after the film of water on the surface of the agar has evaporated completely.

The situation in the other two isolates (BJ–1 and 128–1) when grown on agar presents a contrast. Here, having large zoospores with few lipoid globules per zoospore is apparently correlated with having large sporangia with short exit papillae; thick, contorted rhizoids; a clumping habit of sporangial growth; and failure of the sporangia to release zoospores if a film of liquid does not remain on the agar surface in the plate.

This distinct morphological set of differences is not apparent in the grass cultures and so indicates a possible difference in response to available nutrients by the BJ–1 and 128–1 isolates. A single experiment (in replicate) sheds some light on this idea. A series of dilutions of the agar medium was inoculated with two diverse isolates. After the second subculture, isolate 47–51 (a "normal" isolate) was found to grow poorly and to release few zoospores at a 1/16 dilution. At the same dilution, BJ–1 was greatly stimulated in production of zoospores and in formation of sporangia that approached in appearance those of the majority of the isolates. From this, it appears that BJ–1 is adversely affected by the concentrations of one or more of the components of the agar medium. The morphological differences that distinguish this isolate are at least partially a symptom of the physiological difference in this isolate.

Discussion

The quantitative features of the zoospore of *Rhizophlyctis rosea* investigated in this research have proved to be more variable than previously thought. In the most recent composite diagnosis (Sparrow, 1960, p. 442), the section that pertains to the zoospore is:

> . . . zoospores numerous, oval to spherical, rose-colored, 3.3 to 5.3 microns in diameter, with one or more dark globules or a single colorless globule, and a dense protoplasmic spot. . . .

This portion of the diagnosis should be emended to:

> . . . zoospores numerous, oval to spherical, rose-colored, 2.0 to 8.0 microns in diameter, with 1 to 18 lipoid globules. . . .

The wide range of zoospore sizes and number of lipoid globules found in zoospores of this species raises doubt about the validity of two of the three characteristics used by Karling (1947) for establishing one of his new species, *Karlingia spinosa* Karling [equals *Rhizophlyctis spinosa* (Karling) Sparrow]. *K. spinosa* Karling was considered to have essentially identical development, growth, structure, appearance of sporangia, and behavior of zoospores as *Rhizophlyctis rosea* (deBary and Woronin) Fischer. It was established as a new species on the bases of size of the zoospore body; number of lipoid globules of a zoospore; and the structure of the resting-spore walls. The zoospores of *K. spinosa* were found to be from 3.3 to 4.4 microns in diameter, and Karling considered this to be the "same size" as the zoospore of *R. rosea* but smaller than the zoospore of *K. granulata* Karling (5.5 to 6.5 microns). This larger value is well within the range of zoospore diameter through which *R. rosea* is now shown to extend. Karling cited the presence of one to six golden brown refringent globules in a zoospore. He used this feature to distinguish *K. spinosa* from *R. rosea*. He described the latter as "with a large number of minute granules." Graph II of this paper shows that *R. rosea* frequently has between one and six lipoid globules in a zoospore, and in some isolates this is the number most often found.

The incomplete cleavage of some of the zoospores within a sporangium giving rise to "multiflagellate monsters" was reported in *R. rosea* by Karling (1947). U.N.C. isolates BJ–1 and 128–1 show the same behavior; however, the peculiarity was observed here only when the isolates were grown on agar medium and not on grass. That this formation of "multiflagellate monsters" from agar cultures is at least partially a response to nonoptimal medium is indicated by the experiments of this author with dilute agar medium. When isolate BJ–1 was grown on dilute (1/16 concentration) agar medium, this symptom did not appear. Karling's isolate was grown on cellulosic substrata in water, presumably natural substrata. Abnormal responses in the U.N.C. isolates might be traced to the length of time that the isolates have been maintained in stock culture (BJ–1, eight years; 128–1, seven years). However, one of the abnormal isolates (128–1) was originally isolated from soil from the same location and collected at the same time as isolate 127–1 which shows neither the tendency toward incomplete cleavage nor any of the complex of morphological responses to full-strength agar medium of the two unusual isolates. Mutation and selection in the strict conditions of stock culture is the only explanation that can be offered as to why two isolates from such similar circumstances should show the varying aspects that are presented by 128–1 and 127–1.

Further ramifications of the effect of nutrition on morphology have been presented in the literature on sporangium structure in this species. In his morphological study of *R. rosea*, Willoughby (1958) reported a correlation between sporangium size and exit tube length: small sporangia grew the longer exit papillae, and the converse. The extreme of this condition was very large sporangia with pores instead of tubes, the only surface projection being the gelatinous plugs. This tendency is also apparent in the U.N.C. isolates. Growth of large, short-tubed sporangia is the constant condition in isolates BJ–1 and 128–1 when they are grown on the agar medium (See Figs. 1, 5B, Pl. I). The other habit obtains to a greater or lesser degree in the other six isolates (Figs. 2, 5A, Pl. I). A possible explanation for these phenomena may be deduced from the work of Haskins and Weston (1950). They noted that cellobiose cultures yielded a coarsely granular macroscopic growth of sporangia while glucose cultures were of an even texture. Microscopic examination showed that when their isolate was grown on cellobiose, large sporangia that tended to grow in clumps were formed. On glucose, small sporangia grew evenly dispersed over the surface of the agar. Since the agar medium used in this present study contains both glucose and cellulose, and this size tendency is not found in grass cultures (which produce very irregular and variable sporangia—Figs. 3 and 4, Pl. I), the possibility arises that perhaps some of the U.N.C. isolates use cellulose preferentially. These would show the large-sporangium–clumping habit. The others might use glucose before cellulose resulting in the other mode of growth. A combination of dilution studies and carbon source variation possibly might elucidate the chemical interplay in nutritional requirements of the various isolates of *Rhizophlyctis rosea* and provide an explanation of some of the variability found among clones.

A peculiar structure around the anterior of the nucleus in living zoospores has been observed. It consists of a thin, shiny ring (see Pl. II). In his 1961 paper, Koch included *R. rosea's* zoospore type with a group lacking a nuclear cap. The structure observed here does not resemble the extensive shiny hoods characteristic of nuclear cap structure. Possibly, it is the light microscope view of some nucleus-associated structures observed in Chambers and Willoughby's (1964) electron microscope studies of *R. rosea.* These structures, named extranuclear strands, each connected the nucleus to a mitochondrion. Small dark mitochondria appear close to the structure in question in the observations of this author.

Summary

Eight isolates of *Rhizophlyctis rosea* were grown on Stanier's semi-defined agar medium and *Paspalum* grass in dilute salts solution. The diameters of zoospores and number of lipoid globules in each zoospore were recorded, one hundred per subculture, for five hundred zoospores from each culture medium for each of the isolates. Care was taken to ensure that the sample size was significantly large and random. The two quantitative zoospore characteristics studied here were evaluated as taxonomic criteria.

This study reaffirmed the importance of nutritional effects upon the total reaction of the organism. When the total data of this research are reviewed, it becomes apparent that the cultures grown on agar medium show a consistent constellation of features accompanying the two quantitative characters of the zoospore which were of primary interest. The isolates fell into two groups with respect to zoospore features, sporangium morphology and size, growth habit, and responses to moisture. These distinctions disappeared when the isolates were grown on grass.

Light microscope observations of the cytology of the zoospore body revealed an interesting structure around the anterior of the nucleus.

LITERATURE CITED

CANTINO, E. C., AND M. T. HYATT. 1953. Carotenoids and oxidative enzymes in the aquatic phycomycetes *Blastocladiella* and *Rhizophlyctis.* Amer. Jour. Bot. **40**: 688–694.

CAULFIELD, J. B. 1957. Effects of varying the

vehicle for OsO_4 in tissue fixation. Jour. Biophys. Biochem. Cytol. **3**: 827.

CHAMBERS, T. C., AND L. G. WILLOUGHBY. 1964. The fine structure of *Rhizophlyctis rosea*, a soil phycomycete. Jour. Roy. Mic. Soc. **83**: 355–364.

HASKINS, R. H., AND W. H. WESTON, JR. 1950. Studies in the lower Chytridiales. I. Factors affecting pigmentation, growth, and metabolism of a strain of *Karlingia* (*Rhizophlyctis*) *rosea*. Amer. Jour. Bot. **37**: 739–750.

KARLING, J. S. 1947. Brazilian Chytrids. X. New species with sunken opercula. Mycologia. **39**: 56–70.

KOCH, W. J. 1961. The motile cell in posteriorly uniflagellate Phycomycetes. *In* Recent Advances in Botany. Univ. Toronto Press, Toronto. Pp. 335–339.

MACHLIS, L., AND ESTHER OSSIA. 1953. Maturation on the meiosporangia of Euallomyces. I. The effect of cultural conditions. Amer. Jour. Bot. **40**: 358–365.

PALADE, G. E. 1952. A study of fixation for electron microscopy. Jour. Exptl. Med. **95**: 285.

SPARROW, F. K. 1960. Aquatic Phycomycetes. 2d ed., rev. University of Michigan Press, Ann Arbor. 1187 pp.

STANIER, R. Y. 1942. The cultivation and nutrient requirements of a chytridiaceous fungus, *Rhizophlyctis rosea*. J. Bacteriol. **43**: 499–520.

WILLOUGHBY, L. G. 1958. Studies on soil chytrids. III. On *Karlingia rosea* Johanson and a multioperculate chytrid parasitic on *Mucor*. Trans. Brit. Mycol. Soc. **41**: 309–319.

Studies of the Morphology of *Chytriomyces hyalinus*[1]

LINDA ROANE BOSTICK[2]
Department of Botany, University of North Carolina at Chapel Hill, N. C.

Introduction

The chytrids have proved to be quite variable in culture. They are often peculiar in their response to ordinary culture conditions and techniques. Whenever possible, experimental studies are conducted before new taxa are established and before old taxa are revised or eliminated.

Thus, it is desirable to study a particular chytrid under different environmental conditions to find and describe any variations. Such a study was done with a single-spore, single-genotype isolate of *Chytriomyces hyalinus* Karling.

Very little experimental work has been done on *Chytriomyces hyalinus* since it was first described by Karling in 1945. It was hoped that an investigation might further delimit the species and aid in further establishing its relationship among the other chytrids.

Methods and Materials

The author isolated and purified *Chytriomyces hyalinus* Karling from lake-water collections made at Highlands, Macon County, North Carolina, on 10 July 1963. The isolate was found growing on pollen of *Liquidambar styraciflua* L. (sweet gum), used as bait.

Single-sporangium (thus single spore) cultures were obtained by the "push technique" described by Couch (1939). The isolates were stored at room temperature on corn meal-dextrose-peptone (CMDP) agar slants, and were maintained by transferring to fresh agar slants every six weeks.

Experimental observations were made with a Wild microscope, M–20, with and without a green filter, using fluotar lens, 20X (N.A. 0.60), 40X (N.A. 0.75), and 100X (oil immersion, N.A. 1.30).

Other special methods are described later in the text.

General Observations

An actively discharging five-day-old culture was stained with Janus Green B, and the vesicle with its retained zoospores was observed (Fig. 4).

The apophyses present on this isolate were large and easily seen in one-day-old cultures (Fig. 6). Rhizoids were large and were found to occur inter-, extra-, and intramatrically on pollen (Fig. 5).

The flap operculum was easily seen on empty sporangia and usually occurred as a roughly triangularly shaped, retrorse structure (Figs. 4, 6). It was also infrequently seen on sporangia which were ready to discharge their contents.

Experimental Procedure

Experiment I: *Effects of Inoculum Size and Amount of Available Food on Sporangial Size*

Preliminary observations of *C. hyalinus* growing on *Liquidambar* pollen in sterile distilled water revealed a wide range in diameters of sporangia. To determine whether largeness and smallness separated into two distinct modes or whether there was an intermediate size yielding one mode, two samples were made on two occasions, the material being taken from more than one culture dish. The coefficient of variability obtained from these two samples was so high that it suggested that a curve could not be obtained by sampling.

Because general observation had indicated that pollen quantity per dish seemed to have an influence on sporangium size from each individual dish, a sample was again taken, this time from dishes containing a uniform amount of pollen.

Two sets of conditions were set up: one set of four petri dishes was supplied with an amount of pollen weighing between 11–13 mg. per dish; the second set of four petri dishes was supplied with 45–48 mg. per dish. Inoculation was accomplished by securing *C. hyalinus* actively growing on *Liquidambar* pollen in sterile, distilled water and adding one drop per dish.

After the sporangia had reached maturity, one hundred individuals per dish were measured. The results are summarized in Figure 1. Those plates with fewer pollen grains had a narrower size range and larger sporangia than those with many pollen grains. They also had more individuals occurring in a single size range. The measurements of the sporan-

---

[1] A portion of a thesis submitted to the faculty of the University of North Carolina at Chapel Hill in partial fulfillment of the degree of Master of Arts in Botany.

[2] Present address: Mycology Unit, National Communicable Disease Center, Atlanta, Ga. 30333.

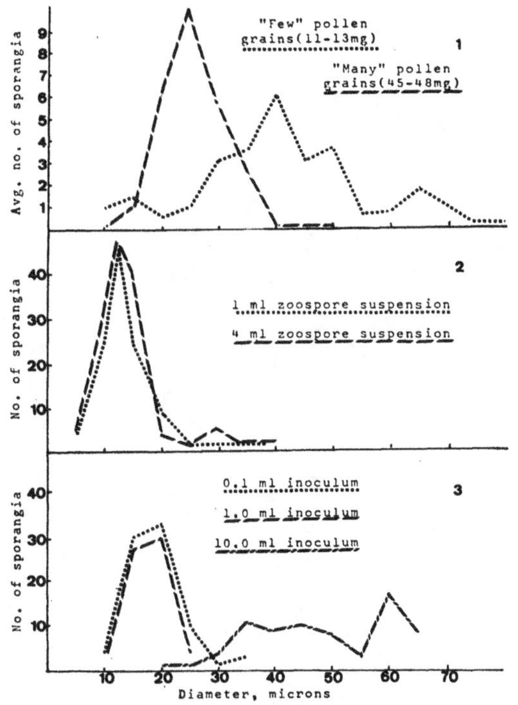

FIG. 1. *Chytriomyces hyalinus.* Comparison of mean diameters of sporangia with varying amounts of pollen.

FIG. 2. *Chytriomyces hyalinus.* Comparison of diameters of sporangia with varying amounts of inoculum.

FIG. 3. *Chytriomyces hyalinus.* Comparison of diameters of sporangia with varying amounts of inoculum.

gial diameters were subjected to the "t" test and were found to be significantly different.

A further experiment was designed in which inoculum size was varied while pollen quantity was held constant. These procedures were followed: Five CMDP plates were inoculated with *C. hyalinus,* the inoculum being taken from stock cultures maintained on slant tubes of the same agar. A small amount of sterile, distilled water was added to each plate. The cultures grew for two days; then the surface of each plate was flooded with sterile distilled water. Within 1½ hours discharge of zoospores had occurred. The zoospore suspensions from each plate were poured off into a sterile flask, in which they were shaken to insure mixing and to achieve a uniform suspension. The suspension was drawn into a sterile syringe. The syringe was shaken before each inoculation to prevent zoospores from settling at the needle end.

Four petri dishes were inoculated with 1 ml. of zoospore suspension and four others with 4 ml. of zoospore suspension. The amount of pollen added to each plate was between 11–13 mg. The mature sporangia were measured, and the results depicted in Figure 2. Essentially, there was no clear difference in the effect of the two conditions on sporangium size.

A further experiment was desirable to determine what effect a smaller or larger inoculum would have. This one was done essentially as that just described except that three plates received 0.1 ml. of zoospore suspensions; three others, 1.0 ml.; and three others, 10.0 ml.

Again, the mature sporangia were measured and the results are shown in Figure 3. The cultures that received 0.1 ml. and 1.0 ml. of inoculum were very similar in sporangium size. These data were submitted to the "t" test and were not found to be significantly different. Data obtained from 10.0 ml. cultures, however, were found to be significantly different from the other two.

Experiment II: *Effects of Temperature on Sporangial Size, Growth, and Development of Resting Sporangia*

Experiments were done with varying temperatures to determine any possible effects on sporangia. Both purified shrimp chitin and *Liquidambar* pollen were used for bait. The amount of chitin added per dish was 43–50 mg., and the amount of pollen added per dish was 45–48 mg. The dishes were inoculated with 1.0 ml. of zoospore suspension each.

The dishes were exposed to temperatures of 10° C., 20° C., 25° C., and 30° C. All cultures were maintained under constant fluorescent illumination of approximately 160 foot-candles

Table I

Effects of temperature and substrata on growth and the occurrence of resting sporangia

Temp. Centigrade	Growth[a]		Resting Sporangia Occurrence[b]	
	Pollen	Chitin	Pollen	Chitin
10........	1	1	0	6
20........	3	4	8	7
25........	3	4	8	6
30........	1	1	0	0

[a]Key:
1 very limited growth
2 limited growth
3 extensive growth
4 very extensive growth

[b]Key:
0 no occurrence
6 occurrence on 6th day
7 occurrence on 7th day
8 occurrence on 8th day

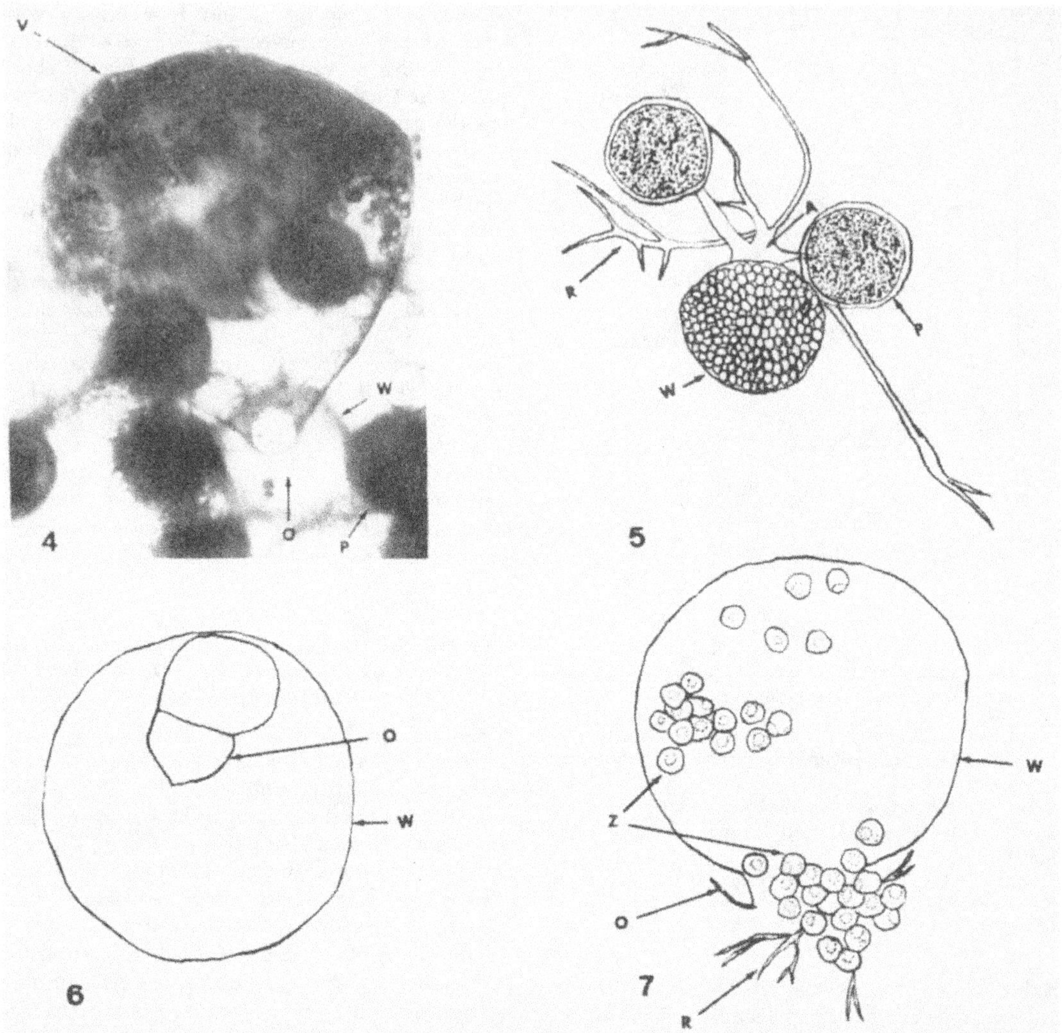

FIGS. 4–7. Photomicrograph and camera lucida drawings of *Chytriomyces hyalinus*. Fig. 4, stained with Janus Green B; Figs. 5–7, unstained.

Fig. 4. Sporangium with vesicle containing zoospores. V, vesicle; O, operculum; W, sporangium wall; P, pollen grain. X890. Fig. 5. Sporangium growing on *Liquidambar* pollen. W, sporangium wall; A, apophysis; P, pollen grain; R, rhizoids. X1097. Fig. 6. Outline of sporangium showing operculum peeled back prior to discharge. W, sporangium wall; O, operculum. X1097. Fig. 7. Sporangium with operculum peeled back. Some of the zoospores failed to emerge and others have germinated at the pore. W, sporangium wall; Z, zoospores; O, operculum; R, rhizoids. X548.

of light. Observations revealed that the growth of cultures and occurrence of resting sporangia varied with temperature and with the type of food available (Table I). Very extensive growth occurred on chitin at 20° C. and 25° C. Very limited growth occurred on either bait at 10° C. or 30° C. Resting sporangia did not form at 30° C., and formed only on chitin at 10° C.

Sporangium size varied with temperature and substrate. The ranges for sporangial size for each substrate and temperature recorded are as follows:

	chitin	*pollen*
30° C.	12.5 μ–44.0 μ	—
25° C.	15.0 μ–75.0 μ	5.0 μ–47.5 μ
20° C.	5.0 μ–45 μ x 47 μ	5.0 μ–31 μ x 35 μ
10° C.	5.0 μ–19 μ x 20 μ	5.0 μ–31 μ x 35 μ

The largest sporangia occurred at 25° C. on both substrates; of the two substrates, sporangia were the largest on chitin. Cultures grown at 25° C. also had wider size ranges than those grown at other temperatures. It was observed that 30° C. and 10° C. were generally unfavorable temperatures for growth to occur.

Experiment III: *Effect of Motion and Aeration on Growth and Morphology*

Experiments were conducted to determine whether motion or aeration had any effect on the growth and morphology of the fungus. A defined medium for *C. hyalinus* was prepared as indicated by Reisert and Fuller (1962) with the following changes in the preparation of the medium. Solution I was added to Solution II with a sterile syringe under a hood. The needle penetrated the cotton plug and decreased the possibility of contamination. The second change involved the growing of cultures to be used for inoculation. The method used, described by Slifkin (1963), permitted an inoculum to be obtained without adherent agar.

Flasks of the medium were inoculated with *C. hyalinus* and exposed to three different conditions. The amount of inoculum and medium per flask were constant. The conditions were: the wrist-action shaker, the swirl-action shaker, and motionless. The still cultures were divided into two groups—one received no aeration, and the second had filtered, compressed air bubbled through.

Under all the described conditions, clumping of the fungal material was quite evident. The least amount of clumping was found in the wrist-action cultures. The largest sporangia occurred in the swirl-action cultures; the smallest, in the wrist cultures. Sporangium shape in all of the cultures was irregular. The rhizoids were matted and thread-like. Resting sporangia were seen to form only in the shaken cultures, occurring earliest in wrist-action cultures. Those that were agitated frequently showed an abnormal type of zoospore discharge. Sporangia could be seen in which not all of the zoospores had escaped at the time of discharge. Germinating zoospores and very young sporangia could be seen inside old sporangial walls or at the mouth of an old sporangium (Fig. 7).

Discussion of Experimental Results and Conclusions

Variations in sporangial sizes and shapes have been noticed and recorded by previous workers. Cox (1939), in describing *Clavochytrium* [now *Blastocladiella stomphilum* (Couch and Cox), Couch and Whiffen, 1942], noted and outlined the many variations in sporangial shape and size. The variations that he figured are quite impressive. Karling (1928, 1931) found variations in sizes of zoosporangia of *Entophlyctis* on different hosts, as well as size variation in resting spores and in number and distribution of rhizoids of the thalli. In 1936, he stated that in a culture of *Diplophlyctis intestina* isolated by zoospore infection from a single zoosporangium, "the zoosporangia may vary tremendously in size and in exceptional cases show almost any conceiveable variation in shape."

In work with *Chytriomyces hyalinus,* it was found that when food supply was limited, more sporangia occurred in a single size range and were smaller in size than when a larger food supply was present. Also, a narrower size range was observed. Varying the amount of inoculum proved to have a similar effect. Sporangia were larger and there was a larger size range when a large inoculum was used than when a small inoculum was used (Fig. 3). McLarty (1941) found that zoosporangia varied in size and shape in *Pseudolpidium* and *Olpidiopsis,* depending upon the amount of nourishment available to the developing thalli. Karling (1936) also found that food supply affected the size and shape of zoosporangia.

In cultures of *Chytriomyces hyalinus* grown on Reisert's defined medium and exposed to varying degrees of aeration, a higher growth rate was observed in aerated cultures than in those not aerated. These results are similar to those of Craseman (1954), working with *Chytridium* and *Macrochytrium.* Craseman found that the yield in shaken cultures was roughly twice that in standing cultures. Such a high increase in yield was not found in *C. hyalinus* by the author.

The appearance and germination of resting sporangia in cultures has been the subject of much speculation and some experimentation. Such workers as Foust (1937), Bartsch (1939), and Johanson (1943) have found that certain cultural conditions may induce the formation of resting sporangia in *Rozella, Chytridium spinulosum,* and *Septochytrium plurilobulum,* respectively. These conditions include shortage of food supply, change in temperature, age of culture, and exposure to light and dark. Sparrow (1933) found no resting sporangia ("bodies") in cultures of *Endochytrium ramosum,* even though a systematic examination of infected *Cladophora* cultures was maintained for some months. Karling (1935), working with *Cladochytrium replicatum,* found no resting sporangia in 1931, but Sparrow found one in the same year. In 1933, however, Karling found them in "unlimited quantity." Karling (1936) found that resting sporangia ("spores") of *Diplophlyctis intestina* formed and germinated under laboratory conditions. He obtained negative results when germination was tried by freezing, heating, and drying. Koch (1959), from collections of *Chytriomyces* made at Highlands, N. C., made the following observations on resting spores that appeared in the cultures: "On chitin . . . they appear to be asexual and formed in the same way as zoosporangia; however, observations of agar cultures reveal that resting spore thalli form sexually at the tips of rhizoids from two contributing thalli. In the type species and in all but one of the subsequently described species, resting spores have been reported."

The occurrence of resting sporangia in *Chytriomyces hyalinus* has been found to be affected by experimental modification of the temperature and substrate. Resting sporangia occur earlier in cultures grown on chitin at 10° C. and 25° C. than in cultures grown on chitin at 20° C. or on pollen at any temperature. Resting sporangia also occur earlier in cultures grown on the wrist-action shaker but in greater abundance in cultures grown on the swirl-action shaker. They form under laboratory conditions but not in great abundance and have never been observed to germinate during a two-year period of study by the author.

The extent of the rhizoidal system was affected by the medium used. Cultures grown in Reisert's defined medium had rhizoids that were stunted in appearance. The rhizoids were matted together and were thin and thread-like. Cultures grown in sterile, distilled water on pollen and on chitin and cultures grown on agar had larger rhizoids. The reason for this is not known.

Cultures grown in Reisert's medium and agitated frequently showed an abnormal type of discharge. Sporangia could be seen in which not all of the zoospores had escaped at the time of discharge. Germinating zoospores and very young sporangia could be seen inside old sporangial walls or at the mouth of an old sporangium. Clearly, the motion disturbs the discharge process in some way.

The variations found to exist in *Chytriomyces hyalinus* by the author are interesting and of value in further defining and understanding the species. Workers such as Antikajian (1949), Koch (1957), and Register (1959) have called attention to variable features in certain other chytridial fungi that place doubt on the true identity of the organism. However, the variability found in *C. hyaliuns* does not seem sufficient to question its validity as a species.

Karling's (1945) original diagnosis needs to be expanded on certain criteria. Sporangia were found that, at maturity, did not measure over 5 μ in diameter, while others were over 100 μ in diameter. In the original description, they are described as ranging from 10 μ to 60 μ. Apophyses as large as 27.5 μ in diameter were found, larger than the range of 3 μ–7 μ specified by Karling. The rhizoidal system is described as being well developed. Under certain cultural conditions described earlier, the author has found the rhizoidal system to look like thin threads and to be matted into a rather small mass. They could hardly be called "well developed." The operculum was flap-like and incompletely delimited, usually remaining hinged to the sporangium following zoospore discharge. The cultures were pigmented, usually being brownish yellow to the naked eye. On the basis of these observations of sifinificant variations between the isolate studied and Karling's fungus, the species is emended as follows:

Chytriomyces hyalinus Karling emend. Sporangia hyaline, smooth pyriform, oblong, spherical, 5–170 μ, appearing brownish yellow in culture; operculum apical or subapical, shallow, saucer-shaped, or flap-like, incompletely delimited, usually remaining hinged to the sporangium following zoospore discharge; 8–16 μ in diam. Zoospores oval, 3–3.5 μ x 5–5.5 μ, with a small (1–1.5 μ), hyaline refractive globule; flagellum 18–20 μ long; zoospores emerging and swarming in vesicle 1–16 minutes before breaking out and swimming away; vesicle continuous with interior of sporangium. Apophysis when present spherical, subspherical, fusiform, or elongate, 3–27.5 μ in diam. Rhizoidal system matted and thread-like to well developed, main axis up to 7 μ in diam.; extending for a distance of 300 μ. Resting spores spherical (10–20 μ), oval (6–8 μ x 10–14 μ), elongate, clavate, pyriform, or slightly irregular, with a smooth, thick (2 μ), light-brown wall; containing a large central refractive globule surrounded by a few to several smaller ones; functioning as prosporangia in germination.

Saprophytic on exuviae of mayflies and on bits of chitin; luxuriant growth on *Liquidambar styraciflua* L. pollen and corn meal-dextrose-peptone agar.

Summary

Experiments were performed on *Chytriomyces hyalinus* Karling to determine the variability of certain characters under various laboratory conditions. Several pertinent observations were made. The vesicle with its retained zoospores was demonstrated using Janus Green B.

The amount of food supply affected size and size range of individual sporangia. With a limited food supply, more sporangia occurred in a single size range and were smaller in size than when a larger food supply was present. The size range was also narrower. Varying the amount of inoculum had a similar effect.

Temperature and substratum also affected sporangium size. Cultures exposed to 25° C. and grown on chitin had larger sporangia than either those exposed to 10° C., 20° C., or 30° C. and grown on pollen or chitin or those grown on pollen at 25° C. All cultures kept under temperature controls and constant illumination had smaller average sporangia diameters than others kept at an average room temperature of 26° C. and exposed to other varying conditions such as light.

Cultures of *C. hyalinus* grown in Reisert's defined medium for *C. hyalinus* and aerated exhibited a narrower size range of sporangial diameters than nonaerated cultures. They also had larger sporangia. More growth was observed in aerated cultures than in those not aerated.

The occurrence of resting sporangia was affected by temperature and substrate used. The cultures grown in Reisert's defined medium had stunted rhizoids that were matted together, thin, and thread-like. Cultures grown in Reisert's medium and agitated frequently showed an abnormal type of discharge. Sporangia were seen in which all the zoospores had not escaped at the time of discharge.

Karling's original diagnosis (1945) of *C. hyalinus* has been emended.

LITERATURE CITED

ANTIKAJIAN, G. 1949. A developmental, morphological, and cytological study of *Asterophlyctis* with special reference to its sexuality, taxonomy, and relationships. Amer. Jour. Bot. **36**: 245–262.

BARTSCH, A. F. 1939. Reclassification of *Chytridium spinulosum* with additional observations on its life history. Mycologia **31**: 558–571.

COUCH, J. N. 1939. Technic for collection, isolation, and culture of chytrids. Jour. Elisha Mitchell Sci. Soc. **55**: 208–214.

COUCH, J. N., AND A. J. WHIFFEN. 1942. Observations on the genus *Blastocladiella*. Amer. Jour. Bot. **29**: 582–591.

COX, H. T. 1939. A new genus of the Rhizidiaceae. Jour. Elisha Mitchell Sci. Soc. **55**: 389–397.

CRASEMANE, J. M. 1954. The nutrition of *Chytridium* and *Macrochytrium*. Amer. Jour. Bot. **41**: 302–310.

FOUST, F. K. 1937. A new species of *Rozella* parasitic on *Allomyces*. Jour. Elisha Mitchell Sci. Soc. **53**: 197–204.

JOHANSON, A. E. 1943. *Septochytrium plurilobulum* sp. nov. Amer. Jour. Bot. **30**: 619–622.

KARLING, J. S. 1928. Studies in the Chytridiales. I. The life history and occurrence of *Entophlyctis heliomorpha* (Dang.) Fischer. Amer. Jour. Bot. **15**: 32–42.

———. 1931. Studies in the Chytridiales. V. A further study of species of the genus *Entophlyctis*. Amer. Jour. Bot. **18**: 443–464.

———. 1935. A further study of *Cladochytrium replicatum* with special reference to its distribution, host range, and culture on artificial media. Amer. Jour. Bot. **22**: 439–452.

———. 1936. Germination of the resting spores of *Diplophlyctis intestina*. Bull. Torrey Bot. Club **63**: 467–471.

———. 1945. Brazilian chytrids. VI. *Rhopalophlyctis* and *Chytriomyces*, two new chitinophyllic operculate genera. Amer. Jour. Bot. **32**: 362–369.

KOCH, W. J. 1957. Two new chytrids in pure culture, *Phlyctochytrium punctatum* and *Phlyctochytrium irregulare*. Jour. Elisha Mitchell Sci. Soc. **73**: 108–122.

———. 1959. The sexual stage of *Chytriomyces*. Jour. Elisha Mitchell Sci. Soc. **75**: 66 (Abstract).

MCLARTY, D. A. 1941. Studies in the Woroninaceae. I. Discussion of a new species including a consideration of the genera *Pseudolpidium* and *Olpidiopsis*. Bull. Torrey Bot. Club **68**: 49–66.

REGISTER, T. E. 1959. Morphological variation in a new species of *Phlyctochytrium*. Unpublished Master of Arts thesis, University of North Carolina, Chapel Hill, N.C.

REISERT, P. S., AND M. S. FULLER. 1962. Decomposition of chitin by *Chytriomyces* species. Mycologia **54**: 647–657.

SLIFKIN, M. K. 1963. Parasitism of *Olpidiopsis incrassata* on members of the Saprolegniaceae. II. Effect of pH and host nutrition. Mycologia **55**: 172–182.

SPARROW, F. K., JR. 1933. Observations on operculate chytridiaceous fungi collected in the vicinity of Ithaca, N.Y. Amer. Jour. Bot. **20**: 63–77.

Observations Concerning Taxonomic Characteristics in Chytridiaceous Fungi

Charles E. Miller
Department of Botany, Ohio University, Athens, Ohio 45701

The stability of certain taxonomic characters frequently used to delineate some taxa of the Chytridiales has been questioned by both past and present workers. Sparrow (1960), in discussing the genus *Rhizophydium*, the largest and most complex of chytridiaceous genera, remarked that "characters which appear from the original description of a species to be fixed, well marked, and distinctive are usually found, upon careful investigation of a great many individuals to be subject to wide variation." Such characters as the presence or absence of an apophysis, type and number of discharge pores, size, shape and color of the sporangium, presence and arrangement of sporangial enations, extent of the rhizoidal system, planospore size, and presence and size of the lipoid bodies in the planospore appear to vary considerably.

Koch (1957) had difficulty using as diagnostic characters the presence or absence of an apophysis, sporangial shape and development, operculation, vesicular discharge, planospore, and relationship of the thallus to the substratum. The instability exhibited by these taxonomic characters strikes at the very heart of chytridiaceous classification. Thus, Koch (1957) found in *Phlyctochytrium punctatum* Koch planospores to be larger when grown on pollen than on agar. Also, planospores from pollen cultures had a greater number of lipoid bodies and were larger and conspicuous. Koch also found that an apophysis may be present or absent in *Phlyctochytrium irregulare* Koch and *P. punctatum* when these chytrids are grown on agar culture media. Miller (1961) found that *Rhizophydium pollinis-pini* (Braun) Zopf, growing on the same pollen grain, exhibited apophysate and nonapophysate thalli.

The state of development of the planospore before and after discharge, including the time of assumption of motility, is presently used as a diagnostic character in taxon delineation and is assumed to exhibit stability in physiology. The physiology or fundamental basis of predischarge and postdischarge development and activity of the planospore, however, is little understood. Thus, except that the presence of oxygen and minimal nutrients frequently initiate sporangial discharge, little else concerning the environmental conditions necessary for this activity are known.

There have been recognized within the Chytridiales (Sparrow, 1958, 1960) two parallel series in which specialization of the thallus has achieved equal complexity. In one series (Inoperculatae), the planospores escape after a deliquescence or rupturing of the tip of the discharge pore or tube; in the other series (Operculatae), the sporangium opens by the dehiscence of an operculum. In *Chytridium sexuale* Koch and *Phlyctochytrium irregulare* Koch, both methods of sporangial dehiscence have been reported, although the former species is predominantly operculate and the latter species predominantly inoperculate. It might be interesting to note here that Whiffen (1944) did not consider inoperculate and operculate sporangial characteristics to be of fundamental importance above the generic level.

During many observations (unpublished) of *Chytriomyces hyalinus* Karling, I have noted that some strains are "operculate" and others "inoperculate." Koch (personal communication) has also studied isolates of *Chytriomyces* and has noted variation in this character. Karling has recorded in several papers his recognition of the lack of stability in some existing taxonomic characters as seen in certain chytrids. Thus in 1945 he stated that ". . . species with occasional tube-like rhizoids show that the presence of an apophysis and the nature of the rhizoids . . . are variable characters and of doubtful importance in distinguishing genera." In 1949 Karling stated, referring to *Rhizidium varians* Karling, ". . . the present species combines the structural and developmental characteristics of a number of genera and even families. . . ."

Recently, Patterson (1963) reported on a study of two species of *Rhizophydium* from Northern Michigan. The results of his study indicated that such characters as the shape and size of the zoospores, shape of sporangia, and number of discharge pores were generally constant on all substrata used. One species occasionally produced angular sporangia resembling those of *Rhizophydium subangulosum* (Braun) Rabenhorst. The sporangial size of both species and the endobiotic part of one species were not constant on all nutrient materials used.

Some chytrids have been described as chitinophilic, keratinophilic, cellulosophilic, etc. The use of such adjectives implies a physiological affinity for or a restriction to a specific substrate. Sometimes convincing supporting evidence to warrant such a conclusion was not included. Much has been made in the past concerning the so-called parasitic or saprophytic nature of certain chytrids. In a few cases, forms present as presumed parasites were assumed to be different taxa if they occurred on nonliving substrates. Perhaps the only reason *Rhizophydium globosum* (Braun) Rabenhorst and *R. sphaerotheca* Zopf are separated is that the former occurs on algae and is supposedly a parasite and the latter occurs as a saprophyte on nonliving substrates. Patterson (1963) found that one of the Michigan chytrids having all the morphological characters of the asexual thallus of both *R. globosum* and *R. sphaerotheca* grew on both nonviable pollen and dead algae. The use of the presumed parasitic nature of some epibiotic and some endobiotic chytrids as a taxonomic character is difficult to apply. It is usually not indicated in the description whether this is an obligate or a facultative phenomenon. There seems to be no general acceptable criterion to measure parasitism. Is an epibiotic chytrid found growing on an algae cell the cause of its death, or is it growing there because the algal cell has died? How does one determine whether the host cell "parasitized" when collected is healthy, moribund, or dead at the particular time the study is made?

Identification of chytrids is frequently very difficult or impossible because there is no basic understanding of the taxonomic significance of the various morphological characters used. Emerson (1950) stated that while variability in morphology is recognized, little attention has been paid to it in many studies, and that "there appears to be greater fascination in describing new species than in determining the precise limits of their variablity. . . . It seems high time to recognize the responsibilities involved in naming new species; the job of critically establishing the true extent of variability and evaluating species . . . cannot be postponed indefinitely for others to undertake. . . ."

The difficulties are evident in a taxon that I first collected and studied in July, 1961. This chytrid, colloquially named by me as "Dentate," because of the tooth-like enations present on the sporangium, was found growing on sweet gum pollen (*Liquidambar styraciflua* L.) and pieces of purified exoskeleton of shrimp, both of which had been added to water collections from several aquatic habitats. The most striking characteristic of "Dentate" is the presence of a varying number of (one to many) sporangial enations (Figs. 1–33). These enations vary also morphologically from a simple pointed, or blunted, broad- or narrow-based spine to a bipartite tooth-like structure, examples of both extremes frequently occurring on a single sporangial thallus (Figs. 21, 22). These enations may be arranged in a random manner on the sporangium or in whorls, apically or laterally situated. The sporangium is usually spherical to subspherical, occasionally pyriform, conical, or irregular in shape.

When growing on sweet gum pollen, "Dentate" is frequently extramatrical, with a long or short epibiotic, tubular rhizoid that may extend through the germ pore of the pollen grain (Figs. 15, 16, 26, 34–36). The endobiotic portion of the rhizoid, generally difficult to observe in pollen, branches slightly. No definite epi- or endobiotic apophysis has been observed. The epibiotic rhizoid may be thicker near the sporangium, tapering gradually to where it enters the pollen grain, or it may remain constant in thickness throughout the epibiotic portion. The rhizoid may attach laterally to the sporangium and have an occasional subsporangial, one-sided bulge (Figs. 15, 26, 28). When chitin (purified exoskeleton of shrimp) is the substratum, the rhizoidal system can be seen clearly (Figs. 1–10). No definite apophysis has been found, and the rhizoid is also a tapering, tubular-shaped structure with branches originating near the base of the sporangium or further removed from it.

Stages in development of the sporangium are similar to those exhibited by many chytrids. The planospore, after settling on or near the substratum, puts forth a germ tube that develops into the rhizoidal system (Fig. 1). Concurrently, the planospore body enlarges to become the sporangium. At first the protoplasm of the young sporangium is homogeneous; then gradually this finely granular aspect disappears as small oil globules, which coalesce, forming larger globules, appear. Now the enations begin to form on the wall of the developing sporangium. When mature, the sporangium contains a number of spherical refractive globules with surrounding protoplasm, each of which indicates the position of a planospore (Figs. 23–28, 34) initial.

The sporangia, containing apparently mature planospores, may remain undischarged for

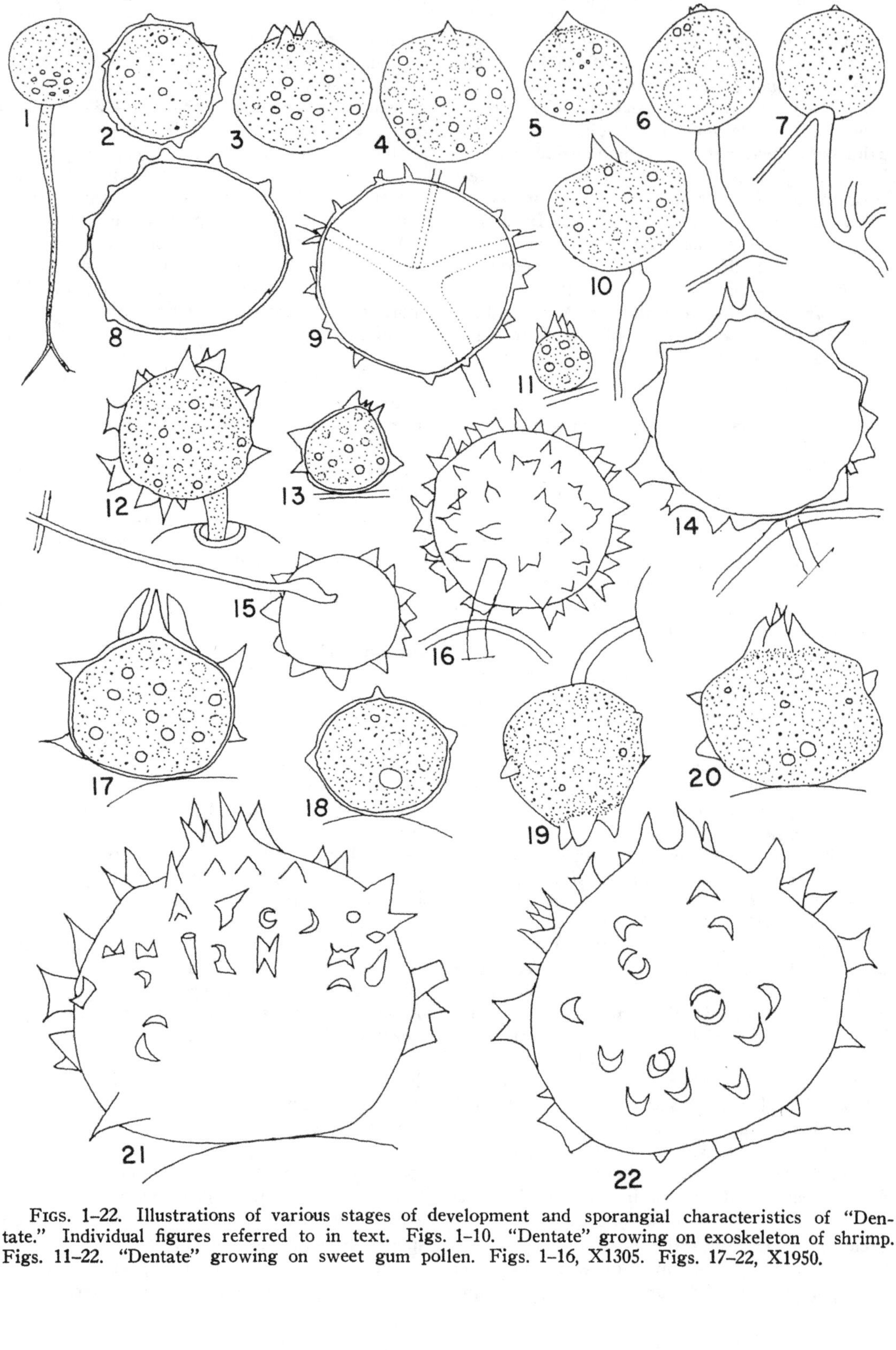

FIGS. 1–22. Illustrations of various stages of development and sporangial characteristics of "Dentate." Individual figures referred to in text. Figs. 1–10. "Dentate" growing on exoskeleton of shrimp. Figs. 11–22. "Dentate" growing on sweet gum pollen. Figs. 1–16, X1305. Figs. 17–22, X1950.

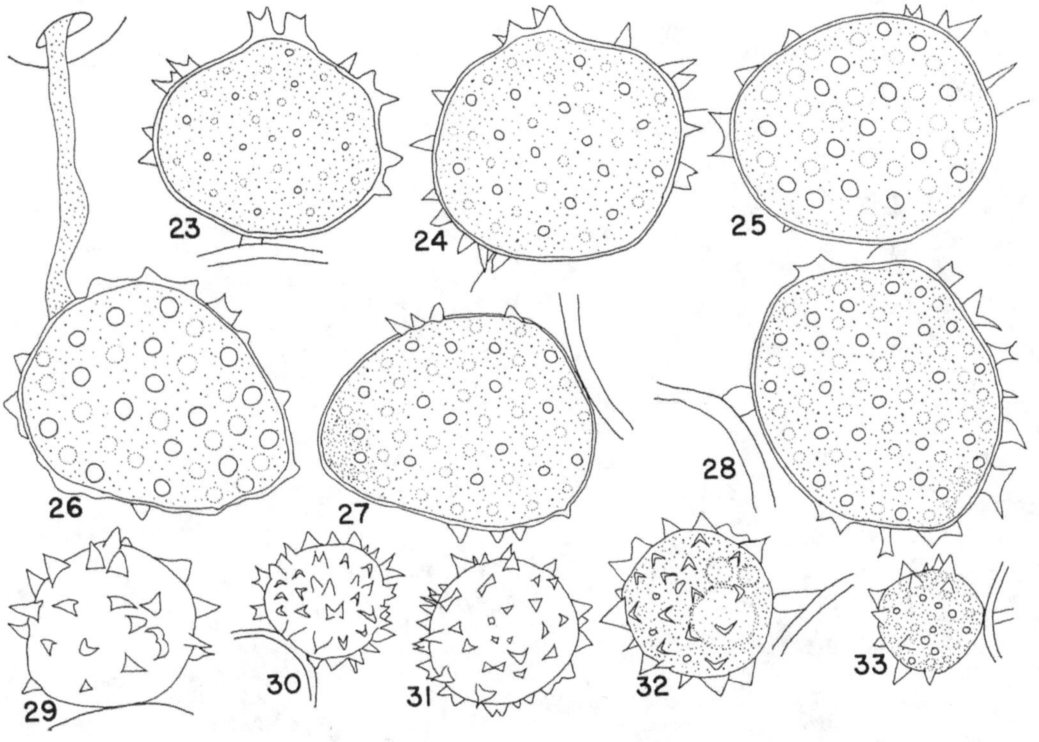

FIGS. 23–33. Illustrations of sporangial variation in "Dentate" growing on sweet gum pollen (except Fig. 24 on corn pollen). Individual figures referred to in text. Figs. 23–29, 32, X1300. Figs. 30, 31, 33, X670.

several hours and sometimes for days. The usual methods of inducing sporangial discharge in chytrids, by adding small amount of distilled or lake water to the microscope slide, works erratically or slowly (up to 3 hours) for this species. Sometimes, long after other species present in mixed collections have discharged, "Dentate" remains apparently unchanged. Planospore emergence is through a single apically or laterally situated pore (Figs. 38–40). The pore is difficult to observe prior to sporangial discharge because of the many sporangial enations. The planospores flow out in a single or double file and form in a loose, irregular or spherical mass. Actual emergence takes about 15 seconds; the zoospores may remain nonmotile in mass at the discharge pore for approximately two minutes, after which they swarm for a few seconds and swim rapidly away.

Emphasizing the inoperculate, predominantly spherical sporangium and the nonapophysate rhizoidal system, "Dentate" might be included in the genus *Rhizophydium*, probably Section I, *sensu* Sparrow (1960). In this section of the genus *Rhizophydium* are two other species with sporangial ornamentation. One species, *Rhizophydium chaetiferum* Sparrow (1939), bearing long, slender, branched or unbranched hairs on the upper sporangial surface, does not appear similar to "Dentate." The other species in this section with sporangial enations is *Rhizophydium keratinophilum* Karling. Karling (1946) described *R. keratinophilum* saprophytic on many keratinized tissues used as baits, including snake skin; feathers; hair of camel, dog, cow, and horse; wool; fingernail, hoof; and horn. *R. keratinophilum* has been collected from a number of localities over the world, primarily on human hair used as bait (Sparrow, 1960).

"Dentate," on the other hand, will grow on corn (Fig. 24), sweet gum (Figs. 11–23, 25–44), pine (*Pinus taeda* L.), and spruce (*Picea abies* Karst.) pollens and purified exoskeleton of shrimp (Figs. 1–10). "Dentate" differs morphologically from *R. keratinophilum* in the following ways: In the former, a single discharge pore, difficult to observe prior to sporangial discharge, situated apically or laterally, is present; in the latter 2–5 fairly prominent exit papillae are found randomly

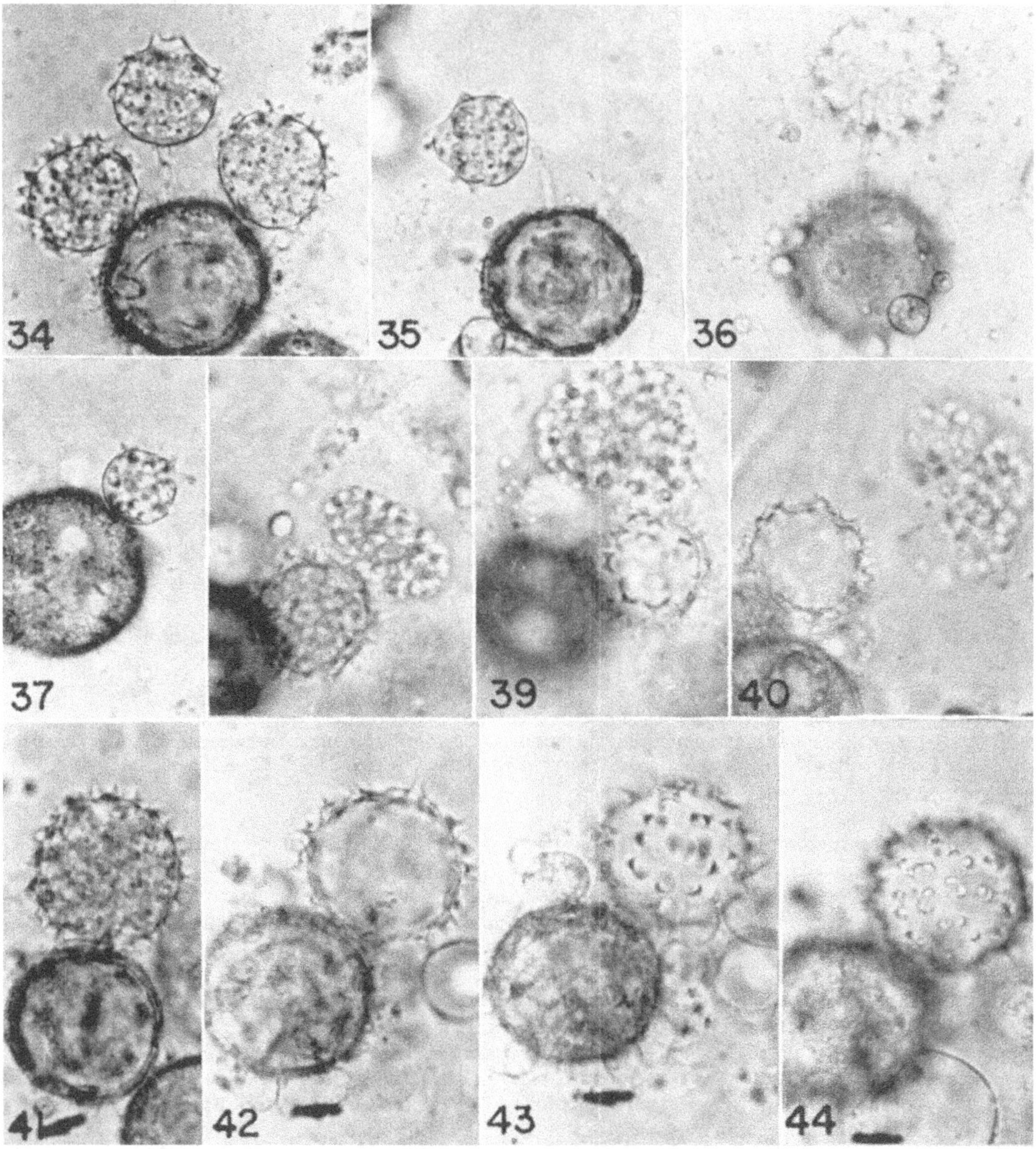

FIGS. 34–44. "Dentate" growing on sweet gum pollen, X1466. Figs. 34–37. Sporangia sessile on substrate and with epibiotic rhizoids. Figs. 38–40. Planospore emergence. Fig. 41. Mature, sessile sporangium. Figs. 42–44. Empty sporangium viewed at three different focal levels.

arranged on the sporangium. The enations of "Dentate" are frequently bipartite, tooth-like structures or single broad-based, fang-like projections, while the enations of *R. keratinophilum* are short, simple, bifurcate, or dichotomously branched spines or long (15–45 µ) simple, branched threads. Planospores of "Dentate" are 4–5 µ in diameter with a single oil globule 1–3 µ in diameter. In *R. keratinophilum*, planospore diameter is 2.5–3 µ, and the oil globule measures 0.3–0.5 µ wide. "Dentate" and *R. keratinophilum* appear to be different taxa. *Rhizophydium verrucosum* Cejp (Section II, Sparrow, 1960), having lemon-shaped sporangia with wart-like, single, toothed enations, is also distinct from "Dentate" (Miller, 1965).

Phlyctochytrium is morphologically very similar to *Rhizophydium* in most characteristics, except that in the former genus an apophysis is present while it is not found in the latter genus. If one has apophysate and nonapophysate specimens expressing this rhizoidal condition at the extremes of the possible variation—that is, either obviously apophysate and obviously nonapophysate—separation of genera may be readily accomplished. However, one must study many specimens showing this relationship to realize that he is not merely measuring the extremes of a continuum from the apophysate to nonapophysate condition, and that the various specimens are better treated as a taxon complex rather than as two or more taxa. Couch (1932) pointed to this problem in his account of *Phlyctochytrium biporosum* Couch. In this species, development of the apophysis is very slight and in some cases may be lacking, which makes generic disposition difficult.

Thus, these two genera (*Rhizophydium* and *Phlyctochytrium*) are probably separated on an artificial basis. This artificial taxonomic arrangement could be responsible for the confusion that has arisen in having very similar (possibly the same) species described as taxa of both genera, depending upon the emphasis placed on the presence or absence of an apophysis, a characteristic very difficult to determine on certain substrata. It is interesting to note that apophysate species in the operculate genus *Chytridium,* which exhibit similar sequences of development as *Phlyctochytrium,* have never been segregated from the genus *Chytridium,* as have the apophysate forms from *Rhizophydium.*

There are fourteen species in the genus *Phlyctochytrium* with varying numbers and expressions of sporangial enations or ornamentations (Sparrow, 1938, 1960, 1966). "Dentate" could be identical with two of them (*P. aureliae* Ajello and *P. mucronatum* Canter) if the apophysate criterion is eliminated. Also, some of these fourteen species are separated primarily on the basis of arrangement, numbers, and shapes of their sporangial enations. Since these characters also vary within these taxa, as well as in "Dentate" (Figs. 2–44), it seems reasonable to question also the significance of such characters in species delineation.

"Dentate" is very similar to *Phlyctochytrium aureliae* in sporangial size, planospores, and planospore oil-globule size (in almost all taxonomic characters), and both taxa are covered with numerous, haphazardly grouped, bipartite (bicornuate), and single-pronged teeth. "Dentate" lacks an apophysis, but *P. aureliae* has a spherical one (3–7 μ in diameter) formed immediately below the sporangium; illustrations of the rhizoidal system of mature ornamented sporangia of *P. aureliae* are, unfortunately, not included with the description (Ajello, 1945). *P. aureliae* was found growing on ". . . decaying vegetation and probably on cast-off integuments of insects" (Ajello, 1945).

It was *Phlyctochytrium mucronatum* with which "Dentate" was first tentatively identified. Except for the apophysis, which is well illustrated (Canter, 1949), it would be impossible to distinguish these taxa. Variations in the number, morphology, and arrangement of the enations are the same for both chytrids. *P. mucronatum* was found by Canter (1949) growing as a saprophyte on *Closterium pritchardianum* Arch. and *C. costatum* Corda. "All attempts to culture the chytrid on dead *Closterium,* on other algae, on pollen grains, and on *Daphnia,* failed." Canter (1949) noted the resemblence to the other dentigerate members of the genus, especially *Phlyctochytrium aureliae.*

"Dentate" may be described as follows: sporangia, epibiotic, sessile or with an epibiotic rhizoid, monocentric, spherical (10–39 [most 23-31] μ in diameter), subspherical, occasionally pyriform, conical or irregular in shape (19–38 μ long x 17–35 μ wide, most 28 μ long x 22 μ wide) with blunted or pointed, narrow or broad-based spine-like, fang-like, or bipartite tooth-like enations (1–5 μ long x 1–5 μ wide at base) varying in number, covering all or limited parts of the sporangium, arranged in a random manner or in whorls apically or laterally placed, with an inconspicuous exit pore (3–4 μ wide); rhizoidal system, extramatrical or intramatrical, the epibiotic portion frequently a tapering, tubular-shaped structure with occasional one-sided bulges or rarely slight epibiotic or endobiotic apophysis-like swellings near or far from the sporangium, the epibiotic rhizoid sometimes attaching laterally to the sporangium; planospores spherical (4–5 μ in diameter) with a single oil globule (1–3 [most 2] μ in diameter), emerging in a single or double file, forming a loose mass or hollow sphere at or beyond aperture of discharge pore; growing on pine, spruce, sweet gun, and corn pollens and purified exoskeleton of shrimp baits added to water and soil collections from various sites at Mountain Lake Biological Station, Virginia (1961–1965), and the University of Maine Forest (1962–1965), Orono, Maine.

What should be done taxonomically with "Dentate?" "Dentate" could be justifiably included in the genus *Rhizophydium,* if the occasional apophysis-like swelling produced were ignored. It might equally as well be placed in the genus *Phlyctochytrium,* if one wished to emphasize that an apophysis-like structure can be produced. When is an apophysis an apophysis? What parameters should be used to judge between a thickened or bulging rhizoid and a very large subsporangial swelling? Is any subsporangial swelling an apophysis? Must this swelling be very close to the sporangium or may it be far removed from it? Can an apophysis be an epibiotic as well as an endobiotic structure? What is the taxonomic significance of the apophysis?

The genera *Rhizophydium* and *Phlyctochytrium,* then, are arbitrarily separated on the basis of a very variable, ill-defined character, the taxonomic significance of which is not known. Erecting "Dentate" as a new taxon in *Rhizophydium* would only add to the existing confusion. Because "Dentate" is morphologically very similar to *Phlyctochytrium aureliae* and *P. mucronatum,* which themselves are separated primarily on the basis of substrate, another taxonomic character loosely defined, it seems logical to ally "Dentate" with *Phlyctochytrium.*

In all taxonomic characters generally used to distinguish taxa (and used also in the case of these three fungi) except the substrate, "Dentate," *P. aureliae,* and *P. mucronatum* are very close. Sporangia are "predominantly" or "more or less" spherical in all three taxa. Sporangia of *P. aureliae* are 12–35 μ in diameter; *P. mucronatum,* 5.7–31 μ in diameter; "Dentate," 10–39 μ in diameter. Nonspherical sporangia of "Dentate" range from 19–38 μ long x 17–35 μ wide. Planospores in all these fungi are 4–5 μ in diameter (*P. aureliae,* 4–4.5 μ in diameter). The enations in all three taxa may be spine-like, fang-like, or bipartite ("bicornuate" or "Y-shaped"), tooth-like structures, 1–5 μ long x 1–5 μ wide at base ("Dentate"); 3.5–4.5 μ long x 3.5–6.5 μ wide (*P. aureliae*); and 1.4–5.2 μ long x 0.9–4.3 μ broad at base (*P. mucronatum*). An oil globule is present in the latter; 1.3 μ (most 2 μ) in diameter in "Dentate"; and 2 μ in diameter in *P. aureliae.*

In *P. aureliae,* the irregular or elongate subsporangial structure is 3–7 μ in diameter and consists of ". . . a swelling or apophysis . . . formed in that portion of the absorbing system immediately below the zoosporangium" (Ajello, 1945). In *P. mucronatum,* "the rhizoid . . . immediately within the algal wall enlarges to form a small (1.3–3.5 μ) spherical or subspherical apophysis. It seems that the apophysis is formed secondarily as a swelling of the germ tube after initiation of the rhizoids" (Canter, 1949). The rhizoid of "Dentate" may exhibit occasional one-sided bulges or rarely a slight, epibiotic or endobiotic swelling adjacent to or far situated from the sporangium.

The substrate on which "Dentate" grows includes sweet gum, pine, spruce, and corn pollens and purified exoskeleton of shrimp; *P. aureliae* grows on decaying vegetation and probably on cast-off insect integuments (chiton?); and *P. mucronatum* grows saprophytically on *Closterium pritchardianum* and *C. costatum.*

Planospore emergence is essentially the same in all three taxa. In "Dentate" the planospores during sporangial discharge form in a loose irregular or spherical mass that may remain near or occasionally float away from the sporangium. In *P. aureliae,* ". . . the zoospores emerge en masse, remaining quiescent for varying periods of time. . . ." (Ajello, 1945). In *P. mucronatum,* "the contents of the sporangium emerge to form a mass outside; although no definite vesicle was observed continuous with the sporangium, the zoospore mass is held together by some invisible substance" (Canter, 1949).

The illustrations of *P. aureliae* and *P. mucronatum* are very similar to each other and to "Dentate"; the diagnostic characters given for each taxon above are essentially the same. The ever so slight differences in the numerical statistics are probably not significant in light of the variations observed. Consequently, it would seem that the present characters are not diagnostic and that on this basis these fungi should be considered members of the same species complex.

Summary

Observations and questions concerning the taxonomic validity of some morphological characters used in taxon delineation in the Chytridiales are presented. An example of the difficulties inherent in using some of these characters in taxon identification in the *Rhizophydium-Phlyctochytrium* complex is given.

ACKNOWLEDGMENTS

This study was supported by grants from The Ohio University Research Fund (G–198), The National Science Foundation (administered by Mountain Lake Biological Station, Virginia), The Coe Research Fund (G–R625–45), University of Maine, Orono, and in part by the National Science Foundation (GB 6855).

LITERATURE CITED

AJELLO, L. 1945. *Phlyctochytrium aureliae* parasitized by *Rhizophydium chytriophagum*. Mycologia **37**: 109–119.

CANTER, H. M. 1949. Studies on British Chytrids VII. On *Phlyctochytrium mucronatum* n. sp. Trans. Brit. Mycol. Soc. **32**: 236–240.

COUCH, J. N. 1932. *Rhizophidium, Phlyctochytrium*, and *Phlyctidium* in the United States. Jour. Elisha Mitchell Sci. Soc. **47**: 245–260.

EMERSON, R. 1950. Current trends of experimental research on the aquatic Phycomycetes. Ann. Rev. Microbiol. **4**: 169–200.

KARLING, J. S. 1945. Brazilian Chytrids VII. Observations relative to sexuality in two new species of *Siphonaria*. Amer. Jour. Bot. **32**: 580–587.

———. 1946. Keratinophilic chytrids. I. *Rhizophydium keratinophilum* n. sp., a saprophyte isolated on human hair, and its parasite, *Phlyctidium mycetophagum* n. sp. Amer. Jour. Bot. **33**: 751–757.

———. 1949. Two new eucarpic inoperculate chytrids from Maryland. Amer. Jour. Bot. **36**: 681–687.

KOCH, W. J. 1957. Two new chytrids in pure culture, *Phlyctochytrium punctatum* and *Phlyctochytrium irregulare*. Jour. Elisha Mitchell Sci. Soc. **73**: 108–122.

MILLER, C. E. 1961. Some aquatic Phycomycetes from Lake Texoma. Jour. Elisha Mitchell Sci. Soc. **77**: 293–298.

———. 1965. Annotated list of aquatic Phycomycetes from Mountain Lake Biological Station, Virginia. Virginia Jour. Sci. **16**: 219–228.

PATTERSON, R. A. 1963. Observations on two species of *Rhizophydium* from Northern Michigan. Trans. Brit. Mycol. Soc. **46**: 530–536.

SPARROW, F. K. 1938. Chytridiaceous fungi with unusual sporangial ornamentation. Amer. Jour. Bot. **25**: 485–493.

———. 1939. Unusual chytridiaceous fungi. Papers. Mich. Acad. Sci., Arts, Letters. **24**: 121–126.

———. 1958. Interrelationships and phylogeny of the aquatic Phycomycetes. Mycologia **50**: 797–813.

———. 1960. Aquatic Phycomycetes. Univ. of Michigan Press, Ann Arbor. xxvi + 1187 pp.

———. 1966. A new bog chytrid. Archiv fur Mikrobiologie. **53**: 178–180.

WHIFFEN, A. J. 1944. A discussion of taxonomic criteria in the Chytridiales. Farlowia **1**: 583–597.

Ecology of *Coelomomyces* Infections of Mosquito Larvae

CLYDE J. UMPHLETT

Department of Botany, University of North Carolina at Chapel Hill, N. C.

Aquatic fungi of the genus *Coelomomyces* are parasites primarily of mosquito larvae. There are a few observations recording their presence in adult mosquitoes, but the effect of the fungus infection on the larvae is usually death before pupation. The possibility of using these fungi as agents for the biological control of mosquitoes deserves serious attention, and the conduct of meaningful experiments to evaluate any control potential the fungus may have should be enhanced by having information concerning the activities and responses of the parasite in its natural environmental setting. After several years of searching, a site has been found nearby that will permit the examination under field conditions of several aspects of the host-parasite relationship. Late in the summer of 1965 an infestation by *Coelomomyces punctatus* Couch of a population of the malaria vector *Anopheles quadrimaculatus* Say, breeding in University Lake at Chapel Hill, North Carolina, was discovered. The level of infection varied from ca. 20–80 per cent of the larvae collected at various times before the mosquitoes stopped breeding for the winter. In 1966 a study was conducted at the lake throughout the breeding season of the host species. During this time data were gathered systematically on the levels of fungus infection in the larval population. In an effort to describe the conditions surrounding infection of larvae in the natural habitat, measurements of selected factors of the physical environment were made.

Materials and Methods

Larvae of *Anopheles quadrimaculatus* were collected with a porcelain dipper from two shallow coves, A and B, in University Lake, a 220-acre impoundment. In taking weekly samples of 100 dips each, the entire portion of each cove supporting larval growth was traversed. It was determined that 100 dips produced an adequate sample of the larval and parasite populations in which the percentage of infected larvae observed was not dependent on the number of larvae in the sample. Larvae were sought in the most likely resting or feeding places in the cove, e.g., in shade near emergent plant stems, along the shore line, and among floating debris. Larvae collected were brought immediately to the laboratory, where they were inspected microscopically for fungus infections not later than the day after collection. The number of infected and uninfected larvae and the percentage of infected larvae for each sample were recorded. Definite identification of the fourth instar larvae was made, and the lesser instars were determined to species insofar as possible. Any larvae of questionable identity were excluded from the data.

Dissolved oxygen was measured at the times of collections with the YSI oxygen meter. Measurements were always taken at the same time of day and same place near the center of the larval-producing portion of the cove in the top six inches of water. The pH was measured with a Beckman pH meter at the same time and site as the dissolved oxygen. Water temperature was measured at collection times, and was monitored continuously (except as noted below) with a maximum-minimum thermometer. Air temperature and relative humidity were measured continuously with a hygrothermograph placed in an elevated, ventilated box located near the shore line of each cove in a partially shaded site.

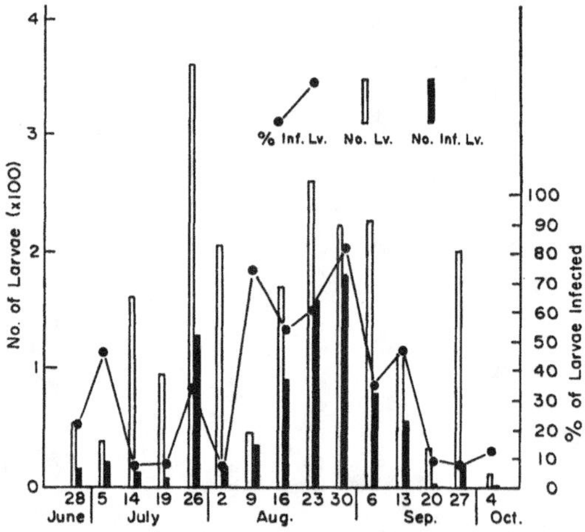

FIG. 1. The number of *Anopheles quadrimaculatus* larvae, and the number and percentage of larvae parasitized by *Coelomomyces punctatus*, in each weekly sample from Cove A.

Results

The number of larvae, infected and uninfected, and the percentage of larvae infected in each weekly sample from Cove A are shown in Figure 1. Infections were noted in larvae from very early second instars through the fourth instar. It is not certain whether any first instars were infected. The period covered by these data includes only the times at which infected larvae were taken. Larvae of *Anopheles quadrimaculatus* were collected in small numbers as early as June 7, but none contained visible infections of *Coelomomyces punctatus* until the collection of June 28. Infected larvae appeared subsequently in every weekly sample from Cove A through October 4, after which larvae were not found at any site in the lake.

Figure 1 shows that the weekly percentage of infected larvae remained relatively low through July, although the host population increased rather sharply during the same period. For most of August the number of host individuals was maintained at a high level, but a much larger percentage of them was infected by *Coelomomyces*. This striking increase in the parasite population is shown graphically in Figure 2, which is a plot of the cumulative number of infected larvae occurring in the weekly samples.

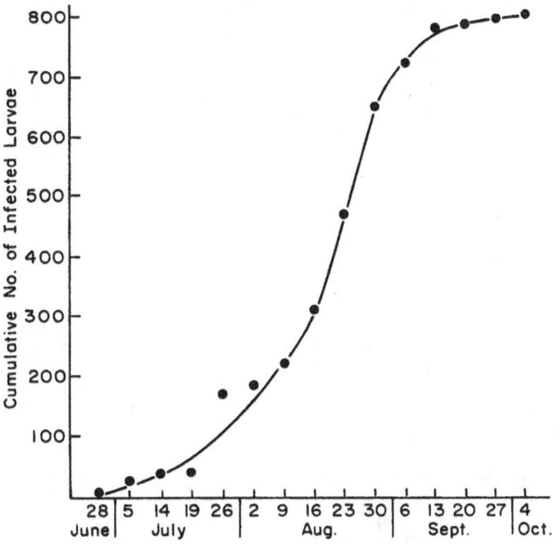

FIG. 2. Plot of the cumulative number of *Anopheles quadrimaculatus* larvae infected by *Coelomomyces punctatus* in weekly samples from Cove A.

Table I gives the number of infected and uninfected larvae, and the percentage of larvae infected, in two instar classes and in total for each weekly sample from Cove A. Fourth instar larvae sustained a higher level of infection than did the lesser instars throughout the

Table I

Number of *Anopheles quadrimaculatus* larvae collected from Cove A, and the number and percentage of these infected by *Coelomomyces punctatus*

Date of Collection	4th Instars			Lesser Instars			Total		
	# Lv.	# Inf.	% Inf.	# Lv.	# Inf.	% Inf.	# Lv.	# Inf.	% Inf.
June 28	17	6	35.5	39	6	15.4	56	12	21.5
July 5	34	17	50.0	5	1	20.0	39	18	46.1
14	71	7	9.9	90	5	5.5	161	12	7.5
19	25	5	20.9	66	2	3.0	91	7	7.7
26	172	83	48.3	189	43	22.8	361	126	34.3
Aug. 2	40	2	5.0	166	11	6.6	206	13	6.3
9	17	17	100.0	30	18	60.0	47	35	74.4
16	12	8	66.7	159	81	50.9	171	89	52.0
23	79	68	86.1	183	90	49.2	262	158	60.3
30	53	47	88.7	171	134	78.4	224	181	80.8
Sept. 6	37	31	83.3	190	47	24.7	227	78	34.4
13	17	14	82.4	98	41	41.8	115	55	47.8
20	4	2	50.0	28	1	3.6	32	3	9.4
27	3	2	66.7	199	15	7.5	202	17	8.4
Oct. 4	1	1	100.0	8	0	0.0	9	1	11.1

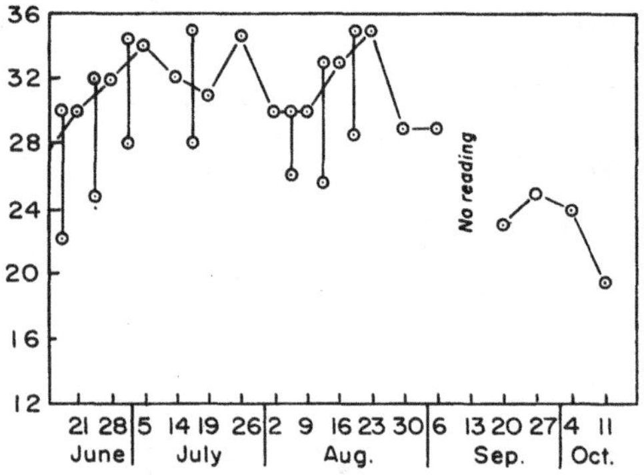

FIG. 3. Water temperatures recorded in Cove A throughout the sampling period.

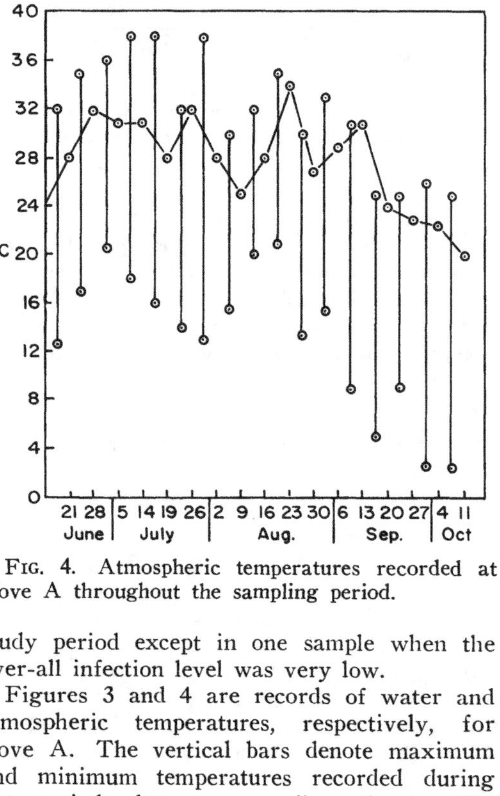

FIG. 4. Atmospheric temperatures recorded at Cove A throughout the sampling period.

study period except in one sample when the over-all infection level was very low.

Figures 3 and 4 are records of water and atmospheric temperatures, respectively, for Cove A. The vertical bars denote maximum and minimum temperatures recorded during the periods between sampling times. The points conected by a line are temperatures recorded at the sampling times. Readings were taken between 2 p.m. and 3 p.m. on the same day of the week throughout the total sampling period. For some periods in July and August maximum and minimum water temperatures are omitted. During these two months the water level in the lake was dropping rapidly, and on some occasions the max.-min. thermometer, being in a fixed position in the water, was left suspended in the air by the receding water.

It is interesting to note that the three samples—July 14, July 19, and August 2—yielded very low levels of infection. Each of these samples was taken following a period of very high maximum air temperature (Fig. 4). Maximum water temperatures were not available for these periods for the reason stated above. The three samples with low percentages of infection taken near the end of the season (Fig. 1) were collected under conditions of lower water and air temperature than any other samples during the summer. The temperature recorded during this late period were similar to those observed in the cove in early April before mosquito breeding had begun.

Dissolved oxygen was present in the water of Cove A in amounts varying from 3.0 ppm to 7.8 ppm, with the lower concentrations occurring under conditions of higher temperatures. In terms of percentage of saturation, the lowest oxygen level was 42 per cent, the highest, 105 per cent. At only two sampling times was the dissolved oxygen at 100 per cent saturation or more.

During the sampling period the pH in Cove A ranged from pH 6.8 to pH 8.4, but at most sampling times pH was near the lower of the two extremes. On only two occasions did the pH exceed pH 8.

The predominant emergent plant present in Cove A was *Peltandra virginica* (L.) Schott and Endl., but *Najas guadalupensis* (Spreng.) Magnus, a submerged aquatic flowering species, grew in the cove in scattered colonies. Occasional large quantities of floating plant debris, e.g., dead twigs and pine needles, were present, and these along with the plants growing there provided an excellent habitat for the breeding of *An. quadrimaculatus*.

Observations were begun in Cove A without the benefit of a suitable area to serve as control for evaluations of environmental effects and of the biological control potential of *Coelomomyces*. Very late in the season, however, a cove, designated Cove B, in University Lake about 0.65 miles from Cove A was found in which *An. quadrimaculatus* was breeding in abundance. The initial collections of larvae from Cove B were not infected with *Coelomomyces*. When this uninfected area was discovered, the same procedures of collecting and screening larvae, and of measuring environmental factors, as in Cove A were initiated. None of the 1,442 larvae taken in the subsequent weekly samples was infected by the fungus. Sampling in Cove B yielded no larvae after the first collection in September because the portion of the habitat suitable for oviposition and larval development was left dry when the water level in the lake dropped.

Cove B contained a few emergent stems of *Cephalanthus occidentalis* L., but the submerged aquatic species, *Najas guadalupensis*, was extremely abundant. In addition to the living plants in the cove a few partially submerged fallen logs and dead branches were present.

Water and atmospheric temperatures were about the same in both coves, but measurements of dissolved oxygen and pH in Cove B were strikingly different from those observed in Cove A (Figs. 5 and 6). Only six weekly

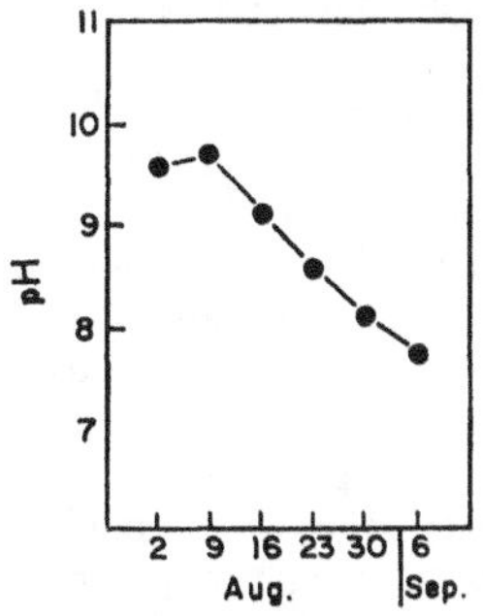

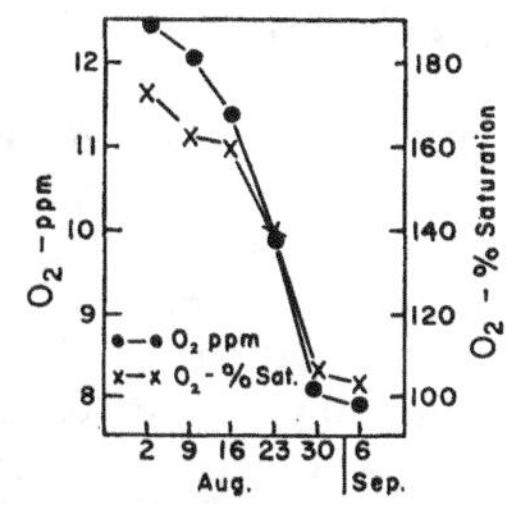

FIG. 5. Record of pH of water in Cove B for the period indicated.

FIG. 6. Dissolved oxygen concentration (in ppm and % saturation) present in water of Cove B for the period indicated.

readings were obtained because of the low water level. The percentage of saturation of dissolved oxygen observed at the first four sampling times was phenomenally high (Fig. 6), and the pH was more than two units above the usual pH observed in other places in the lake (Fig. 5). It is not known, of course, how long these conditions prevailed in Cove B before the measurements were begun. As the water level dropped in the cove, however, the pH and dissolved oxygen content became progressively lower (Figs. 5 and 6).

With both infected and uninfected larvae of possibly the same genetic strain now available from natural habitats, laboratory observations were initiated to determine the effect of *Coelomomyces punctatus* infections on the emergence of adults from larvae of *An. quadrimaculatus* collected in the field but reared under insectary conditions. The insectary was maintained at 82° F. and relative humidity of ca. 70 per cent. Samples were taken from Coves A and B on August 30, brought into the laboratory, screened for infection, and counted. The larvae were put into rearing pans containing the lake water in which the larvae had been brought to the laboratory and were placed in the insectary. Both samples were reared to maturity according to insectary routine. The sample from Cove A consisted of 224 larvae, of which 181 (80.8 per cent) were infected by the fungus when collected. The Cove B sample contained 146 larvae, none of which was infected. As the larvae pupated they were removed and handled as usual. Emerging adults were collected and checked for identification and infection. Any pupae not producing adults were checked for infection also. Of the 146 larvae from Cove B, 111 (76 per cent) emerged as adult mosquitoes. None of these was infected by the fungus. The 224 larvae from Cove A yielded 14 adults (6.3 per cent), one of which contained *C. punctatus.*

An attempt was made to carry out similar observations on another collection of larvae, but Cove B was dry. Larvae were collected from Cove A anyway, and were treated as described above. This sample contained 115 larvae, 55 (47.8 per cent) of which were infected when collected. Twenty-seven (23.5 per cent) of these emerged as adults. None of the adults were infected, but two dead pupae contained the parasite.

Discussion

In attempts to obtain experimental infection of larvae of *Anopheles gambiae* Giles by *Coelomomyces*, Muspratt (1946a) enjoyed some measure of success, but was unable to obtain infections repeatedly. He stated that although his experiments did not show that continuous, unlimited infection of *An. gambiae* could be obtained in a confined space, he felt that, "given suitable climatic conditions, this would be the case." In view of the sporadic, or sometimes negative, results of workers who have attempted experimental infection of mosquito larvae with *Coelomomyces* (Walker, 1938; Couch and Dodge, 1947; Laird, 1959; Couch and Umphlett, 1963; Madelin, 1964, 1965), the statement by Muspratt appears to be particularly significant. Since the conditions surrounding *Coelomomyces* infections in nature have not been described before, the environmental data presented in this paper, although representing observations made in only a single season, may be an important step toward performing definitive infection experiments in the laboratory, at least where *C. punctatus* and *An. quadrimaculatus* are concerned.

It seems unlikely that Cove B, which produced no infected larvae during the study, is barren of resting sporangia of *Coelomomyces punctatus,* for larvae infected with this fungus have been collected both up- and downstream from this cove. That resting sporangia do occur as free-floating bodies in the water, and are thereby subject to distribution by the currents, was shown by Umphlett (unpublished) when he isolated resting sporangia of *C. punctatus* by continuous-flow centrifugation of water samples from University Lake. It seems more likely that one or more factors of the physical environment may be acting to prevent zoospore production by the resting sporangia present in the water. Observations in this laboratory point, although not conclusively, to the zoospores as the most likely agents of infection (Couch, 1960). There are indications from preliminary observations of J. N. Couch (personal communication) that very low dissolved oxygen concentrations stimulate zoospore release. Cove B contained extremely high levels of dissolved oxygen during the period when it was supporting a large, uninfected host population. The possibility exists, then, that discharge of zoospores from resting sporangia likely to be in this cove was inhibited. The very large quantity of *Najas guadalupensis* in Cove B were undoubtedly contributing greatly through photosynthesis to the oxygen content of the water. Observations on this species growing in tanks in the greenhouse are that the oxygen level of the water in the tanks containing *Najas* is always considerably higher than that in tanks without this plant. Small, isolated colonies of *Najas* were present in Cove A from which infected larvae were taken regularly, but the dissolved oxygen level there was usually between 5 ppm and 6 ppm. The highest concentration of dissolved oxygen noted in Cove A was 7.8 ppm.

The possibility of an effect on resting sporangium germination by the relatively high pH observed in Cove B for at least a part of the observation period may warrant further investigation, but laboratory experiments on resting sporangium germination in *Coelomomyces punctatus* indicate that these bodies are not particularly sensitive to pH. Zoospores were released in abundance from resting sporangia over a range from pH 3 to pH 9.

That temperature may have an effect on the level of infection by *Coelomomyces* in a larval population was indicated by observations made in Cove A. As noted earlier, two collections in July and one in August, all with very low levels of infection, were made following periods when air temperature rose to 38° C. Maximum water temperatures were not obtained for these times, but it seems likely that in this shallow cove, well lighted by direct sunlight much of the day, the water temperature probably reached a level comparable to the high air temperature. A comparison of water and air temperatures taken at sampling times shows that on all but two occasions the water temperature was 1°–5° C., higher than air temperature. At the two times when this was not so, air and water temperatures were the same. Laboratory experiments by J. N. Couch (personal communication) to determine the temperatures at which resting sporangia of *C. punctatus* release zoospores showed that upward from about 30° C., a progressive reduction in the number of resting sporangia germinating occurred. At about 35° C., little, if any, zoospore release was observed. No germination occurred at 38° C. With the likely achievement of temperatures similar to these in the water of Cove A during the periods of high air temperature, it is possible that germination of resting sporangia was inhibited. Periods such as these without zoospore production could lower the concentration of inoculum (zoospores) in the water so that the potential for numerous infections would be reduced. Furthermore, the zoospores already released into the water would probably be affected adversely by the high temperature. Although there is no experimental evidence for this presumption about the zoospores of *Coelomomyces,* experience in this laboratory with other aquatic fungi indicates strongly that it may be so.

The observation that a higher percentage of infection in fourth instar larvae than in the lesser instars may be interpreted to be a result simply of longer exposure of the fourth instars to the inoculum in the water. However, the possibility exists that the infection seen in a fourth instar larva might have occurred initially in a lesser instar, and the fungus had not developed in the lesser instar to a detectable stage at the time of sampling. Involved here is a question as to which aquatic stage in the life cycle of a mosquito is susceptible to attack by *Coelomomyces.* While Table I does not show which of the lesser instars was infected, it was noted during the screening of the larvae that some very early second instars contained advanced infections consisting of many resting sporangia with very little mycelium, and some fourth instars had infections consisting of only scanty amounts of mycelium. The data and observations seem to support the idea that invasion of mosquitoes by *Coelomomyces* may not be limited to any specific larval stage.

The levels of infection found in Cove A compare favorably with or exceed those noted by other authors. Walker (1938), Muspratt, (1946b), Manalang (1930), and Shemanchuk (1959) have reported infection levels of *Coelomomyces* in various species of mosquitoes ranging from 1 per cent to 95 per cent of samples. Umphlett (1961), collecting in Thomasville, Georgia, found *C. dodgei* Couch in *An. crucians* Wiedeman at levels varying from 4 per cent to 38 per cent. *Culex erraticus* (Dyar and Knab), collected at the same time and site as above, was infected by *Coelomomyces pentangulatus* Couch in one collection in October at a level of 7 per cent, and in two collections in September at 79 per cent. Data of this nature that have been reported usually cover short periods of time. Estimations and evaluations of the biological control value of *Coelomomyces* obviously will be of greater meaning when information from long-term field and laboratory studies is available.

Adult mosquitoes were considered only briefly in one phase of this study. In future work involving parasitism and biological control by *Coelomomyces,* however, the imago must receive more attention, since ultimately this is the medically, economically, and biologically important stage in the life of a mosquito. The control mechanism, in this case a fungus, should be evaluated relative to its impact on the reproductive phase of the host population. In the one experiment on adult emergence reported here, there appears to be a considerable effect of the parasite upon the numbers of host individuals reaching the reproductive stage. Because of its brevity, however, this experiment should not be considered conclusive.

Adult mosquitoes must be included in studies similar to this one for still another reason. The reports of Manalang (1930), Walker (1938), and others, including the present paper, of the occurrence of *Coelomomyces* in adult mosquitoes implicate the adult as a probable means of dispersal of the parasite. *An. quadrimaculatus* is reported to have a flight range of about one mile. Transportation of potential *Coelomomyces* inoculum for such a distance, or even shorter distances, could contribute greatly to the local and geographical distribution of the parasite. How far a mosquito with "coelomomycosis" can fly, however, is a quesion, but some infected adults have been taken by some workers using light traps, which indicates that they are able to fly to some extent.

Summary

Weekly samples of larvae of *Anopheles quadrimaculatus* were taken during the mosquito breeding season in 1966 from two coves, A and B, in University Lake, Chapel Hill, N. C. Larvae from Cove A were infected regularly by *Coelomomyces punctatus* in quantities varying from 6.3 per cent to 80.8 per cent of the samples. No infected larvae were found in Cove B. Environmental conditions in both coves were described from measurements of water and air temperatures, pH, and dissolved oxygen. The dissolved oxygen concentration and pH were much higher in Cove B than in Cove A. These factors, and the higher temperatures observed at times in Cove A, seemed to be exerting some influence on the infection potential of the fungus. In one brief laboratory experiment on emergence of adult mosquitoes from field larval collections, 76 per cent of the larvae from Cove B (uninfected) emerged while 6.3 per cent of those from Cove A (infected) emerged as adults. One infected imago emerged. A role for infected adult mosquitoes in the dispersal of *Coelomomyces* is possible.

ACKNOWLEDGMENTS

The very able assistance throughout this investigation of James M. Roane, research assistant, is gratefully acknowledged. It was he who first collected *Coelomomyces punctatus* from University Lake in 1965. I am grateful, too, to Dr. J. N. Couch for the access granted to his data, and to him, Dr. W. J. Koch, and Dr. J. F. McCormick for criticizing the manuscript. This work was supported by PHS Grant No. AI 03235–06 from the National Institute of Allergy and Infectious Diseases, U.S. Public Health Service.

LITERATURE CITED

COUCH, J. N. 1960. Some fungal parasites of mosquitoes. *In* Biological Control of Insects of Medical Importance. AIBS Tech. Rept. (Unnumbered). Washington, D. C.

COUCH, J. N., AND H. R. DODGE. 1947. Further observations on *Coelomomyces*, parasitic in mosquito larvae. Jour. Elisha Mitchell Sci. Soc. **63**: 69–79.

COUCH, J. N., AND C. J. UMPHLETT. 1963. *Coelomomyces* infections. *In* E. A. Steinhaus, ed., Insect Pathology, an Advanced Treatise. Vol. 2. pp. 149–188. Academic Press, New York.

LAIRD, M. 1959. Parasites of Singapore mosquitoes, with particular reference to the significance of larval epibionts as an index of habit pollution. Ecology **40**: 206–221.

MADELIN, M. F. 1964. Laboratory studies on the

infection of *Anopheles gambiae* Giles by a species of *Coelomomyces*. Bull. World Health Organization, WHO/EBL/17, WHO/Mal/438, WHO/ Vector Control/64. Mimeo., 23 pp.

———. 1965. Further laboratory studies on a species of *Coelomomyces* which infects *Anopheles gambiae* Giles. Ibid., WHO/EBL/52.65, WHO/Mal/530.65. Mimeo., 22 pp.

MANALANG, C. 1930. Coccidiosis in *Anopheles* mosquitoes. Philippine J. Sci. **42**: 279.

MUSPRATT, J. 1946a. Experimental infection of the larvae of *Anopheles gambiae* (Dipt., Culicidae) with a *Coelomomyces* fungus. Nature **158**: 202.

———. 1946b. On *Coelomomyces* fungi causing high mortality of *Anopheles gambiae* larvae in Rhodesia. Ann. Trop. Med. Parasitol. **40**: 10–17.

SHEMANCHUK, J. A. 1959. Note on *Coelomomyces psorophorae* Couch, a fungus parasitic on mosquito larvae. Can. Entomologist **91**: 743–744.

UMPHLETT, C. J. 1961. Comparative studies in the genus *Coelomomyces* Keilin. Unpublished Ph.D. Dissertation. University of North Carolina. Chapel Hill, N. C.

WALKER, A. J. 1938. Fungal infections of mosquitoes, especially of *Anopheles costalis*. Ann. Trop. Med. Parasitol. **32**: 231–244.

Studies on the Infection by *Coelomomyces indicus* of *Anopheles gambiae*

M. F. Madelin
Department of Botany, The University, Bristol, England

Introduction

Species of the genus *Coelomomyces* are aquatic fungi parasitic in mosquitoes and a few other insects. The genus has been comprehensively reviewed by Couch and Umphlett (1963), and recent contributions are noted by Madelin (1966). Characteristically the fungus develops in larvae but sometimes also in pupae (Walker, 1938) and even in adults (Manalang, 1930; Gibbins, 1932; Walker, 1938; Haddow, 1942; Van Thiel, 1954; Kellen, *et al.*, 1963; Brady, 1963). No one has yet observed how the fungus enters its host. As Couch and Umphlett (1963) remark, the difficulty in making such observation lies partly in rearing parasitized larvae in the laboratory. Success in experimentally infecting larvae has been limited. Walker (1938) obtained larvae of *Anopheles gambiae* infected with *Coelomomyces africanus* in experiments in a pool and in a shallow concrete tank out of doors in the tropics. Muspratt (1946b) also used a concrete tank in the open and obtained about 15 larvae of *A. gambiae* infected with *Coelomomyces* "type (a)" out of a batch of a hundred larvae, and a few more in successive tests. Couch and Dodge (1947) exposed larvae of *Anopheles crucians* to zoospores of *C. dodgei* in petri dishes, but none became diseased. Muspratt (1946a) also failed in attempts to infect mosquito larvae with zoospores, as did Lum (1963), who exposed larvae of *Aëdes taeniorhynchus* to zoospores of a variety of *Coelomomyces psorophorae*. Couch and Umphlett (1963) conducted experiments in which sporangia of *Coelomomyces* and mosquito larvae were placed together in petri dishes of water. They used *Coelomomyces* sp. with *Aëdes taeniorhynchus* and a variety of *C. psorophorae* with *Psorophora howardii* but obtained no infections. Laird (1959) reported that larvae of *Aëdes aegypti* became infected in pans of distilled water buffered to pH 6.6 and containing dried sporangia of *Coelomomyces stegomyiae* from *Aëdes albopictus* plus sediment from the latter's container. A few infection experiments in the field have been successful. Laird and Colless (1962) introduced *C. stegomiae* into a parasite-free population of *Aëdes polynesiensis* in a South Pacific atoll, and Muspratt (1963a) successfully introduced *C. ? indicus* into a healthy population of *Anopheles gambiae* in a pool in Zambia.

The following report describes experiments with *Anopheles gambiae* Giles, one of the most important vectors of malaria in Africa, and *Coelomomyces indicus* Iyengar, which is recorded by Muspratt (1963a) as an important pathogen of this mosquito in at least the area around Livingstone, Zambia. The continuous parasitization of laboratory-bred *A. gambiae* by this fungus in comparatively small vessels has hitherto not been accomplished (Muspratt, 1963a). The experiments reported here have led to recurrent outbreaks of *Coelomomyces* infection, to a partial definition of the conditions for infection, and to the provision of a small amount of fungus material for use in small-scale field trials, as well as for distribution under World Health Organization auspices to workers in the field to familiarize them with the appearance of *Coelomomyces*-infected mosquito larvae. Observational and experimental studies on this host-parasite combination in the field are described by Muspratt (1946a, 1963a).

Materials and General Methods

Material of *Coelomomyces* was derived from larvae of *Anopheles gambiae* which contained thick-walled resting sporangia (referred to hereafter as RS) of the parasite. These larvae were collected by Mr. J. Muspratt from breeding pools near Livingstone, Zambia, in the rainy seasons early in 1962 and 1963 and stored at room temperature (15–25° C.), dried (see Muspratt, 1963a) on filter papers or on microscope slides, or in small jars of mopane clay. This is the characteristic dark brown sandy clay from the area in Zambia in which diseased larvae are found (Muspratt, 1946a). The species of *Coelomomyces* is "Muspratt's type (a)" (Muspratt, 1946a; Couch and Umphlett, 1963) and is evidently conspecific with *C. indicus* Iyengar (= *C. indiana* Iyengar, *nomen nudum*) (Iyengar, 1962).

Additional quantities of mopane clay were collected from the sites of breeding pools of *A. gambiae* near Livingstone and stored dry at room temperature.

Eggs of *A. gambiae* were derived from laboratory populations kept in the Ross Institute of Hygiene, London. Two strains were employed, one from Kano, the other from Lagos, Nigeria. Eggs were added directly to the water used in infection tests or were hatched and the first instar larvae transferred to tests one day later. Larvae in tests received carefully regulated amounts of either finely ground yeast tablets, "Bemax" (a proprietary cereal food prepared from wheat germ), or "Farex" (a proprietary cereal baby food).

The opinion was soon formed that infection was most likely to occur under conditions when the potential hosts themselves flourished. This view was based on the experience of successful tests and the supposition that because the fungus—as far as is known—is a highly specialized obligate parasite, it would probably resemble its host in its tolerances of physical environmental factors. A guiding principle was adopted that conditions in all tests should be adjusted to secure as nearly normal rates of development of the mosquitoes as possible.

The small polythene and large porcelain basins used for infection tests are described in the text. In them, conditions prevailing in ephemeral rain pools in Zambia were simulated. Natural and/or artificial illumination was provided, the latter for alternate 12-hour periods because continuous illumination led to the development of larvae at uneven rates. Generally the water temperature for most of the day was between 28° and 32° C.; in the large basins it fell to ca. 25° C. by night. Muspratt (1963b) records that the temperature of natural breeding pools in Zambia may rise to over 37.8° C. on hot days. The reaction of the water was close to pH 8, a value near that in natural breeding pools (pH 8–8.5, and exceptionally 9.2: Muspratt, 1963a). This reaction was achieved spontaneously in basins to which soil was added. The soils used, either singly or in mixtures, were mopane clay, a physically similar slightly alkaline soil from near Bristol, and a greenhouse potting soil. Unrestricted evaporation was allowed; in some vessels periodic replenishment of the water was required.

The water in test vessels evaporated in, usually, 2–4 weeks. During this period eggs or newly hatched larvae were added, generally on several occasions, to ensure a succession of overlapping larval populations. Most larvae pupated 6 or 7 days after hatching and were then removed. Infected larvae were recognized by microscopic examination or by the distinctive orange color they assumed when resting sporangia were present.

Further details of particular infection tests are given below.

Results

Only a small proportion of the infection tests were successful. These are described below, but circumstances in tests that failed are noted when they contribute to an understanding of the conditions influencing infection. The experiments comprised three series.

1. Infection Tests in Small Basins

Eleven circular polythene basins—9 cm. deep, 25 cm. diameter at the base, and 27 cm. at the top—were employed. These "small basins" are

Table I

Summary of experiments on the infection of *Anopheles gambiae* with *Coelomomyces indicus* in small basins

Basin No.	Strain(s) of larvae[a]	Presence of mopane clay	Special features	Tests conducted and results[b]
S.B.1. . .	KL	+	—	*1* **2** 3 *4* 5 6 7
S.B.2. . .	KL	−	—	*1* 2 ***3*** 4 5
S.B.3. . .	K	+	Single strain of mosquito with or without mopane clay	*1* ***2*** 3
S.B.4. . .	L	+	"	*1* ***2*** 3
S.B.5. . .	K	−	"	*1* ***2*** 3
S.B.6. . .	L	−	"	*1* ***2*** 3
S.B.7. . .	KL	−	Foundation of sterilized local soil	*1* ***2*** 3
S.B.8. . .	K	−	No foundation of soil	*1*
S.B.9. . .	L	−	"	*1*
S.B.10. .	KL	+	Similar to S.B.1.	*1* 2
S.B.11. .	KL	+	No RS deliberately added	1 2

[a]K=Kano strain, L=Lagos strain
[b]Test numbered in bold type yielded infected larvae.
Tests numbered in italics received an inoculum of RS at their beginning or during their course.

referred to subsequently as S.B.1, S.B.2, etc. Most were used for two or three successive tests, between which they were fully dried. Successive tests with the same small basin are referred to as S.B.1/1, S.B.1/2, S.B.1/3, etc. Table I summarizes the principal circumstances of 33 separate tests. One alone (S.B.1/2) was successful, having been prepared as follows.

A layer 2.5 cm. deep, consisting of a mixture of equal parts of potting soil and white sand and rather less than an equal part of mopane clay, was placed in a basin located 75 cm. away from two 4-feet, 80-watt, "warm white" fluorescent tubes operating during alternate 12-hour periods in a 31 ± 1° C. constant temperature room. Resting sporangia dried in mopane clay were scattered on the soil surface, rain water muddied with potting soil was gently added to a depth of ca. 2.5 cm., and further large numbers of RS dried on filter paper were introduced. The temperature of the freely evaporating water was ca. 28° C. During S.B. 1/1, in which ca. 150 first instar larvae were added, there was no infection; in about ten days the soil in the basin was dry. Two days later the basin was refilled with water to 4 cm. depth and again ca. 75 of each strain of larva were added. Eight days later about ten infected fourth instar larvae were present, most of which were left to reinforce the inoculum. The day after the infections were first noted, the water was at pH 7.9, and flame photometric analysis of a centrifuged and filtered sample demonstrated calcium at 45 p.p.m., potassium at 25 p.p.m., and sodium at 123 p.p.m. Water fleas which were not a native species and so presumably were derived from the mopane clay appeared during this test and had multiplied rapidly. They characteristically appeared also in other infection tests to which mopane clay was added (see below), and were a feature of all tests in which mosquito larvae became infected with *Coelomomyces.*

Although S.B.1 was used for five more tests (Table I) there was no recurrence of infection despite further addition of RS dried in mopane clay. Other tests in small basins under the same conditions of temperature and illumination were negative (Table I). These comprised tests with single strains of mosquito, tests with

Table II

Summary of experiments on the infection of *Anopheles gambiae* with *Coelomomyces indicus* in large basins

Basin No.	Location of tests	Presence of mopane clay	Initial inoculum[a]	Later additions of inoculum[a]	Tests conducted and results[b]
L.B.1	1-3 outdoors 4-6 greenhouse	+	Massive (see text)	D.L. dried on filter paper in tests 2 and 4; and 3 D.L. from test 2 left in basin	**1 2 3** 4 5 6
L.B.2	Greenhouse	+	ca. 50 D.L. dried in soil, and 3 D.L. from experiment	D.L. dried on filter paper in test 2	1 2 3
L.B.3	Greenhouse	+	ca. 25 D.L. dried in soil, and 3 D.L. from experiment	D.L. dried on filter paper in test 2	1 2 3 1 2 3
L.B.4	1-5 in greenhouse 6-23 in laboratory	+	25 D.L. dried in soil, and ca. 25 D.L. dried after storage in water	Some D.L. from tests left in basin (see Table III)	**1 2 3** 4 **5** 6 **7 8 9** **10 11 12 13** 14 **15** 16 17 18 19 20 21 22 23
L.B.5	Laboratory	–	ca. 20 D.L. dried on filter paper	none	1 2 3 4 5 6 7
L.B.6	Laboratory	+	none (test of infectivity of mopane clay)	none	1 2 3 4 5 6 7
L.B.7	Laboratory	–	100 laboratory-infected larvae	none	1 2 3 4 5 6
L.B.8	Laboratory	+	ca. 20 laboratory-infected larvae	D.L. dried on filter paper in test 6	1 2 3 4 5 6 7
L.B.9	Laboratory	+	ca. 20 laboratory-infected larvae	D.L. dried on filter paper in test 6	1 2 3 4 5 6 7
L.B.10	Laboratory	+	ca. 20 laboratory-infected larvae	D.L. dried on filter paper in test 7	1 2 3 4 5 6 7 8

[a]D.L. = diseased, sporangium-laden larvae
[b]Tests numbered in bold type yielded infected larvae.

and without mopane clay in the soil foundations, tests without a soil foundation or with one of sterilized soil, a test to investigate whether the foundation layer of mopane clay alone was an effective source of infection, and a test in which conditions in the original basin were reproduced except that the inoculum comprised RS which had been stored in pool water since their collection in Zambia two months previously.

2. Infection Tests in Large Basins

These tests were conducted in rectangular porcelain wash basins. These "large basins" are referred to subsequently as L.B.1, L.B.2, etc., and successive tests with the same large basin as L.B.1/1, L.B.1/2, L.B.1/3, etc. The largest (L.B.4) was 98 x 51 x 17 cm. deep and the smallest 54 x 33 x 18 cm. deep. Of 77 tests in ten basins, 15 led to infection of larvae (Table II). Three successful tests were in L.B.1, and 12 in L.B.4. All tests with the other eight basins were negative. The early tests with L.B.1 which were successful (L.B. 1/1–3) were conducted in the open at Bristol from August to early October, 1963. As in all "large basin" tests, a 100-watt aquarium-type, thermostatic, flexible immersion heater prevented water temperatures from falling below 25° C. as they otherwise would do by night. Supplementary radiation from dawn to sunset from a 250-watt clear-glass infrared lamp 46 cm. above the basin helped keep daytime temperature of the water at ca. 30° C. In the third outdoor test (L.B. 1/3) a red-filter infrared bulb was operated *continuously* to offset the lower ambient temperatures. To control the environment more closely, the later tests with this basin, all tests with L.B.2 and L.B.3 and the first five with L.B.4, were conducted inside a greenhouse. There either one or two clear-bulb infrared lamps installed and a ca. 40 cm. above each basin horizontally mounted 40-watt tubular "dark light" (Philips TL 40 W/08), all of which operated for alternate 12-hour periods (6 a.m.–6 p.m.), partially simulated the intense solar radiation in natural habitats of *A. gambiae*, which prefers to breed in small natural collections of water completely or partially exposed to direct sunlight. The "dark light" radiated maximum power in the near ultraviolet (320–390 mμ). Because early tests in which supplementary radiation was provided showed vigorous and normal development of larvae and in some instances led to outbreaks of disease, this supplementary radiation was adopted as standard, but no experiments have conclusively proved that it enhances infection. All other tests in large basins were conducted inside a laboratory under the same conditions of illumination and temperature control as were provided in the greenhouse.

Of the successful tests, those with L.B.1 were prepared as follows. Local soil which resembled mopane clay was banked at each end of the basin (64 x 33 x 14 cm. deep) and overlaid with a layer of genuine mopane clay 12–25 mm. deep. Rain water was added and replenished every few days. A massive inoculum of fungus collected from the field was introduced, comprising ca. 60 diseased (i.e., RS-filled) larvae stored in slightly damp mopane clay, an aqueous suspension of RS from ca. 150 diseased larvae, and a large number of RS from larvae slowly dried on filter paper. In addition, aqueous suspensions of RS from ca. 25 and 75 diseased larvae respectively were added 10 and 18 days after the first test began. During the first three outdoor tests—between each of which the basin dried out—9, 3, and 21 diseased larvae respectively were observed, but none appeared in three further tests after transfer indoors.

L.B.4 was prepared similarly but with a very much smaller inoculum (see Table II). During 23 indoor tests (Table III) the temperature lay for most of the daytime period between 27° and 32° C. and at night fell to ca. 25° C. The overhead supplementary radiation provided for 12 hours a day kept the surface water by day generally 1–2° warmer than that below—mostly at ca. 31° C. Only occasionally did it rise above 34° C. In tests L.B.4/1, 3, 6, 7, 8, 12, and 20, the soil foundation was kept wet by sprinkling with water during respectively the 4, 2, 1, 1, 5, 3, and 4 days immediately before the basin was filled to simulate the rain showers which precede complete irrigation of a natural pool. In the first three tests first instar larvae were added, but in all others eggs were added directly. Both strains of mosquito were employed. The water depth in each test was initially ca. 16 cm., which left the tops of the soil banks just exposed, and fell by evaporation to zero in about three weeks. If the level became low while larvae were still present, more water was sometimes added. The essential circumstances and results of 23 successive tests are presented in Table III, but a number of additional features merit comment.

In L.B.4/1 the reaction of the water on day 9, which was probably close to the time that the larvae became infected (see Discussion), was pH 8.16. On day 15 it was almost the same, at pH 8.3. In L.B.4/2 it was pH 7.7 on day 10, when most diseased larvae were to be seen. In this test, 182 of the 271 insects which

Table III
Summary of experiments on the infection of *Anopheles gambiae* with *Coelomomyces indicus* in basin L.B.4.

Test No.[a]	Water in test[b]	Date basin was filled	Duration of test (days)[c]	Approximate duration of preceding drought	Number of diseased mosquitoes observed	Day when first diseased larva observed	Number of diseased larvae returned as inoculum
1	R	18 Nov. 63	20	—	7	14	ca. 2
2	R	12 Dec. 63	ca. 19	3 days	182	7	0
3	T	30 Apr. 64	>26	17 wks	30	11	5
4	R	1 Aug. 64	ca. 21	9 wks	0	—	0
5	T	7 Sep. 64	>16	2 wks	>43	12	12
6	T	8 Dec. 64	>17	10 wks	0	—	0
7	T	5 Jan. 65	31	1 wk	921	9	100
8	T	22 Mar. 65	27	6 wks	4	17	0
9	T	4 May 65	24	2 wks	>33	10	few
10	T	31 May 65	19	1 day	>30	9	0
11	T	25 Jun. 65	>16	2 days	2	11	2
12	T	16 May 66	22	43 wks	26	12	1
13	T	13 Jun. 66	22	1 wk	7	11	0
14	T	11 Jul. 66	19	5 days	0	—	0
15	T	9 Aug. 66	>17	1 wk	2	13	0
16	T	20 Sep. 66	15	3 wks	0	—	0
17	T	11 Oct. 66	>11	4 days	0	—	0
18	T	1 Nov. 66	16	3 days	0	—	0
19	T	23 Nov. 66	13	1 wk	0	—	0
20	T	13 Dec. 66	>12	1 wk	0	—	0
21	T	13 Jan. 67	16	1 day	0	—	0
22	T	7 Feb. 67	15	1 wk	0	-	0
23	T	28 Feb. 67	15	1 wk	0	—	0

[a]First five tests conducted in greenhouse, and remainder in laboratory; environmental conditions similar in both sites (see text).
[b]R=rain water, T=tap water.
[c]Symbol > means observations were terminated on specified day when there were no longer any larvae present, but more days elapsed before basin dried out.

developed became infected—i.e., 67.2 per cent. The diseased larvae which appeared on successive days 7–12 inclusive ranged from second to fourth instars and numbered 32, 1, 3, 109, 10, and 25 respectively, totaling 180. Only 19 of a sample of 41 examined on day 10 appeared to be without infection; they were not returned to the basin. Seventy other larvae pupated; of these one was seen to be diseased. Two additional pupae hatched into adults, of which one escaped and one (a female) was infected, dying as it strove to free itself from the pupal skin. This adult contained in its hemocoel sporangia with smooth, thin walls and a small number of what appeared to be immature RS that, although thin walled, bore traces of the characteristic ribbed patterning of RS. Few normal RS were present.

LB.4/7 led to 921 larvae's becoming diseased. No infections were seen among the ca. 180 pupae which appeared during the test. The infection rate was therefore ca. 84 per cent of the hatched insects. Because some eggs were added after the outbreak had largely run its course (e.g., on day 22; see Table IV), the rate of infection in the earlier part of the test was probably even higher than 84 per cent and may have compared closely with Muspratt's (1963a) figure of 95 per cent in natural pools. The numbers of visibly diseased larvae which appeared and were collected each day are tabulated in Table IV; Figure 1 portrays the form of the outbreak. Most of the diseased larvae were in third or fourth instars. It was not possible to recognize to which strain the diseased insects belonged.

In L.B.4/13, in which 7 larvae became infected (Table III), the reaction of the water

Table IV

The course of the seventh test of infection of *Anopheles gambiae* by *Coelomomyces indicus* in basin No. L.B.4., showing the times of introduction of mosquito eggs, of appearance and removal of pupae, and of appearance of visibly diseased larvae

Day	Addition of eggs[a]	Pupae present[b]	Diseased larvae	Cumulative total of diseased larvae
1	L	—	0	0
2	—	—	0	0
3	—	—	0	0
4	—	—	0	0
5	K	—	0	0
6	—	—	0	0
7	K	++	0	0
8	L	++	0	0
9	K	+	4	4
10	—	++	8	12
11	K	—	48	60
12	—	—	92	152
13	—	— }	441 }	593
14	—	— }		
15	KL	+	240	833
16	—	+	30	863
17	K	++	34	897
18	—	+	10	907
19	—	+	6	913
20	—	+	0	913
21	—	+	3	916
22	KL	+	0	916
23	—	+	4	920
24	—	+	0	920
25	—	—	0	920
26	—	+	0	920
27	—	—	0	920
28	—	+	1	921
29	—	+	0	921
30	—	—	0	921
31	—	—	0	921

[a] L = Lagos strain of *A. gambiae*
K = Kano strain of *A. gambiae*
[b] + = <10 pupae
++ = >10 pupae

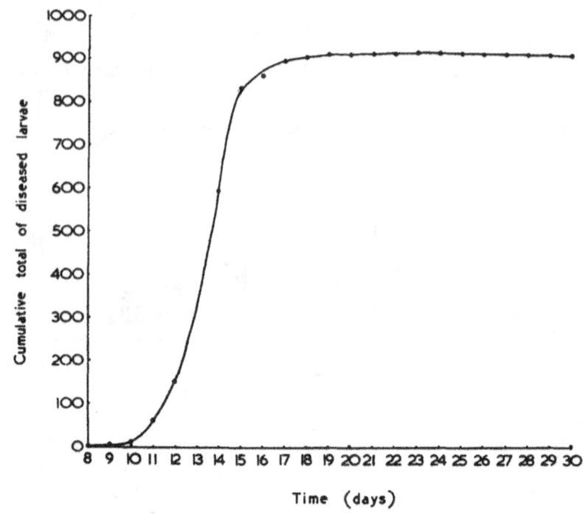

FIG. 1. The cumulative numbers of larvae of *Anopheles gambiae* recognizably infected with *Coelomomyces indicus* which were collected from basin L.B.4 during the course of the seventh test.

was determined electrometrically each day. Up to day 11, when the first diseased larvae appeared, it varied between pH 7.8 and 8.0. In L.B.4/15, in which one diseased larvae appeared on each of the thirteenth and fourteenth days, the reaction of the water up to this time was pH 7.8 ± 0.2 after wider variations up to pH 8.76 in the first four days had subsided.

L.B.4/13 involved a subsidiary experiment to test whether larvae prevented from browsing on the submerged surface of the soil layer could become infected. A series of 6-cm.-diameter polythene cups with nylon net bases with 0.2 x 0.2-mm. perforations were attached to a tethered float so that they provided bodies of water 3 cm. deep within which larvae could develop. Six such cups contained one strain of larva and six the other, while the basin at large contained a mixture. On day 11 of the test, one Lagos-strain larva in an isolation cup was heavily infected. The six other diseased larvae that subsequently appeared were all in the main basin.

After five successive tests (L.B.4/16–20) had failed to lead to infection, the basin was equipped with a gently bubbling aerator for L.B.4/21, but still there was no infection. The inoculum in the basin was reinforced at the beginning of L.B.4/17 and L.B.4/22 by adding small quantities of RS that had been stored dry for three years at room temperature, but also without effect. The possibility that natural breeding pools might be contaminated with urine or feces of animals and that these might enhance infection prompted the decision to add at the beginning of L.B.4/23 ca. 500 ml. of a mixture of slightly weathered cow dung and mud from near a cattle drinking trough, but still no infection ensued.

All tests with other large basins were negative, but some points may be noted. L.B.2 and L.B.3 were prepared and used in the greenhouse in rather similar ways to that used with L.B.1. Apart from their being smaller, they resembled L.B.4. L.B.5 contained an inoculum of sporangia from a field collection but no mopane clay, and L.B.6 contained mopane clay but no deliberate addition of RS. L.B.7 also lacked mopane clay but was seeded with laboratory-produced RS from 100 larvae (from L.B. 4/7). L.B.8, L.B.9, and L.B.10, were provided with a uniform basal layer of mopane clay and inclined at 6° C. so that the soil foundation became progressively exposed as the water evaporated. They were each seeded with RS from ca. 20 diseased larvae which had been produced in L.B.4/7 and stored dry in soil. Together with L.B.7, these tests constituted attempts to complete a full infection cycle in the laboratory with laboratory-produced RS.

3. Infection Tests in Other Vessels

A number of experiments were conducted in which first instar larvae or eggs were placed in glass distilled water at 31 ± 1° C. in diurnally illuminated crystallizing or petri dishes together with sporangia from field collections of diseased larvae but without a soil layer. Carefully controlled feeding avoided fouling of the water. The tests were repeated after initial unsuccessful ones had dried out, but none led to infection. Examination of the added RS revealed none that had dehisced.

In a similar experiment a live laboratory-produced diseased larva was added. It contained more or less mature RS as well as thin-walled smooth sporangia whose precise nature as immature RS or functionally distinct structures could not be conclusively determined. About twenty hatched larvae were added initially and others later to replace those that pupated. The diseased larva soon died and disintegrated, but no infections appeared in the other larvae during two weeks of incubation.

Discussion

It is possible from these results to define at least some of the conditions which are conducive to infection even though some may not be wholly necessary and unresolved factors may still exercise an overriding restraint. The following features were shared by the successful tests:

(1) A freely evaporating body of water.
(2) A reaction in the water of ca. pH 8.
(3) A water temperature for much or all of the day of ca. 28–32° C., not falling below 25° C. and seldom rising above 34° C.
(4) Natural and/or artificial illumination for about half of each day.
(5) A foundation of soil containing a proportion of mopane clay.
(6) An inoculum of sporangia of *C. indicus* collected from field sites in Zambia.
(7) A population of *Anopheles gambiae* introduced either as eggs or first instar larvae.
(8) The presence of water fleas which probably originated from the mopane clay.

The association of the above features in one test was, however, not in itself sufficient to ensure infection, for several tests which presented these conditions failed. Unresolved factors are clearly limiting. Further, some features appear to be uncritical. For example:

(1) Whether the temperature does rise and fall diurnally (cf., S.B.1 [Table I] with L.B.1 and L.B.4 [Table II]).
(2) Whether mosquitoes are introduced into test basins as first instar larvae (as in L.B.4/1–3; see Table III) or as eggs (as in all other successful tests with this basin).
(3) Which of the two strains of mosquito is used. Although tests with single strains of mosquito all gave negative results, the attainment of 67.2 per cent infection in L.B.4/2 indicates that at least the Kano strain is susceptible because more than two-thirds of the larvae added in this particular test were of that strain. The occurrence of a single infection among an isolated population of Lagos larvae in L.B.4/13 shows that this strain, too, is susceptible. The possibility that there are quantitative differences in degree of susceptibility is, however, totally unexplored.
(4) Whether basins are filled with rain or tap water (see Table III). However, the eventual cessation of infections in L.B.4 after 15 tests might have resulted from a progressive concentration of salts by evaporation of successive volumes of tap water.
(5) Whether the soil layer is banked at the ends of the vessel (e.g., in L.B.1 and L.B.4), and so becomes increasingly exposed as the water level falls, or is flat and completely submerged (as in S.B.1).
(6) Whether the soil foundation is sprinkled with water a few days before the basin is filled.

There are some factors whose significance cannot be determined with certainty from the results at hand. The quality of radiation provided is one. While S.B.1 received only "warm white" light from a fluorescent discharge tube (radiation from ca. 300 to 750 mμ with a major peak near 580 mμ), L.B.4 received daylight supplemented with infrared and near ultraviolet radiation of intensities comparable with those in natural habitats in the tropics. Both led to outbreaks of disease, the former to one (S.B. 1/2) and the latter to twelve. Although infection can occur with artificial "warm white" illumination alone, it is possible that the stronger radiation in L.B.4 may have helped predispose the inoculum in the basin to cause infection, and so may account for the lengthy run of successes achieved. A major obstacle in these researches has been inability to observe dehiscence of sporangia even though very many tests, involving—among other things—attempts to break sporangial dormancy, have been made. The RS appear to be constitutively dormant in the sense of Sussmann (1966). Muspratt (1946a) reported dehiscence of RS of this species—his "type (a)"—on rare occasions. It seems almost certain that it must have occurred in the course of successful infection tests here, but always went unnoticed because of the complexities of the experimental system. Possibly the conditions experienced during the long, dry season in the natural environment, when the temperature of the mud in dried-out breeding pools may rise to 50° C. (Muspratt, 1963b), serve to break the constitutive dormancy of at least some RS. The intense radiation in L.B.4 may have favored infection not so much by its effects during the course of infection tests as during the periods of drought that intervened.

The importance of drying of the basins between tests is hard to evaluate. S.B.1 yielded infected larvae in only its second test, i.e., after drying out once. Muspratt (1946b), with the same fungus, similarly succeeded in his second test after a first one was negative. On the other hand, there were outbreaks in the first tests with L.B.1 and L.B.4 (Tables II and III). It may be significant that most of the RS added to these as inoculum had been stored dry. Therefore no exception has occurred to the generalization that all successful tests have been inoculated with at least some RS which have been thoroughly dried at least once beforehand. That the RS of some *Coelomomyces* species, notably those that parasitize mosquitoes which inhabit temporary pools, cannot germinate until dried was suggested by Couch (1945).

Only a single outbreak of disease was observed in any one successful test; at most there was but a sporadic appearance of diseased individuals after the peak of infection had passed (Fig. 1). This suggests that after those RS which were able to germinate had done so, the rest remained constitutively dormant until the next period of drought rendered some more ready to germinate. Such a characteristic would be biologically advantageous to the species in its natural habitats, where, as Muspratt (1963a) describes, pools may sometimes fill with water yet contain no larvae of *A. gambiae*. It must be noted that Muspratt (1963a) found that a nearly 100 per cent infection rate was sustained in some natural breeding pools which happened to remain continuously filled for six weeks. Nevertheless in those that were not replenished by rain, all the larvae became infected at about the same time (Muspratt, 1946a), a situation which resembles that revealed in these laboratory tests.

The occurrence of only single outbreaks of disease in tests despite the overlap of generations of larvae suggests also that there was no secondary spread of infection. None of the thin-walled, smooth sporangia from diseased larvae have even been seen to dehisce despite many tests, even though Muspratt (1946a, 1963a) has reported that some do so in 3–6 days if the larval remains which contain them are left in water from breeding pools. The very late appearance of infection (day 17) in L.B.4/8 (Table III) suggests that the late sporadic infections in L.B.4/7 (Table IV) might also have derived from sporangia of the primary inoculum rather than represent secondary infections.

The first appearance of sporangium-filled larvae has varied in different tests from day 7 to day 17 (see e.g., Table III). It was most commonly about day 11. The earliest detected stage of infection has been that in which small, round thalli (Muspratt's "buds" [1963a]) are present in the anal gills of the larvae. These thalli soon become lobed. One larva in which only a single rounded thallus in an anal gill was to be seen proceeded to the stage in which the hemocoel contained large numbers of RS in only 28.5 hours at ca. 31° C. In view of this rapid rate of development, it seems likely that infection of larvae occurs only about two or three days before they become laden with RS. If, as seems probable, infection requires the dehiscence of sporangia, then a number of days must have elapsed after filling a test basin before any RS dehisced. Indeed, many or most of the first population of larvae commonly pupated before any infections appeared.

Larvae are clearly susceptible to infection because infections have resulted when only first instar larvae have been added. Development of infected insects is usually halted at the third or fourth instars, though occasionally earlier (second instar) and sometimes later (i.e., at pupal and adult stages). No first instar larvae in these tests have ever been recognized as diseased, though Muspratt (1946a) has reported their natural infection, but in any case one might expect the symptoms to be slight at this stage. Thus it is concluded that infection can occur at least as early as the second instar. It might occur even in the first instar, or at the first moult, but certainly does not *necessarily* occur at or prior to hatching. Because the parasite develops so rapidly and normally arrests its host's development, the existence of occasional diseased pupae and adults probably testifies to the occurrence of infection during the fourth instar and even pupal stages, or at the intervening moult.

Infection does not require larvae to browse at the submerged soil surface, a fact which suggests that zoospores and not RS themselves are the infective units.

The role of mopane clay in these tests is obscure. No test has succeeded in which it has been absent, but its presence has not assured success. Its contribution to the experimental system could be physical, chemical, or biological. The water fleas, which appeared in tests to which it was added, were not noticeably involved in any direct way with the biology of the fungus. The mopane clay itself might even have been the source of effective inoculum of the fungus, for example, of fortuitously present RS whose dormancy has been broken naturally. On the other hand, no test succeeded in which mopane clay was added unaccompanied by collected RS. Apart from any effect on the fungus, a layer of soil in rearing pans is reported by Armstrong and Bransby-Williams (1961) to improve development of larvae of *Anopheles gambiae,* so its effects in the present experiments may be several.

A curious feature of the experiments has been the wide variations in numbers of larvae which became diseased in different tests (Table III). Because host populations comprised overlapping generations and because their numbers were not standardized, the numbers of diseased larvae in different tests (Table III) cannot be compared directly. Nevertheless, they clearly fall into several orders of magnitude (0–10, 10–100, 100–1000), and differences between these are almost sure to be significant; yet they have not been correlated with any observed circumstance. For example, the magnitude of an outbreak was not directly related to the duration of the preceding period of drought, nor to the practice of gradual remoistening of the soil layers.

The single diseased female adult obtained in these tests contained sporangia widely distributed in its hemocoel. The development of sporangia of *Coelomomyces* in the ovaries of adult females so that they replace the eggs has been reported by Gibbins (1932), Haddow (1942), Van Thiel (1954), Kellen, et al. (1963), and Brady (1963). The records of Gibbins, Haddow, and Brady referred to specimens of *Anopheles gambiae. A. gambiae* is an anautogenous mosquito; therefore because the adult in the present tests died before it received a blood meal, its ovaries would not have been developed anyway.

Although these tests have yielded quantities of RS-filled larvae sufficient to inoculate laboratory infection tests and indeed to distribute to other workers for various trials, it has not so far proved possible to secure infections using these RS alone as inocula; nor has their dehiscence been observed. There is, in fact, no certain proof that infection in any of the above tests resulted from RS gathered directly from diseased larvae, whether from the field or from laboratory tests. All infections might have stemmed from inoculum of unknown history within the mopane clay that was also present.

Because RS have never been seen to dehisce in the present study, it has not been possible to discriminate between those features of successful tests which contributed toward making RS germinate and those which directly influenced the infection process. It would appear necessary to solve the problem of how sporangial dehiscence may regularly be induced in *C. indicus* before rapid progress can be made toward a detailed analysis of the circumstances which influence infection.

ACKNOWLEDGMENTS

The author is deeply indebted to Mr. J. Muspratt for much advice and direct assistance in the early stages of these investigations and for providing material of *Coelomomyces* and mopane clay from sites near Livingstone, Zambia. Thanks are also due to Mr. G. Davidson of the Ross Institute of Hygiene, London, for supplying eggs of *Anopheles gambiae* on very many occasions; to Mr. M. V. Angel for information on the water fleas mentioned; and to Misses P. Bees, M. Messenger, and J. Weber for technical assistance from time to time. These researches were supported by a grant from the World Health Organization, Geneva. My thanks are also due to Dr. Marshall Laird, late of that organization, for his sustained interest and help.

REFERENCES

ARMSTRONG, J. A., AND BRANSBY-WILLIAMS, W. R. 1961. The maintenance of a colony of *Anopheles gambiae*, with observations on the effects of changes in temperature. Bull. World Health Organization **24**: 427–435.

BRADY, J. 1963. Results of age-grouping dissections on four species of *Anopheles* from southern Ghana. Bull. World Health Organization **29**: 147–153.

COUCH, J. N. 1945. Revision of the genus *Coelomomyces* parasitic in insect larvae. Jour. Elisha Mitchell Sci. Soc. **61**: 124–136.

COUCH, J. N., AND DODGE, H. R. 1947. Further observations on *Coelomomyces* parasitic in mosquito larvae. Jour. Elisha Mitchell Sci. Soc. **63**: 64–79.

COUCH, J. N., AND UMPHLETT, C. J. 1963. *Coelomomyces* infections. *In* E. A. Steinhaus, ed., Insect Pathology, Vol. II, pp. 149–188. Academic Press, New York and London, 689 pp.

GIBBINS, E. G. 1932. Natural malaria infection of house frequenting *Anopheles* mosquitoes in Uganda. Ann. Trop. Med. Parasitol. **26**: 239–266.

HADDOW, A. J. 1942. The mosquito fauna and climate of native huts at Kisumu, Kenya. Bull. Entomol. Research **33**: 91–142.

IYENGAR, M. O. T. 1962. Validation of two species of *Coelomomyces* described from India. Jour. Elisha Mitchell Sci. Soc. **78**: 133–134.

KELLEN, W. R., CLERK, T. B., AND LINDEGREN, J. E. 1963. A new host record for *Coelomomyces psorophorae* Couch in California (Blastocladiales: Coelomomycetaceae). J. Insect Pathol. **5**: 157–166.

LAIRD, M. 1959. Parasites of Singapore mosquitoes, with particular reference to the significance of larval epibionts as an index of habitat pollution. Ecology **40**: 206–221.

LAIRD, M., AND COLLESS, D. H. 1962. A field experiment with a fungal pathogen of mosquitoes in the Tokelau Islands. Proc. Intern. Congr. Entomol. 11th., Vienna, 1960. **2**: 867–868.

LUM, P. T. M. 1963. The infection of *Aedes taeniorhynchus* (Wiedemann) and *Psorophora howardii* Coquillet by the fungus *Coelomomyces*. J. Insect Pathol. **5**: 157–166.

MADELIN, M. F. 1966. Fungal parasites of insects. Ann. Rev. Entomol. **11**: 423–448.

MANALANG, C. 1930. Coccidiosis in *Anopheles* mosquitoes. Philippine J. Sci. **42**: 279.

MUSPRATT, J. 1946a. On *Coelomomyces* fungi causing high mortality of *Anopheles gambiae* larvae in Rhodesia. Ann. Trop. Med. Parasitol. **40**: 10–17.

———. 1946b. Experimental infection of the larvae of *Anopheles gambiae* (Dipt., Culicidae) with a *Coelomomyces* fungus. Nature **158**: 202.

———. 1963a. Destruction of the larvae of *Anopheles gambiae* Giles by a *Coelomomyces* fungus. Bull. World Health Organization **29**: 81–86.

———. 1963b. Progress report (May 1963) on investigations concerning three mosquito pathogens at Livingstone, Northern Rhodesia. World Health Organization, Geneva: WHO/EBL/12, 24 July, 1963, 7 pp. (mimeographed).

SUSSMANN, A. S. 1966. Types of dormancy as represented by conidia and ascospores of *Neurospora*. *In* M. F. Madelin, ed., The Fungus Spore, pp. 235–256. Proceedings of the 18th Symposium of the Colston Research Society. Butterworths, London. 338 pp.

VAN THIEL, P. H. 1954. Trematode, gregarine and fungus parasites of *Anopheles* mosquitoes. J. Parasitol. **40**: 271–279.

WALKER, A. J. 1938. Fungal infections of mosquitoes, especially of *Anopheles costalis*. Ann. Trop. Med. Parasitol. **32**: 231–244.

Life History of the Motile Spore of *Blastocladiella emersonii*: A Study in Cell Differentiation[1]

Edward C. Cantino, Louis C. Truesdell, and David S. Shaw[2]
Department of Botany and Plant Pathology, Michigan State University, East Lansing, Mich.

We are grateful to have the opportunity to participate in this joint tribute to John N. Couch, and pleased by the fortuitous coincidence that our current research, and thus the subject of our contribution to this commemorative volume, is especially appropriate for the occasion. We are reminded that it was a full quarter of a century ago, and some six years before we had even begun our own 18 years of continuous work with *B. emersonii,* that John Couch and Alma Whiffen (1942) had already published the first comparative study of motile cells derived from several members of the then newly discovered genus *Blastocladiella*—including, in fact, some new species which they themselves described. In the light of what has been learned since then with benefit of better optics, phase and electron microscopes, and lots of hindsight, it is instructive and quite good for one's humility to contemplate upon the penetrating insight into the structure of the blastocladiaceous zoospore that Couch and Whiffen were able to achieve with instruments of the time; the keenness of their observations will speak for itself in the following chronicle on *Blastocladiella.*

Form and Fabric in the Spore of Blastocladiella emersonii

During the thirty years which have now followed Miss Matthews' discovery of the genus *Blastocladiella,* enough observations have been made to provide convincing evidence that its motile cells are exceptionally—and in certain ways, even uniquely—suitable for experimental studies of a variety of provocative biological phenomena. We shall take advantage of this occasion, therefore, to trace chronologically how this has come about.

(1937). V. D. Matthews drew up a rather short account of the spores of *B. simplex,* and—to sum up her over-all impression, it would seem—concluded that they were similar in structure to the spores of *Blastocladia* as described by Cotner (1930). By and large, Miss Matthews' brief generalization is reflected in her drawings. Her sketch of a large tripartite structure in the swarmer of *B. simplex* is almost identical with the corresponding body depicted by Cotner in the spore of *Blastocladia*; he had interpreted it to be—as Thaxter did before him (1896)—a subtriangular nucleus consisting of a dense chromatic portion (presumably the nucleolus), a clear zone (probably the nucleus), and a broad, dense, distal end containing irregular masses of chromatin just inside the "nuclear" membrane (undoubtedly the nuclear cap, and so called by Miss Matthews). Cotner also noted the presence of a highly refractile body, the blepharoplast, at the point of insertion of the flagellum on the spore's membrane, and said that it was joined to the "nucleus" by a thread-like structure; Matthews also depicted such a connecting link in one of her three figures, although she wrote nothing about it.

But in other ways, Miss Matthews' conclusion and her sketches are not quite in harmony. Among his numerous drawings of the zoospores of *Blastocladia,* Cotner pictured swarmers with a dense granular mass of cytoplasm between the broad end of the "nucleus" and the posterior surface of the cell; in this same region, Matthews showed a large vacuole. Cotner drew in collections of granules, varying in number from a few to very many, in the cytoplasm opposite both sides (a two-dimensional view, of course) of the "nucleus"; Miss Matthews depicted linear arrays of granules and localized them on only one side of the spore, adjacent to the nucleus—where, in fact, one might now expect to find a "side body" (see below).

(1938). R. Harder and G. Sörgel noted the presence of a saddle-shaped "food body" (i.e., a nuclear cap) in the spores of *Rhopalomyces variabilis* (=*B. variabilis*; cf. Couch and Whiffen, 1942).

(1939). H. Stüben, too, in his commentary on the internal morphology of the spore of *Sphaerocladia variabilis* (=*B. stübenii*; cf. Couch and Whiffen, 1942) pointed to the same principal ingredients: a homogeneous "food

[1] All unpublished work on *B. emersonii* reported herein was supported by research grants to E.C.C. from the National Science Foundation and the National Institutes of Health.

[2] N.A.T.O. Post-Doctoral Fellow; the present address of D.S.S. is Department of Botany, University College of North Wales, Bangor, Wales.

body" (nuclear cap) with strong affinity for basic dyes; a nucleus; the flagellum, connected at its point of insertion to a refringent granule; and finally, a strand connecting this granule to the nucleus. But in addition, Stüben highlighted for the first time what he considered to be a new organelle in the spore of *Blastocladiella*: a side body, shaped like a teardrop, and positioned with one end alongside the "food body," its other end extending into a region between nucleus and refractive granule. *In vivo*, its homogeneity and refractive properties were similar to those of the "food body"; it was strongly stainable with hemotoxylin but not with other dyes suitable for coloring the "food body" (i.e., Safranin, gentian violet, etc.).

(1942). J. N. Couch and A. J. Whiffen, in a study of several species of *Blastocladiella*, wrote: "In the living zoospores of *B. simplex*, *B. laevisperma* and *B. asperosperma* we are unable to find with certainty any structures which would be interpreted as Stüben's side body. However, when the zoospores of these species are killed with the fumes of osmic acid and weakly stained with an aqueous solution of gentian violet and studied immediately without drying, a very conspicuous curved body is visible. In *B. laevisperma* and *B. asperosperma* this body extends from near the rhizoplast over the nucleous and nuclear cap usually almost to the opposite end of the spore . . . the side body . . . appears to be hollow or vesicular in structure. . . ." And about *B. simplex*, in particular, they wrote (italics added by us; compare, below, with details of fine structure described in 1963 and later): "A particularly good view of the side body and indeed of the entire contents is shown in the spore in figure 64 in which the parts were considerably swollen and the plasma membrane had begun to disintegrate. The side body was much enlarged, purplish and clearly vesicular. *Adhering to the side body* were the darkened fat bodies. *There seemed to be an intimate association between the rhizoplast and the side body*. Indeed, it seemed that the *cilium was attached both to the side body and the rhizoplast*."

(1953). E. C. Cantino and M. T. Hyatt described a new species of *Blastocladiella*, *B. emersonii*, whose spores possessed a "typically blastocladiaceous" nuclear cap and some 3 to 16 (generally 6–8) conspicuous refractive granules inserted near the flagellum and to the side of the nucleus.

(1956). E. C. Cantino and E. A. Horenstein made out a new cytoplasmic body in the spores of *B. emersonii*, estimated its size at ca. 0.5 μ, and christened it a gamma particle. Its frequency (range: ca. 8–16/spore) was related to the previous history of the parent plant (from which the spores had been derived), the presence or absence of endogenous carotenoid therein, and the environmental conditions which had been imposed upon it.

(1963). E. C. Cantino, J. S. Lovett, L. V. Leak, and J. Lythgoe showed by electron microscopy that the spore of *B. emersonii* contained, in addition to its nucleus and nucleolus: (a) a single eccentrically disposed mitochondrion, occupying an extraordinarily large proportion of the cell's volume, which was situated in the same position as—and almost certainly corresponded to—Stüben's "seintenkorper" and the side body of Couch and Whiffen; (b) a group of prominent, strongly osmiophilic, lipid-like organelles, localized in a sac-like region by a double-layered membrane, and situated along the outer edge of the mitochondrion, which clearly corresponded to the darkened fat bodies of Couch and Whiffen; (c) a flagellum, with a "9 + 2" internal structure, whose fibrils seemed to end within the spore in a differentiated basal structure; this, in turn, terminated abruptly at the nucleus; (d) the basal structure was completely surrounded by the mitochondrion and apparently attached to it by one or more banded "rootlets" which extended deep into lateral mitochondrial canals; it appeared to be anchored, as well, by nonbanded structures extending upward between the nucleus and the mitochondrion; (e) a flagellar sheath confluent with the outer membrane of the spore; (f) a massive nuclear cap separated from the nucleus by a double-layered membrane bearing pores ca. 100 mμ in diameter; and (g) a cytoplasm essentially devoid of endoplasmic reticulum but containing about a dozen organelles, bound by single-layered membranes, which seem to consist of a spherical body incompletely girdled by a band of strongly osmiophilic material. These organelles, possibly lysosome-like in character, were undoubtedly the gamma particles previously seen by Cantino and Horenstein.

(1963). J. S. Lovett isolated nuclear caps from the spores of *B. emersonii* in sufficient quantity to permit their chemical and physical characterization. The intact cap, strongly basophilic, was composed of 60 per cent protein and 40 per cent RNA; it represented 18 per cent of the dry weight of the spore, and contained 69 per cent of its RNA and essentially all of its small (25–30 mμ) electron-dense particles. More than 95 per cent of the cap contents had a sedimentation coefficient of 83S in 0.005 M Mg, the 83S particles aggregating at higher Mg concentrations and dissociating to yield

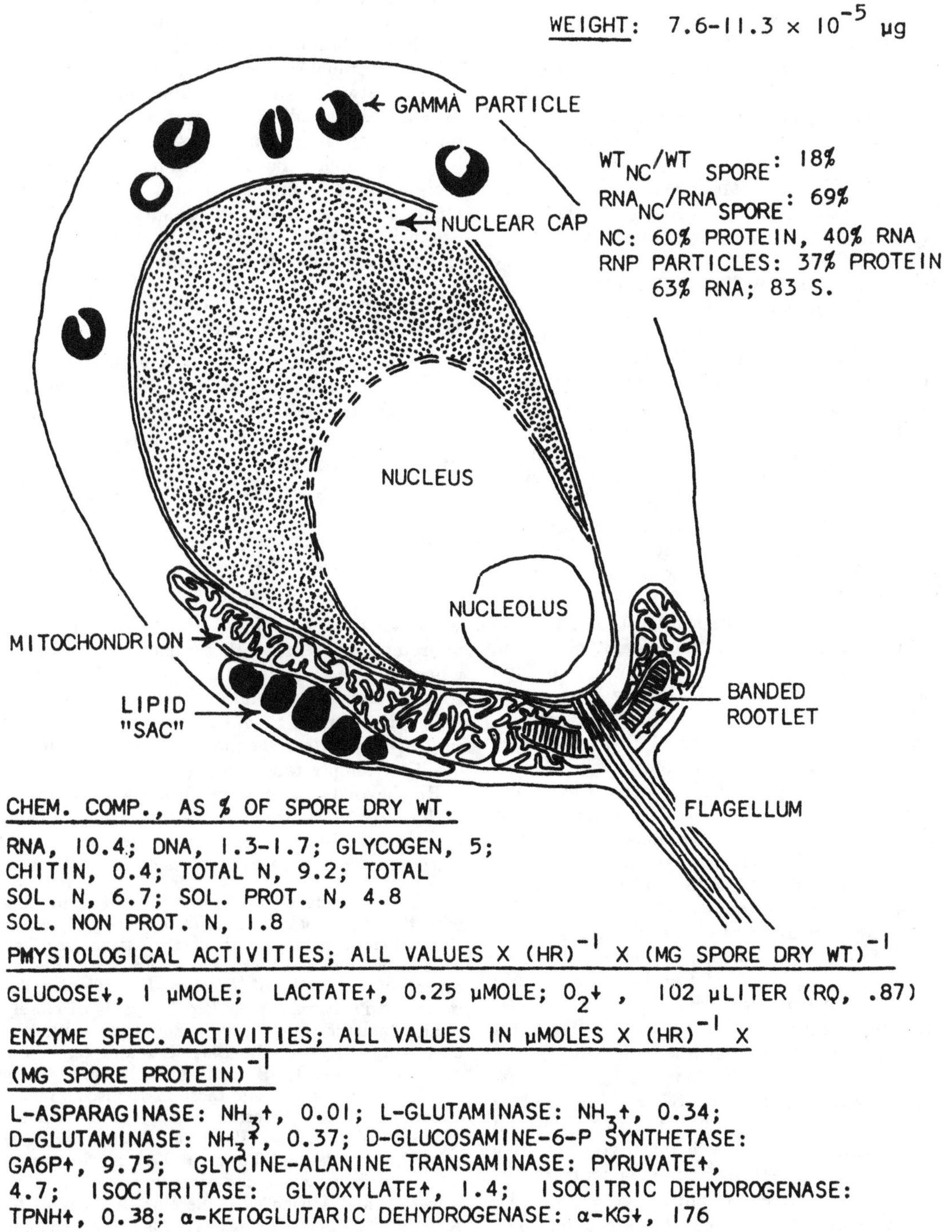

Fig. 1. A summary of structure and function in the spore of *B. emersonii*. For particulars, see references in text; for recent, new details on fine structure, see Fuller (1966), Reichle and Fuller (1967), and Lessie and Lovett (1968).

63S and 41S particles at lower Mg concentrations; cap particles contained 37 per cent protein and 63 per cent RNA. For the first time, *direct chemical evidence* pointed strongly to the conclusion that the nuclear cap in the spore of *B. emersonii* was a package of particles which, in their chemical and physical properties, resembled typical ribosomes.

(1965–1959). H. McCurdy, J. S. Lovett, E. A. Horenstein, A. Goldstein, A. Domnas, and E. C. Cantino, in a series of studies (references in Cantino, 1966) established the dry weight of the spore ($7.6–11.3 \times 10^{-5}$ µg) and part of its chemical makeup (protein-, non-protein-, and total N, chitin, soluble polysaccharide, RNA, DNA, etc.), quantified on a per-cell basis its high endogenous oxygen consumption (Q_{O2}= ca. 100), some of its other *in vivo* physiological activities (capacities for lactic acid production, glucose consumption, etc.), and many of its *in vitro* enzymatic activities (D- and L-glutaminase, L-asparaginase, D-glucosamine-6-phosphate synthetase, glycine-alanine transaminase, isocitratase, isocitric dehydrogenase, and α-ketoglutaric dehydrogenase).

(1966). A diagrammatic digest of structure and function in the motile spore of *B. emersonii*, as visualized solely from our own first-hand studies with it, is shown in Figure 1.

(1966). Lastly, M. S. Fuller (and just recently, Reichle and Fuller, 1967) sharpened and significantly enlarged our view of fine structure in the spore of *B. emersonii* by showing that: (a) each pair of fibrils in the flagellar axoneme possesses a pair of arms ca. 16 mµ long and directed clockwise; (b) the two single-structured central strands of the axoneme terminate close to the spore's perimeter, not beside the nucleus as reported by Cantino et al. (1963); (c) the kinetosome (the "differentiated basal structure" of Cantino et al., 1963; i.e., a "blepharoplast") displays a cartwheel-like substructure in cross section, is closed off by a terminal plate near the point of contact of the axoneme with the spore body, and is associated with a nearby centriole lying next to the nuclear membrane; (d) the mitochondrial canals which house the banded rootlets may number three rather than only one or two as previously described (Cantino et al., 1963), and some may be plugged at their outer ends by vesicle-like bodies (Reichle and Fuller, 1967); (e) each banded rootlet, surrounded by its own unit membrane, appears to be separated from the mitochondrion; (f) the "nonbanded structures of Cantino et al. (1963) which originated at the kinetosome may have been sections of penetrating double membranes which can extend back, either along or within folds of the double membrane surrounding the nuclear cap; (g) microtubules are detectable in the cytoplasm between the mitochondrion and the nuclear cap; and (h) the membrane of the nuclear cap is continuous with a membrane which envelopes the mitochondrion and its associated lipid granules; in fact, a double unit-membrane system seems to surround all cell organelles except gamma particles, vacuoles, and a few membrane fragments.[3]

"Life History" of the Spore of **Blastocladiella emersonii**

The origin, life, and fate of a viable zoospore of *B. emersonii* come to pass in quick succession along the following path:

SPORE SWIMS AWAY ⟶ SPORE STOPS SWIMMING ⟶ SPORE RETRACTS FLAGELLUM
↓
Spore germinates ⟶ Sporangial plant ⟶ Spore is born

With virginal simplicity, the spore's birth is unpretentious. And with comparable propriety, when its life time is at an end, it retracts its sole appendage and simply ceases to exist—yet it is not returned to dust. For it can be truly said of *Blastocladiella* that its old spores never die, they simply fade away (into germlings).

But beyond this, how much knowledge do we really have about this sexless cycle of events in *B. emersonii*—or, for that matter, in other water molds where it applies? Miss Water-

[3] The foregoing chronological account dealt only with the spores of *Blastocladiella* except, of necessity, where conclusions about their structure had been expressed in terms of observations made by others of the cells of *Blastocladia*. The reader interested in the motile uniflagellate cells of closely related aquatic fungi may want to consult the pioneering studies of the Blastocladiales–e.g., those of Butler (1911), Barrett (1912), and Kniep (1929) on *Allomyces*, and those of Reinsch (1887), Thaxter (1896), Petersen (1910), von Minden (1915), and Kanouse (1927) on *Blastocladia*; the paper by Blondel and Turian (1960) on fine structure in *Allomyces* and the preceding work of Kellenberger and Turian cited therein; the chapter by Kole (1965) in *The Fungi* for references to early studies of Cotner, Couch, etc., and the subsequent work of Manton et al., Kole and co-workers, Koch, and others on flagellar structure in aquatic Phycomycetes; and finally, the recent publications by Renaud and Swift (1964), Fuller and Reichle (1965), and Fuller (1966) on fine structure in several types of uniflagellate spores.

house (1962) posed the same sort of question in pleasingly direct fashion, thus: "After all its activity the zoospore apparently gets worn out and come to a halt. We all know the familiar text-book story—movement becomes more sluggish, it settles down, withdraws or sheds its flagella, rounds off, and encysts. But how much do we really know? Why does it round off? Why not keep its oval, pear, or bean shape. The answer to that probably is that the sphere presents the least surface to the bleak outside world. But what changes occur inside to cause the protoplasm to contract to the minimum volume?" Coming back to *Blastocladiella,* then, it is in these very things that we, too, are interested; we, too, would like to learn how, to what degree, and at which stages during its existence the motile spore undergoes change—in what way, precisely, does its own "ontogeny" unfold? To accomplish this, we have begun a study of the swimming spore and all its doings. By way of introduction, let us first consider briefly what is known of sporogenesis in *B. emersonii*; and then we will close in upon the three central stages (solid arrows) which can perhaps most properly be considered to constitute the spore's life history.

Sporogenesis in Blastocladiella emersonii

Over the years, many mycologists have watched sporogenesis in water molds; examples can be found in Sparrow's treatise (1960) on aquatic Phycomycetes. As for *Blastocladiella,* Miss Matthews (1937) observed the process in thin-walled sporangia of *B. simplex,* illustrated it (Fig. 2A), and described it thus: ". . . some enlargement of the sporangium now takes place and the refractive granules become smaller and scattered throughout the very fine cytoplasm. . . . A few minutes after this stage these granules are much smaller and arranged in a more or less circular manner. . . . A few minutes later the origins of the individual spores become visible with a group of granules in each. Five to ten minutes later the conspicuous papilla suddenly seems to be pulled inward, movement of the spores begins, and the papilla is pushed outward to form a very thin vesicle, which lasts only a few seconds. No cap is cut off. After the vesicle bursts, the spores separate and swim away with their cilia propelling them from behind. The zoospores remaining in the sporangium come out rather rapidly and swim away. . . ."

Later, Couch and Whiffen (1942) also wrote of sporogenesis in the conspicuously vacuolated and densely granular developing sporangium of another species, *B. laevisperma*: "Just before the zoosporangium reaches its mature size the numerous fat bodies break up into small particles and become evenly dispersed through the protoplasm. . . . Soon after the basal wall is formed, the tiny globules become arranged in a reticulate fashion. . . . About this time one or more emergence papillae are formed and simultaneously the tiny globules collect into small groups, a group in the center of each spore origin. . . . A few minutes later the outlines of the zoospores appear as rounded, clear areas each with several small globules. . . . In some sporangia just after the spores are completely formed the spore mass near the center may start cyclic motion which increases in vigor until most of the spores have been discharged. In other sporangia there is no autonomous movement of the spores until after they emerge."

During the quarter-century that followed these early descriptions of sporogenesis by Matthews and Couch and Whiffen, nothing was published that added significantly to our understanding of the origin of organelles and their architectural arrangement in the spore of *Blastocladiella.* But now, research along this front has begun again; preliminary unpublished work by Murphy (1964: cf. Murphy and Lovett, 1966) suggests that sporogenesis in *B. emersonii* may be similar to gametogenesis in *Allomyces* as described by Blondel and Turian (1960) and Renaud and Swift (1964). Of particular interest, therefore, are: (a) observations by these four workers (in contrast to an earlier report by Ritchie, 1947) that flagella are detectable before deposition of the membranes eventually laid down to delimit the coenocytic gametangial protoplast into gametes; and (b) the conclusion of Renaud and Swift that the inductive event in flagellum formation is genesis (*via* pinocytosis from membranes) of a vesicle which makes contact with a basal body (the larger of a pair of centrioles on the nuclear membrane), inducing it to invaginate into the vesicle and thereby grow into a flagellum. How closely these observations—as well as those of Blondel and Turian on the origin of the nuclear cap, the increase in size and number of vesicles and their redistribution at sites of membrane formation, etc.—on gametogenesis in *Allomyces* correspond to events which occur during sporogenesis in *B. emersonii* must now await publication of the details of recent work along these lines (Lessie and Lovett, 1968).

In contrast to our present deficiency of descriptive detail about the origin of organelles and fine structure, biochemical investigations of sporogenesis in *B. emersonii* have begun to yield interesting facts. Cantino (1962) compared chemical and physiological features of mature RS plants (i.e., capable of sporogenesis) with corresponding properties of spores derived from OC cells. Tacitly assuming that RS and OC spores are essentially identical, the inference was drawn that sporogenesis in an RS cell is associated, per unit of protoplasm, with: (1) a 7.4-fold increase in total RNA, and simultaneous utilization of an insoluble protein fraction thought to have been associated with large aggregates of insoluble RNA in the RS cell; (2) a 2.4-fold increase in soluble protein-N; (3) a 186-fold increase in both endogenous Q_{O_2} and lactic acid; and (4) a 5- to 30-fold increase in the specific activities of several enzymes. In conclusion, indirect evidence suggests that a surprisingly high level of synthetic activity is associated with sporogenesis in an RS cell.

Of greater import, however, are the more definitive data of Murphy and Lovett (1966) on synthesis of RNA and protein derived directly from synchronized populations of OC cells induced to undergo sporogenesis; in these cultures, cleavage of the coenocytic protoplast into spores was visible about an hour after papillae had formed, and was completed in ca. 10 minutes. This experimental system, ideally suited for labeling experiments with ^{14}C-uracil, provided evidence that: (a) the nuclear cap is formed by aggregation of pre-existing ribosomes, not from *de novo synthesis*; (b) aggregation, as well as formation of the cap's membrane, is completed in 10–15 minutes; and (c) all of this occurs about 2½ hours after the OC cell has ceased to manufacture ribosomes. Although a considerable fraction of the sporangial RNA is degraded during sporogenesis, very little new RNA is synthesized. Experiments with inhibitors also led to tentative though highly interesting interpretations about temporal relationships between sequential synthesis of different species of short-lived m-RNA and the subsequent onset of morphological events during sporogenesis (i.e., formation of papillae, protoplasmic cleavage) which they controlled.

To conclude, our presently limited knowledge about the origin of gross morphological features and fine structure in the spore of *B. emersonii* has now been overtaken—indeed, outdistanced—by new facts and fresh insight into the origin of biochemical differences that ensue prior to and during sporogenesis. But as work progresses, and especially in view of Fuller's (1966) recent work as well as that of Reichle and Fuller (1967) and the oncoming publication by Lessie and Lovett (1968) of an extensive series of pictures showing additional details of fine structure in the spore of *B. emersonii*, the accumulating information about structure and function will surely blend together soon to provide an exceptionally informative collage, in modern hues, of sporogenesis in *B. emersonii*.

The Swimming Spore of Blastocladiella emersonii

We have come to look upon the interval between the time when a motile swimming spore is born and then released from its sporangium and the time when it has just finished retracting its flagellum as a period in the life of *B. emersonii* that is most critical but still essentially uncharted. Our long-range goal is to establish what kinds of changes occur along the line and how plastic they may be, to find out whether "points of no return" are detectable, and to learn something of the mechanisms involved therein at any level of organization that we can handle. In the following pages, we will present an account of some of our observations, experimental data, and interpretations about the spore; we intend that it be viewed as a progress report of studies that will later be published in full elsewhere.

Preparation of Uniform Suspensions of Swimming Spores

Experience has repeatedly revealed (see Cantino and Lovett, 1964; Cantino, 1966; Lovett, 1967, for reviews) the practical utility of synchronized, single-generation cultures of *B. emersonii* for developmental studies. Since such cultures can be exploited for purposes of spore production (cf., Murphy and Lovett, 1966), and since all that follows in this essay is concerned with spores, it becomes important to review the whole procedure briefly.

Swarmers are germinated on PYG agar (Difco Labs., Detroit, Mich.); the first generation of OC cells derived therefrom is flooded with water and the second generation of spores collected. The cycle is repeated until a suitably dense population of swarmers is obtained. The suspension finally selected for starting liquid cultures is passed through filter paper to remove germlings and any plants dislodged from the agar surface; they are trapped and held effectively among the fibers while motile spores pass through unhindered. The absolute density of the spore suspension is established with a Coulter Counter, and the number of viable spores therein is established by making transfers to PYG agar and scoring for mature OC plants. The age of a population of swimming spores can also be controlled precisely; for example, if swarmers are selected which have been discharged within a ten-minute period, the synchrony of the suspension is defined with certainty, for no spore can be more than ten minutes old. Finally, these spores of known age and viability are used to produce submerged, synchronized, single generations of OC cells, either (a) in flasks containing PYG and phosphate (PYG-P) or phosphate-citrate (PYG-PC) buffer (Goldstein and Cantino, 1962; Cantino and Goldstein, 1967), or (b) in water-jacketed spinner flasks (Bellco Glass, Inc., Vineland, N. J.) agitated by a rotating magnet (Murphy and Lovett, 1966). Using spinner flasks, an air stream supplemented with O_2, and PYG-P, Murphy and Lovett showed that thalli developed uniformly and differentiated synchronously into mature OC plants; the latter were then induced to produce and release spores by replacement of spent media with a salts solution.

Questions of Variability in Populations of Swimming Spores

We have stressed the demonstrable usefulness of single-generation culture techniques for synchronous production of motile spores. However, what can be said about the *uniformity* of the swarmers so produced? There was little reason for Murphy and Lovett (1966) to ask this question; but for our purposes—wherein the spore itself is central—it may have to be posed with every new experiment.

Actually, it was shown years ago (Cantino and Hyatt, 1953) that spores derived from all four phenotypes of *B. emersonii* (Resistant sporangial (RS)-, Late colorless (LC)-, Orange (O)-, and ordinary colorless (OC) plants) did exhibit variability, both in their viability and in the frequency with which they produced these four thallus types in subsequent generations. Other investigations (Cantino and Horenstein, 1956) revealed that the spore's internal population of gamma particles was also variable: i.e., that the average number of gamma particles/spore was not constant but, instead, was related to the parental phenotype from which the spores were born; that even within a population of swarmers from any one type of parent, the distribution curve for gamma particles/spore was broad; and that the number of gamma particles/spore could be altered by prior treatment of parental spore progenitors with metabolic inhibitors. Thus, zoospore populations from *B. emersonii*, randomly collected, should be expected to possess potential for some variation in their internal architecture, viability, and capacity for genesis of progeny with differing phenotypes.

Now it must be emphasized that some of this potential heterogeneity is not manifested by spores prepared according to the preceding section. For example, peptone tends to suppress formation of O cells; also, because of their longer generation times, both LC and O cells are effectively bypassed when OC cells on PYG are flooded to provide suspensions of spores with which synchronous liquid cultures are started. Presumably, the same argument then applies to such submerged cultures when they, in turn, are used as the source of spores; the fact is, however, that direct evidence for this has been lacking. And even if we limit our attention to one type of spore progenitor, some pressing questions still arise. Do properties such as size, weight, fine structure, chemical composition, and physiological activities of spores derived solely from OC cells follow a normal distribution curve? Or, alternatively, might such spore populations consist of two or more distinct classes and thus yield bi- or even multi-modal distribution curves for some of these characteristics?

Table I

Viability of the spores of *B. emersonii* after two hours in liquid media

Viable spores/total spores (%)			
Medium PYG		Medium PYG-PC*	
Light	Dark	Light	Dark
55	57	54	60

*PYG-PC: Difco PYG containing 1.1 × 10^{-2}M Na_2HPO_4 and 3 × 10^{-3}M citric acid, pH 6.6 (Cantino and Goldstein, 1967).

We have begun to study this question and are accumulating evidence that spores, derived from OC cells grown to maturity in either solid or liquid media, are in fact *not* all alike. Consider the following experiment on viability. OC cells were grown on PYG agar and flooded with water when ca. 30 per cent had discharged. The spore suspension was pre-chilled at 1° C. for two hours; its density was estimated with a Coulter Counter. Liquid PYG-PC and ordinary Difco PYG at 24° C. were inoculated (6.96 x 10^6 spores/600 ml.), incubated in light (Fluorescent, 500 f.c.; Gold-

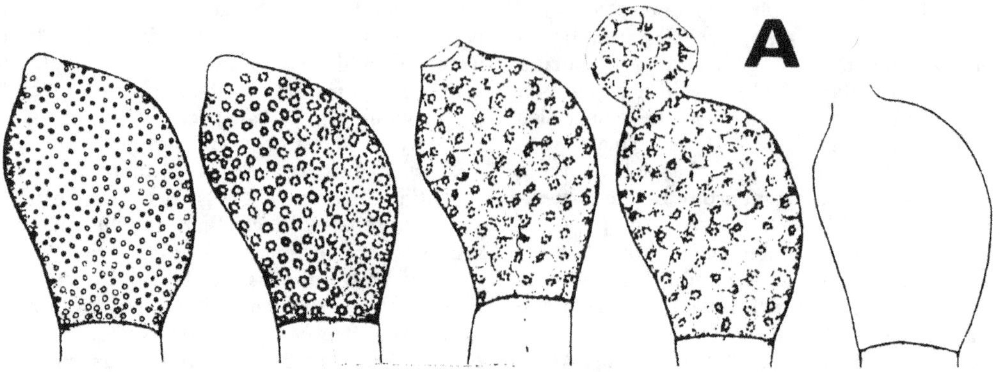

FIG. 2A–2B. Fig. 2A. Sporogenesis in the zoosporangia of *B. simplex* as sketched by Miss Matthews in her original description of the genus *Blastocladiella* (redrawn from Matthews, 1937).

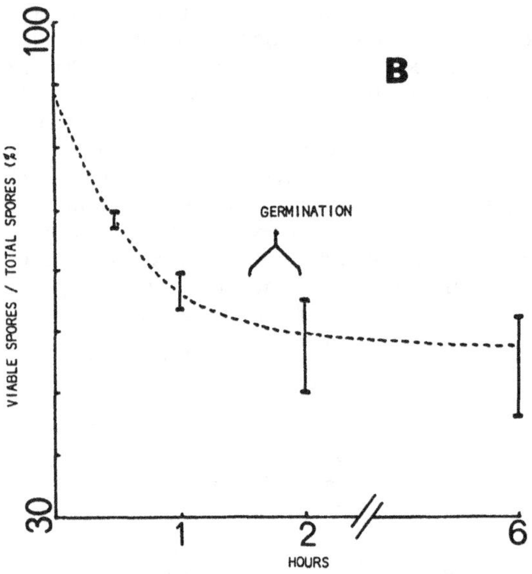

FIG. 2B. Kinetics of the loss in viability of spores of *B. emersonii,* derived from OC cells grown in spinner flask cultures and inoculated into media PYG-P and PYG-PC. Vertical bars represent ranges of values in various experiments. See text for details.

---

stein and Cantino, 1962) and darkness, and gassed with 3600 ml. air/minute. After two hours, samples were transferred to plates of PYG and checked for viable spores by counting the first generation OC cells obtained thereon. In all instances, only ca. half the spores survived (Table I).

The kinetics of this loss in viability was then established. Synchronous cultures of OC cells were grown in spinner flasks according to Murphy and Lovett (1966); spore formation was induced by replacing spent medium with their "½ DS Salt Solution." The spore suspension was pre-chilled at 1° C. for two hours and inoculated into new spinner flasks (8 x 10^4 spores/ml.) containing either PYG-P or PYG-PC. Immediately after inoculation and at intervals thereafter, cultures were sampled and checked for total cells (Coulter Counter), viable cells (on PYG agar), and the condition of the spores (microscopically). During the first two hours, about half of the spores disintegrated and were no longer scored by the Coulter Counter (Fig. 2B); the remaining half began synchronous growth in normal fashion. This was true for both media. The viable and Coulter counts corresponded almost exactly at zero, two, and six hours. Judging solely from this particular criterion of viability, all members of a population of pre-chilled spores of *B. emersonii* are *identical* if transferred to solid PYG either directly or after a quick dip in liquid PYG containing phosphate or phosphate-citrate; yet, this same population consists of *two classes*—viable and nonviable, roughly half and half—when forced to germinate in these two media. Clearly, spores produced by OC cells grown in submerged, synchronized, single-generation cultures *are not all identical*! The unavoidable question then arises: are the details of fine structure described by Cantino et al. (1963; see also, Fig. 1), Fuller (1966), Reichle and Fuller (1967), and Lessie and Lovett (1968) characteristic features of *all* the spores of *B. emersonii*? Or, does some organelle exhibit an alteration in structure, position, or number in some of the spores derived from OC cells? The issue is provocative enough to motivate the needed search.

Flagellar Activity and Anaerobic Metabolism in Swimming Spores

By what mechanical means do zoospores of *B. emersonii* push themselves about with their flagella? Probably the most that can be said at the moment must be based on Couch's (1941) conclusion that the spores of its close relative, *Allomyces*, are propelled by transmission of waves through the flagellum in one plane and one direction with respect to the spore's body. As for energetics, what transduces chemical energy into motion? In the spore of *B. emersonii*, the intimate tie-in between its single mitochondrion and the sheathless end of its flagellum strongly suggests that the transducing "device" must be centered in this linkage. But once again, we lack experimental evidence that these two structures are functionally connected by an energy-transfer mechanism. And third, since the spores of *B. emersonii* can swim about continuously for quite a while with little apparent relaxation, an associated question also arises: what source of chemical energy provides its long-range maintenance requirements, and what enzymatic machinery makes it available? Here, perhaps, we know enough to justify a good guess: namely, that the important substrate for energy production is a sizeable pool of glycogen-like polysaccharide (see Cantino and Lovett, 1964, for references in this paragraph). The spore has considerable capacity for endogenous formation of lactic acid (ca. 0.25 μmole x mg. dry wt.$^{-1}$ x hr.$^{-1}$). Although metabolism of the polysaccharide has not been followed in the spore itself, release of lactic acid is related mole for mole to utilization of this polymer by RS plants. The specific activity of the spore's glucose-6-phosphate dehydrogenase is greater than that at any stage during the exponential growth of plants. And exogenous glucose does not increase a spore's endogenous Q_{O_2} or output of lactic acid, even though glucose is consumed (ca. 1.0 μmole x mg.$^{-1}$ x hr.$^{-1}$). All of this is consistent with our suspicion that the spore's glycogen-like polysaccharide serves as the main source of phosphorylated hexose units and energy metabolism for the motile swarmer, and that "Zwischenferment" may constitute one route for initial disposition of the carbohydrate. Finally, to the arguments above, we can add the results of a simple yet informative experiment. Equal quantities of spores were placed in each of two vessels containing PYG-PC at 24° C. One was agitated continuously with air, the other with N_2. In the former, viable spores germinated after ca. 90-120 minutes (as in Fig. 2B); in the latter, all spores that did not die or explode were still swimming some six hours later!

Although there are numerous reports (with some exceptions; see Sleigh, 1962, for discussion and references) that oxygen is necessary for ciliary activity in various kinds of cells, we must tentatively conclude from the available facts that in the spores of *B. emersonii*: (a) anaerobic metabolism of a glycogen-like polysaccharide, probably *via* glucose-6-phosphate dehydrogenase, can provide most if not all the energy needed for flagellar activity; (b) that partial anaerobiosis *per se* cannot cause the spores to stop their swimming; and (c) that continuous agitation under low oxygen tensions promotes continued swimming (i.e., prevents germination) by removing something from the spore's surface which is normally required for flagellar retraction under these semianaerobic conditions. The significance of this last speculation will become evident later on in this report.

From Swimmer to Cyst; Flagellar Retraction in **Blastocladiella emersonii**

Flagellar Retraction

The central issue to which we will address ourselves in the remaining pages is simply posed, though far from simply answered: how does the swimming spore of *B. emersonii* transform itself into a rounded cyst? In 1953, Cantino and Hyatt emphasized that it did not simply shed its tail but, in fact, retracted it. Ten years later, after more extensive study, we said (Cantino et al., 1963) that when the spore stopped swimming and settled down, it became rather spherical, vibrated its flagellum, and then straightened it out such that it took on the aspect of a rigid structure; that then the flagellum swept around in a wide arc, and when this happened, the nuclear apparatus—the nucleus and its associated cap—rotated within the cell through almost one full turn; and that simultaneously, the flagellum got progressively shorter and eventually disappeared within the cell. Due stress was placed upon a critical observation: that during rotation of the nuclear apparatus, the spore itself did *not* rotate.

When these results were published, we had not seen the retracted flagellum in thin sections of the spore. But a short time later (oral presentation, symposium on Morphogenesis in Fungi, International Botanical Congress, Edinburgh, Scotland, 1964) Lovett showed unpublished pictures of the flagellar axoneme—for much of its full length—within the spore. Later, Fuller (1966) published pictures showing "9 + 2" arrays in sections of spores which had taken in their flagellar fibers. Finally, since then we too have seen the full lengths of retracted tail pieces within spores *in vivo*, using phase microscopy. There is no question,

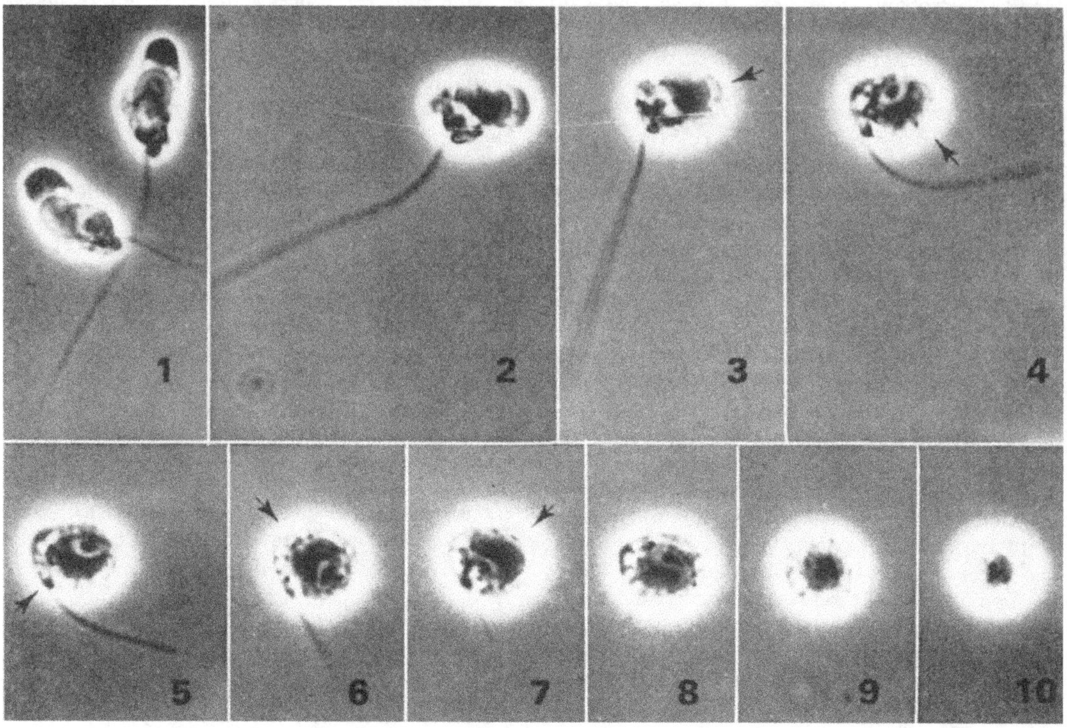

FIG. 3. Rotation of the nuclear apparatus during flagellar retraction by the spore of *B. emersonii*. Spores were prepared according to a standard procedure we devised for synchronization of spore "ontogeny." The method is defined and discussed later on in this report. Conditions employed (see Table III for procedures and definition of terms) were: (S) = H_2O; (W) = O; (T) = 60; (T') = 2.5; (P) = 10^6. Pictures were taken ca. 15 seconds apart; all spores are in their original positions on the slide. The arrow points to the broad top side of the nuclear cap, which lies anteriorly adjacent to the nucleus (light colored) with its nucleolus (dark spherical body therein). As the amoeboid spore (upper cell, Fig. 3–1) begins to round up, its flagellum vibrates without propelling the spore away (Fig. 3–3); immediately thereafter, the nuclear apparatus begins to rotate clockwise (Fig. 3–4) while the flagellum sweeps around in an arc and simultaneously begins to move into the spore (Fig. 3–4); as rotation continues, the flagellum gets progressively shorter (Figs. 3–5 through 3–8) until it is withdrawn *in toto* (Figs. 3–9, 3–10). The progressive change in the position of the rotating nuclear cap, and that of the nucleolus opposite it, is most clearly seen in Figs. 3–3 through 3–7. Note particularly, the point at which the flagellum makes contact with the periphery of the spore; it remains fixed at the same position throughout the sequence (Figs. 3–2 through 3–7), thus serving as an important "marker" which shows that the spore's external surface does not rotate during the internal gyration of the nuclear apparatus. Pictures were taken with a Wild Phase Microscope (40 X objective), B & L ribbon filament lamp, Panatomic-X film, and exposures of 1/10 second. Magnification as shown, approximately X1380.

then, that the axoneme finds its way into the spore of *B. emersonii* just before it germinates. With this conclusion, all of us—Fuller at California, Lovett at Purdue, and we at Michigan State—agree.

However, apparently unable to observe rotation of the nuclear apparatus during flagellar retraction, Reichle and Fuller (1967) questioned the reality of this phenomenon. We suspect that it was simply an unsuitable environment around the spores which inhibited nuclear rotation and thus prevented Reichle and Fuller from verifying our conclusion, for we know that unsatisfactory ionic balance can interfere with the process (see subsequent sections of this paper for a treatment of our methods for working with the spores). We take this opportunity, therefore, to provide pictorial evidence (Fig. 3) that rotation of the nuclear apparatus does, indeed, occur, and trust that this time-lapse series of photographs of our whirligig in action will be sufficiently convincing to bring our friends at California back into our camp.

In any case, the punctual occurrence of nuclear rotation in spores exhibiting flagellar retraction has now provoked so many questions that it has become a major focal point in our research. In the original report (Cantino et al., 1963), we suggested—with partial tongue in cheek—that rotation of the nuclear apparatus without concomitant movement of the spore itself might be interpreted to mean that the nuclear assemblage was winding in the tail. We would like to say, now, that we do not believe such a pulley mechanism is the answer. But neither can we accept Fuller's (1966) suggestion that the flagellum is retracted by absorption along the spore's surface. We have repeatedly watched the full length of the flagellum enter and then finally disappear *at a single and fixed point* on the spore's surface—i.e., under conditions where the flagellum, albeit curved in the form of an arc, *did not make contact with the spore periphery* at any point along its (the flagellum's) entire length except where it was originally attached. This point is well illustrated by the time-lapse series in Figure 3.

What, then, is our position? We believe that the envelope of the spore—continuous with the membranous sheath around the flagellum itself—by decreasing in surface area, *forces* the axoneme to be translocated to a more central

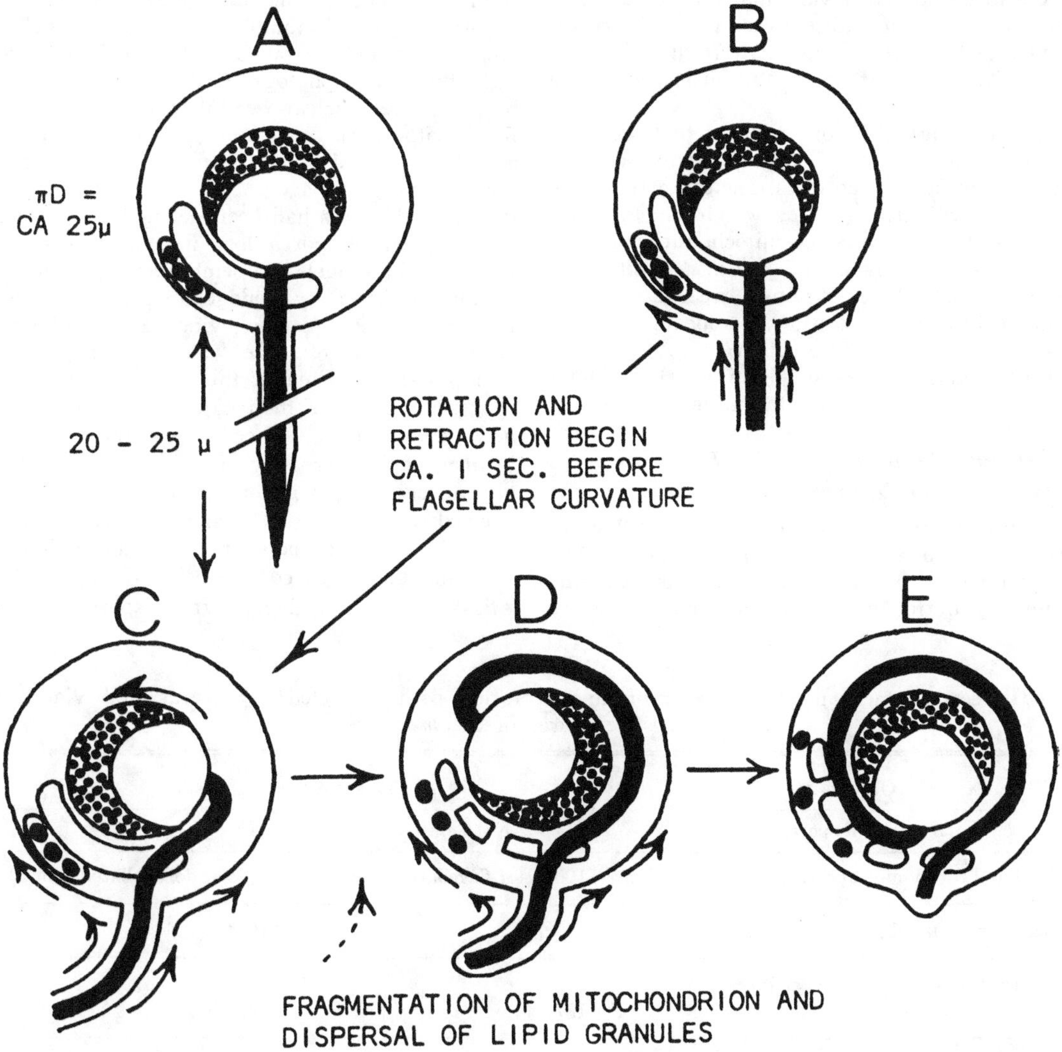

FIG. 4. Hypothetical mechanism of flagellar retraction by the spore of *B. emersonii*. The elongated colorless body represents the mitochondrion; the short body containing three dark spheres, the lipid sac. Note that the circumference of a rounded spore about to retract its flagellum is approximately equal to the length of the flagellum. See text for details.

position within the spore, and that it is the resultant *push* of the basal flagellar fibers (kinetosome) upon the nucleus, to which these fibers are connected, that causes rotation of the nuclear assembly. This point of view is illustrated in Figure 4. It is important to re-emphasize that during revolution of the nucleus and its cap, the spore itself does not turn; the arrows along the periphery of the spore in Figure 4 signify the supposed contraction of the membrane, not pivoting of the cell. The process can be likened, loosely speaking, to what would occur if a strong but flexible wire, extending through the neck of a 500-ml. Florence flask and firmly attached to a tennis ball therein, were to be pushed inward into the vessel. The ball, especially if bathed in a viscous lubricant and free to move, would rotate if the wire was not kinkable yet flexible and firm enough to push the ball around.

In our original report, we said that the lipid granules alongside the mitochondrion do not move significantly until the nuclear assemblage has finished turning, after which the granules become dispersed and the mitochondrion begins to change in shape and texture. With improved optical conditions, we have now modified this interpretation: it appears (Fig. 4) as if mitochondrial fragmentation and dispersal of lipid can commence slightly earlier, about midway during nuclear rotation or even sooner.

Nuclear Behavior and Viability in Deflagellated Spores

If the postulated mechanism of flagellar retraction is correct, a spore whose flagellum has been removed by means which do not alter the integrity of its other structures and activities should never display rotation of its nuclear apparatus; to demonstrate this experimentally would offer direct evidence for our premise. Indeed, this has been done. The spore's flagellum can be knocked off mechanically by bubbling air vigorously—i.e., with a turbulence much greater than that associated with routine aeration of a culture—through a few ml. of a thick suspension of spores. The results of one such experiment are shown in Table II. Spores derived from OC cells on PYG agar were filtered, chilled to 1° C., centrifuged two minutes at 1300 x G, and resuspended in ca. 10 ml. of 10^{-3}M $MgCl_2$, $CaCl_2$, and NaCl, and 5 x 10^{-3}M KCl. Half of the suspension was used as a control; the other half was aerated vigorously three minutes to remove flagella. Replicate samples from both suspensions were (a) studied and photographed *in vivo* to determine whether nuclear revolution occurred; (b) fixed with osmic acid and scored for normal and deflagellated spores; and (c) tested for viability on PYG agar. About 61 per cent of the aerated spores had been deflagellated; none of the control spores lost flagella. *In vivo,* rotation of the nuclear assembly *never occurred* in the mechanically deflagellated spores (see Fig. 5 for photographic evidence); it *always* occurred in control spores (as shown previously in Fig. 3). Finally, the viability of these spores (in terms of their capacity to produce a new generation of OC cells) and the viability in turn, of their progeny (in terms of the number of second-generation clones of OC cells derived from the first generation OC cells) were the same for populations of deflagellated cells and unaerated control cells. Thus, mechanical removal of flagella from spores of *B.*

Table II

The effects of mechanical deflagellation upon rotation of the nuclear apparatus and viability among the spores of *B. emersonii*

Observation	Treatment of spores	
Number of spores which:	Not aerated	Aerated
Exhibited rotation of nuclear apparatus.	100% (of flagellated spores)	0% (of deflagellated spores)
Possessed no flagella.	0%	61%*
Yielded first generation OC cells on medium PYG.	137/plate	131/plate
Yielded second generation clones (from first generation OC cells above).	93%	92%

*Up to ca. 100% of a population of spores have been deflagellated by this procedure in subsequent experiments.

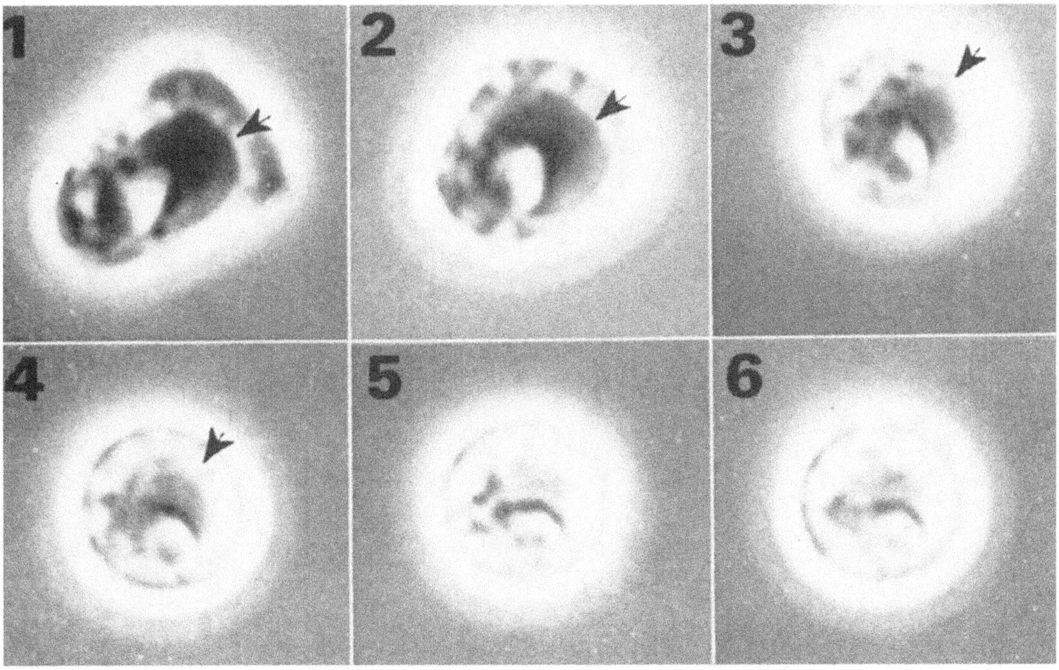

FIG. 5. The absence of rotation of the nuclear apparatus in a demonstrably deflagellated spore of *B. emersonii* during its "rounding up" stage prior to germination. See text for explanation and legend to Fig. 3 for photographic and microscopic methods, significance of arrows, structural details in the spores, etc. Magnification as shown approximately X2630.

emersonii does not affect their capacity to germinate nor their potential for producing normal progeny for at least two successive generations, but it does destroy their capacity to rotate the nuclear apparatus. These results provide substantial evidence for our theory about the causal mechanical relationship between nuclear rotation and flagellar retraction.

Release of Binding Sites; a Prerequisite for Retraction of Flagella?

If the flagellum is retracted as visualized in Figure 4—whereby the mitochondrion and its associated lipid sac are left behind as the axoneme is translocated inward because of its attachment to the nucleus *via* the kinetosome—the mitochondrial-lipid complex must be held back in its original position by some mechanical connection. Whatever its exact nature and location, we shall call it "Binding Site L" (Fig. 6). By the same token, since the kinetosome is translocated from its original position in a mitochondrial canal, we shall assume that a second "Binding Site M" (Fig. 6)—perhaps at the banded rootlets, perhaps elsewhere—must be broken. Thus, a rupture at Binding Site M but not at Binding Site L would characterize what we shall arbitrarily label the "normal path" (sketched in Fig. 6; illustrated photographically in Fig. 3).

However, an "alternate path" (Fig. 6) is possible. For example, if spores are treated with sulfonated compounds such as Biebrich Scarlet or sodium lauryl sulfonate (see legend, Fig. 6), the mitochondrial-lipid complex moves along with the kinetosome (Fig. 6A–6F). In this instance, Binding Site L would have been broken while Binding Site M remained intact. Our working hypothesis is: first, that Binding Site L involves the membrane network (cf. Reichle and Fuller, 1967) which ramifies about the lipid sac and mitochondrion; and second, that the strength of this network can be so weakened by sulfonate-type detergents and related compounds that it will break preferentially over Binding Site M. Much further work on this phenomenon is needed; some of it is now in progress.

Spore Flocculation—a Reflection of Membrane Change Prior to Retraction?

If the spore's membrane must contract to effect flagellar retraction, then does the spore possess this capability from the moment it is born, or does it acquire this capacity only at some later stage in its life history? If it is

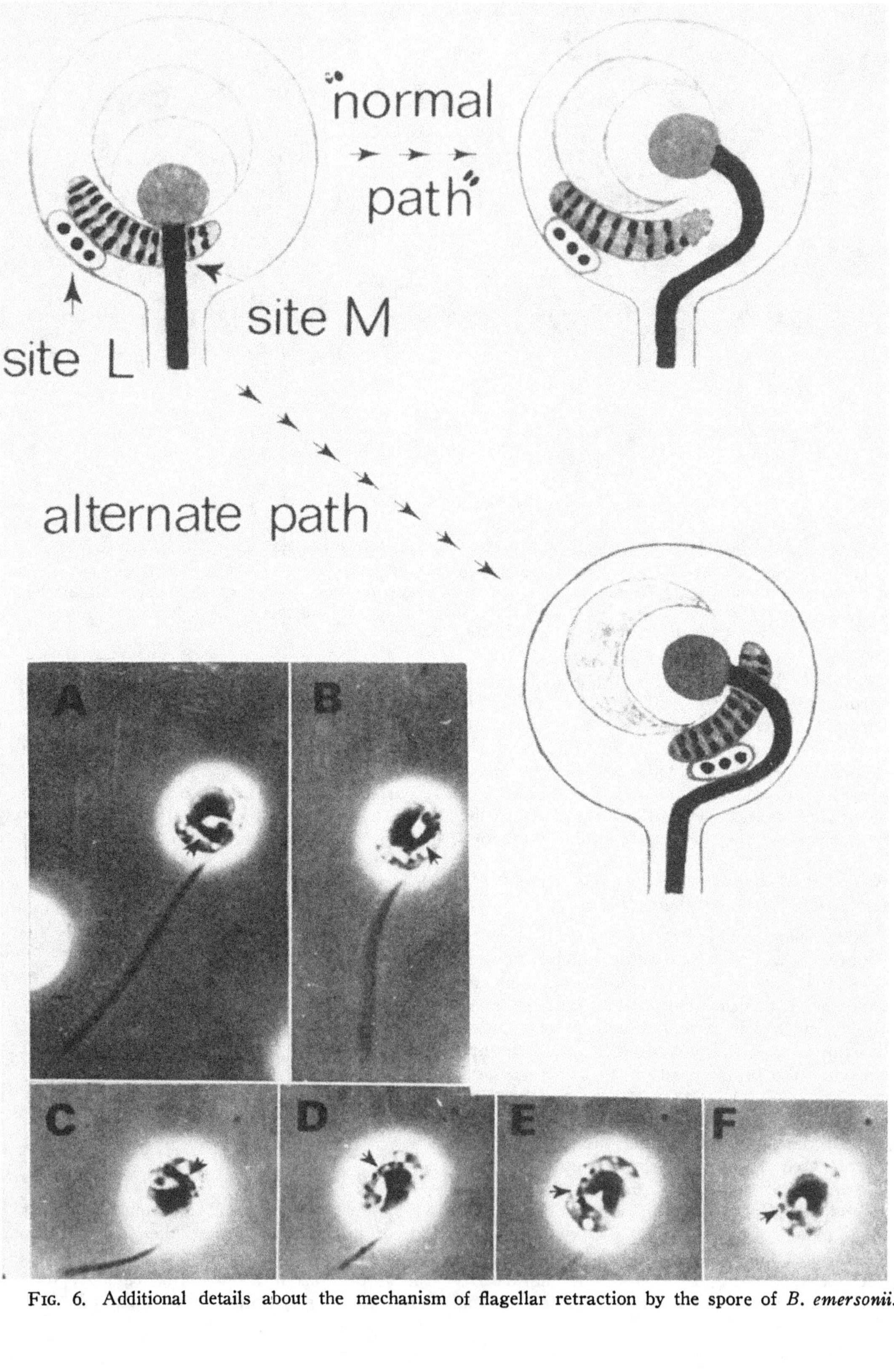

FIG. 6. Additional details about the mechanism of flagellar retraction by the spore of *B. emersonii*.

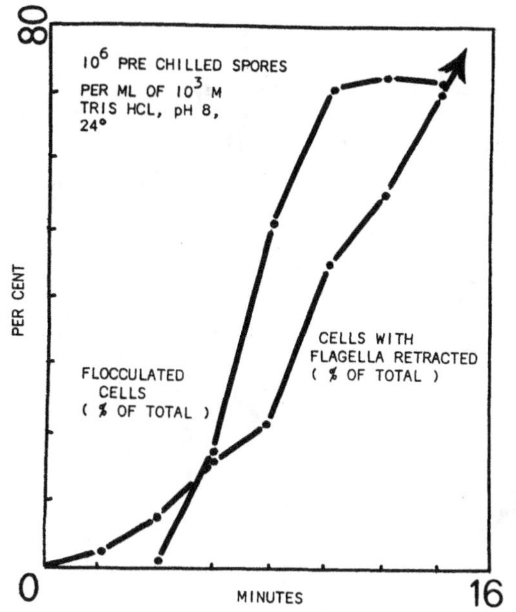

FIG. 7. The time course for cell flocculation and flagellar retraction in a population of spores of *B. emersonii*; see text for details.

the latter, a detectable change in quality of th swarmer's membrane might ensue. We have gotten some indirect evidence which suggests that this may in fact be so. At some point in time before a spore becomes immobile preparatory to retracting its flagellum, a change occurs—as seen microscopically *in vivo*—which imparts upon it a capability to agglutinate with sister spores. This observation was quantified by incubating pre-chilled spores at 24° C. in TRIS.HCl, sampling at intervals, and scoring the population for cells that had flocculated and/or retracted their flagella. The results (Fig. 7) showed that, except for an initial period when flocculation-competent spores are still so few in number that chances of collision among them are very small, flocculation precedes flagellar retraction. We must now determine more precisely the temporal aspects of this change in the spores of *B. emersonii,* for the appearance of visible attraction among spores may be due to genesis of surface binding sites, and fabrication of the latter could reflect a membrane change associated with the onset of competence for retraction.

Standardized Procedure for Studies of a Spore's Life History

The numerous questions that arose out of the work discussed in the preceding pages had also underlined our growing need for a procedure with which the spore could be studied, experimentally and under controlled conditions, along the full course of its life history. Such a method was developed and, although we doubt that it has reached its final stage of evolution, it has permitted us thus far to make a population of spores proceed (see Fig. 8) in essentially synchronous fashion from swimming stage (A) through a spherical stage (a), a transient ameoboid stage (B), and thence—by way of a "rounding up" process (C through E)—to retraction of flagella (C, D, E) and germination (F). The essential steps in the procedure are outlined in Table III. Most of the sequence (a to F) is followed *in vivo* on the glass slide controlled at constant temperature with a water-cooled heat sink (Fig. 9A). The method can, of course, be scaled upward; by substituting a larger chamber for the glass slide, chemical, enzymological, and other assays can be made and put on a per-cell basis, and thus related to any stage in the spore's development.

With the slide procedure, we established that the degree of synchrony among the "differentiating" spores—i.e., the time interval over which the spores pass through stages B to F in Figure 8—is, in large measure, a function of (T). For example, with (S, S′) = TRIS. HCl, 10^{-3}M, pH 8, (P) = 5 x 10^5/ml., (T′) = 15, and (C°) = 24, none of the spores will pass beyond stage (B) in Figure 8 if (T) approaches zero; increasing numbers of them do so as the pre-chill period is progressively increased. Thus, by making plots of (T) vs. % (R) at (T′) on X and Y axes, respectively, families of standard curves can be established whose slopes and X-axis intercepts are a function of (S), (S′), (T), and (C°). With this many variables, each system will have its own

Fig. 6, top. Two alternate paths for the disposition of the lipid sac-mitochondrion complex during flagellar retraction. See text for details and explanation of binding sites L and M. Fig. 6, bottom. Pictorial documentation for the "alternate path" shown in Fig. 6, top: the series reveals *simultaneous* rotation of the lipid sac (arrows point to a chain of lipid bodies within the sac which serve as markers) and the nuclear apparatus during flagellar retraction. Spores were prepared according to a standard procedure for synchronization of spore ontogeny, defined and discussed in a later section of this paper, whose parameters (see Table III for definition of terms) were: (S) = H_2O; (W) = O; (T) = 60; (T′) = 2.5. See text for further explanation, and legend, Fig. 3, for photographic and microscopic methods. Magnification as shown, ca. X1380.

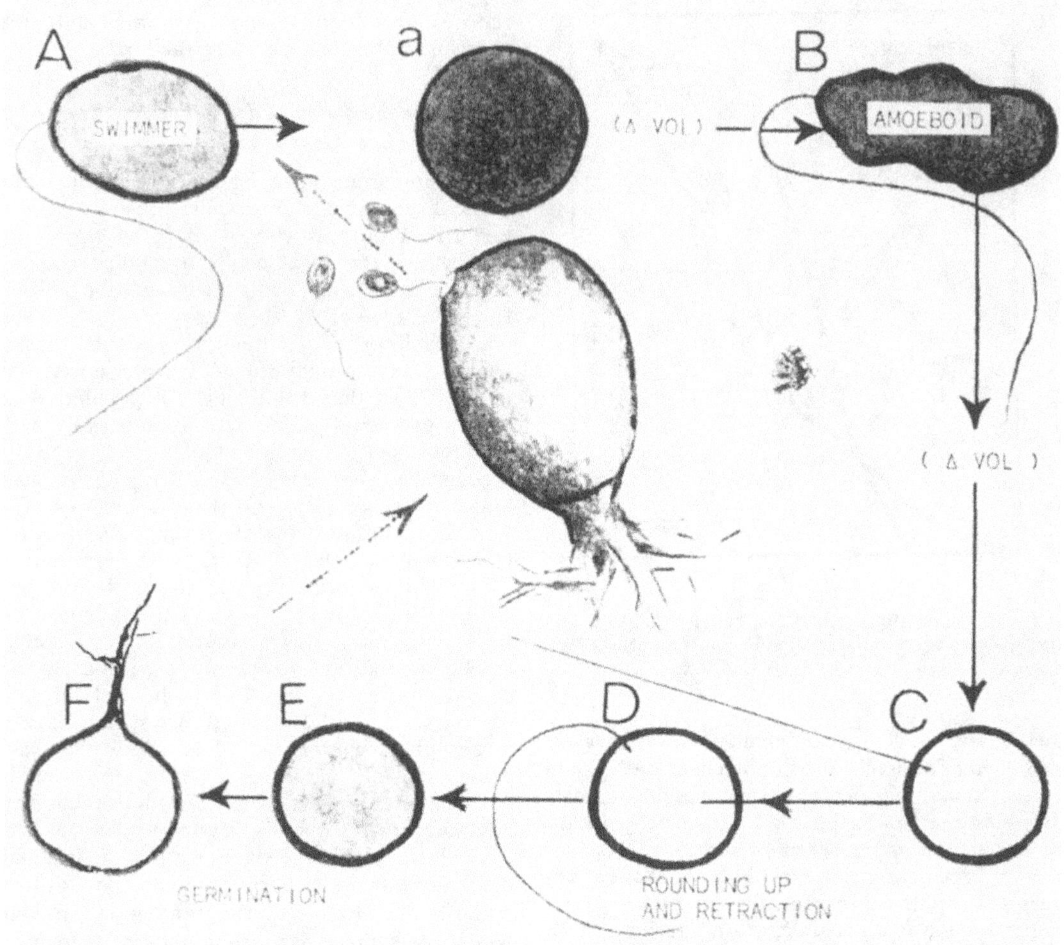

FIG. 8. The principal stages in the "ontogeny" of a spore of *B. emersonii*, as deduced from studies of populations of spores treated by the standard procedure for synchronization of spore ontogeny (see Table III).

particular sensitivities and must be worked out empirically in detail. To illustrate: the number of permissible washes with (S′) depends upon the system chosen; with (S′) = NaCl and KCl at 2 x 10^{-3}M, (W) can be 0 or 1, whereas with (W) = 2, many spores swell and burst. Or, to use one more case in point, the composition of (S)—especially if it contains organic substances—can adversely affect the late stages in the spore's ontogeny; with (S) = H_2O, (S′) = 0.8% sucrose, 2 x 10^{-4}M $MgCl_2$, 10^{-3}M KCl and NaCl, (W) = O, and (T) = 5, spores round up satisfactorily but then exhibit obvious "difficulty" in retracting their flagella.

We have begun to exploit the system; some results obtained with it thus far are summarized below.

Other Events That Precede Flagellar Retraction—Reflections of Changes in Membrane and/or Mitochondrial Activity?

We have assumed that in the process whereby a spore becomes competent to retract its flagellum, its membrane is also altered for this special task. Achievement of flocculation competence may be but one reflection of this commitment. If the membrane which surrounds the spore of *B. emersonii* resembles that of other organisms—i.e., is composed of a bimolecular phospho-lipid layer with a leaf of protein on each side and probably other macromolecules as well—it should exhibit some chemical or structural alterations. This might, in turn, provoke (or permit) other things to happen at

Table III
Standard procedure for synchronization of spore development

(1) Grow OC cells on solid medium PYG.
(2) Flood OC cells with appropriate solution (S) to induce spore discharge.
(3) Filter suspension of spores in (S) through filter paper and collect over ice bath.
(4) Centrifuge chilled spore suspension 2.5 minutes at 1300 x g.
(5) Resuspend spores in solution (S'); (S') may or may not be the same as (S).
(6) Wash spores one to three times (W=1, 2, or 3) with (S) or (S') by centrifugation, or use spores directly without wash (W=0).
(7) Pre-chill spores in ice bath for (T) minutes.
(8) Determine number of spores/ml (P) with Coulter Counter.
(9) Transfer pre-chilled spores to glass slide on temperature-controlled stage.
(10) Incubate for (T') minutes at known temperature (C°).
(11) Examine spores *in vivo*, and/or fix with osmic acid and score for % cells "rounded up" (R), or for other stages (Fig. 8) in development.

its surface: changes in permeability, enzyme components, metabolic activities, membrane plasticity, and perhaps even the volume of the spore.

With our standard procedure, some effects of exogenous substances on spore ontogeny have been established. For example (Fig. 9B), in a TRIS.HCl system, K^+ augments the beneficial effect of pre-chilling upon synchronous development of the spore; the degree to which it augments is inversely related to the duration of the pre-chill period; the optimum concentration of K^+ is ca. 10^{-3}M; and K^+ cannot be replaced by Na^+.

With another medium (Table IV), Ca^{++} (as contrasted with Mg^{++}) decreases the rate at which a population of spores will pass through the amoeboid stage and round up. In fact, with (T') less than 60, and (P) suitably high, we are using this approach to prepare pure populations of amoeboid cells for comparative physiological and chemical studies.

Also, 2, 4 dinitrophenol (5 x 10^{-5} M if added during pre-chill), EDTA (if added at 10^{-3}M immediately after pre-chill), and anaerobiosis, all inhibit rounding up; they block the (B)-to-(C) transformation (Fig. 8) and, in effect, provide a means for synchronizing the conversion of a population of pre-chilled spores into amoeboids.

At this stage in our work, the foregoing results are individually interpretable in various ways. The effect of cations could be related to a Na- and K-activated, Ca-inhibited (Rossini, et al., 1966), or Mg-activated (Whitaker, 1966) microsomal ATPase, to respiratory activity associated with (and affected by) a mitochondrial type of DNP-sensitive Ca (but not Mg) uptake (Lehninger, 1966), to differential association of cation-sensitive enzymes with inner and outer membrane layers (Wolman and Bubis, 1966), etc. But, taken all together, the effects of Ca (vs. Mg), K (vs. Na), 2, 4 DNP, EDTA, and anaerobiosis upon the spore of *B. emersonii*, along with preliminary observations (Table V) which suggest that O_2 consumption by amoeboids may be much lower than that of spores which are rounding up and germinating, lead us to speculate that the swarmer's single giant mitochondrion plays some important role in a spore's morphogenetic transformations.

Other Changes Associated with "Rounding Up"—Further Reflections of Membrane Change?

The spore of *B. emersonii* has neither a rigid wall to hold its membrane in nor a central vacuole. We presume that its single outer boundary is subjected to very little hydrostatic pressure from within for maintenance of the cell's rigidity. From microscopic measurements, we conclude that a decrease in spore volume is normally associated with both conversion of pre-chilled swarmer to amoeboid cell (a to B; Fig. 8) and "rounding up" (C to E; Fig. 8).

Other observations (see Fig. 10) are also more or less consistent with this thought about a membrane change. We have detected an unknown fluorescent material which is liberated from spores at a stage preceding flagellar retraction. But we do not have direct evidence that it is localized on the cell surface or that it is released as a result of a change in the membrane.

The onset of flagellar retraction is also preceded by a structural alteration which we have followed with phase microscopy; namely, the formation of "vacuole"-like bodies (Fig. 10) which originate near the nuclear cap, migrate to the cell surface, and there seem to "break through" it, apparently release their contents, and then disappear. We have the impression that some of these migrating "vacuoles" arise from gamma particles. The phenomenon is somewhat reminiscent of the ways in which internal zymogen sacs, vesicles, etc. have been seen to travel to the outer membrane in other

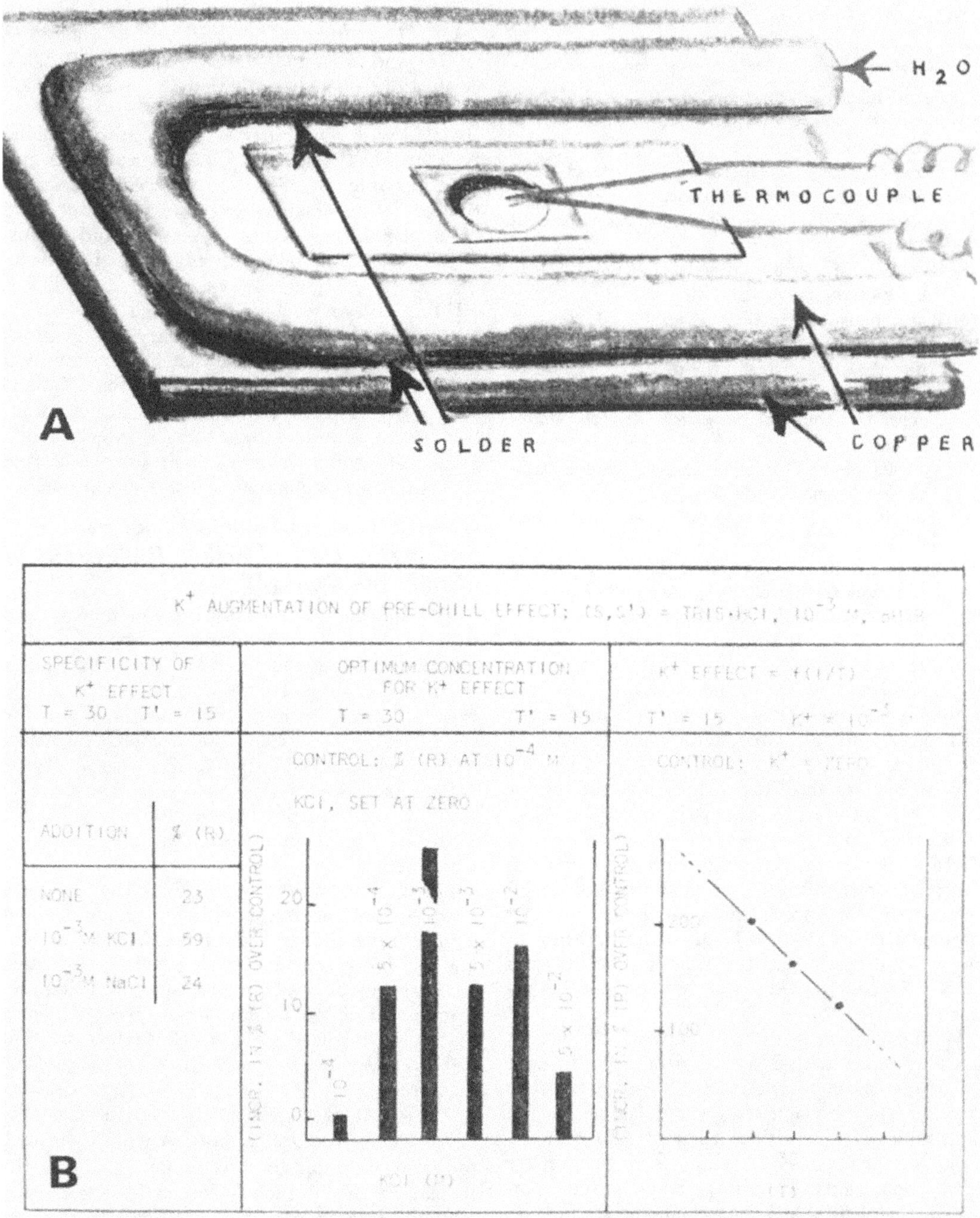

FIG. 9A–9B. Fig. 9A. Heat sink (slide holder) for temperature control of spore "ontogeny." Water of known temperature is circulated through a copper tube (0.8 cm. I.D.) slightly flattened and heavily soldered to a copper plate (11.5 x 6.5 x 0.1 cm.) containing a hole (1.3 cm. diam.) for transmission of the light beam through a microscope slide carrying a very fine copper-constantan thermocouple and a cover slip. Fig. 9B. The effect of potassium upon "rounding up" of the spores of *B. emersonii*, using the standard procedure for synchronization of "spore ontogeny" (see Table III); see text for details.

Table IV

Inhibitory effect of calcium on "rounding up," using the standard procedure for synchronizatioı of spore development in *B. emersonii* (see Table III)

Constants
(S) = H_2O (S') = 5 × 10^{-4}M Na_2HPO_4, 5 × 10^{-3}M KCl, 10^{-4}M Glutamate
(W) = 1 (T) = 0 (T') = 60 (C°) = 24 (P) = ca. 10^7

Counts made for:	Additions:	
	10^{-3}M $MgCl_2$	10^{-3}M $CaCl_2$
Total cells	326	372
(R) cells	109	12

Table V

Oxygen uptake by amoeboids of *B. emersonii*

Constants
(S, S') = 10^{-3}M TRIS.HCl, pH 8. (W) = 2 (T) = 0
(T') = 15 (C°) = 24 (P) = 1.8 × 10^7

Addition:	Q_{O2}*
None	21.0 14.9 14.1 (Ave. = 16.7)**
1% PYG	12.6 14.8 13.3 12.4 13.7 (Ave. = 13.4)**

*These measurements, made with an oxygen electrode, were completed before our standard procedure for synchronization of spore "ontogeny" had been worked out in detail. Although microscopic observations were not made at the time the measurements were taken, we feel reasonably sure that most if not all cells must have been in their amoeboid state under the environmental conditions employed.

**In previous manometric work (see Cantino and Lovett, 1964, for references), where spores were probably "rounding up" and/or germinating during Warburg runs of relatively long duration (as compared to the oxygen electrode runs above, where (T') = 15 at a maximum, the endogenous Q_{O2} was ca. 100; with added 1/2% PYG, ca. 200.

organisms whereupon they presumably spill their contents.

We also have evidence (Fig. 10) that rounding up is accelerated by some cytoplasmic dyes—for example, Nigrosin, Amido Black, and Biebrick Scarlet.

Finally (see Fig. 10), "rounding up" is also accelerated by thionine and methylene blue; notably, the oxidation-reduction potentials of their half-reactions are greater than for the sulfhydryl-disulfide couple. Conversely, "rounding up" is decelerated by anaerobiosis, $HgCl_2$, and mercurochrome, and by Janus Green and Vital Red; the oxidation-reduction values for the latter are *less* that for the sulfhydryl-disulfide couple. We are inclined to take the position, therefore, that a change in sulfhydryl content of the surface on *Blastocladiella's* motile spore is probably associated with the onset of its capacity to retract its flagellum. Perhaps the synchronous "rounding up" of the motile spore of *B. emersonii*—in effect, induced by increasing the temperature of pre-chilled spores—shares a good deal in common with the DNP-sensitive "synchronous rounding" (Tamura et al., 1966) of *Tetrahymena*, which is also induced by increase in temperature and which seems to involve changes in its sulfhydryl proteins.

In summary, the 2, 4 DNP-, K-, Ca-, EDTA-, and O_2-sensitive "rounding up" process which precedes flagellar retraction in the spore of *B. emersonii* is closely linked on a temporal basis with: (a) release of unidentified fluorescent substance; (b) genesis of flocculation competence; (c) formation of "vacuoles" which migrate to the surface of the spore and apparently release their contents; (d) a decrease in cell volume; and (e) a probable change in the sulfhydryl components on the spore surface.

Non-"Linear" Approaches to Biological Investigations; A Point of View

Many biologists before us have made the plea, in one form or another, that to understand

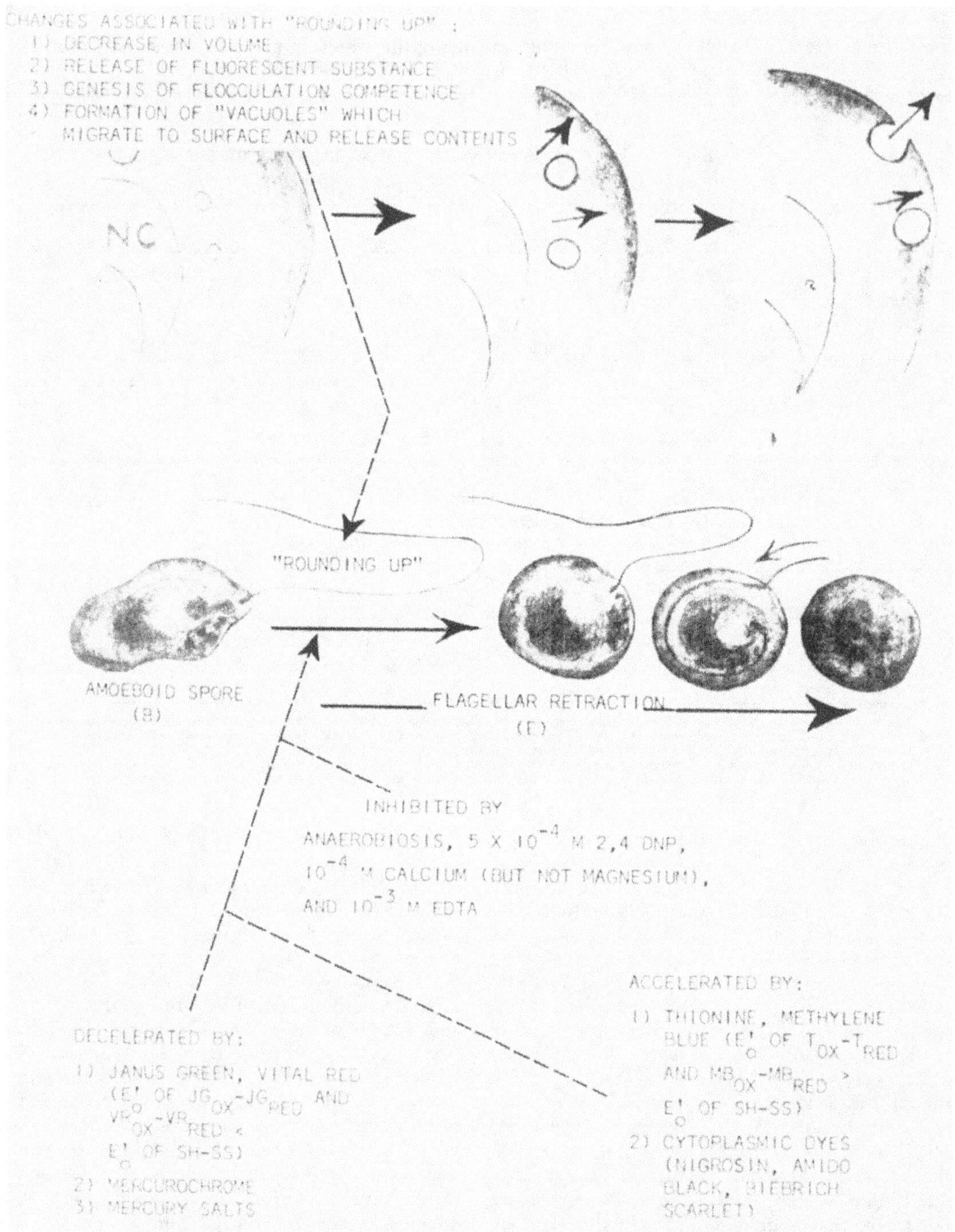

FIG. 10. A summary of various other demonstrable in vivo changes associated with spore "ontogeny," and the effects of some exogenous substances thereon; see text for discussion.

the nature of an intact living organism fully, it must be studied as it is, in its entirety. In our current work with spores of *Blastocladiella*, we have obviously cast our lot with this philosophy. Since our approach has been non-"linear"—to use a term somewhat in vogue in other quarters —we would like to take this opportunity to elaborate briefly upon the reasons why we think such an approach is important.

If it were our task to describe van Gogh's "The Orchard" for a listener, perhaps we'd first block out its grove of trees, their trunks arising in a grass of green and orange-red, and their boughs with blossoms white and pink extending skyward into a heaven blended of blue and horizontal clouds. We could then move on to define its other components, bit by bit, down to their finest details. But no matter how meticulous and accurate our verbal sketch, we would simply have transformed van Gogh's blend of strokes and color into a "linear" array of phonetic symbols; our listener could never have achieved the understanding that he would have had he himself looked at the painting. Indeed, if we had arbitrarily cut up the picture into its apparent components—trees, blossoms, clouds, grass—and then allowed our listener to view the pieces one by one in "linear" fashion, he could not have had the same experience that he would have had from a direct view of the intact masterpiece.

From the foregoing, there stem two important conclusions that deserve explicit emphasis: first, *with every interaction there is automatic synthesis*, the sort of synthesis which makes a painting a painting, a swimming spore a swimming spore; and second, with every attempt to understand something by a "linear" approach, there is *automatic loss of the effect of interaction* and, thus, automatic loss of the essence and uniqueness of the entity in question.

To homogenize a living cell destroys its very nature; yet, today, biologists frequently take this as an early step if not the first approach in efforts to understand the nature of the cell they have destroyed. Intrinsically inconsistent, this course is patently unreasonable. Why, then, has the incongruity evolved? It is because in recent decades, biology has been profoundly influenced by physical science, not only by its concepts but also by the success with which it has applied a "linear" approach to its own problems. The physical scientist works with natural phenomena in which complexities, ensuing from interacting systems, generally fall far short of those inherent in a living cell; this allows him much more freedom to use his "linear" approach. When the properties of some isolated system are modified by interaction with a second, he can often deal with the isolated interaction itself by circumscribing it within arbitrary limits; therefrom, he may eventually deduce a principle or law which closely approximates the "truth" (whether or not such laws stand the test of time is immaterial for our argument).

But in biology, primary errors first arise when we fail to recognize the details of some single interaction among the units of our system. These errors are then multiplied many fold by the great numbers of interactions which occur; the total error is increased well beyond the resolvable level for the system. Thus, when it is applied to experimental investigations of biological phenomena, the utility of the "linear" approach becomes limited at a very early stage; not to recognize this fact is to proceed along a dead-end street—and, moreover, without even having seen the road sign! Therefore, it seems to us that if his main concern is the living cell, the biologist has no choice but first to place his emphasis upon the totality and simultaneity of his system. He must experiment with the intact organism in its environment. In the process, answers then become the problems for subsequent experiments, and he can move successively to smaller and smaller units of organization with a minimum of approximations.

We would be among the very last to deny the great power and accomplishments of "molecular biology" in its sphere of influence. But we think, too, that there is increasing need to re-emphasize the continuing necessity for acute *in vivo* observations of the sort that Professor Couch has made so often in his long and productive career. We feel certain that he, in turn, would be among the first to agree that in this age of automatized science and "instant data," a biologist cannot afford to forget that a living cell is more than just a supernatant and a pellet.

LITERATURE CITED

Barrett, J. T. 1912. The development of *Blastocladia strangulata* n. sp. Bot. Gaz. **54**: 353–371.

Blondel, B., and Turian, G. 1960. Relation between basophilia and fine structure of cytoplasm in the fungus *Allomyces macrogynus* Em. J. Biophys. Biochem. Cytol. **7**: 127–134.

Butler, E. J. 1911. On *Allomyces*, a new aquatic fungus. Ann. Bot. **25**: 1023–1035.

Cantino, E. C. 1962. Transitional states of ribonucleic acid and morphogenesis in synchronous single generations of *Blastocladiella emersonii*. Phytochem. **1**: 107–124.

———. 1966. Morphogenesis in aquatic fungi. *In* G. C. Ainsworth and A. S. Sussman, eds., The

Fungi. Vol. II, 283–337. Academic Press, New York.

CANTINO, E. C., AND GOLDSTEIN, A. 1967. Citrate-induced citrate production and light induced growth of *Blastocladiella emersonii*. J. Gen. Microbiol. **46**: 347–354.

CANTINO, E. C., AND HORENSTEIN, E. A. 1956. Gamma and the cytoplasmic control of differentiation in *Blastocladiella*. Mycologia **48**: 443–446.

CANTINO, E. C., AND HYATT, M. T. 1953. Phenotypic "sex" determination in the life history of a new species of *Blastocladiella, B. emersonii*. Antonie v. Leeuwenhoek **19**: 25–70.

CANTINO, E. C., AND LOVETT, J. S. 1964. Non-filamentous aquatic fungi: Model systems for biochemical studies of morphological differentiation. Adv. Morphogen. **3**: 33–93.

CANTINO, E. C., LOVETT, J. S., LEAK, L. V., AND LYTHGOE, J. 1963. The single mitochondrion, fine structure, and germination of the spore of *Blastocladiella emersonii*. J. Gen. Microbiol. **31**: 393–404.

COTNER, F. B. 1930. Cytological study of the zoospores of *Blastocladia*. Bot. Gaz. **89**: 295–309.

COUCH, J. N. 1941. The structure and action of the cilia in some aquatic Phycomycetes. Amer. J. Bot. **28**: 704–713.

COUCH, J. N., AND WHIFFEN, A. J. 1942. Observations on the genus *Blastocladiella*. Amer. J. Bot. **29**: 582–591.

FULLER, M. S. 1966. Structure of the uniflagellate zoospores of aquatic Phycomycetes. Proc. 18th. Sympos. Colston Res. Soc. **18**: 67–84.

FULLER, M. S., AND REICHLE, R. 1965. The zoospore and early development of *Rhizidiomyces apophysatus*. Mycologia **57**: 946–961.

GOLDSTEIN, A., AND CANTINO, E. C. 1962. Light-stimulated polysaccharide and protein synthesis by synchronized, single generations of *Blastocladiella emersonii*. J. Gen. Microbiol. **28**: 689–699.

HARDER, R., AND SÖRGEL, G. 1938. Über einen neuen plano-isogamen Phycomyceten mit Generationswechsel und seine phylogenetische Bedeutung. Nachrichten Gesell. Wiss. Göttingen. **3**: 119–127.

KANOUSE, B. B. 1927. A monographic study of special groups of the water molds. I. Blastocladeaceae. Amer. J. Bot. **14**: 287–306.

KNIEP, H. 1929. *Allomyces javanicus*, n. sp., ein anisogamer Phycomycet mit Planogameten. Berichte Deutsch. Bot. Gesell. **47**: 199–212.

KOLE, A. P. 1965. Flagella. *In* G. C. Ainsworth and A. S. Sussman, eds., The Fungi. Vol. I, 77–93. Academic Press, New York.

LEHNINGER, A. L. 1966. Dynamics and mechanisms of active ion transport across the mitochondrial membrane. Ann. N.Y. Acad. Sci. **137**: 700–707.

LESSIE, P. E., AND LOVETT, J. S. 1968. Ultrastructural changes during sporangium formation and zoospore differentiation in *Blastocladiella*. Am. J. Bot. (in press).

LOVETT, J. S. 1963. Chemical and physical characterization of "nuclear caps" isolated from *Blastocladiella* zoospores. J. Bacteriol. **85**: 1235–1246.

———. 1967. Aquatic Fungi: *Allomyces* and *Blastocladiella*. *In* F. Wilt and N. Wessels, eds. Experimental Techniques of Development. T. Y. Cronwell Co. (in press).

MATTHEWS, V. D. 1937. A new genus of the Blastocladiaceae. Jour. Elisha Mitchell Sci. Soc. **53**: 191–195.

MINDEN, M. VON. 1915. Chytridiineae, Ancylistinaea, Monoblepharidineae, Saprolegniineae. Kryptogaenfl. Mark. Brandenburg. **5**: 193–630.

MURPHY, SR. M. N., AND LOVETT, J. S. 1966. RNA and protein synthesis during zoospore differentiation in synchronized cultures of *Blastocladiella*. Dev. Biol. **14**: 68–95.

PETERSEN, H. E. 1910. An account of Danish freshwater Phycomycetes, with biological and systematical remarks. Ann. Mycologici. **8**: 494–560.

REICHLE, R. E., AND FULLER, M. S. 1967. The fine structure of *Blastocladiella emersonii* zoospores. Amer. J. Bot. **54**: 81–92.

REINSCH, P. F. 1878. Beobachtungen über einige neue Saprolegnieae, über die Parasiten in Desmidienzellen und über die Stachelkugeln in Achlyaschläuchen. Jahrb. wiss. Bot. **11**: 283–311.

RENAUD, F. L., AND SWIFT, H. 1964. The development of basal bodies and flagella in *Allomyces arbusculus*. J. Cell Biol. **23**: 339–354.

RITCHIE, D. 1947. The formation and structure of the zoospore in *Allomyces*. Jour. Elisha Mitchell Sci. Soc. **63**: 168–206.

ROSSINI, L., COHEN, H. P., HANDELMAN, E., LIN, S., AND TERZUOLO, C. A. 1966. Measurements of oxidoreduction processes and ATP levels in an isolated Crustacean neuron. Ann. N.Y. Acad. Sci. **137**: 864–876.

SLEIGH, M. A. 1962. The biology of cilia and flagella. MacMillan, New York.

SPARROW, F. K., JR. 1960. Aquatic Phycomycetes. Univ. Mich. Press, Ann Arbor.

STÜBEN, H. 1939. Über entwicklungsgeschichte and Ernährungsphysiologie eines neuen niederen Phycomyceten mit Generationswechsel. Planta (Archiv. wiss. Bot.) **30**: 353–383.

TAMURA, S., TOYOSHIMA, Y., AND WATANABE, Y. 1966. Mechanism of temperature-induced synchrony in *Tetrahymena pyriformis*. Japan J. Med. Sci. Biol. **19**: 85–96.

THAXTER, R. 1896. New or peculiar aquatic fungi. 3. *Blastocladia*. Bot. Gaz. **21**: 45–52.

WATERHOUSE, G. M. 1962. Presidential address. The zoospore. Trans. Brit. Mycol. Soc. **45**: 1–20.

WHITTAKER, V. P. 1966. Some properties of synaptic membranes isolated from the central nervous system. Ann. N.Y. Acad. Sci. **137**: 982–998.

WOLMAN, M., AND BUBIS, J. J. 1966. The relation of various enzymes to cellular membranes. Histochemie **7**: 105–115.

Fine Structure of Mycota.
13. Zoospore and Nuclear Cap Formation in *Allomyces*

ROYALL T. MOORE
Department of Botany, North Carolina State University, Raleigh, N. C.

The genus *Allomyces* has had a long history of popularity with aquatic phycomycetologists (cf. Emerson, 1941; Skucas, 1966), and recently the electron microscope has inaugerated a new phase of research into these fungi (Blondel and Turian, 1960; Moore, 1964a, 1964b; 1965a, 1965b; Renaud and Swift, 1964; Skucas, 1967; Turian and Kellenberg, 1956). The present communication reports certain observations, presented in an interpreted ontogenic sequence, of zoospore initiation and maturation for *Allomyces* × *javanicus* Kniep.

When grown on agar this fungus produces quantities of thin-walled zoosporangia that develop only as far as pre-cleavage. These sporangia are hyaline and multinucleate, bear several discharge papillae, and are delimited from the somatic hyphae by abscissional septa (Fig. 1). Zoospore formation is induced by immersing the sporangia in water. Cleavage takes about an hour, after which time numerous posteriorly uniflagellate zoospores, each with a single nucleus containing a prominent nucleolus and bearing a large nuclear cap, are released.

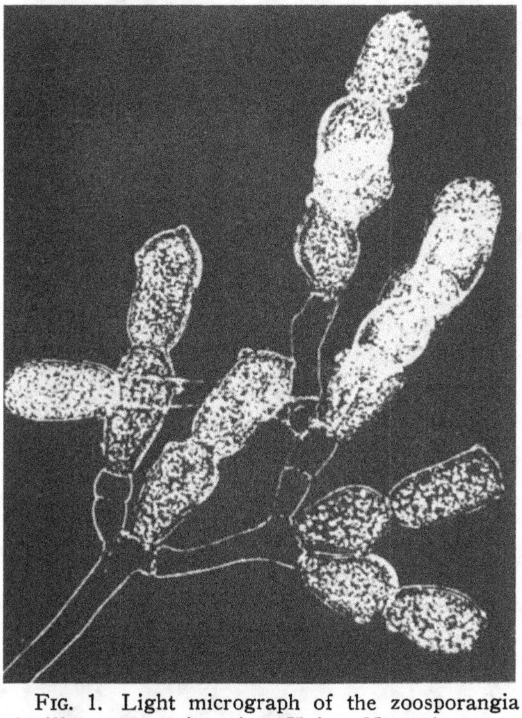

FIG. 1. Light micrograph of the zoosporangia of *Allomyces* × *javanicus* Kniep. Note the evacuated somatic hyphae and the developing discharge papillae. Such sporangia have a size range of 60–80 x 27–50 µ. Courtesy of R. Emerson.

Although material for study was originally selected at ten-minute intervals from time of initation, it was found, not surprisingly, that cleavage synchrony was only approximate and that all but the earliest stages could be found in the one-hour material. Harvested sporangia were fixed in 1.5% aqueous unbuffered $KMnO_4$, dehydrated in an acetone series (interrupted by overnight staining in 1% uranyl nitrate in 70% acetone), and embedded in Epon. Sections were examined in a Siemens Elmiskop I electron microscope.

The first indications of cleavage are the appearance of aggregations of small vesicles in the cytoplasm (Figs. 2, 4). These cleavage vesicles are about 20–30 mµ in diameter and are filled with an electron-dense material that sets them in marked contrast to the nuclei, mitochondria, and endoplasmic reticulum of the surrounding cytoplasm. The source of these vesicles is presently unknown, but the complete absence of the Golgi complex (Moore, 1965a) denies postulating that they might derive from the vesicles of this organelle. (This is in contrast to *Phytophthora*, in which Golgi vesicles are reported to be the source of the cleavage vesicles (Hohl, Hamamoto, and Uehara, 1966). Also prominent are a number of filled vesicles confluent with the endoplasmic reticulum (Figs. 2, 3). These vesicles occur at only one end of an ER-element and contain a material of comparable appearance with that composing the matrix of the ribosome-rich nuclear cap (Turian, 1955; Turian and Kellenberger, 1956) (Fig. 9 *et seq.*).

The process of cleavage appears to progress from the initial appearance of the pre-sporolemma vesicles (Fig. 2), to their linear orientation (Fig. 4), to their subsequent fusion into developing cleavage planes (sporolemma anlagen) (Fig. 5). These sporolemma anlagen continue to expand at their margins until they approach one another (Fig. 5); subsequent anastomosization with themselves and the sporangial plas-

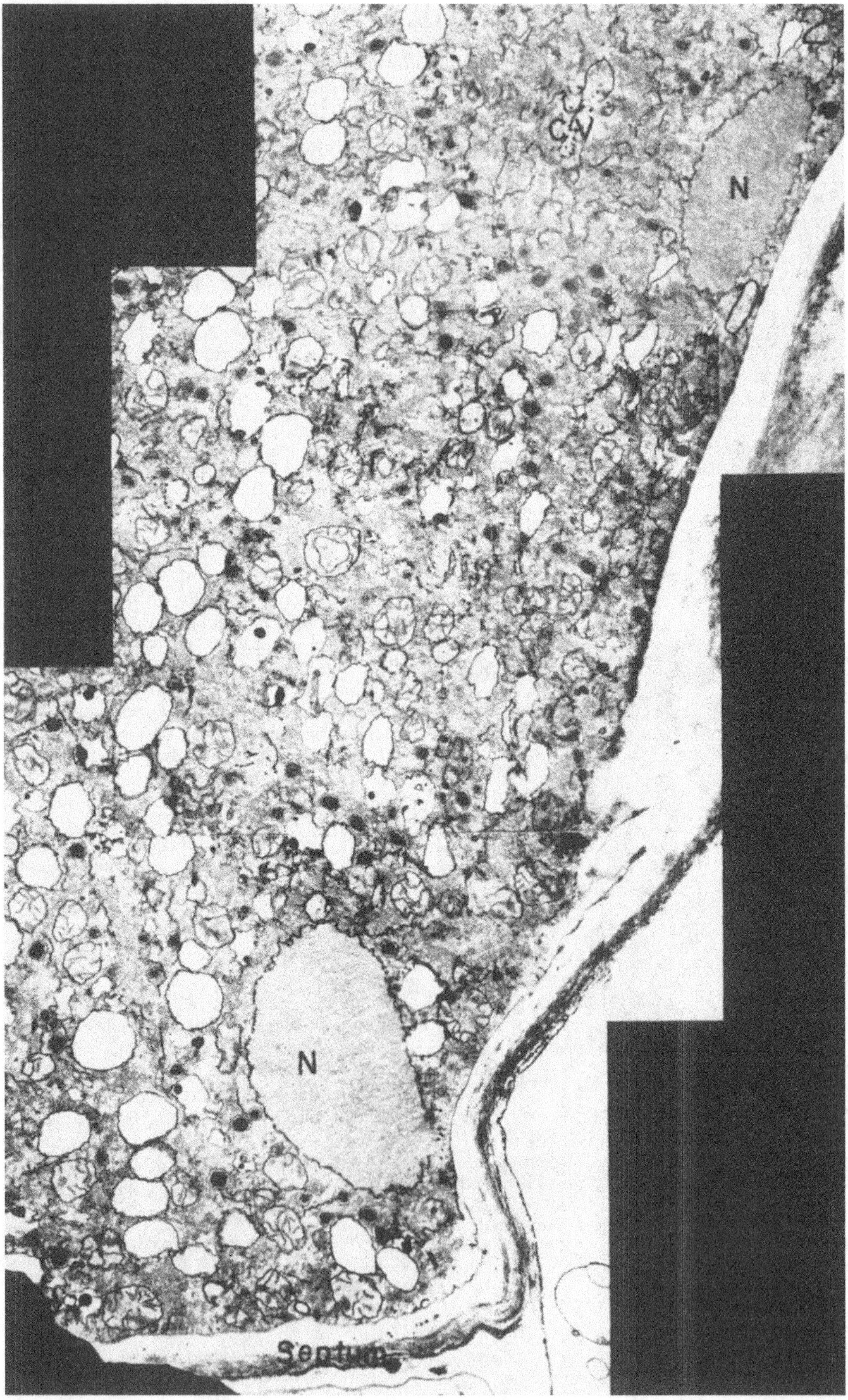
CV
N
N
Septum

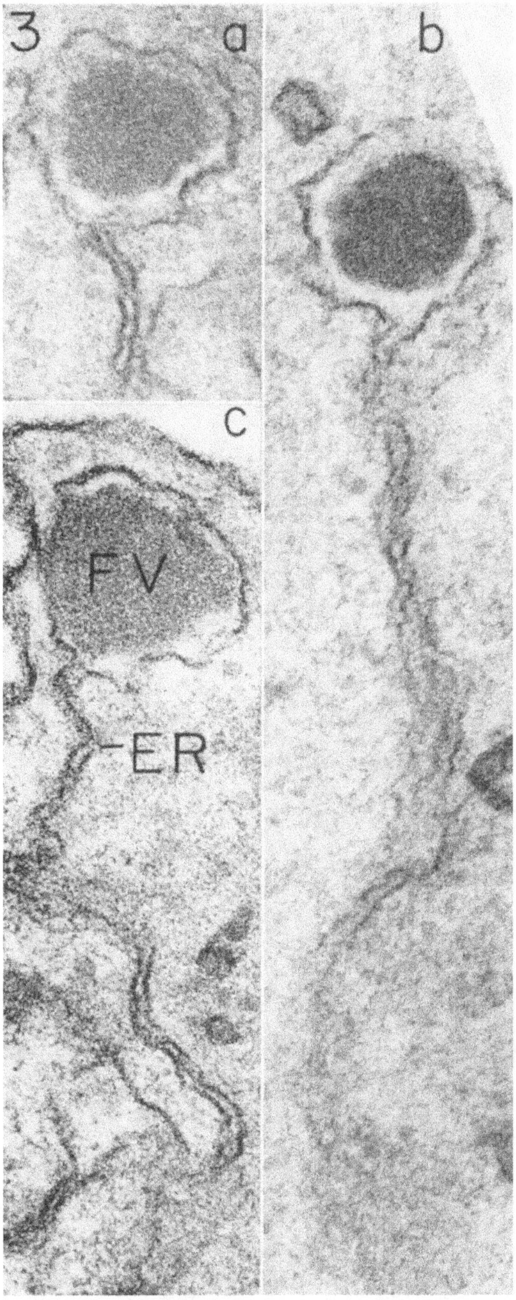

FIG. 3a, b, c. Examples of the filled vesicles (FV) that constitute one end of the elements of the endoplasmic reticulum (ER) (compare with Fig. 2). X120,000.

ma membrane completes sporolemma formation and marks out the uninucleate regions of the future zoospores (Fig. 6).

Through the early stages of blocking out these uninuclear domains, the mitochondria remain scattered (Fig. 6). But as cellulation progresses, nuclear cap membranes begin to be organized from the endoplasmic reticulum, and the mitochondria vacate the ground cytoplasm (Figs. 7–9).

The completion of cleavage finds the sporangium divided into a number of uninucleate polyhedral-shaped cells (Figs. 9, 14). At the same time each nucleus has become invested by a prominent nuclear cap that is bounded by a continuous double membrane partially attached to the outer membrane of the nuclear envelope (Fig. 12) and upon which impinge numerous mitochondria (Figs. 9, 10, 14, 16, 20, 21); concomitantly the extranuclear cytoplasm has become coarsely granular and devoid of both mitochondria and most of the endoplasmic reticulum with its attached vesicles (Figs. 9, 15). It seems probable that the nuclear cap membrane is derived from the endoplasmic reticulum, that the source of its contents is from the ER-attached vesicles, and that this transformation is effected by the action of the mitochondria. This differentiation also seems to cause a redistribution of nuclear pores to that portion of the envelope adjacent to the cap (Fig. 11).

It seems worthwhile to inject at this point a comparison with *Callimastix cyclopsis.* This hemolymphic parasite of *Cyclops* was described by Weissenberg in 1912 as the type of a new zooflagellate genus, but a recent ultrastructure study by Vavra and Joyon (1966) clearly shows it to be a member of the Blastocladiales. Characteristic of the order, *C. cyclopsis* lacks a Golgi system (Moore, 1965a) and has a prominent nuclear cap which has, however, a turbinate rather than the more usual napiform profile. Further, as in *Allomyces* (Renaud and Swift, 1964), there evidently are paired centrioles, only one of which develops into a flagellum. It is singular in that the numerous mitochondria remain discrete and dispersed in the zoospore cytoplasm.

Soon after the polyhedral stage the future zoospores begin to round up—a process that begins at the intersections of the adjoining

FIG. 2. Survey electron micrograph montage of a peripheral portion of a zoosporangium soon after the initiation of cleavage. The abscission septum, nuclei (N), and one group of cleavage vesicles (CV) are identified. These CV, even at this low magnification, are readily distinguished from the numerous mitochondria and the abundant endoplasmic reticulum with its attached filled vesicles (compare with Figs. 3 and 4). Note the complete absence of the Golgi organelle. The clear areas are believed to represent unpreserved lipid inclusions. X10,000.

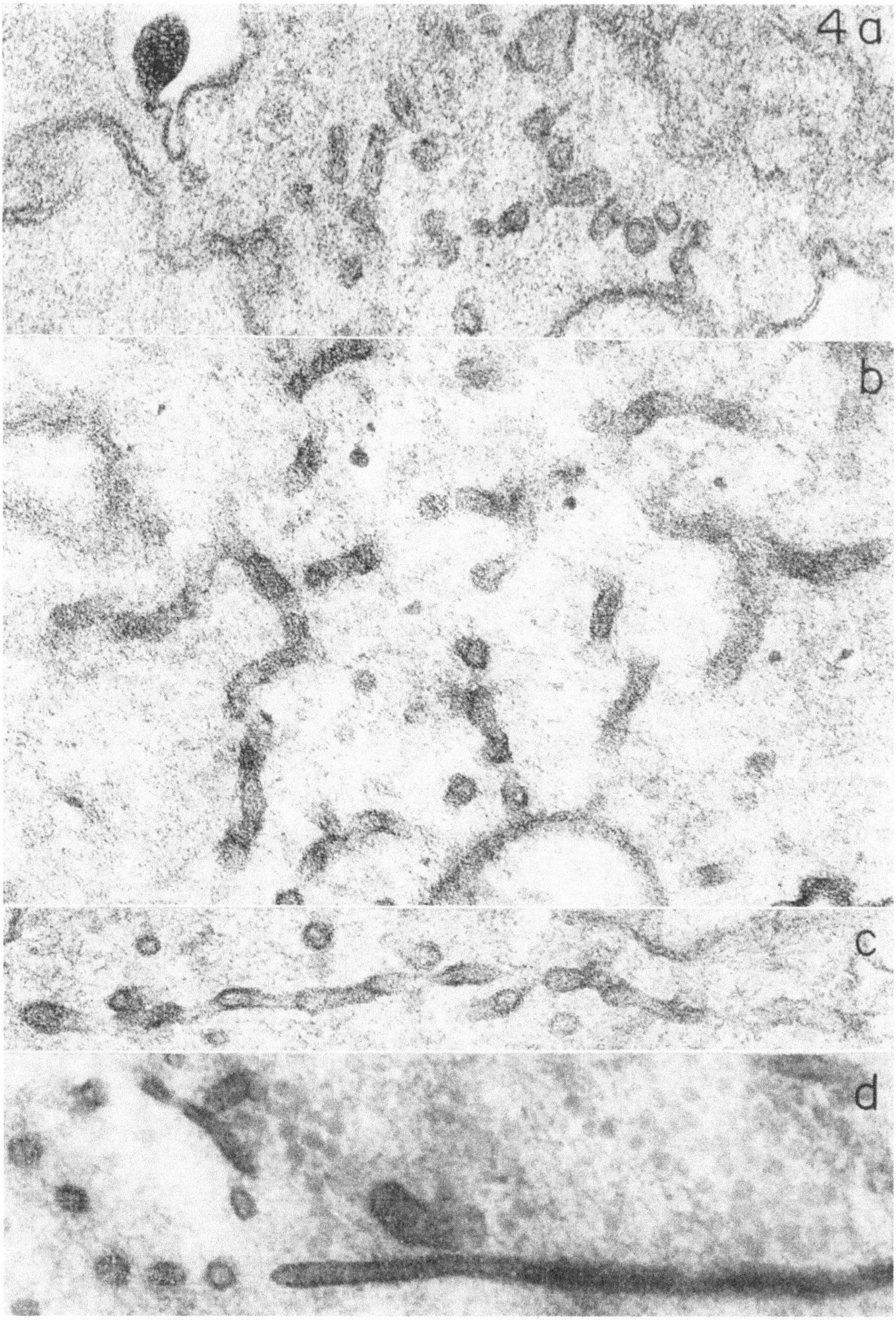

FIG. 4. Cleavage vesicles. Note that they are bounded by a unit membrane and filled with an

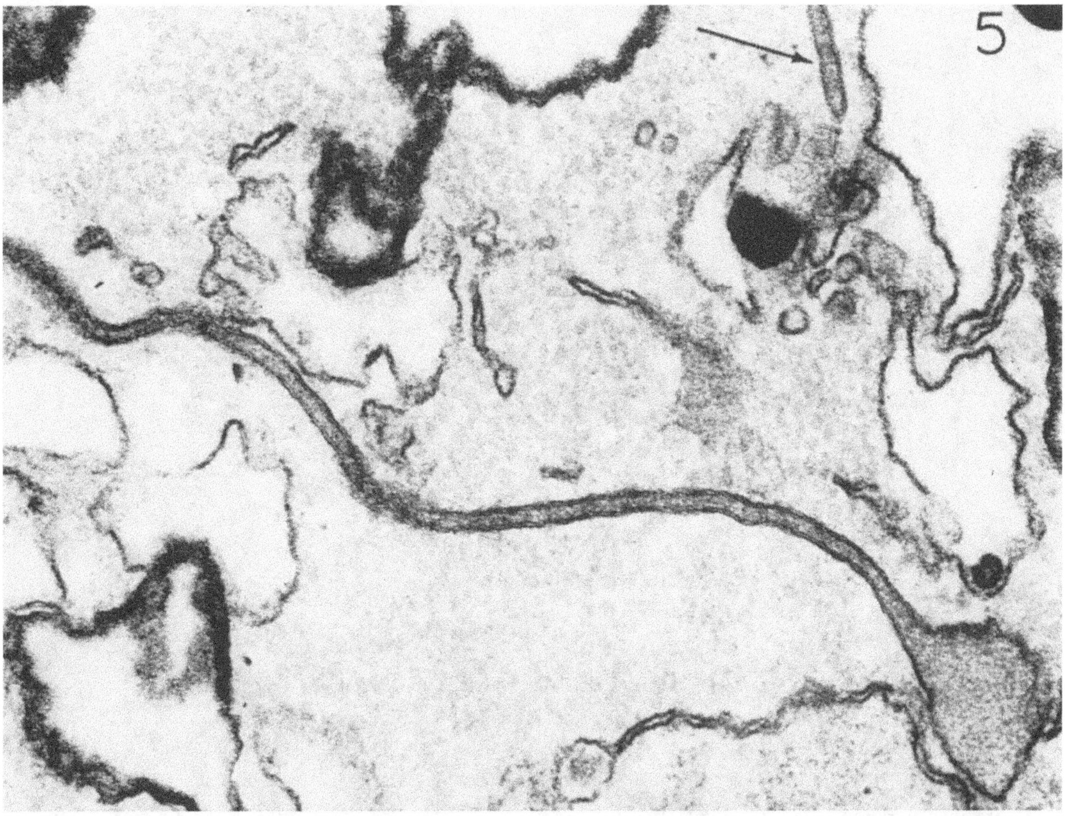

FIG. 5. Beginning of cellulation of the sporangium. The lower cleavage plane, whose sheet-like nature is evident in the tangential transect in the lower right, is being approached by the margin of another cleavage plane from the top of the micrograph (arrow) (compare with Fig. 6). X90,000.

cells. Rounding up continues and proceeds to separate the zoospores. This separation is achieved by a splitting of the sporolemma to form the respective plasmalemmas of the zoospores, by the probable dissolution of what was originally lumen material (Figs. 15b, 16, 17).

Zoospore emergence is preceded by the dissolution of the pulley-shaped plugs (Skucas, 1966) (Fig. 18). The mature zoospores (Figs. 20, 21) that then exit are bounded by a flexible cell membrane that is either wholly derived from the vesicle-plane complex (Fig. 15b) or, if the spore is of peripheral origin, is partially composed of the original sporangial plasma membrane (Fig. 15a). The mature zoospore is also prominently characterized by a posterior flagellum (Fig. 21). This last-mentioned micrograph also offers suggestive evidence that at the base of the flagellum there is a particularly large mitochondrion which is similar to that observed in *Blastocladiella* by Cantino et al. (1963) and Reichle and Fuller (1967) and is probably the "side-body" in Koch's (1961) diagrams based on light microscopy. (*Blastocladiella*, however, has only the one mitochondrion, and thus its nuclear cap is smooth; the attachment of the paired cap membranes in this genus is interpreted by these investigators as being totally continuous with the outer nuclear membrane. Further, though Reichle and Fuller (1967) do not comment upon it, their micrographs indicate that in *Blastocladiella* also the nuclear pores in the mature zoospores are limited to that portion of the envelope adjacent to the ribosome-rich nuclear cap.)

electron-dense material (compare with Fig. 1). a) Early appearance when single and randomly distributed; b) later configuration as the vesicles begin to become organized into cleavage planes; c) still later as linear orientation becomes pronounced; d) end of vesicle stage when they have mostly coalesced into membrane sheets that in turn will anastomose with other such membrane sheets to form the sporolemma. X120,000.

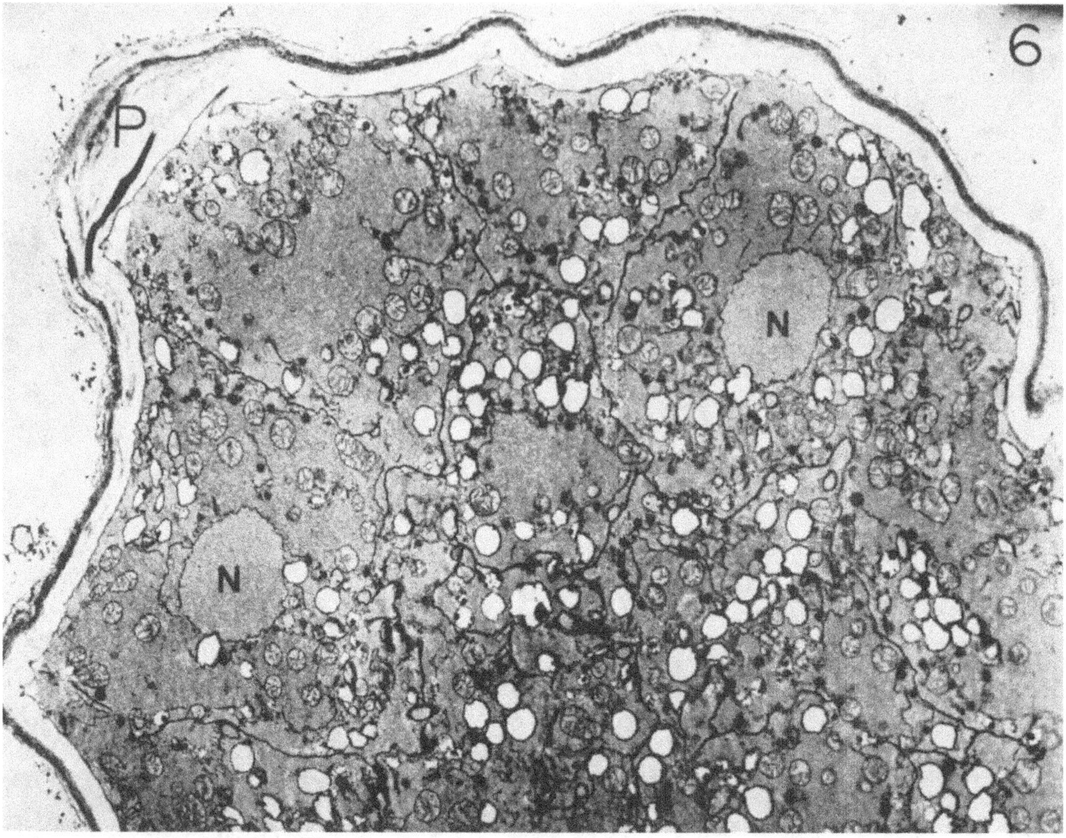

FIG. 6. Early cellulation. The domains of the future uninucleate zoospores have become blocked out, but the cleavage planes have not yet completely fused to form the sporolemma (compare with Figs. 5–14). Note that the mitochondria and most of the ER with attached vesicles are still scattered, but around each nucleus (N) the nuclear cap is beginning to be organized (these nuclei are seen at higher magnification in Figs. 7 and 8). Also evident is the beginning of a discharge papilla (P) (compare with Fig. 18). X 2,500.

Though the development of the flagellum was not observed in the present study, it has been well described and diagrammed for gametogenesis in *Allomyces arbusculus* by Renaud and Swift (1964), and the two ontogenies might be expected to be comparable. The gametangia, like the zoosporangia, develop up to pre-cleavage on agar. During this initial phase paired centrioles are observed in the cytoplasm; but during this early growth only one develops, and it increases to a length three time its original size. Transferring the gametangia to distilled water promotes gamete formation and the final maturation of the "chosen" member of the centiolar pair to become the single, posterior, whiplash flagellum. (These observations, it would seem, would offer strong support for the long available hypothesis of the derivation of the Chytridiomycetes from a biflagellate saprolegnid archetype by the evolutionary suppression of the anterior tinseled flagellum. What an insight into phylogenetic time might be gained if a way could be found to induce the second centriole to produce its flagellum!)

Finally, in the words of Turian (1962), "Dedifferentiation is the necessary corollary of differentiation in the developmental cycle of microorganisms. In *Allomyces* (and Blastocladiales in general) it manifests its rejuvenating effects with particular clarity during the first phase of germination of zoospores or zygotes where disintegration of the nuclear cap is the most spectacular dedifferentiative event." That is, one might anticipate that germination would effect, in essence, a time reversal, and that an ultrastructural study of germination would elicit comparable micrographs to those shown here for nuclear cap formation the sequence of which would be reversed.

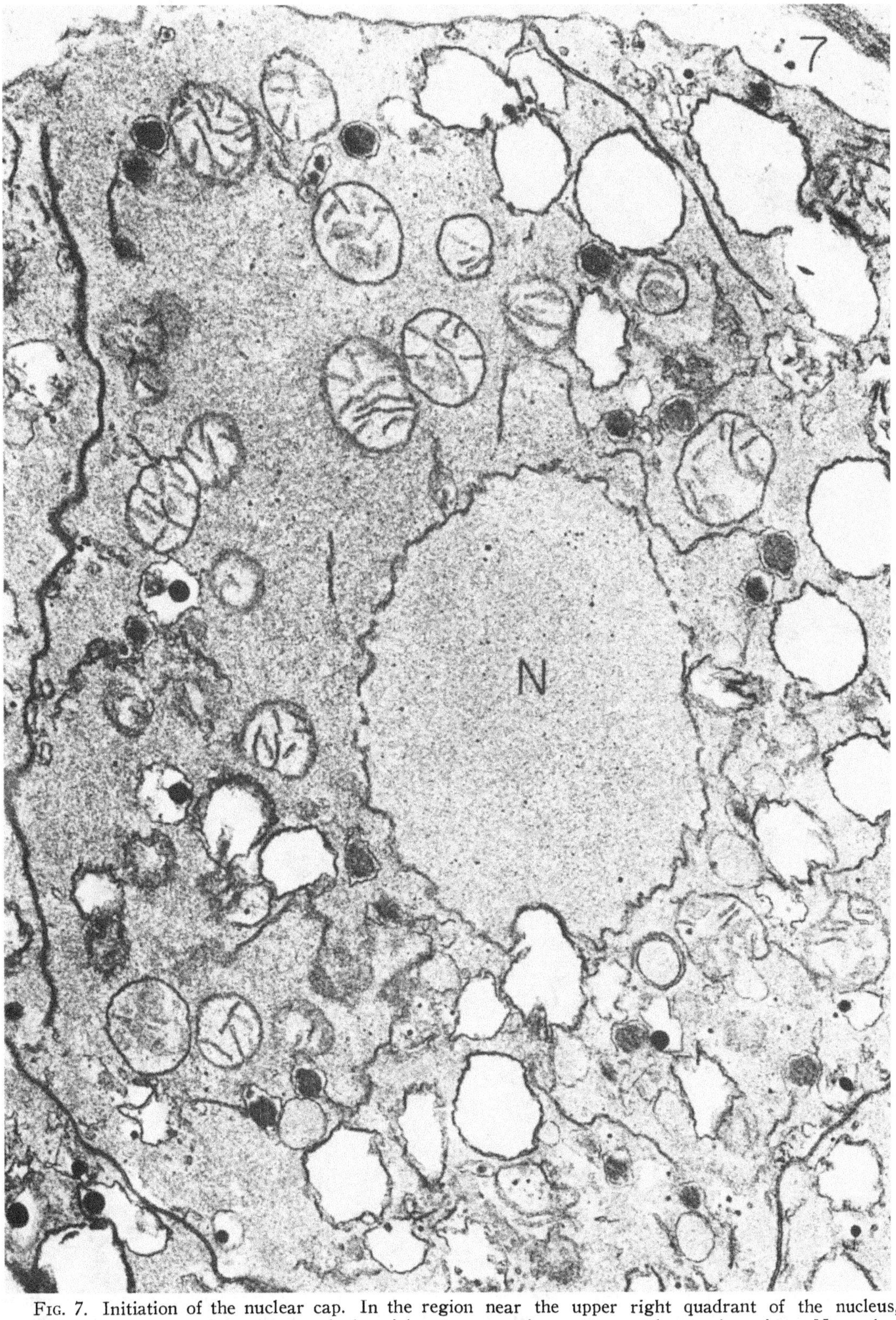

FIG. 7. Initiation of the nuclear cap. In the region near the upper right quadrant of the nucleus, several elements of the ER with attached vesicles are assuming a paranuclear orientation. Note that cellulation is still incomplete. (This is the right nucleus in Fig. 6). X10,000.

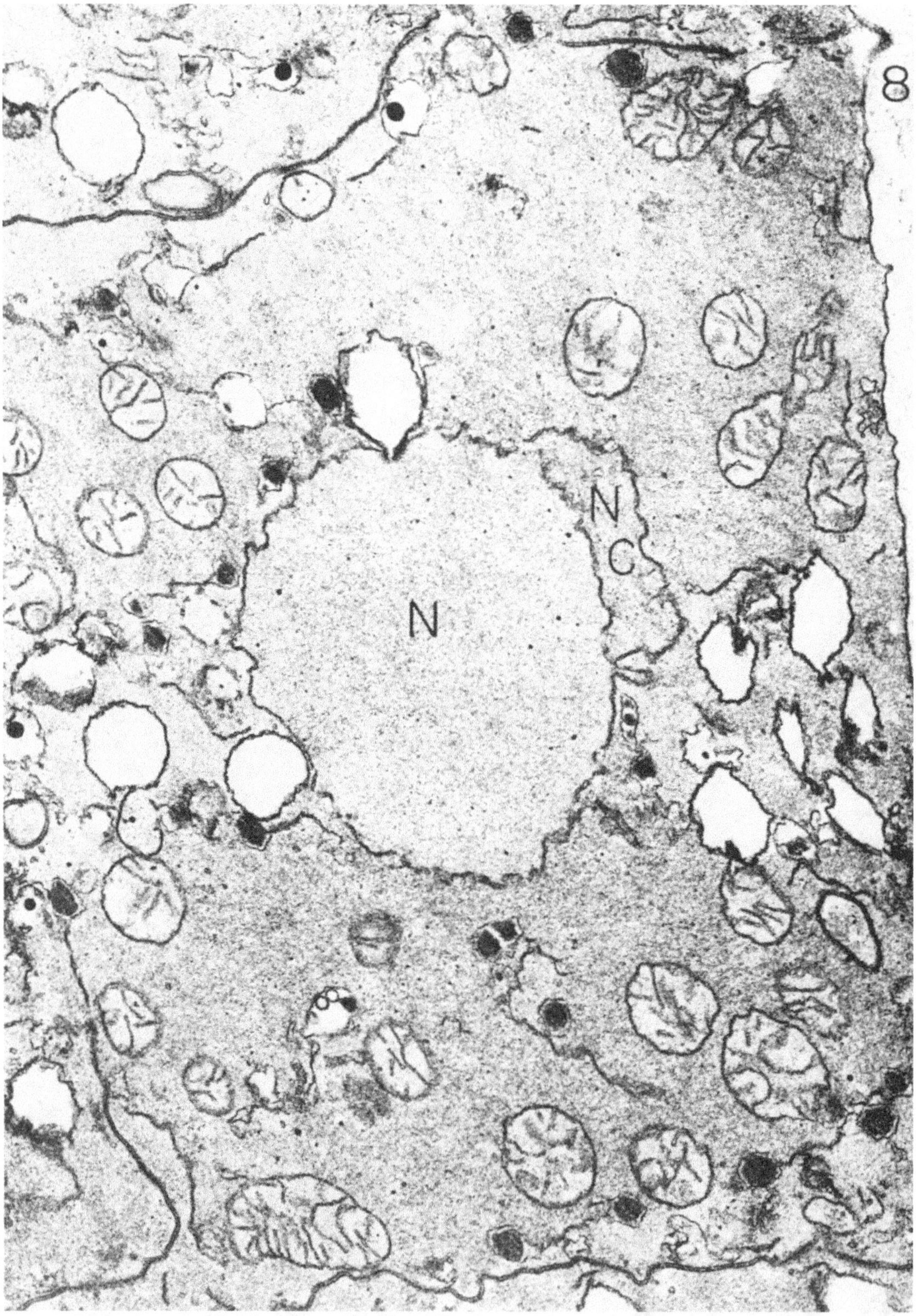

FIG. 8. Nuclear cap (NC) anlage. The first ER elements have joined the nuclear envelope to establish the location of the future ribosome reservoir. Note that the nuclear pores are still regularly distributed and that cellulation has not been completed. (This is the left nucleus in Fig. 6). X10,000.

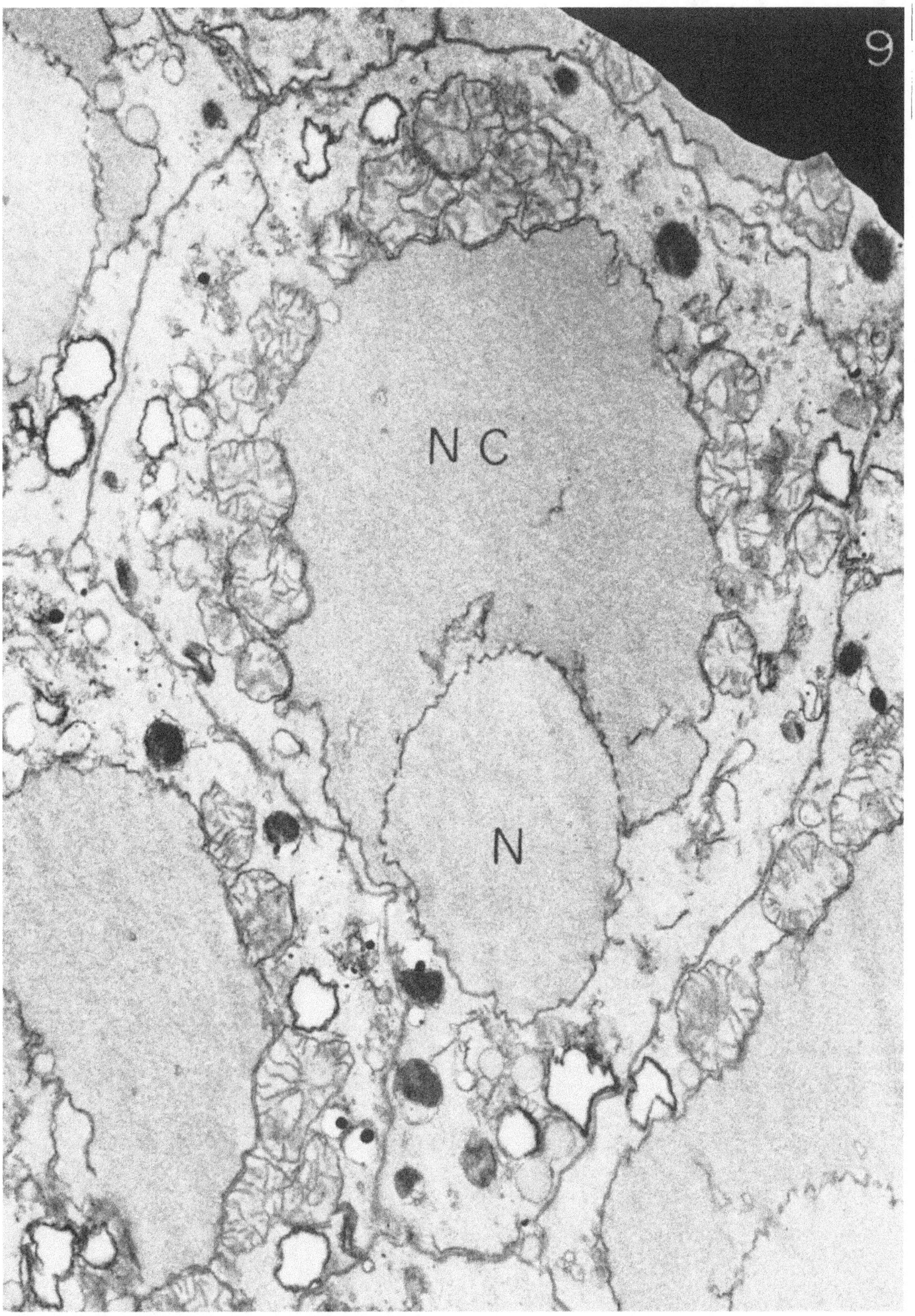

Fig. 9. Mature nuclear cap. The nuclear cap (NC) has nearly reached its full size; the mitochondria have lost their random distribution and now impinge in upon the cap membranes; the nuclear pores have become limited to the interface between the nucleus (N) and the hooding cap. Further, by the time of this stage of development, the cleavage planes have completely interfused and the sporolemma is nearly complete (compare with Figs. 6 and 14). X11,000.

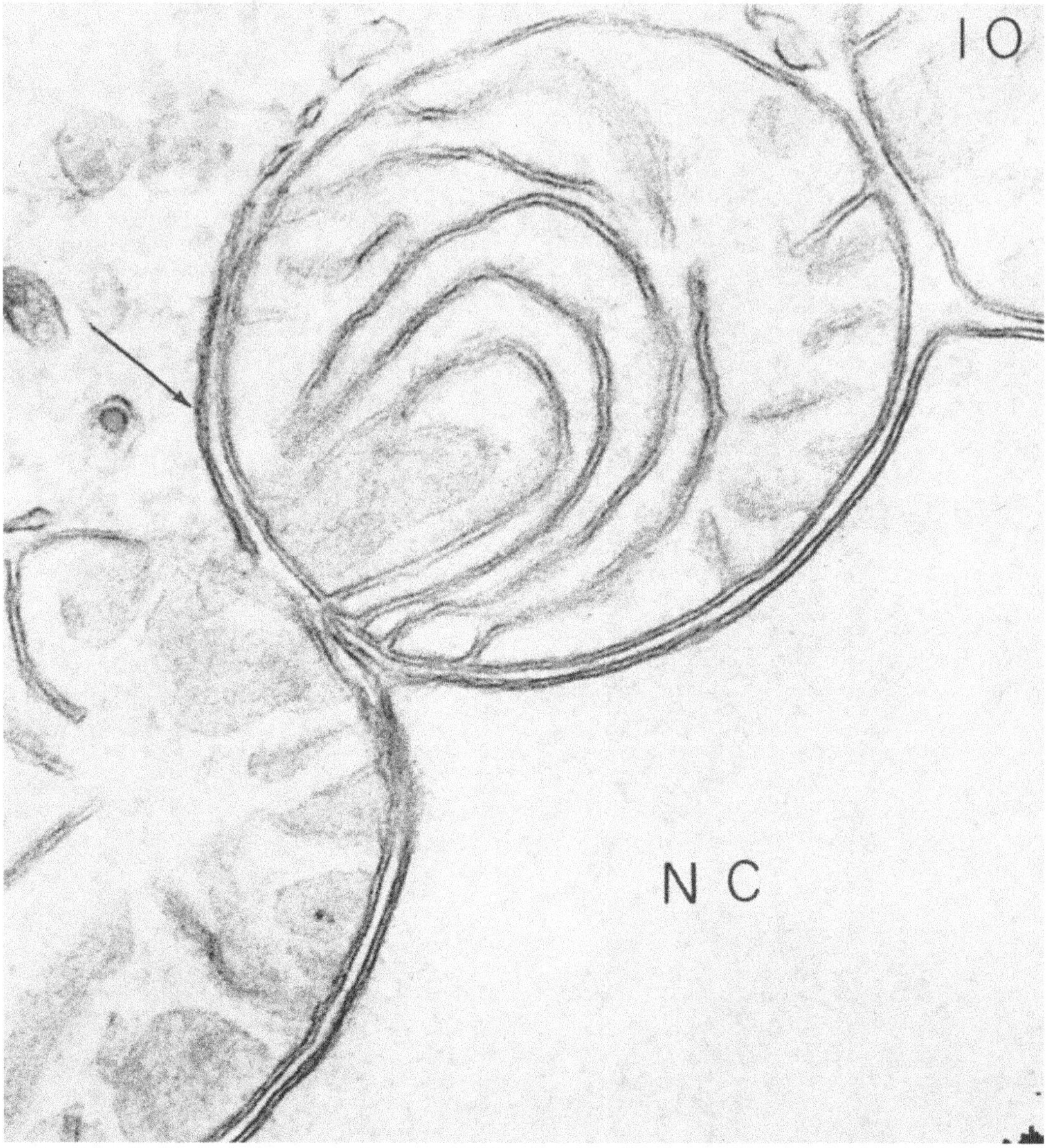

FIG. 10. Portion of a mature nuclear cap (NC) with impinging mitochondria. Note that, like the mitochondria, the cap is bounded by pairs of continuous membranes. Also evident are some fragments of unincorporated ER (arrow) and the granular nature of the ground cytoplasm (upper left). X100,000.

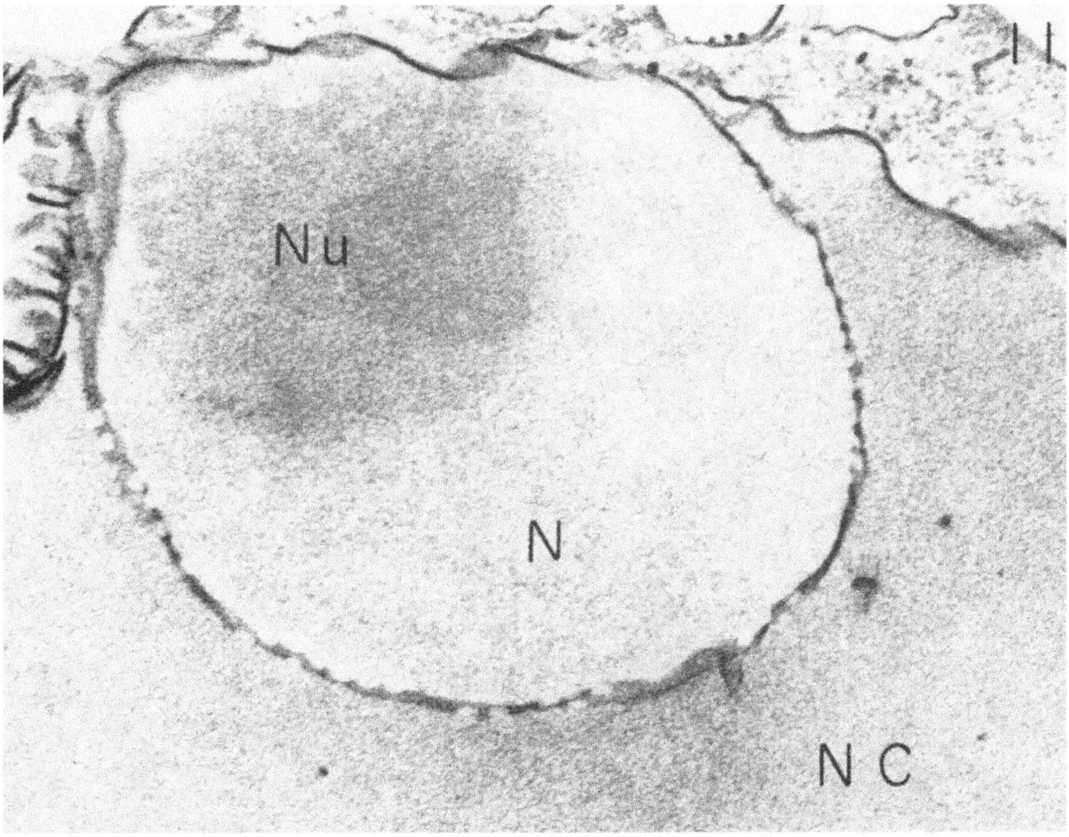

FIG. 11. Showing the limitation of the pores in the nuclear envelope to the interface between the nuclear cap (NC) and the nucleus (N). Note—the flagellum (cf., Fig. 21) will emerge out the upper left corner and be attached to the nucleolar (Nu) portion of the nucleus. X6,000.

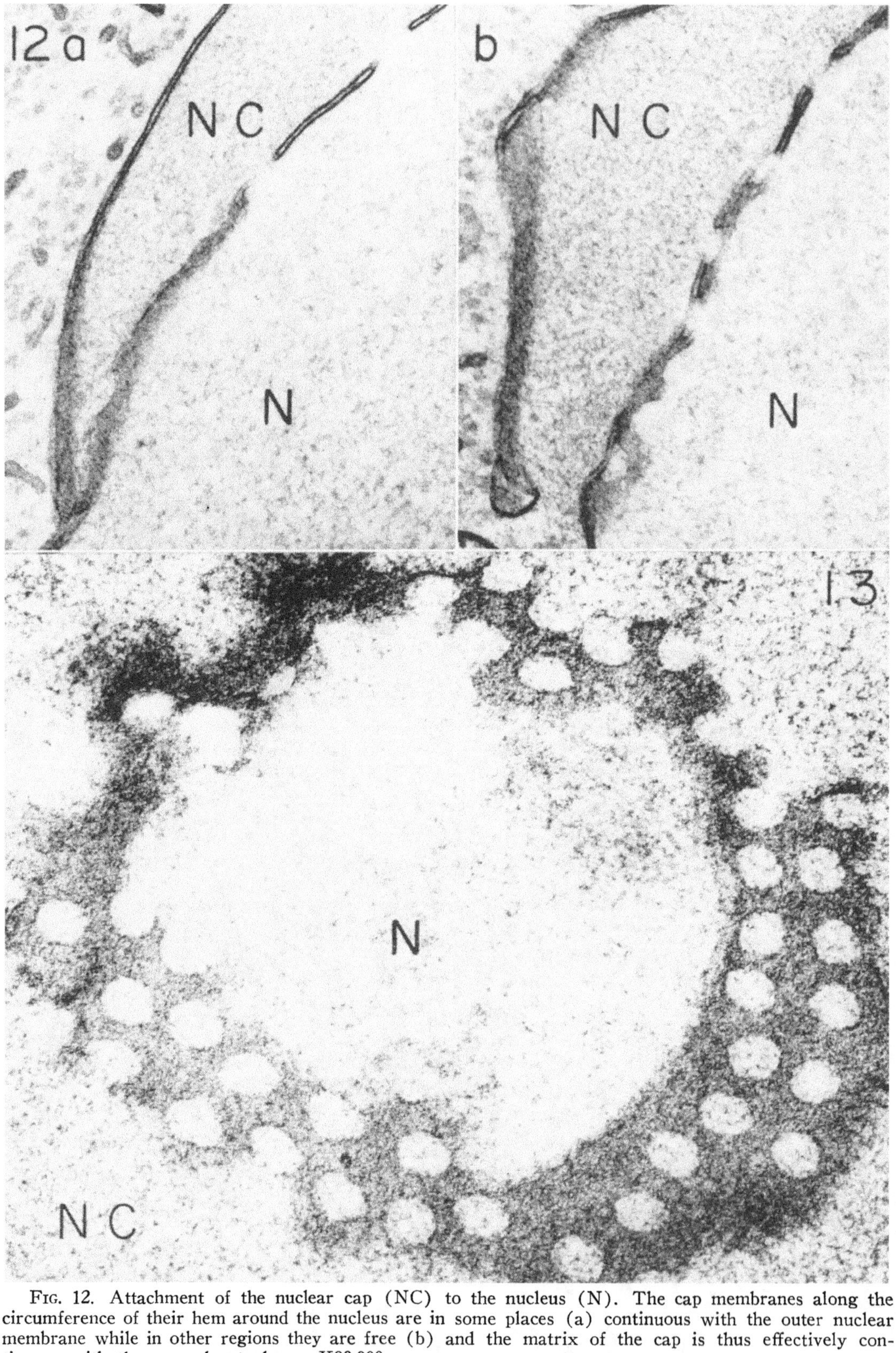

FIG. 12. Attachment of the nuclear cap (NC) to the nucleus (N). The cap membranes along the circumference of their hem around the nucleus are in some places (a) continuous with the outer nuclear membrane while in other regions they are free (b) and the matrix of the cap is thus effectively continuous with the ground cytoplasm. X90,000.

FIG. 13. Tangential section through the interface of the nucleus (N) and the nuclear cap (NC) showing the prominent and quite regular arrangement of the pores in the nuclear envelope. X120,000.

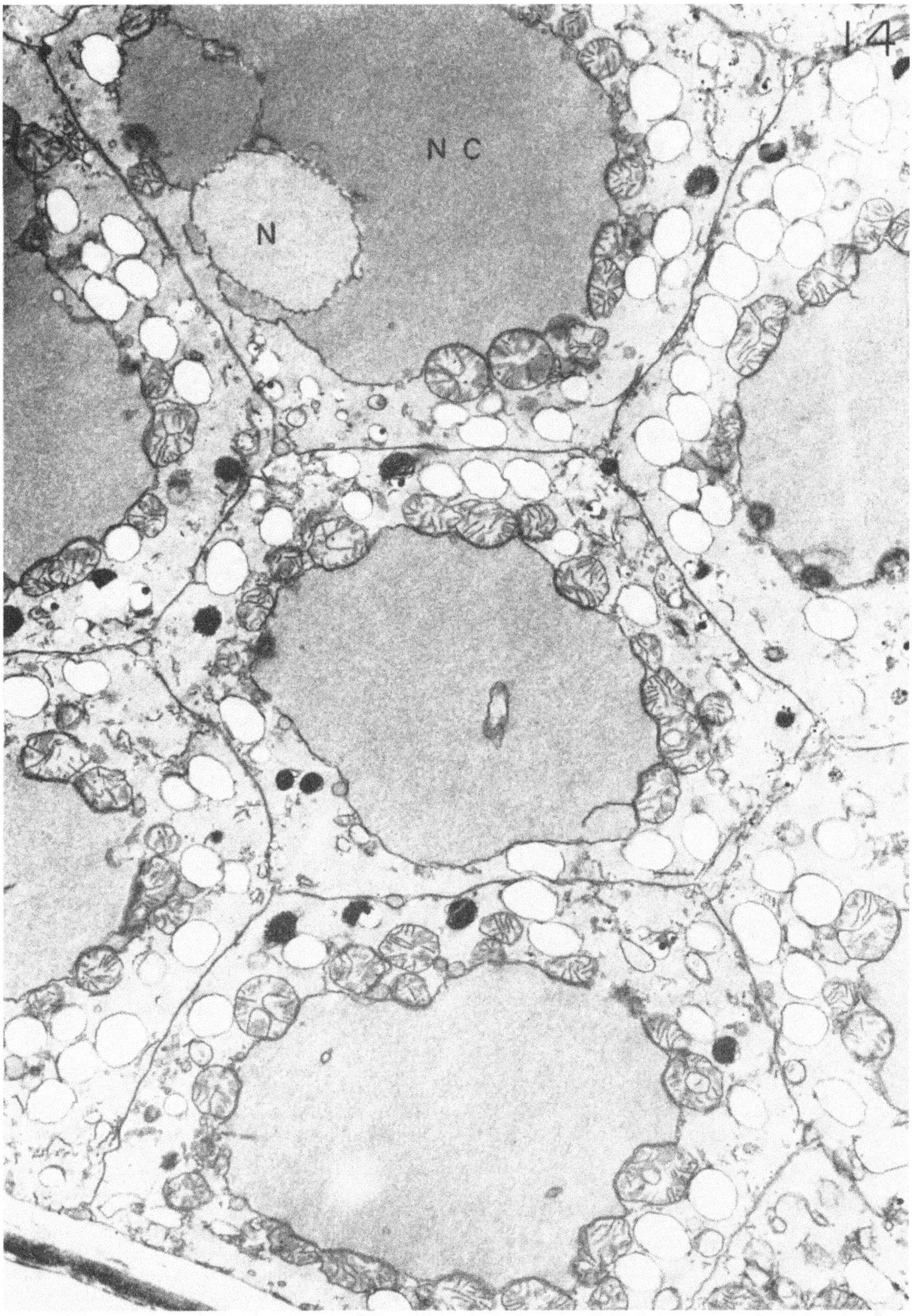

FIG. 14. Completion of sporolemma. The sporangium is now divided into a number of uninucleate polyhedral cells that will round up, starting from the corners, to become the future zoospores. Note the distribution of the mitochondria and the prominent relative size of the nuclear cap (NC) to the nucleus (N) and also the distribution of the nuclear pores. X2,000.

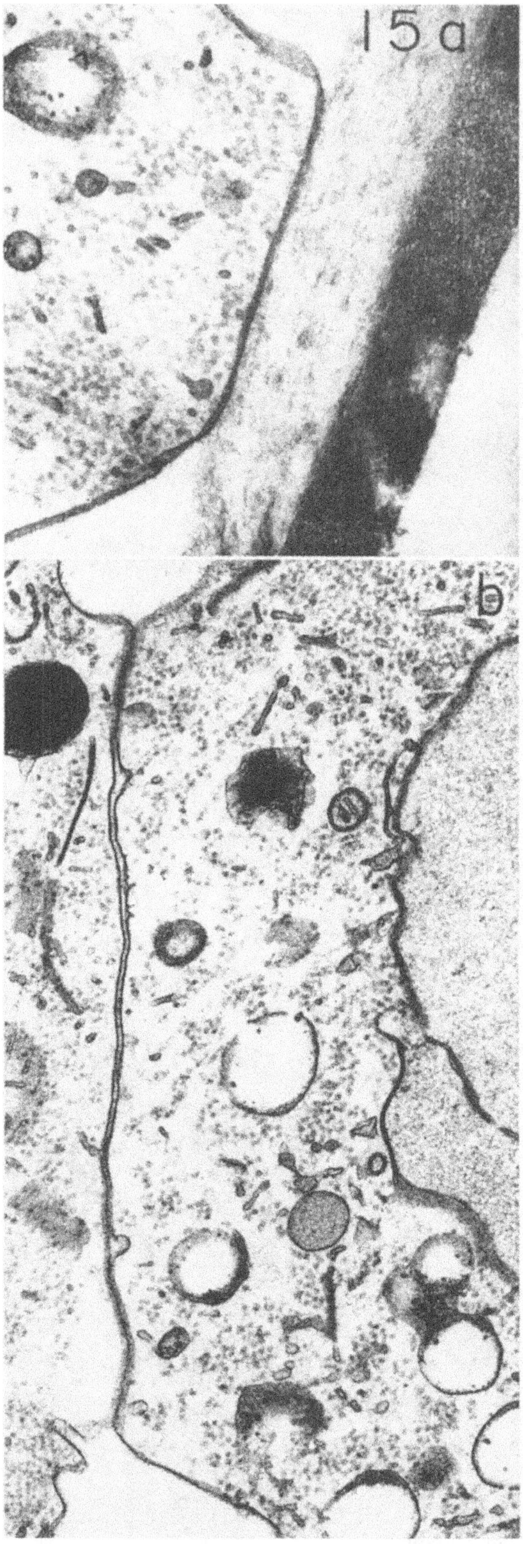

FIG. 15. Composition of the zoospore plasmalemma a) showing the incorporation of the plasma membrane of the sporangium into a portion of the plasma membrane of a peripheral zoospore (X50,000); b) internal pair of zoospores whose respective plasmalemmas are being formed by the splitting of the *de novo* originated sporolemma (compare with Fig. 16). X45,000. Note the absence of ER in the ground cytoplasm.

ACKNOWLEDGMENTS

This study was supported by grants from the National Institute of Allergy and Infectious Diseases (AI-05514-01 and 1 RO 1 AI-07122-01 BMA). It gives me deep pleasure to recognize the kind assistance of Dr. J. H. McAlear and his capable staff. Acknowledgment is also gratefully made to Mr. Robert Berman for supplying the cultures, to Prof. R. Emerson for making available his isolate Java 1 and for permission to use Fig. 1, and to Mr. J. R. McCurry for the preparation of the photographs used in the present report.

LITERATURE CITED

BLONDEL, B., AND G. TURIAN. 1960. Relation between basophilia and the fine structure of cytoplasm in the fungus *Allomyces macrogynus* Em. Jour. Biophys. Biochem. Cytol. **7**: 127–134.

CANTINO, E. C., J. S. LOVETT, L. V. LEAK, AND J. LYTHGOE. 1963. The single mitochondrion, fine structure and germination of the spore of *Blastocladiella emersonii*. Jour. Gen. Microbiol. **31**: 393–404.

EMERSON, R. 1941. An experimental study of the life cycle and taxonomy of *Allomyces*. Lloydia **4**: 77–144.

FULLER, M. F. 1967. The ultrastructure of motile cells in fungi. Seminar, October 16, Univ. of North Carolina, Chapel Hill.

HOHL, H. R., S. HAMAMOTO, AND M. UEHARA. 1966. Bacteriol. Proc. **1966**: G74.

KOCH, W. J. 1961. Studies of the motile cells of Chytrids. III. Major types. Amer. Jour. Bot. **48**: 786–788.

MOORE, R. T. 1964a. The nuclear cap of the phycomycete *Allomyces*. Jour. Cell Biol. **23**(2): 62A.

———. 1964b. Zoospore formation in the phycomycete *Allomyces*. Jour. Cell Biol. **23**(2): 108A–109A.

———. 1965a. Distribution and characterization of the Golgi complex in the phycomycetes. Jour. Cell Biol. **27**(2): 69A.

———. 1965b. The ultrastructure of fungal cells. *In* The Fungi. Vol. I. G. C. Ainsworth and A. S. Sussman, eds. Academic Press, New York.

REICHLE, R. E., AND M. S. FULLER. 1967. The fine structure of *Blastocladiella emersonii* zoospores. Amer. Jour. Bot. **54**: 81–92.

RENAUD, F. L., AND H. SWIFT. 1964. The development of basal bodies and flagella in *Allomyces arbuscula*. Jour. Cell Biol. **23**: 339–354.

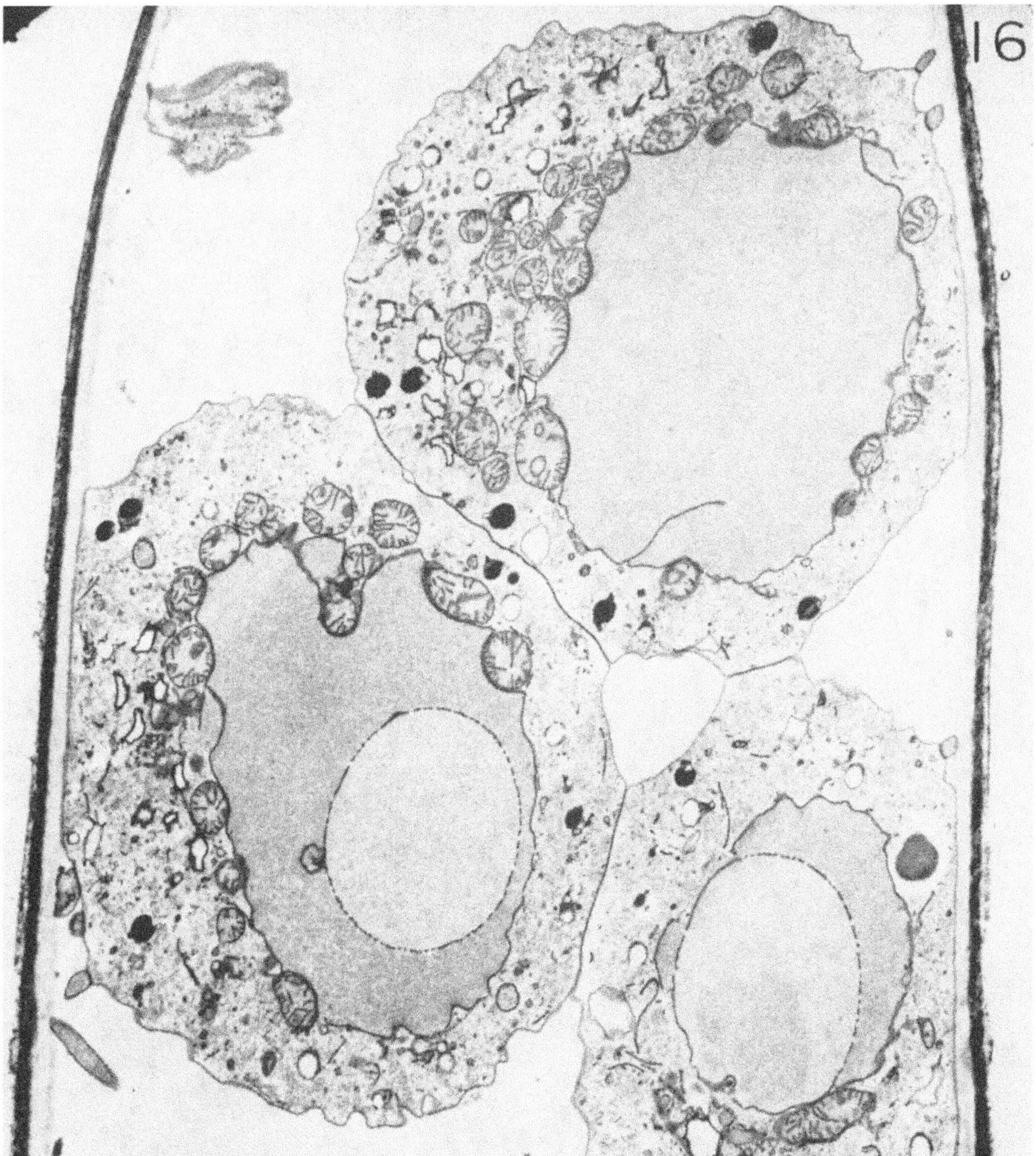

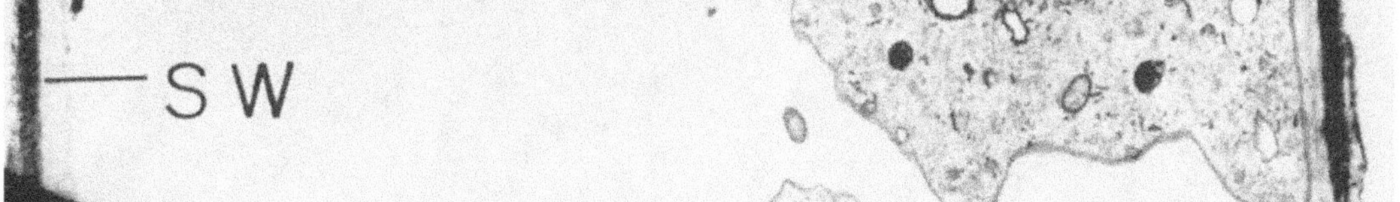

FIG. 16. Late cleavage in which the future zoospores have nearly finished rounding up and separating. Note the sources of the plasma membranes (cf. Fig. 15), the distribution of mitochondria and nuclear pores, and the absence of ER. (SW—the two-layered sporangial wall.) X5,000.

FIG. 17. Separation of the sporolemma between two zoospores. The separation appears to be effected by the dissolution of an evident lumen material. X200,000.

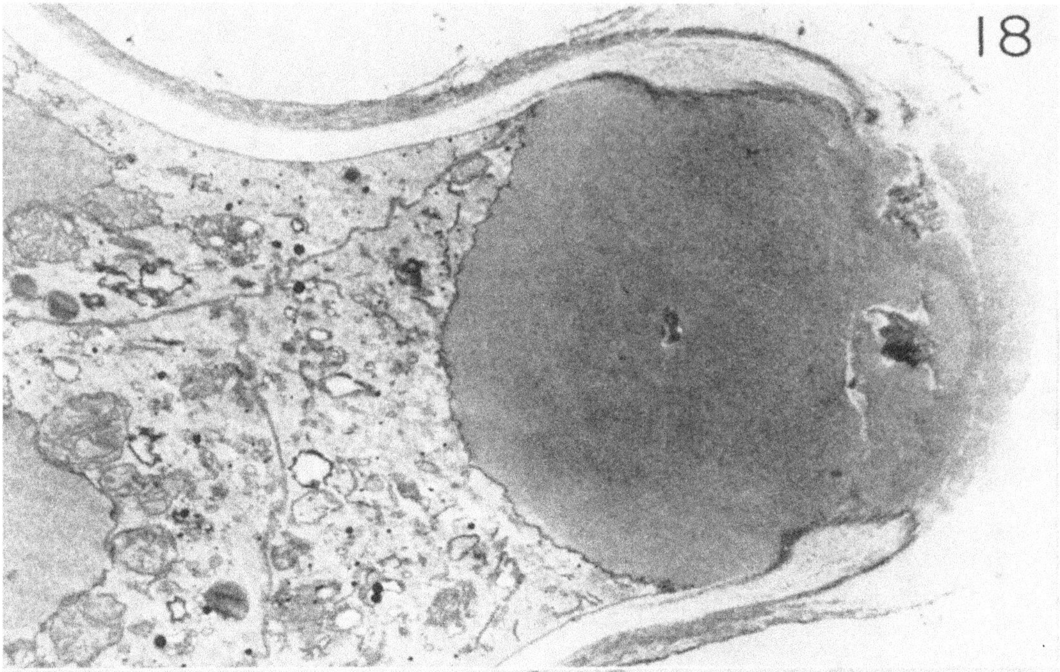

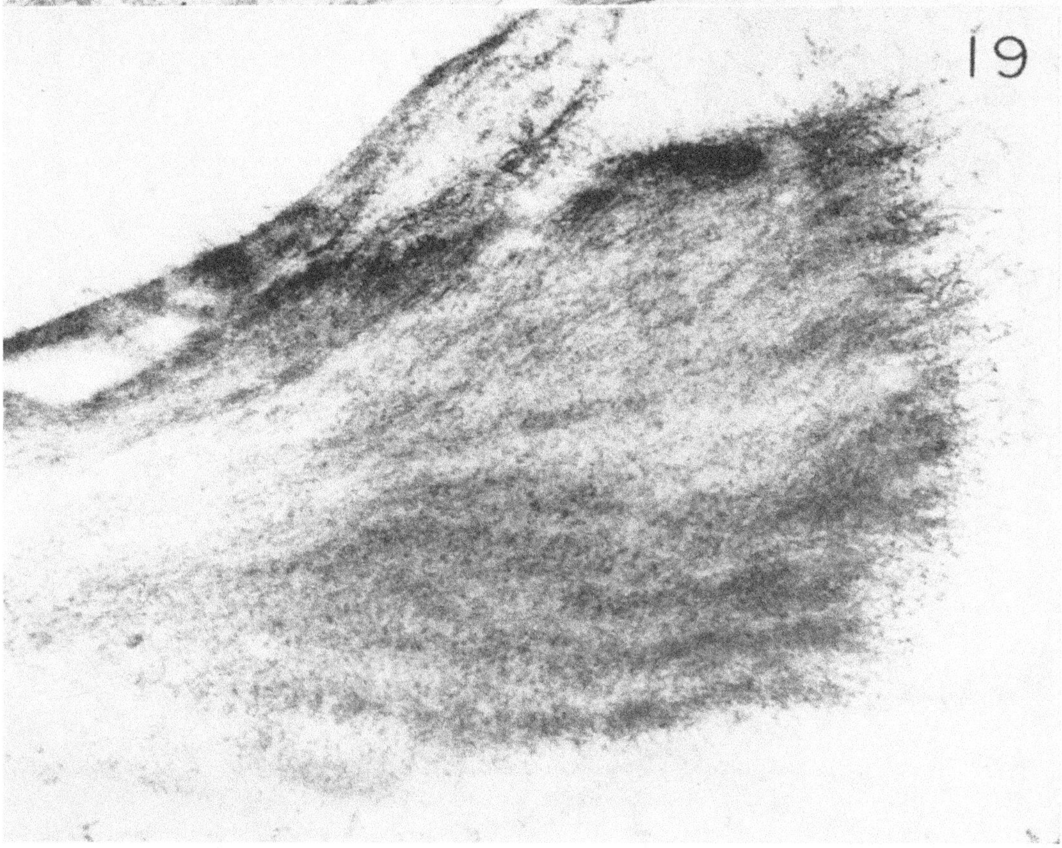

FIG. 18. Mature discharge papilla at about the same stage as Fig. 9. In the interval since the stage shown in Figure 6 the wall has differentiated into a pulley-shaped plug corking an opening in the sporangium wall. Note that the outer wall has already ruptured. X25,000.

FIG. 19. Margin of a discharge pore after the plug has blown out. The frayed edges well show the fibrous nature of the sporangial wall. Note also the major separation between the inner and outer walls. X90,000.

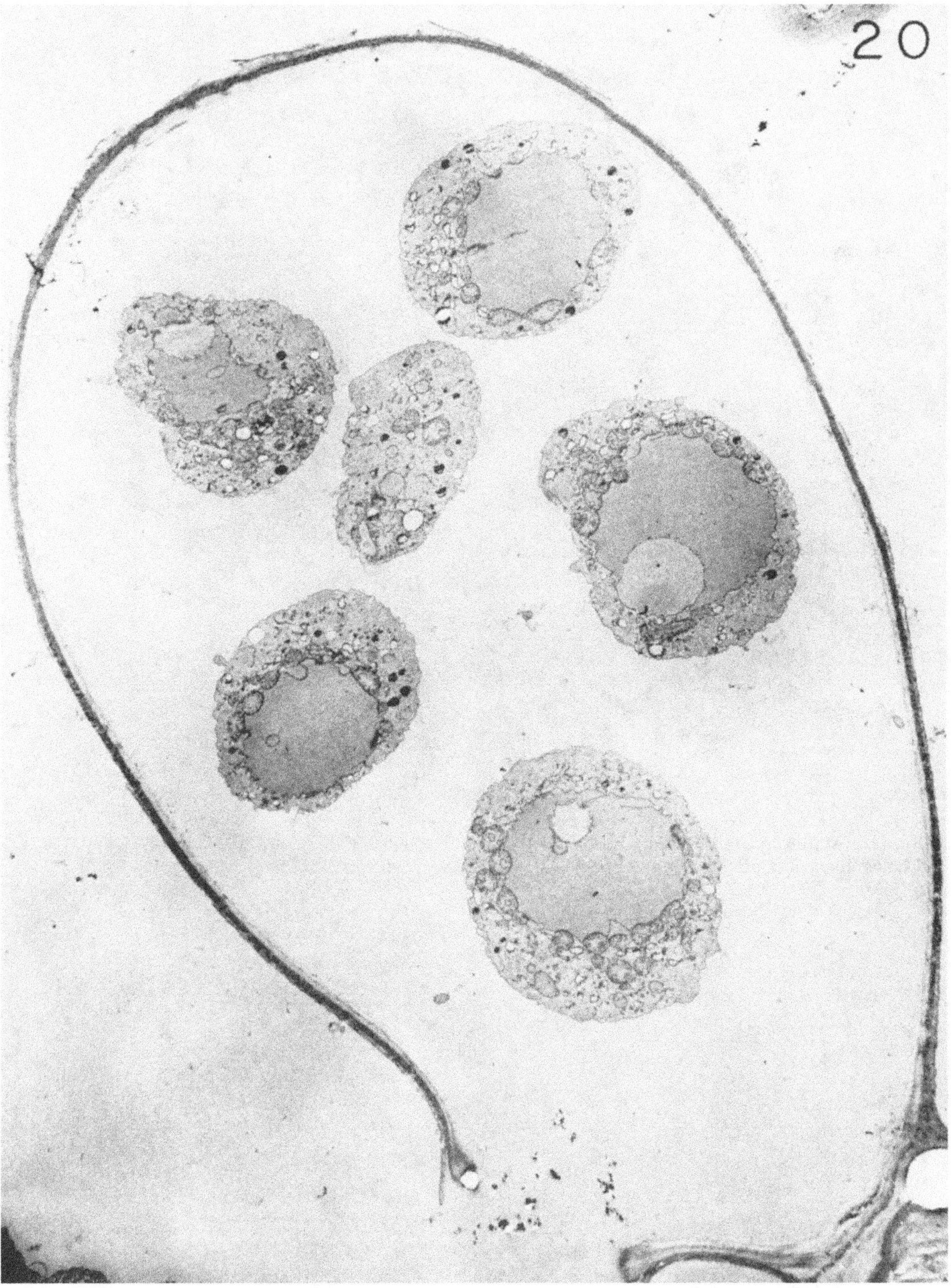

FIG. 20. Mature sporangium from which most of the zoospores have been discharged. X1,000.

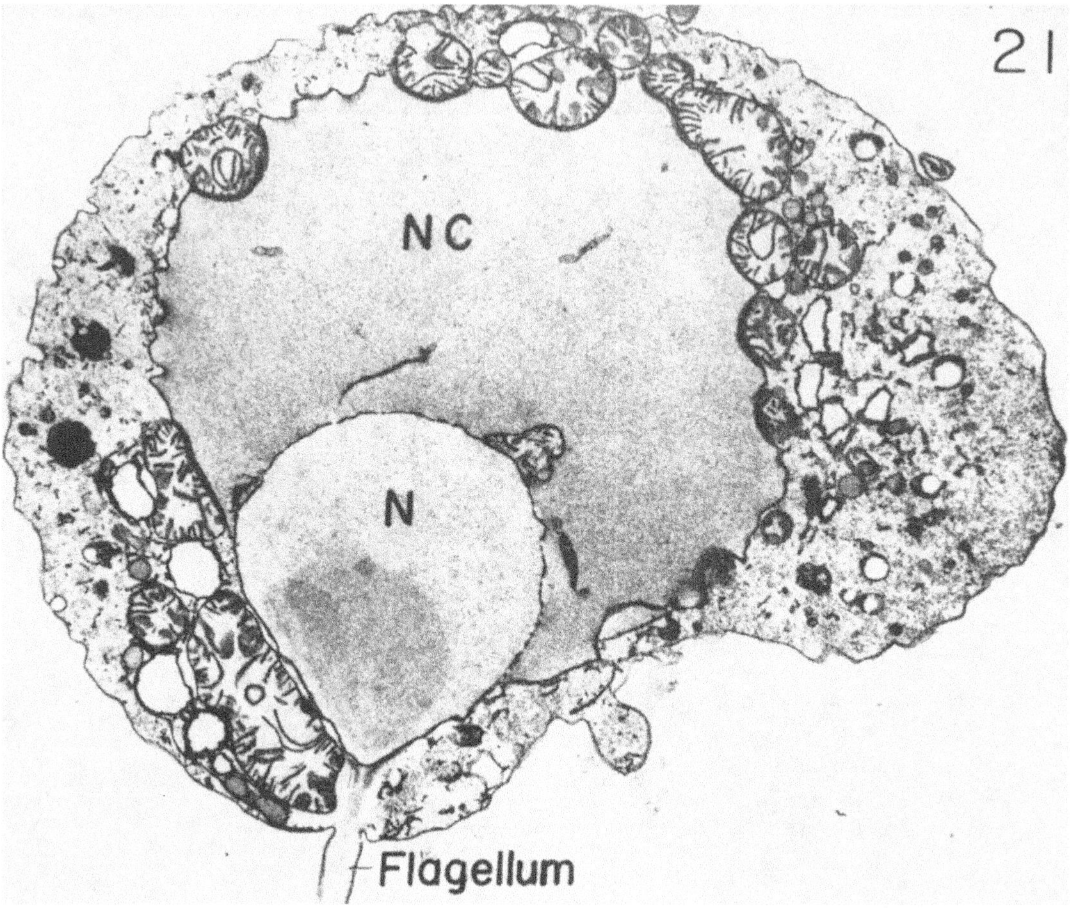

FIG. 21. Mature zoospore. Note the large basal mitochondrion that may have a similar relationship to the flagellum here as the single mitochondrion in *Blastocladiella*. X2,500.

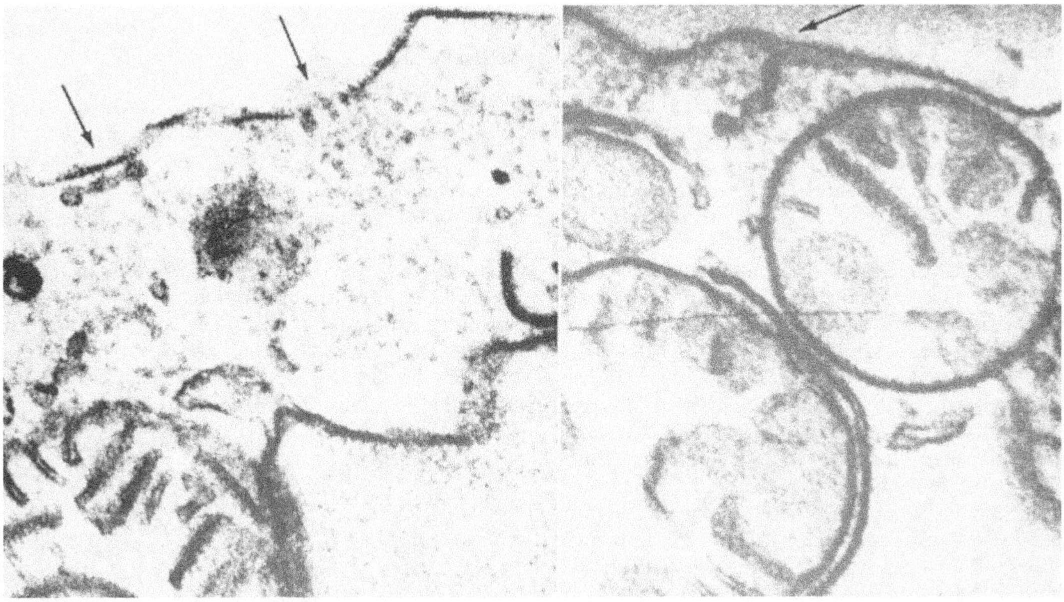

FIG. 22. Two areas of Fig. 21 extra-enlarged to show what may represent terminal ends of flagellar sheath microtubules (arrows). Left, a portion of the peripheral cytoplasm directly above the upper-left single mitochondrion; right, a portion of the peripheral cytoplasm right of center of the uppermost edge of the micrograph. Both approximately X15,000. Fuller (1967) has shown that the encircling microtubules of the flagellum, after entering the body of the cell and becoming a typical fascicle of nine triplets at the base of the flagellum, continue into the spore cytoplasm and flare out equidistantly around the nucleus. Both the termination and function of these tubules are undetermined, but perhaps they are part of the aquadynamic design of the spore.

SKUCAS, G. P. 1966. Structure and composition of zoosporangial discharge papillae in the fungus *Allomyces*. Amer. Jour. Bot. **53**: 1006–1011.

———. 1967. Structure and composition of the resistant sporangial wall in the fungus *Allomyces*. Amer. Jour. Bot. **54**: (in press).

TURIAN, G. 1955. Sur la nature ribonucléique du corps paranucléaire et ses relations avec la différentiation du sexe chez *Allomyces javanicus*. Compt. Rend. Acad. Sci. (Paris) **240**: 2344–2346.

———. 1962. Cytoplasmic differentiation and dedifferentiation in the fungus *Allomyces*. Protoplasma **54**: 323–327.

TURIAN G., AND E. KELLENBERGER. 1956. Ultrastructure du corps paranucléaire, des mitochondries et de la membrane nucléaire des gametes d'*Allomyces macrogynus*. Exptl. Cell Res. **11**: 417–422.

VAVRA, J., AND L. JOYON. 1966. Étude sur la morphologie de la cycle évolutif et la position systématique de *Callimastex cyclopsis* Weissenberg 1912. Protistologica **2**: 5–16.

Zoosporic Fungi of Oceania. I[1]

John S. Karling

Department of Biological Sciences, Purdue University, Lafayette, Indiana

Except for species that parasitize economic plants, the zoosporic fungi of the Pacific Islands have not received much attention from mycologists and plant pathologists, and relatively few studies have been made on the occurrence and distribution of such fungi. Gregory and Wentworth (1937) and Bessey (1943) reported *Dictyuchus* and *Achyla* from Hawaii, and Sparrow (1948) described three chytrids from soils on the Bikini, Eniwetok, Rongerik, and Rongelap atolls in the Marshalls but found no filamentous Phycomycetes. Later (1965), he reported 30 chytrids and 14 species of the Blastocladiales, Monoblepharidiales, Saprolegniales, Peronosporales, and Hyphochytriales from Hawaii. In the meantime Johnson (1950) described a new species of *Brevilegnia, B. longicaulis,* from New Caledonia. Persiel, Scholz, and Harder (1966) made a study of 33 soil samples from New Caledonia, Fiji, Samoa, Tonga, and Tahiti, and identified one species each of the Hyphochytriales and Blastocladiales and 17 species of the Chytridiales from such samples.

During the author's tenure as a Fulbright Research Fellow in New Zealand in 1965 and 1966, he had opportunity to visit several of the neighboring islands. During these visits soil samples were collected at various places for study of the zoosporic fungi that occur in them. The location of these collections and the types of soil are described in the appendix of this paper and referred to by alphabetical symbols and numbers. The techniques employed in trapping and studying the fungi were the same as described by the author (1965) for study of the New Zealand species. The present contribution concerns the Anisochytridiales or Hyphochytriales; subsequent papers will deal with species in other orders and families.

Anisochytrids are almost as abundant as species of *Pythium* in oceanic soils, according to the author's experience, and most of the keratinic and cellulosic substrata used as baits as well as cleared pollen grains of *Pinus sylvestris* became densely infected when they were floated on watered soil samples. In fact, the surface of bleached corn leaves and strips of snake skin were almost completely covered with sporangia in many instances, precluding the development of most other zoosporic fungi except *Pythium.* So far, nine species have been identified, among which *Anisolpidium saprobium, Rhizidiomyces bulbosum,* and *R. coronum* are described as new species.

Anisolpidium saprobium sp. nov.

Sporangia 1–5 in cella; globosa (8–32 μ diam.) vel ovoidea (12–15 x 18–22 μ); tubis emissariis osteolaris 4–7 x 20–60 μ. Zoosporae ovales vel oblongatae, 3.5–4 x 5–6 μ. Sporae perdurantes ignotae.

Sporangia 1 to 5 in a cell, spherical, 8–32 μ diam., ovoid 12–15 x 18–22 μ, with a hyaline smooth wall; developing one or two tapering, 4–7 x 20–60 μ exit tubes for discharge of protoplasm. Zoospores usually delimited outside of sporangium, ovoid to slightly elongate, 3.5–4 x 5–6 μ, with several small, hyaline refractive globules; encysting after a motile period to become spherical, 5–5.8 μ diam., cystospores that may enlarge, develop exit tubes, and produce 1 to 8 zoospores directly, or enlarge further to become normal-sized sporangia, or germinate by a tube and infect the substratum. Resting spores unknown.

Saprophytic in cleared pollen grains of *Pinus sylvestris* in soil samples P16 and P17, from Pitcairn.

This species occurred in great abundance, and most of the pollen grains floating on the surface of watered soil samples harbored one to five thalli and sporangia of various sizes and degrees of maturity. When first observed it was tentatively identified as a species of *Olpidium,* but its method of sporogenesis and the formation of anteriorly uniflagellate zoospores soon showed it to be an anisochytrid. After this identification had been made, the possibility was considered that it might relate to reduced monocentric and holocarpic thalli of a *Hyphochytrium* species such as the one Persiel (1960) found in pollen grains of *Corylus.* Her unidentified species (H.28) elongated markedly when transferred to agar media and produced a

[1] Oceania is regarded by many geographers as including the central and south Pacific Islands, i.e., Micronesia, Melanesia and Polynesia and sometimes New Zealand, Australia, and the Malay Archipelago. Others restrict it to include only Polynesia, Melanesia, and Micronesia. The author is interpreting it here in this restricted sense.

complex of more than 100 sporangia. Accordingly, attempts were made to grow *A. saprobium* on agar media to determine whether it would become polycentric in a similar manner. Finally, after many difficult trials a few thalli were isolated and grown in slush agar, but in this environment they merely enlarged and became spherical to ovoid sporangia. The zoospores produced by such sporangia usually came to rest as groups of cystospores near the discharge tube. From such limited results the author concluded that the thallus remains monocentric and holocarpic. This view was supported later by the discovery of such thalli in water between numerous adjacent pollen grains. The development of *A. saprobium* is similar to that reported for other species of this genus. The content of the cystospore flows into the pollen grain through a short infection or germ tube (Fig. 16), and after 18 to 32 hours becomes visible as a small thallus (Fig. 17) containing numerous large and small refractive globules. Such thalli enlarge progressively to become spherical or ovoid sporangia (Fig. 1). In the slush agar noted above and in water mounts, the cystospores merely enlarge to become sporangia (Fig. 18). In water surrounding films of agar on cover slips, the cystospores may develop exit tubes when they are quite small and produce 1 to 8 zoospores (Fig. 11–13), depending on their relative sizes.

Mature sporangia in pollen grains develop 1, rarely 2, tapering exit tubes in 2 to 6 hours after mounting in fresh water (Figs. 2, 3). These may branch occasionally (Fig. 4) or attain a length of 60 μ, but in such cases the protoplasm is "used up" in the extensive growth, with the result that no zoospores are formed. Normally, the exit tube begins to enlarge at its tip (Fig. 5) as the protoplasm moves into it, and in this process its wall thins out and becomes imperceptible at the apex or tip. Suddenly, the protoplasm begins to flow out with increasing rapidity to form a globular, naked mass of protoplasm (Fig. 6) which cleaves into zoospores (Fig. 7) within a few minutes as in many other anisochytrids. During this process the mass of wiggling zoospores may float away, and it is not uncommon to find such masses of zoospores completing maturation at considerable distances from the infected pollen grains.

Quite often the zoospores come to rest at the tip of and surrounding the exit tube (Fig. 14) after a brief period of wiggling around by their flagellum. Also, if the tip of the tube fails to expand and the protoplasm does not flow out, cleavage occurs within the sporangium and exit tube, and as the tip of the tube bursts later, the actively motile zoospores escape in succession (Fig. 15). Thus, cleavage may occur within as well as on the outside of the sporangium. Nevertheless, the latter type of sporogenesis is the predominant one. As in most of the anisochytrids studied by the author, a portion of the protoplasm in *A. saprobium* may fail to emerge, and it may develop into another sporangium within the old one or undergo cleavage into zoospores. These may swim out or come to rest as cystospores within the sporangium (Fig. 14).

Anisolpidium saprobium is the first known saprophytic species of the genus. All other known species are parasites of algae (Karling, 1943), and with the exception of *A. stigeoclonii* Canter (1950), they are reported to delimit their zoospores within the sporangia. From descriptions in the literature, Canter separated the anisochytrids into two groups on the basis of whether the zoospores are delimited within or outside of the sporangia, and Sparrow (1960) created a new genus, *Canteriomyces,* for Canter's *A. stigeoclonii* because its zoospores are formed outside of the sporangia. On this basis *A. saprobium* should be included perhaps in Sparrow's new genus. However, the place of cleavage in this species as well as in other anisochytrids to be described below varies markedly, and the author doubts that the place of cleavage is a criterion that merits generic rank. In this connection it may be noted that Canter (1950, Fig. 3b) also showed that portions of the protoplasm remaining in the sporangium and exit tube may undergo cleavage and form zoospores in the sporangium.

Rhizidiomyces bulbosus sp. nov.

Sporangia, hyalina, laevia, globosa, 15–62 μ diam.; raro ovoidea; 1–3 bulbosis, subglobosis, clavatis (8–12 x 13–16 μ) papillis. Zoosporae ovoideae (3–3.8 x 4–4.5 μ) vel subproductae; nonnullis globulis parvis et hyalinis; flagello 12–16 μ longo. Sporae perdurantes ignotae.

Sporangia nonapophysate, spherical, 15–62 μ diam., or rarely ovoid, with a hyaline smooth wall; usually developing 1 to 3, rarely 4, exit papillae at maturity. Rhizoids somewhat hypha-like, often irregular in contour, and ending bluntly, usually reduced and limited in extent, arising usually at the base of the sporangium, rarely from 2 to 3 places on the periphery. Exit papillae growing out and enlarging to become bulbous, subspherical, or broadly clavate, 8–12 μ diam. by 13–16 μ high. Place of cleavage and maturation of zoospores variable; all, part of, or none of protoplasm emerging through exit

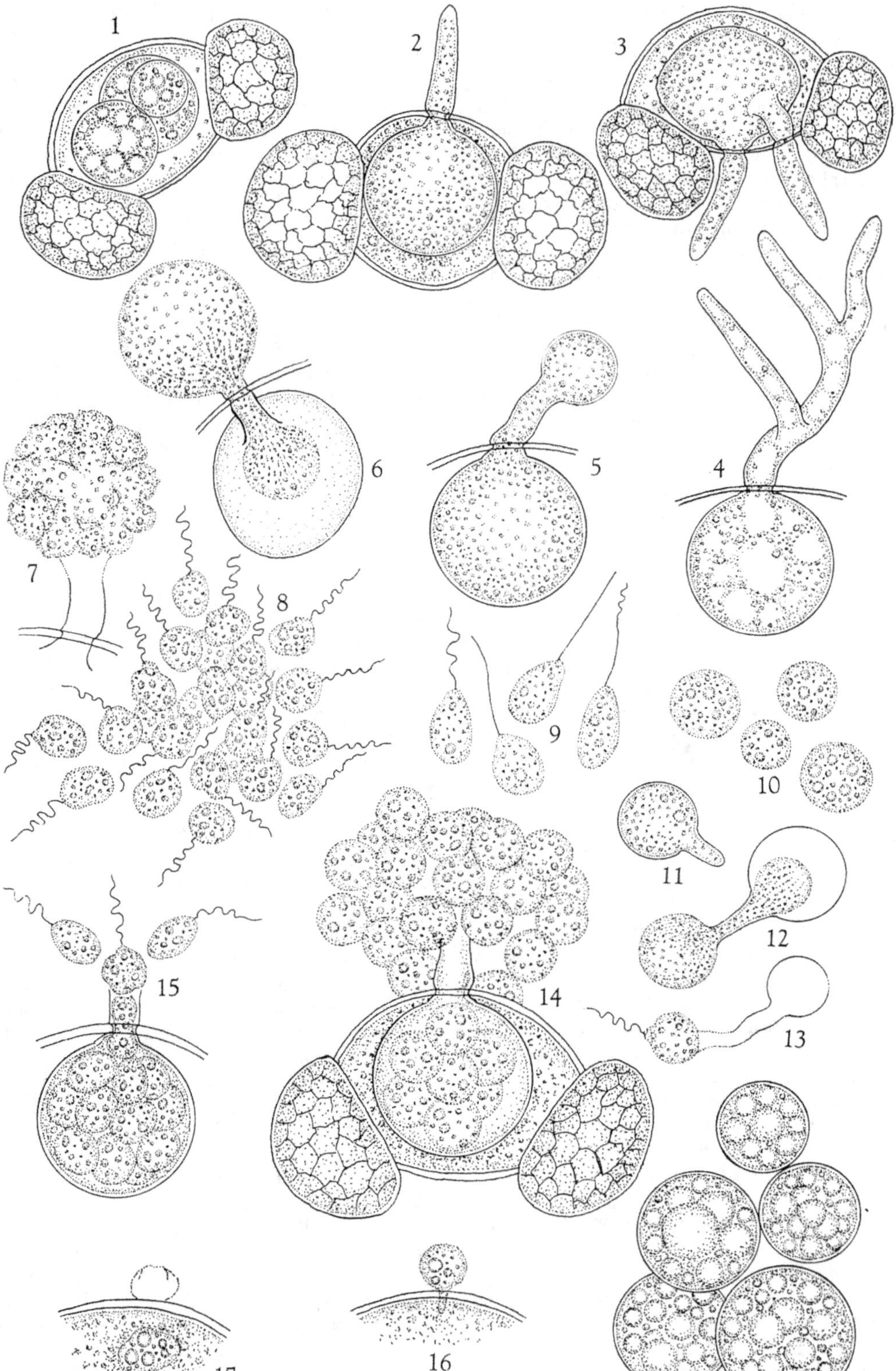
1
2
3
6
5
4
7
8
9
10
11
12
13
15
14
17
16
18

papillae; cleavage usually occurring outside of sporangium but often within the sporangium to form zoospores which swarm and emerge singly in succession, or come to rest in the sporangium as cystospores. Zoospores ovoid, 3–3.8 x 4–4.5 μ, to slightly elongate with several small, hyaline refractive globules; tinsellate flagellum 12–16 μ long. Cystospores spherical, 3.5–5 μ diam., frequently enlarging to become small sporangia, 8–15 μ diam., and forming 2 to 8 zoospores directly without germinating with a germ tube. Resting spores unknown.

Saprophytic on strips of bleached corn leaves, snake skin, and pollen grains of *Pinus sylvestris* in soil samples NI2, Niue; P10, Pitcairn; and AI1, Aitutaki, Cook Islands.

This species differs primarily from other smooth-walled members of *Rhizidiomyces* by its slightly smaller zoospores and the development of 1 to 3, rarely 4, relatively short, bulbous, subspherical or broadly clavate exit papillae instead of long, tapering necks for the discharge of the protoplasm and zoospores. Also, its rhizoids may be reduced to short or long pegs, or become branched and more extensive, irregular, contorted, and somewhat hypha-like. In addition it exhibits marked variations in the place of cleavage and maturation of the zoospores.

It occurred very abundantly on floating strips of corn leaves and snake skin, and in many instances the surface of these substrata was almost completely covered with sporangia in various stages of development. In 25 areas taken at random on such substrata, the number of sporangia per square 100 μ varied from 19 to 52, with an average of 26. It was isolated and grown in axenic cultures on PDY agar, where it developed only sparingly as successive generations of zoospores were liberated. Frequently, the zoospores did not become actively motile but encysted at the apex of the exit papillae, where they subsequently developed into sporangia. As a result, heaped-up groups of sporangia occurred on the agar cultures which eventually became up to 1.5 mm. diameter.

In such cultures as well as in the substrata cultures noted above, the zoospores remain actively motile for periods of 20 minutes to 3 hours, after which they slowly come to rest, round up, and become cystospores. During these periods the small refractive globules and granules coalesce to form a large and conspicuous one (Figs. 19, 20), and immediately after they are formed the cystospores are approximately 3.5–4.5 μ in diameter (Fig. 21). Their subsequent development and behavior in water surrounding a thin film of agar on cover slips may vary greatly. They may remain in this state (Fig. 21) for several hours, germinate by a broad germ tube (Fig. 25), or enlarge up to 15 μ in diameter without germinating. Such small and enlarged cystospores may develop broad and elongate (up to 24 μ long [Fig. 22]), or bulbous (Figs. 23, 24) exit papillae through which the protoplasm emerges and subsequently cleaves into zoospores. The number of zoospores formed depends on the size of the cystospore. In a few cases only 2 developed while in larger cystospores as many as 8 were formed. Accordingly, the cystospores may function directly as sporangia instead of developing into thalli with rhizoids. In agar film mounts such as noted above, zoospores may continue to be formed for up to 6 days as successive generations of cystospores function as sporangia.

In rare instances observed, the incipient bulbous exit papilla did not emit protoplasm, but enlarged to form a larger swelling than the papilla, and from this second swelling a long tube grew out. The end of this tube subsequently enlarged to form a third swelling. Thus, an elongate thallus with 3 swellings was formed, each of which later emitted protoplasm or zoospores. *Rhizidiomyces bulbosus*, accordingly, rarely exhibits a tendency to form an elongate thallus with several swellings somewhat similar to but much less extensive than that in *R. hirsutus*, to be described below.

On corn leaves, snake skin, and films of agar the majority of cystospores form a broad germ tube that penetrates the substratum (Fig. 20) and becomes the rudiment of the rhizoidal or absorbing system. It may develop a few (Fig. 25), or several short, blunt-ending branches (Fig. 26), or it may remain as a short or elongate, up to 25 μ long, unbranched peg.

FIGS. 1–18. *Anisolpidium saprobium*. Fig. 1. Pollen grain with three thalli. Fig. 2. Pollen grain with one sporangium which has developed a tapering exit canal. Fig. 3. Sporangium with two exit canals. Fig. 4. Sporangium with a long, branched exit canal. Fig. 5. Beginning of emission of protoplasm. Fig. 6. Later stage. Fig. 7. Cleavage of protoplasm on outside of sporangium. Fig. 8. Loose group of wiggling zoospores. Fig. 9. Zoospores. Fig. 10. Cystospores of various sizes. Fig. 11. Enlarged cystospore with a developing exit canal. Fig. 12. Emission of protoplasm from a larger cystospore. Fig. 13. Cystospore which formed one zoospore. Fig. 14. Encystment of zoospores outside and within sporangium. Fig. 15. Zoospores emerging in succession from sporangium. Figs. 16, 17. Infection of pollen grain. Fig. 18. Group of incipient sporangia growing in slush agar.

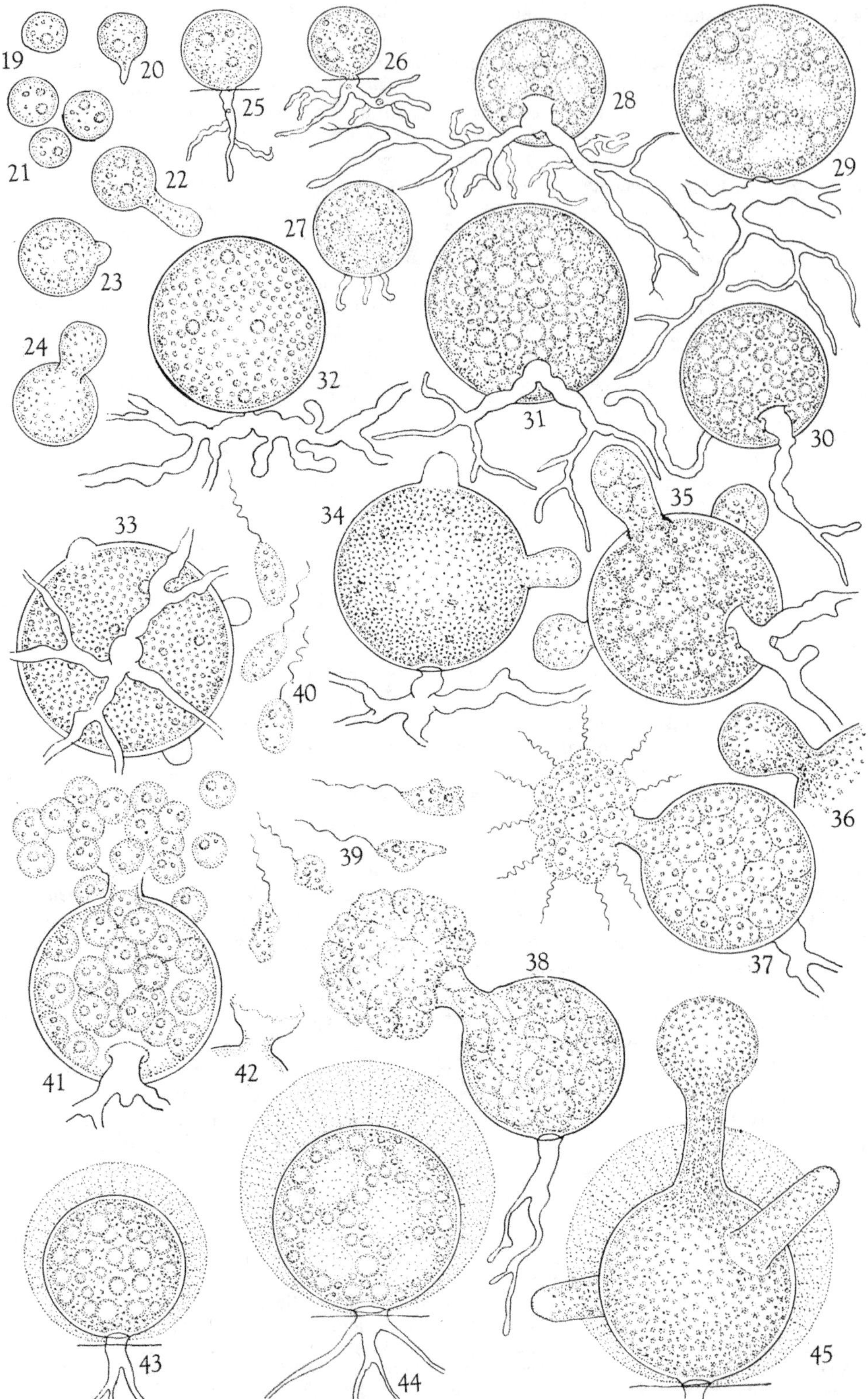
19
20
21
22
23
24
25
26
27
28
29
30
31
32
33
34
35
36
37
38
39
40
41
42
43
44
45

Figure 27 shows a small thallus with 3 short pegs arising at separate points at the base of the incipient sporangium. On agar the rhizoids may be reduced to irregular filaments such as shown in Figure 30. In corn leaves and snake skin, the rhizoids may be more extensively developed and branched (Figs. 28, 29, 31, 32–34), and frequently irregular and nodular in contour.

As in other species of *Rhizidiomyces* and *Hyphochytrium* studied by the author (1939, 1944, 1945, 1964, 1966b, 1967), the young and fully grown sporangia of *R. bulbosus* are filled with large and small conspicuous refractive bodies (Figs. 28–34), and these often appear to be heaped up (Fig. 34), imparting a rather dark and dense appearance to the sporangia in transmitted light. However, as the sporangia mature, the refractive material appears to become more evenly and finely dispersed, and the protoplasm takes on a more evenly coarse, granular appearance (Figs. 32–34).

When *R. bulbosus* was first discovered on corn leaves, a large number of sporangia were observed with zoospores swarming within them, and these emerged singly in succession. From such observations it was considered that this fungus might be a species of the doubtful genus *Rhizidiomycopsis* Sparrow (1960), and not until it had been isolated and grown in axenic cultures was the swarming of zoospores in sporangia explained. In such cultures as well as in sporangia on corn leaves, only a portion of the protoplasm may emerge as a naked mass and cleave on the outside (Figs. 37, 38). The portion remaining in the sporangium also cleaves into zoospores which subsequently swarm and escape through the exit orifice. Accordingly, as noted in the diagnosis, the place of cleavage and maturation varies markedly in this species. In the majority of sporangia, most or all of the protoplasm emerges to the outside, and this appears to be the normal sequence of development. But in other sporangia cleavage and zoospore development may occur before expansion and opening of the exit papillae occurs. In such cases the zoospores may glide around slowly in the sporangium. Sometimes the papillae fail to open, and the zoospores encyst to become cystospores, with the result that a sporangium may be filled with polyhedral cystospores. The peripheral ones may germinate and develop germ tubes through the wall of the sporangium, and in such cases the latter becomes surrounded by a weft of radiating, continuous, or branched germ tubes.

Very frequently in grass and axenic agar cultures, the zoospores outside as well as within sporangia become only feebly motile and creep around as flagellate amoebae (Fig. 39). They usually round up in a short while and become cystospores. As a result, groups of cystospores frequently occur around and within the exit orifices and within the sporangia, as shown in Figure 41. Often, when only a few zoospores and subsequent cystospores remain in the sporangium, they may enlarge and become sporangia within the primary one. As they enlarge and fill the old sporangium, they become polyhedral from contact and pressure. Such sporangia may develop and produce zoospores without the formation of rhizoids.

The zoospores that become actively motile are ovoid to elongate in shape (Fig. 40) and usually contain one larger and several smaller hyaline refractive bodies. Such zoospores are slightly smaller than those of other smooth-walled species of *Rhizidiomyces,* but resemble most closely those of *R. hansonii* Karling, 4–4.5 x 5.5–6.5 μ, and *R. parasiticus* Karling, 3.5–4.2 x 5–7 μ, in size.

As noted in the diagnosis above, *R. bulbosus* usually forms 1 to 3, rarely 4, bulbous, subspherical to broadly clavate exit papillae instead of long, tapering tubes for the emission of the protoplasm. These begin as relatively hyaline protrusions (Fig. 33), the time of development

FIGS. 19–42. *Rhizidiomyces bulbosus.* Fig. 19. Cystospore. Fig. 20. Germination of cystospore by a tube on corn leaf. Fig. 21. Enlarging cystospores. Fig. 22. Cystospore, developing an exit canal, which later formed two zoospores; germ tube or rhizoids lacking. Fig. 23, 24. Stages in formation of a bulbous exit papilla in a larger cystospore which formed six zoospores. Figs. 25, 26. Young thalli on corn leaves; blunt-ended rhizoids reduced and sparingly branched. Fig. 27. Young thallus on agar film; rhizoids reduced to three pegs. Figs. 28, 29, 31, 32. Stages in development of thalli and sporangia on corn leaves, with more extensive rhizoids. Fig. 30. Thallus from agar culture; rhizoids reduced to two filaments. Figs. 33, 34. Stages in development of bulbous exit papillae. Fig. 35. Mature sporangium shortly before dehiscence. Fig. 36. Protoplasm beginning to flow into expanding tip of papilla. Figs. 37, 38. Cleavage of protoplasm within and outside of sporangium. Fig. 39. Amoeboid zoospores. Fig. 40. Motile zoospores. Fig. 41. Cystospores within and outside of sporangium. Fig. 42. Exit papilla after dehiscence.

FIGS. 43–45. *Rhizidiomyces coronus.* Figs. 43, 44. Young and older incipient sporangia enveloped by a corona in which radial striations occur. Fig. 45. Sporangium with three exit canals; central one emitting protoplasm.

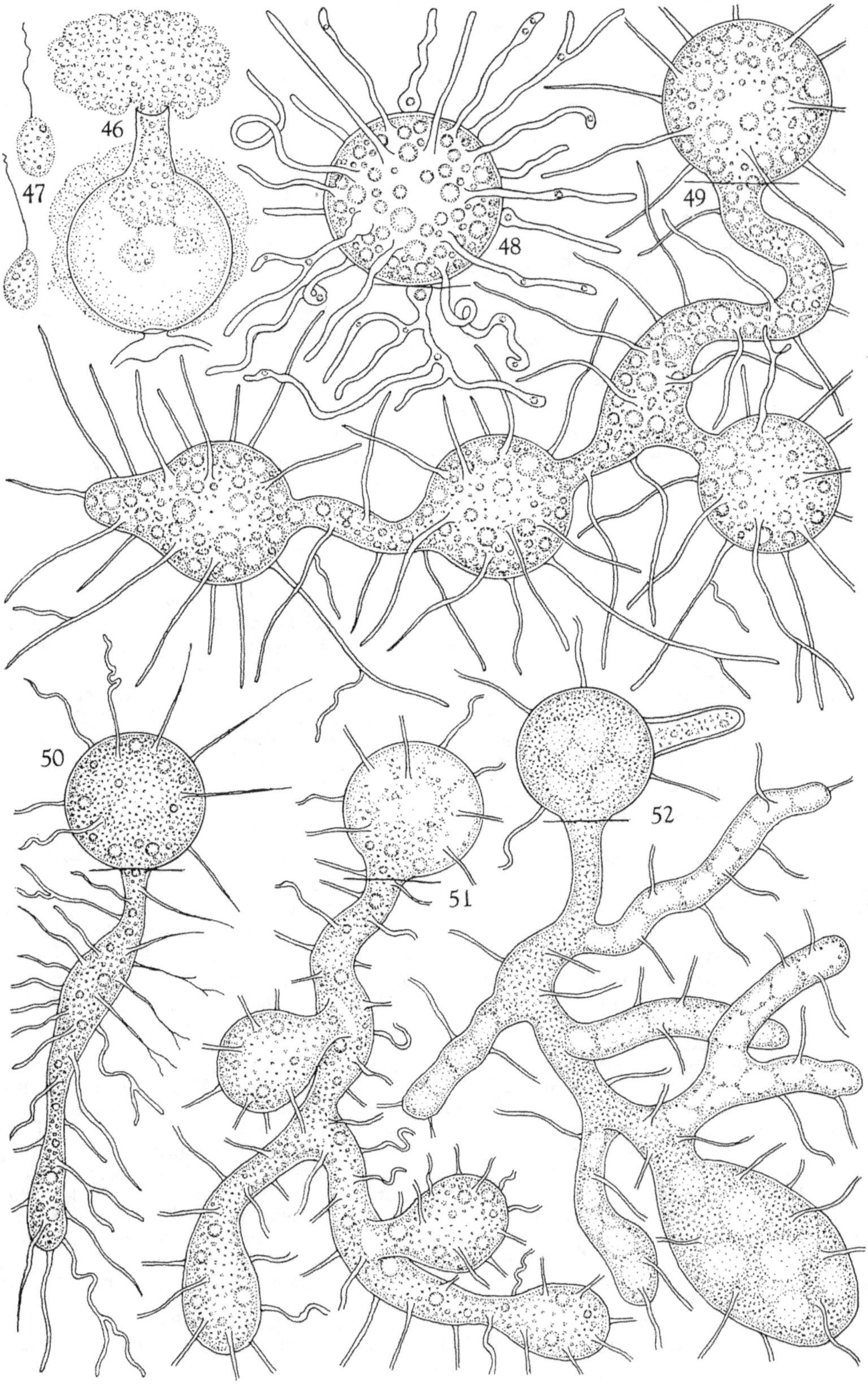
46
47
48
49
50
51
52

of which may vary greatly. In grass cultures in the presence of bacteria, amoebae, and other protozoa, their development may take 2 to 6 hours from the time of their first appearance to the emission of the protoplasm. In axenic agar-culture sporangia mounted in tap water, their development may take only 30 to 70 minutes. As they grow out (Fig. 34), their apex begins to expand as the granular protoplasms moves into them, and within this 30- to 70-minute time span, they become bulbous (Figs. 35, 36) or broadly clavate in shape. As their wall thins out, the protoplasm surges up into them and they expand suddenly and almost explosively in appearance. After the protoplasm has emerged and undergone cleavage and the zoospores have dispersed, the remnants of the exit papillae (Fig. 42) become more sharply visible. When two or more papillae develop and expand, one of them may do so more rapidly than the others, and as the protoplasm surges into it, that in the other papillae may flow back into the sporangium and out into the predominating papilla.

As observed in grass, snake skin, and axenic agar cultures and as described above, *R. bulbosus* exhibits marked variations in structure, development, and, particularly, in the place of cleavage of the protoplasm and the maturation of the zoospores. Although zoospores are usually delimited outside of the sporangium, they are often formed and swarm within the sporangium and emerge singly and in succession from one or more exit orifices. These variations in zoosporogenesis seriously raises the question whether genera of the anisochytrids may be established and recognized on the basis of whether the zoospores are delimited within or outside of sporangia, as Sparrow (1960) has done in the case of *Rhizidiomycopsis* and *Canteriomyces*. Present studies on *R. bulbosus* and also *R. hirsutus*, to be described below, in axenic cultures show that this process may be highly variable and is probably influenced greatly by the microenvironment.

Rhizidiomyces coronus sp. nov.

Sporangia, hyalina, laevia, globosa, 20–63 μ diam.; corona omne sporangium circumdante. Zoosporae ovalovoideae vel productae, 3–3.6 x 5–5.5 μ; flagello 12–15 μ longo. Sporae perdurantes ignotae.

Sporangia predominantly spherical, 20–63 μ diam., hyaline, smooth, each surrounded by 8–12 μ thick, semi-hyaline halo or corona in which faint radial striations occur; usually non-apophysate, sometimes apophysate; apophysis subspherical to ovoid, 6–9 μ diam. Rhizoids extensive and branched. Protoplasm of sporangia emerging usually through 3 slowly developing and tapering exit tubes and undergoing cleavage into zoospores on the outside. Zoospores ovoid to elongate, 3–3.6 x 5–5.5 μ, with several refractive granules; flagellum approximately 12–15 μ long. Resting spores unknown.

Saprophytic on bleached corn leaves and pollen grains on *Pinus sylvestris* in soil sample (AI1) from near Peperao, Aitutaki, Cook Islands.

This species is characterized primarily by the presence of a halo or corona around the sporangia. The corona is a constant accompaniment of the sporangia and appears to be a distinctive characteristic. At least, this is substantiated indirectly by the lack of a corona around the sporangia of species of *Nowakowskiella*, *Rhizophydium*, and *Chytriomyces* that were present in same soil cultures and growing on the same substratum with *R. coronus*. Were the corona a mere accumulation of slime and bacteria around the sporangia, it probably would not be as constant in occurrence and uniform in structure and shape. Also, if it were an accumulation of slime, its presence could be expected occasionally, at least, around the sporangia of the chytrids present in the same cultures.

This species was found only once and in limited quantities in slowly draining fertile Tongaruta clay derived from basalt alluvium, rich in P and K and with a pH range of 5.7 to 5.8. Unfortunately, at the time of its discovery facilities were not available for isolating and growing it in axenic culture in synthetic media. Also, it has not been recovered from a part of the same soil sample that was brought back to the United States. Consequently, the present observations are based on its occurrence and development on corn leaves and pollen grains in a mixed culture with several chytrids. Its development, so far as it is known, is similar to that of other species of *Rhizidiomyces* except that the corona becomes

FIGS. 46, 47. *Rhizidiomyces coronus.* Fig. 46. Cleavage of protoplasm. Fig. 47. Zoospores.

FIGS. 48–52. *Rhizidiomyces hirsutus.* Fig. 48. Spherical sporangium with relatively short, simple or branched, straight or curved and coiled, blunt-ended hairs or rhizoids. Fig. 49. Elongate tubular thallus growing down into agar with three secondary swellings; primary swelling on surface of agar. Fig. 50. Relatively young stage in the development of tubular thallus. Figs. 51, 52. Variations in tubular, branched thalli.

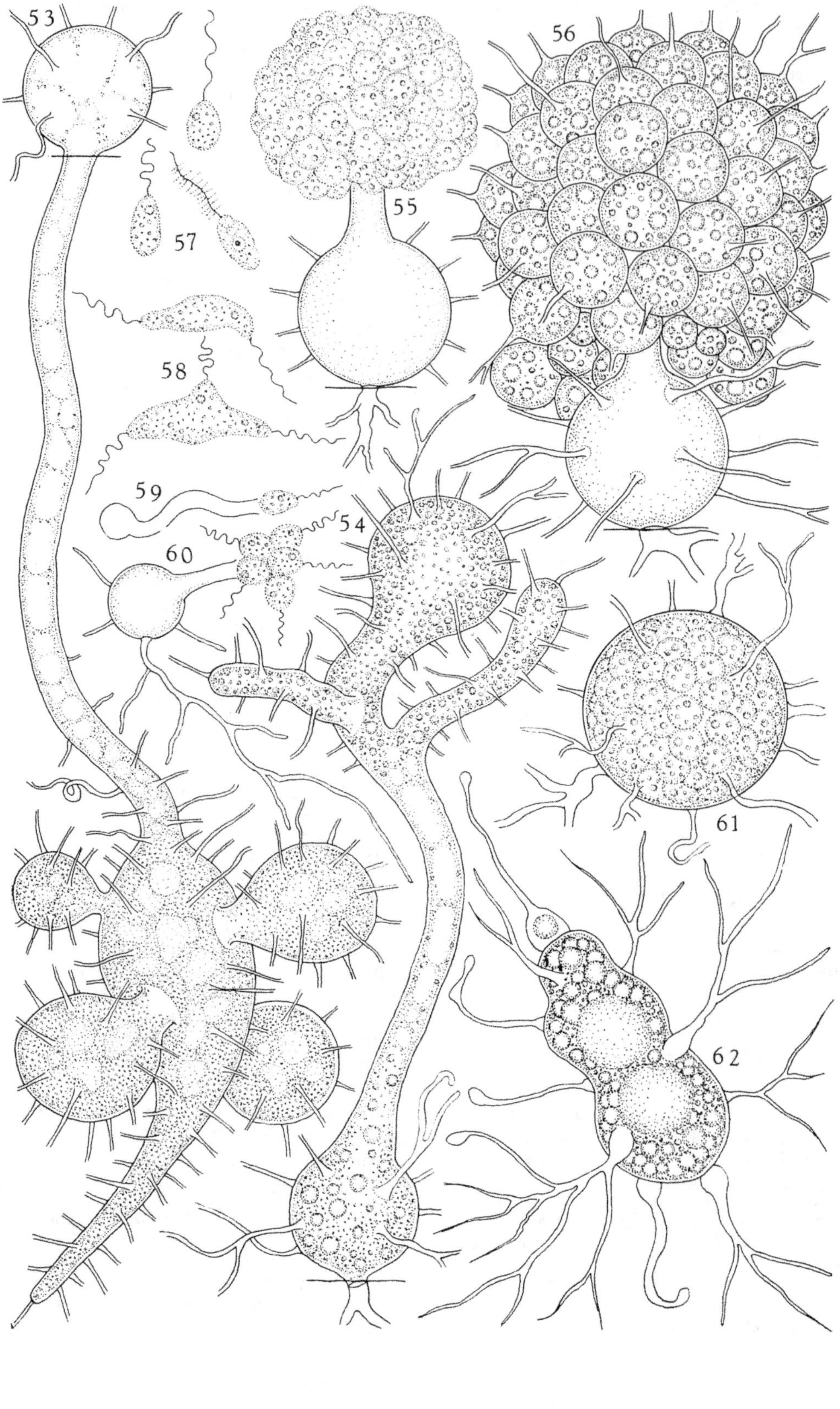
53
54
55
56
57
58
59
60
61
62

visible after the incipient sporangia have attained about ⅓ mature size (Fig. 43). The radial striations in the corona are barely discernible at first, but they become more evident as the sporangia enlarge (Figs. 44, 45) and mature. As in *R. bulbosus* and other species, the sporangia contain numerous refractive globules and bodies of various sizes, but in preparation for sporogenesis these decrease in size as the refractive material that composes them becomes more evenly dispersed. Thus, shortly before the exit tubes begin to develop, the protoplasm has a grayish, coarsely but evenly granular appearance in transmitted light.

The exit canals begin as protrusions and elongate slowly at first, but by the time their tips have extended through the corona, they expand at the apex (Fig. 45) as the protoplasms flows outward. Cleavage and maturation of the zoospores usually occur at the tip of the exit canal (Fig. 46), but in this species also, a portion of the protoplasm may remain in the sporangium and cleave into zoospores. During and after dehiscence of the sporangium the corona may become distorted in shape and reduced in thickness (Fig. 46), but in several dehisced sporangia observed, it was still intact. After the zoospores (Fig. 47) have come to rest and become cystospores, the latter may enlarge and give rise to a few zoospores directly or germinate with a broad germ tube, as in *R. bulbosus* and other species.

Rhizidiomyces hirsutus **Karling, 1945.**

Bull. Torrey Bot. Club **72**: 47, Figs. 1–19

Saprophytic on strips of bleached corn leaves, snake skin, fibrin film, shrimp chitin, bits of hempseed and pollen grains of *Pinus sylvestris* in soil samples AI1, Aitutaki, RI1, RI2, Rarotonga, and MI1, Mangaia Islands, in the Cook group; NI1, NI2, Niue; P6, P10, P17, Pitcairn; MI, Mana Island, Fiji group; WS, Western Samoa; and AS1, Pago Pago, American Samoa.

As indicated above, this species appears to be widely distributed in soils of Oceania. A collection of it from Pitcairn (P17) was isolated and grown in axenic culture on PDY agar; on this medium it varied so markedly in development and thallus structure that the author was inclined at first to regard it as a different species. However, from several months of study of axenic cultures and the transfer of zoospores from such cultures back to corn leaves and snake skin, he is convinced that the Pitcairn isolation is *R. hirsutus*, or possibly a variant of it. When zoospores from axenic cultures were transferred back to corn leaves and snake skin, they usually developed into fairly typical *R. hirsutus* thalli. Only occasionally did elongate ones develop as on agar. However, on corn leaves the tips of the rhizoidal branches did not run out to fine points as the author illustrated them (1945, Fig. 1), but ended more bluntly. But in snake skin the terminal branches were more finely divided and pointed at the tips.

The majority of sporangia on agar were ovoid to spherical (Fig. 48) and bore numerous simple or sparingly to frequently branched filaments or hairs, which were relatively short or elongated, straight or curved, occasionally coiled, and slightly inflated at the tip. Usually, one or more of these hairs became enlarged, more frequently branched, and extensive and served as anchoring rhizoidal axes. They developed at the base of the sporangium (Figs. 55, 56) or from several points on its periphery.

A large percentage of the thalli grew down into the agar and developed into elongate, simple or branched tubular structures in which one to several swellings occurred as shown in Figures 49 to 54. Many other variations in thallus structure occurred in agar cultures, but Figures 49 to 54 are sufficient, in the author's opinion, to illustrate the extent of these variations. Some of the thalli attained a length of 2.4 mm. and a diameter up to 24 μ, and bore as many as five swellings. In the development of such thalli the cystospores on the surface of the agar enlarged to sporangium size and developed numerous hairs, after which they formed 1, or occasionally 2 or 3, tubes which grew down into the agar (Fig. 50). Hairs or simple to branched rhizoids usually developed from the periphery of the tubes; when first formed these hairs were relatively narrow or delicate with finely pointed tips, as Fig. 50 illustrates for

FIGS. 53–61. *Rhizidiomyces hirsutus*. Fig. 53. Greatly elongate thallus with a terminal carrot-like swelling from which four secondary swellings are budding out. Fig. 54. Tubular thallus growing up into the air from the surface of agar. Fig. 55. Monocentric thallus on agar with cystospores at exit orifice. Fig. 56. Similar thallus with cystospores which are developing into sporangia. Fig. 57. Living and fixed and stained zoospores. Fig. 58. Bi- and tri-flagellate zoospores. Fig. 59. Small cystospore in water which developed a long exit canal and one zoospore. Fig. 60. Small thallus in water which formed five zoospores. Fig. 61. Sporangium on agar within which cleavage occurred.

FIG. 62. Thallus of *Hyphochytrium oceanum* with radiating, branched rhizoids with fine tips and blunt-ended hypha-like rhizoids.

some of them. Later, they became coarser and blunt at their tips. Such hairs extended for distances up to 210 μ with a diameter of 2 to 3 μ and were frequently undulating in contour. For the sake of economy of space they are not drawn to full length in Figures 49 to 54.

As in the incipient sporangia, the tubes contained numerous refractive globules and bodies of various sizes (Fig. 49). As the tubes elongated, they remained simple or branched once to several times (Figs. 51, 52), and the tips of the broad branches often enlarged into ovoid to subspherical swellings (Fig. 51). Figure 52 shows a frequently branched thallus in which only the terminal branch had enlarged while the other branches were highly vacuolate. Other thalli formed one to several swellings along the length of the tube (Fig. 49). Figure 53 shows a thallus which attained a length of 2.2 mm. and formed a carrot-like swelling at its end, from the surface of which 4 ovoid secondary swellings budded out. In this thallus hairs or rhizoids were lacking on the upper portion of the vacuolate tube. Usually, the enlarged cystospore or swelling on the surface of the agar remained full of protoplasm (Figs. 49, 50, 52) as the tube grew downward, but sometimes it became vacuolate and almost empty (Figs. 51, 53). Occasionally, instead of growing down into the agar, the tubes elongated upward into the air and developed branches and swellings (Fig. 54), particularly if the relative humidity in the petri dishes was high.

The tubes that grow down into the agar or up into the air and become as extensive as described above appear to be modified exit tubes which proliferate and develop hairs and swellings under favorable nutritional conditions. At least, this is suggested by the fact that when young thalli such as the one shown in Figure 50 were mounted in water, the tip of the tube enlarged fairly rapidly and functioned as an exit tube as the protoplasm flowed out and divided into zoospores. In the Brazilian collection, the author (1945) noted that large sporangia sometimes developed long (170 μ) exit tubes which branched once to several times without emitting protoplasm, and it is not improbable that such tubes are comparable to those which developed in agar in the Pitcairn isolate.

The primary swelling on the surface of the agar, as well as the secondary and tertiary ones within, developed exit canals and usually emitted naked protoplasm when a tubular thallus was mounted in fresh water. In this respect such thalli might be regarded as polycentric, but the swellings are not delimited by septa, as are the sporangia of the Cladochytriaceae and Catenariaceae. The tubular thalli are coenocytic so far as the author has been able to determine. Part or all of the protoplasm of the whole thallus emerged through the exit canals which developed on the swellings and divided into zoospores. That which remained within also divided into zoospores, or encysted as a spherical mass of protoplasm. Occasionally, the development of the exit canal on the primary sporangium was arrested and became thick walled and nonfunctional (Fig. 52). Also, in old cultures the wall of the tubular portion as well as that of the swellings became markedly thick, with numerous tylose-like plugs projecting inward.

On relatively hard (2–2.2%) agar media, irregular clumps of sporangia frequently occurred and projected up on the surface. Such clumps became up to 7 mm. in diameter and consisted of sporangia in various stages of development. These clumps developed when the masses of zoospores formed at the mouth of the exit canals failed to become motile and encysted as cystospores (Fig. 55). Subsequently, they enlarged to become sporangia (Fig. 56). These in turn, apparently, gave rise to zoospores which also encysted, and thus built up the large, irregular masses of sporangia. When such masses were flooded with a film of sterile water, motile zoospores were formed and spread over the entire surface of the agar media in the petri dish. As the agar gradually dehydrated in such cultures, irregular raised clumps of sporangia developed again, and these imparted a grayish white appearance to the surface of the agar plate.

Like *R. bulbosus, R. hirsutus* may vary with regard to the place of cleavage and maturation of the zoospores. In the Pitcairn isolate the protoplasm normally and usually emerged as a naked mass and divided into zoospores on the outside as described previously by the author (1944, 1945). Fairly often, however, part of the protoplasm remained in the sporangium and cleaved into zoospores which became motile and emerged in succession. If they failed to emerge, they lost their motility and became cystospores which subsequently developed into secondary sporangia within the old one. When no exit canals developed, the protoplasm occasionally divided into zoospores which did not become motile or glided about slowly in the sporangium before becoming cystospores (Fig. 61). This occurred, also, infrequently in the tubular thalli such as those shown in Figures 49 to 54. In such cases most of the cystospores degenerated, but the peripheral ones developed rhizoids through the sporangium or thallus wall. As a result, these structures were enveloped by wefts of radiating hairs and rhizoids.

In water surrounding agar films on cover slips, the cystospores frequently enlarged in size and gave rise directly to zoospores, as in *R. bulbosus*, or germinated with a broad germ tube which developed into a rhizoidal axis (Fig. 60). Sometimes, a small cystospore developed a relatively long exit tube and formed a single zoospore (Fig. 59). From other thalli odd numbers such as 5 (Fig. 60), 7, 11, and 14 zoospores were formed, and in some cases the zoospores were unequal in size.

In connection with the extensive growth of *R. hirsutus* described above, mention should be made again of Persiel's *Hyphochytrium* sp. H28, which developed monocentric holocarpic thalli in pollen grains of *Corylus*. However, when it was transferred to agar media, it grew out to form a complex of more than 100 sporangial rudiments. Her discovery and description of this species are additional indications of the markedly variability which anisochytrid species may exhibit.

***Rhizidiomyces bivellatus* Nabel, 1939.**
Arch. f. Mikrobiol. **10:** 537, Figs. 1–7

Saprophytic on bleached corn leaves in soil sample NI1, Niue.

***Rhizidiomyces hansonii* Karling, 1944**
Amer. J. Bot. **31:** 396, Figs. 35–64

Saprophytic on bleached corn leaves and snake skin in soil samples P16, Pitcairn; RI1, Rarotonga, and AI1, Aitutaki, Cook group; and NI1, Niue.

***Hyphochytrium catenoides* Karling, 1939**
Amer. J. Bot. **26:** 512, Figs. 1–18

Saprophytic is bleached corn leaves and snake skin in soil samples P17, Pitcairn; NI1, Niue; RI1, Raratonga, AI1, Aitutaki, MI1, Mangaia, Cook Islands; and VL1, Fiji Islands.

***Hyphochytrium oceanum* Karling, 1967**
Sydowia **20:** (in press)

Saprophytic in bleached corn leaves, onion skin and snake skin in soil samples RI1, AI2, MI1, Cook Islands; NI1, Niue; P6, P8, P16, P17, Pitcairn; VL2, VL6, YI, TI2, Fiji Islands; and AS, American Samoa.

This is one of the most widely distributed species found by the author, and it occurred in abundance in strips of corn leaves and onion skin. In a previous study of this species from New Zealand, the author (1967) reported that the rhizoids or absorbing organs were usually hypha-like and ended bluntly or were sometimes inflated at the tips. In additional studies on this species, the rhizoids were found to branch frequently and run out to fairly fine points, as shown in Figure 62. Dehiscence of the sporangia occurred sporadically and only now and then in fresh water mounts of infected grass leaves, although new infections occurred abundantly when leaf strips were added to the watered soil cultures. Because dehiscence occurred so seldom under the conditions noted above, all attempts to isolate and grow *H. oceanum* in axenic cultures on synthetic media have failed so far, but the efforts are being continued.

Most of the sporangia observed were monocentric and eucarpic like species of *Rhizidiomyces,* but occasionally polycentric ones were also found. Whether this species should be regarded as a member of *Rhizidiomyces* or *Hyphochytrium* remains to be established. If it is placed in the former genus, it will be first known intramatrical member of this taxon. The fact that *R. hirsutus* and *Hyphochytrium* H28 Persiel may vary so markedly in thallus structure and organization shows that the categories of holocarpy, eucarpy, monocentricity, and polycentricity are not sharply defined and constant criteria of classification in some anisochytrids.

Summary

Anisochytrids are commonly and widely distributed in the soils of Oceania and develop in great abundance when favorable substrata are floated on watered soil samples. So far nine species have been isolated and identified, among which *Anisolpidium saprobium, Rhizidiomyces bulbosus,* and *R. coronus* are described as new species. In axenic cultures on agar media *R. hirsutus* varied markedly in development and structure, often becoming extensively elongate and tubular and forming several globular swellings which function as sporangia.

ACKNOWLEDGMENTS

The author is grateful for the support of this study by a grant from the National Science Foundation, and also to Drs. Margaret De-Minna and S. A. Widdowson of the Soil Bureau, Taita, N.Z., and Mr. Stanley Stewart Cameron, Department of Geography, University of Otago, Dunedin, N.Z., for their assistance in collecting soil samples.

Thanks are also due to Dr. H. M. Martin, Jr., of the Department of Classics at the University of North Carolina, who translated the Latin diagnoses.

Index to Oceanic Soils Examined

Cook Islands

RI1. Rarotonga Island. Tamarua clay loam, waterlogged soil, pH. 5.5–6.5, from taro field.
RI2. Te Manga clay loam, pH. 5.8–6.1, from steep to moderately steep slope.
AI1. Aitutaki Island. Tongaruta clay, near Peperao, derived from basalt alluvium, pH. 6.8–7.3, slowly-draining, fertile, rich in P and K.
MI1. Mangaia Island. Ivarna clay loam, pH. 4.5–4.6, from alluvium, deeply weathered, infertile.

Niue

NI1. Shallow soil, pH. 7.1, from under *Crotalaria* near Vaiea; a complex of Fonaukula silty loam with 8 per cent calcium carbonate.
NI2. Shallow soil similar to NI1. pH. 7.4, under *Digitaria decumbens*, near Vaiea with 3 per cent calcium carbonate.
NI3. Deep soil, pH. 6.4, under rye grass.

All of these niue soils had been fertilized and were quite high in organic phosphorus and low in potassium.

Fiji Islands

VL1. Viti levu. Red volcanic soil from flower bed at Mocambo hotel, Nadi.
VL2. Viti levu. Sandy loam from sugar cane field between Nadi and Latouka.
VL3. Viti levu. Soil from park in front of Latouka hotel, Latouka.
VL4. Viti levu. Soil from steep hill about 10 miles west of Latouka along the road to Mba.
VL5. Soil from rice paddy near Mba.
VL6. Viti levu. Soil from sugar cane field at Rarawai near Mba.
VL7. Viti levu. Red volcanic soil from hills above Latouka near the radio station.
VL8. Similar soil under mango tree in the same general vicinity.
YI. Yunaya Island. Sandy soil from under tapioca and banana plants.
CI. Castaway Island. Leaf mold and sandy soil under trees.
MI. Mana Island. Sandy soil and leaf mold under trees and vegetation.
TI1. Tavua Island. Sandy soil in garden in a village.
TI2. Tokoriki Island. Sandy soil from under vegetation.

Western Samoa

WS5. Clay loam from slopes near Apia airport.

American Samoa

AS1. Red volcanic soil above Pago Pago.
AS2. Red volcanic soil in vicinity of TV telecasting towers above Pago Pago.

Pitcairn

Thirty-five collections of soil were made at various localities, and these are numbered consecutively. Unfortunately the data on the locations, types of soil, origin and pH were lost.

REFERENCES

BESSEY, E. A. 1943. Notes on Hawaiian fungi. Papers Mich. Acad. Sci., Arts, Letters **28**: 3–8.
CANTER, H. M. 1950. Studies on British chytrids, IX. *Anisolpidium stigeoclonii* n. comb. Trans. Brit. Mycol. Soc. **33**: 335–344, Figs. 1–6. Pls. 24–26.
GREGORY, H. E., AND C. K. WENTWORTH. 1937. General features and glacial ecology of Mauna Kea Hawaii. Bull. Geol. Soc. Amer. **48**: 1719–1742.
JOHNSON, T. W., JR. 1950. A study of an isolate of *Brevilegnia* from New Caledonia. Mycologia **52**: 242–252, Figs. 1a–1q.
KARLING, J. S. 1939. A new fungus with anteriorly uniciliate zoospores: *Hyphochytrium catenoides*. Amer. J. Bot. **26**: 512–519, Figs. 1–18.
———. 1943. The life history of *Anisolpidium ectocarpii* gen. nov. et sp. nov., and a synopsis and classification of other fungi with anteriorly uniflagellate zoospores. Amer. J. Bot. **30**: 637–648, Figs. 1–21.
———. 1944. Brazilian anisochytrids. Amer. J. Bot. **31**: 391–397, Figs. 1–64.
———. 1945. *Rhizidiomyces hirsutus* sp. nov., a hairy anisochytrid from Brazil. Bull. Torrey Bot. Club **72**: 47–51, Figs. 1–19.
———. 1964. Indian anisochytrids. Sydowia **17**: 193–196, Figs. 1–6.
———. 1965. Some zoosporic fungi of New Zealand. I. Sydowia **19**: 213–226, Pl. XLVI.
———. 1966. The chytrids of India with a supplement of other zoosporic fungi. Beihefete zu. Sydowia **6**: 1–125.
———. 1967. Some zoosporic fungi of New Zealand. IX. Hyphochytriales or Anisochytridiales. Sydowia **20**: (In press).
NABEL, K. 1939. Über die Membran neiderer Pilze, besonders von *Rhizidiomyces bivellatus* nov. spez. Arch. f. Mikrobiol. **10**: 515–541, Figs. 1–7.
PERSIEL, I. 1960. Über die Verbreitung niederer Phycomyceten in Böden verschiedenen Höhenstufen der Alpen und an einigen Standorten subtropischer und tropischer Gebirge. Arch. f. Mikrobiol. **36**: 257–282, Figs. 1–14.
PERSIEL, I., E. SCHOLZ, AND R. HARDER. 1966. Chytridiales und einige niedere Phycomyceten aus Ozeanien. Arch. f. Mikrobiol. **53**: 173–177.
SPARROW, F. K., JR. 1948. Soil Phycomycetes from Bikini, Enirwetok, Rongerik and Rongelap atolls. Mycologia **40**: 445–453, Figs. 1–16.
———. 1960. Aquatic Phycomycetes. 2d ed. vii–1187 pp. Univ. of Michigan Press, Ann Arbor.
———. 1965. The occurrence of *Physoderma* in Hawaii, with notes on other Hawaiian Phycomycetes. Mycopath. Mycol. Appl. **25**: 119–143, Figs. 1–78.

Aquatic Fungi of Iceland: Introduction and Preliminary Account

T. W. JOHNSON, JR.
Department of Botany, Duke University, Durham, N. C.

Save for a report of *Saprolegnia ferax* (Gruith.) Nees, four species of *Synchytrium* and five of *Physoderma* published in 1931 by Larsen, and a report of *Rozella marina* Sparrow by Johnson (1966), the aquatic fungi of Iceland and its territorial waters are virtually unknown. In 1964, with the volcanic upthrust Surtsey as the focal point, a taxonomic-ecological study of aquatic fungi in Iceland was begun. The investigation centered on the marine lignicolous and alga-inhabiting Ascomycetes and Deuteromycetes, the Chytridiomycetes and Oömycetes associated with marine filamentous algae and phytoplankters, and the Chytridiomycetes and Oömycetes of fresh water and soils. This paper is an account of some of the fungi collected during the period from March, 1964, to September, 1966. A brief account of the development of Surtsey, the volcanic island, is included by way of introduction to the Surtsey Research Project.

Surtsey

On 14 November 1963 a submarine volcanic eruption began on the fissured submarine ridge of the North Atlantic at 63° 18′ 30″ N, 20° 36′ 30″ W, three nautical miles west-southwest of Geirfuglasker, the southernmost island of Iceland. Initially the eruption was characterized by steam and vapor, but one day after the eruption began a small island some 10 m. high appeared. During the early stages of its development, Surtsey consisted of a single tephra cone. Subsequently two to four vents appeared and the island began to increase in size. By mid-December, 1963, only a single vent was active, and the island became nearly circular. With sea water flooding the vent as the volcanic sand was eroded by wave action, tephra continued to build up until the sea was blocked from access to the crater vent. When this occurred, the effusive phase began and lava flow commenced. This flow was chiefly to the south and southwest, forming a semicircular lava dome. Surtsey is thus a basaltic shield volcano resting on a tephra socle.

The island enlarged to an area of 2.3 $km.^2$ above the ocean level and some 300 m. above the ocean floor. The highest point was some 170 m. above sea level. Because of the predominant lava-flow direction, the southern and western coast consists of hardened lava sheets and lava blocks at the water's edge, but is largely volcanic sand along the northern and northeastern coasts. A small saline lake was trapped on the northeastern side of the island during the effusive stage of the eruption.

In 1965, two new eruptions occurred northeast and east-northeast of Surtsey. These eruptions built into small tephra islands, but did not develop as the submarine fissures became closed. Subsequently, wave action obliterated these new small islands. The history of the development of Surtsey is described and illustrated in detail by Thorarinsson (1964).

The newly developed land mass is admirably suited to biological studies, chiefly those dealing with colonization and succession. Surtsey, however, has significant geological and oceanographic implications as well, serving as a site for the study of tectonic forces, seismic refraction, lava viscosity, the volatile constituents of the magma, and the sublimates and encrustations of the lava flow. Surtsey also gave an opportunity to explore the unique aspect of the relationship of the upthrust of the Palagonite Formation in Iceland itself.

Recognizing the biological, geophysical, and geochemical implications of the eruption, scientists from Iceland and the United States formed the Surtsey Research Society to organize and stimulate research on the various aspects of the island and the adjacent waters and land masses. The Society, founded in 1965, is a chartered, independent, international organization charged with coordinating the research on Surtsey and related land masses including the mainland of Iceland. Biological research includes studies of colonization by birds, establishment of algal populations, and investigations of the composition of drift deposits. Long-range ecological investigations of the terrestrial and marine flora and fauna are planned, and initial or preliminary studies are in progress.

Mycoflora

It is not the purpose of this paper to report the total aquatic mycoflora known to occur in Iceland. Rather, in the following account some common representatives are listed as are a few fungi of less common occurrence. Where ap-

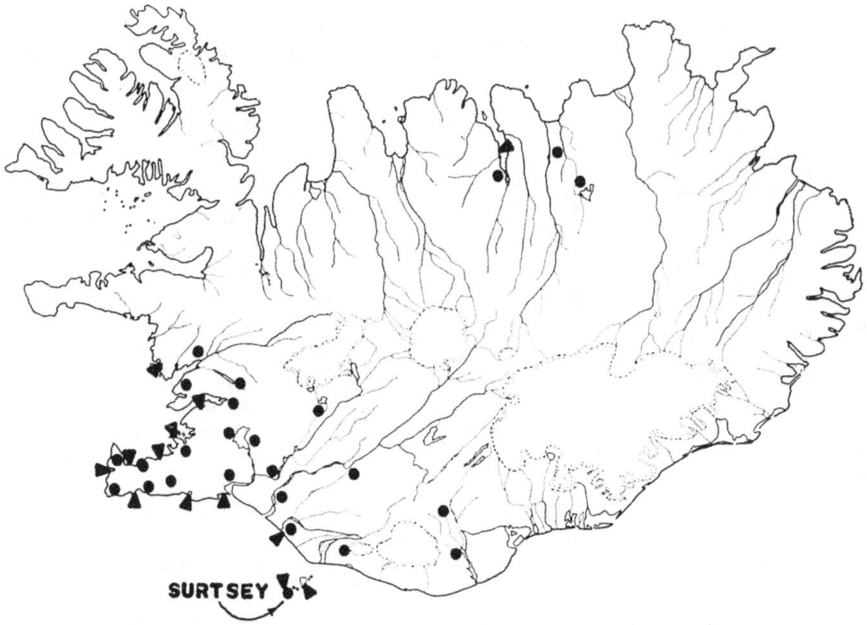

FIG. 1. Outline map of Iceland. The centers of the major collecting areas for fresh-water and soil-inhabiting fungi are shown as solid circles; marine fungi have been collected in the general coastal areas indicated by the solid triangles.

propriate, some descriptive notes are added, or comments are made on speciation. All of the species included are new records for Iceland. Distributional records are purposely brief since collecting sites are regularly being extended, and knowledge of distribution is therefore changing. Figure 1 shows the chief areas of collecting. Vast areas of the island still remain unexplored mycologically.

Chytridiomycetes

1. *Rhizophydium globosum* (Braun) Rabenhorst. Common in volcanic soils of pasture lands. This complex species is represented by many variants, hence the taxonomy of the specimens at hand needs additional consideration.

2. *Rhizophlyctis rosea* (deBary and Woronin) Fischer. Although this species is perhaps the most widespread of all Chytridiomycetes, it is not common in Icelandic soils. The sporangia are predominantly a pale rose-red.

3. *Phlyctochytrium punctatum* Koch.

4. *Phlyctochytrium irregulare* Koch. This and the previous species are found in lava soils below moss tussocks. The characteristics agree well with the descriptive matter given by Koch (1957).

5. *Olpidium pendulum* Zopf. Collected once in pine-pollen bait in water from a roadside stream near hot-spring vents. The collection is small; the few resting spores with persisting infection tubes seem to confirm the identity of this species.

6. *Nowakowskiella elegans* (Nowakowski) Schroeter. Found once in bits of decaying grass leaves, from a small pasture pond. None of the sporangia in the rather sparse material possessed discharge tubes.

Oomycetes

7. *Achlya spiracaulis* Johnson. This species occurs frequently in waters from small ponds and lakes in southern Iceland, and may be the most common of the water molds. Its only other records include Michigan and Louisiana. Some isolates tend to show variation in the degree of coiling in the oogonial stalks, and in a few specimens, the tendency toward coiling diminishes with subculturing.

8. *Achlya conspicua* Coker. In lava soil and water. Although the oospores seldom mature, they are clearly eccentric when they do.

9. *Saprolegnia turfosa* (Minden) Gäumann. In pasture soil and water. Coker's illustrations (1923: pl. 20, figs. 8, 11–13) are probably of this species.

10. *Saprolegnia hypogyna* Pringsheim. In pasture and lava soils and in water; common.

11. *Aphanomyces irregulare* Scott. In water. This species now includes some isolates pre-

viously reported under the name *Aphanomyces scaber*.

12. *Pythium* spp. Species of this genus are commonly collected on hempseed bait in water. A number of isolates are as yet unidentified, but the following have been confirmed: *P. torulosum* Coker and Patterson, *P. rostratum* Butler, *P. dissotocum* Drechsler, and *P. monospermum* Pringsheim. All are known to occur in the west-central inland portion of Iceland.

13. *Ectrogella perforans* Petersen. Known from a single collection of *Striatella unipunctata* from the southwest coast of the Reykjanes Peninsula.

14. *Aphanomycopsis bacillariacearum* Scherffel. Parasitic in an unidentified species of *Pinnularia* from the southern coast near Grindavik. There were no oöspores in this rather sparse, single collection.

15. *Thraustochytrium* sp. An unidentified species in this genus occurred on pine pollen in a salt-water sample from a tidal pool. It is possibly a variant of *T. proliferum* with globose rather than obpyriform sporangia. It is not *T. globosum* Kobayasi and Ookubo, since the sporangia proliferate internally.

16. *Olpidiopsis saprolegniae* var. *saprolegniae* (Braun) Cornu. The host species, a *Saprolegnia*, is unidentified since the sexual apparatus was absent. Ornamentations on the resting spores are on occasion warts rather than the spines that are characteristic of the species. Save in one instance, companion cells are absent from the resting spores; none of the sporangia is spiny.

17. *Pontisma lagenidioides* Petersen. A single collection in dead cells of *Ceramium* (possibly *rubrum*) from a rocky shore line near Alftanes in southwestern Iceland.

18. *Sirolpidium bryopsidis* (deBruyne) Petersen. Occurs as linear, unbranched chains of sporangia which disarticulate. A single collection in dead *Bryopsis plumosa*.

19. *Lagenidium* sp. A single collection in a species of *Pythium* collected on hempseed from a water sample taken in a pool in old lava beds. In most major respects the fungus resembles *L. pythii* Whiffen, but the organism failed to develop oogonia in culture, and the host plant did not survive.

20. *Leptolegniella keratinophilum* Huneycutt. Common on bits of snakeskin bait in water and pasture soil samples.

In addition to the foregoing "Phycomycetes," a large number of specimens remain unidentified. Among genera represented by such material are *Sorosphaera, Chytridium, Hyphochytrium, Olpidium, Entophlyctis, Cladochytrium, Rhizidium, Petersenia,* and *Pleotrachelus*.

The southern and southwestern coastal waters of Iceland seem reasonably rich in lignicolous marine fungi. The subsequent sections deal with some Fungi Imperfecti and Ascomycetes collected on driftwood from various shore-line locations in southwest Iceland (Fig. 1).

Fungi Imperfecti

21. *Phoma* spp. Several collections are extant. None of the lignicolous marine species in this genus has been formally circumscribed; hence the identities of the various collections remain unknown. In driftwood on Surtsey, among other locations.

22. *Cirrenalia macrocephala* (Kohlmeyer) Meyers and Moore. In its usual aspect, the recurved conidia of this fungus have a dark- or reddish-brown terminal cell and yellowish to light-brown subterminal ones. In the material at hand, the conidia are concolorous.

23. *Culcitalna achraspora* Meyers and Moore. The conidia of this fungus are structurally like those of species of *Trichocladium*. Whether the species should be retained in *Culcitalna* or transferred to *Trichocladium* turns on the presence or absence of sporodochia. No such fructification type is found in the Iceland material, although conidiophores may be densely clustered and thereby give the superficial appearance of sporodochia. The fungus presumably produces sporodochia in culture; until intensive culture work confirms this for the Iceland material, a generic reassignment is not in order.

24. *Piricauda pelagica* Johnson. Uncommon. The identity of this species with *Piricauda* remains in doubt.

25. *Cremasteria cymatilis* Meyers and Moore. This species may well be merely the conidial stage of an ascomycete. It has been observed in connection with fructifications of *Halosphaeria mediosetigera* and on wood devoid of ascocarps. Evidence from culture work points to the conidia of *C. cymatilis* as being nothing more than chlamydospores in vegetative hyphae.

26. *Helicoma-Zalerion* complex. Occasionally, on driftwood from the south coast, dematiaceous fungi assignable to genera having species with coiled conidia are found. Interspersed with these are conidia that are branched or irregular as well. Indeed, in one collection, small colonies producing *Helicoma-, Zalerion-* and *Dictyosporium*-like conidia were common. The entire complex must be studied compara-

tively, and in culture, before positive assignments can be made.

27. *Diplodia orae-maris* Linder. The conidia of the single collection are, like those reported by me in 1956, larger than described by Linder: 8–12 x 5–7 μ. This does not constitute a sufficient difference on which to base another species, as the Kohlmeyers (1964) suggest.

28. *Dinemasporium marinum* Nilsson. A single collection on driftwood washed ashore on Surtsey. The conidia are 11–15 x 3–6.5 μ, thus having a slightly greater diameter than described by Nilsson from the Denmark material. The slender, terminal setae on the conidia mark this species.

29. *Nia vibrissa* Moore and Meyers. Collected once, in an embayment near Reykjavik. According to the authors of the species, the conidia are produced on a sclerotium-like fructification, but Kohlmeyer believes the fungus to be sphaeropsidaceous. The few conidia observed in the Iceland material were found associated with what may have been broken pycnidia.

Ascomycetes

30. *Ceriosporopsis halima* Linder. Common on driftwood.

31. *Ceriosporella calyptrata* (Kohlmeyer) Cavaliere. Since the fructifications of this species are stromatic, Cavaliere (1966a) was correct in removing it from the genus *Ceriosporopsis*.

32. *Corollospora maritima* Werdermann. Found twice, on driftwood. The equatorial appendages are stiff and setose.

33. *Halosphaeria appendiculata* Linder. A single collection with characteristics closely paralleling those given by Linder. The Kohlmeyers (1964) list *Remispora ornata* Johnson and Cavaliere as a synonym. Linder's type material, which Kohlmeyer examined, is poorly preserved, and does not show more than vague, general similarities to the *Remispora*. Certainly ascocarp wall structure in the two species is very different.

34. *Halosphaeria mediosetigera* Cribb and Cribb. In one collection, the ascospore diameter is predominantly 11–15 μ. This size range clearly overlaps the diameters reported for *H. mediosetigera* var. *grandispora* Kohlmeyer, a variety reduced by Johnson and Sparrow (1961) but retained by the Kohlmeyers (1964).

35. *Amphisphaeria maritima* Linder. Collected on drifted segments of bamboo from Surtsey and the Vestmannaeyjar, as well as along the south coast of Iceland. Cavaliere (1966b) has shown this species to be stromatic, in which case it probably cannot be retained in *Amphisphaeria*. It most assuredly is not assignable to *Microthelia*, a valid genus of lichens, as the Kohlmeyers (1964) maintain.

36. *Lulworthia* spp. Several collections are extant, but the identifications remain unconfirmed.

Summary

Six species of Chytridiomycetes, seventeen Oömycetes, nine marine Fungi Imperfecti and six marine Ascomycetes are reported for the first time from Icelandic soils and waters. In this preliminary account of the aquatic mycoflora of Iceland eleven additional genera are mentioned. The development of the submarine volcanic upthrust, Surtsey, and the coordinated Surtsey Research Project are defined.

ACKNOWLEDGMENTS

This study was supported chiefly by the U.S. Atomic Energy Commission through Contract AT-(40-1)-3556. Partial support from the Office of Naval Research, Contract No. 1181 (15) NR 104–820, initiated the mycological efforts. The assistance and encouragement of the National Research Council of Iceland and the Surtsey Research Society (Surtseyjarfélagid) are acknowledged with gratitude. Many individuals contributed of their time and facilities to insure that the work could be carried out with a maximum of efficiency and a minimum of effort. In particular, I wish to thank Mr. Steingrimur Hermannsson, Chairman of the National Research Council, Dr. Eythor Einarsson of the Museum of Natural History, and Dr. Unnsteinn Stefánsson, chemical oceanographer with the Marine Research Institute. Professor A. R. Cavaliere, Gettysburg College, and Dr. Roland Seymour and Mr. K. L. Howard of Duke University assisted in the collection and identification of some specimens.

LITERATURE CITED

CAVALIERE, A. R. 1966a. Marine Ascomycetes: ascocarp morphology and its application to taxonomy. I. *Amylocarpus* Currey, *Ceriosporella* gen. nov., *Lindra* Wilson. Nova Hedwigia **10**: 387–398.

———. 1966b. Marine Ascomycetes: ascocarp morphology and its application to taxonomy. IV. Stromatic species. Nova Hedwigia **10**: 438–452.

COKER, W. C. 1923. The Saprolegniaceae. Univ. North Carolina Press, Chapel Hill. 201 pp.

JOHNSON, T. W., JR. 1966. *Rozella marina* in *Chytridium polysiphoniae* from Icelandic waters. Mycologia **58**: 490–494.

JOHNSON, T. W., JR., AND F. K. SPARROW, JR.

1961. Fungi in Oceans and Estuaries. J. Cramer, Weinheim. 668 pp.

KOCH, W. J. 1957. Two new chytrids in pure culture, *Phlyctochytrium punctatum* and *Phlyctochytrium irregulare*. Jour. Elisha Mitchell Sci. Soc. **73**: 108–122.

KOHLMEYER, J., AND ERIKA KOHLMEYER. 1964. Synoptic Plates of Higher Marine Fungi. 2d ed. J. Cramer, Weinheim. 64 pp.

LARSEN, P. 1931. Fungi of Iceland. *In* The Botany of Iceland, Vol. II, Part 3.

THORARINSSON, S. 1964. Eyjan nýja i Atlantshafi: Surtsey. Almenna Bókafélagid, Reykjavik. 55 pp.

The Effect of Colchicine on *Olpidiopsis incrassata* and Its Host, *Saprolegnia delica*

MIRIAM K. SLIFKIN[1]
Department of Botany, University of North Carolina at Chapel Hill, N. C.

Introduction

The reaction of all stages in the life cycle of *Saprolegnia delica* Coker to the chemical colchicine was described in an earlier work (Slifkin, 1967). Colchicine inhibited the growth and development of the sexual stages, but it did not affect the vegetative stage. On the basis of this fact, it was assumed that nuclear division in the somatic hyphae of *S. delica* is not of the classical mitotic form, whereas nuclear division in the antheridia and oogonia is either mitotic or meiotic. There may be some question, however, as to whether the mycelial wall may be different from the oogonial. Can it be penetrated by the chemical? In hopes of answering this question, *Olpidiopsis incrassata* Cornu, a parasite that infects the mycelium of species of *Saprolegnia* and *Isoachlya*, was studied.

O. incrassata is an endobiotic, holocarpic fungus that has zoospores with flagella similar in structure to those found in the Saprolegniaceae. The genus *Olpidiopsis* was shown by McLarty (1941) to have mitotic-type divisions.

These studies were also made to determine the reactions, if any, of *O. incrassata* to colchicine.

The isolate of *O. incrassata* used in this study was collected along with its host, *S. delica* from a hempseed-baited trap set in Bolin Creek, Chapel Hill, N. C., in the spring of 1964. The two-membered culture was purified and maintained in the laboratory by procedures described by Slifkin (1962).

Results

When the zoospore of *O. incrassata* settles on a susceptible host such as *S. delica*, it encysts and sends out a germ tube that pushes the spore up from the mycelium. The exact mechanism of penetration is not known, but the protoplast of the parasite enters the host through the wall, and protoplasmic streaming of the host carries the parasite away from the site of penetration. It could be assumed that since the parasite is internal, if the host-parasite complex is put into colchicine, the chemical would have to penetrate the host wall before it could have any direct effect on the parasite.

This study was designed to evaluate the effect of colchicine on the infection of *S. delica* by *O. incrassata*. Its purpose was to determine whether (1) the parasite will infect the host which was grown in colchicine; (2) the parasite could infect while both members are in the chemical; (3) once established in the host, the parasite will continue to grow in colchicine; and (4) if the parasitic growth is inhibited while in colchicine, it could resume growth when put into water.

Six different tests of five replicates each were made. Hempseeds on agar blocks containing *S. delica* were placed into petri dishes with water or 0.1% colchicine. These were allowed to grow for 24 hours at 20° C. During the last 4 hours of this period, discharging sporangia of *O. incrassata* were introduced into each of the dishes. In the meantime another six tests with five replicates each were grown and incubated in the same manner except that the parasite was not added. These were the secondary hosts. In the next 24-hour period the primary hosts that were grown in water were placed in colchicine, two tests of those grown in colchicine were put in water, and the other two grown in colchicine were placed in other dishes with colchicine. To each of these was added a secondary host so that every test had a secondary host from either water or colchicine. At a concentration of 0.1%, colchicine inhibited the sexual cycle, and zoospores were encysted, but there was no evidence of abnormalities in the mycelial growth of *S. delica* (Slifkin, 1967). For this reason 0.1% colchicine was used throughout these studies.

The results of the experiment are condensed in Table I. The extent of infection of both primary and secondary host by the parasite is also given.

At the end of the second 24-hour period, the original hosts, grown in water and then put in colchicine, were lightly infected. No infection was observed in either the secondary hosts grown in water or those grown in colchicine. These results are shown in tests 1 and 2 of Table I. The original hosts that were grown

[1] Present address: National Environmental Health Science Center, P.O. Box 12233, Research Triangle Park, N. C. 27709.

Table I

The effect of colchicine on the infection of *Saprolegnia delica* by *Olpidiopsis incrassata*

Test No.	Medium for Primary Host First 24 hrs.		Medium for Secondary Host First 24 hrs.		Medium for Second 24 hrs.		Infection Hosts	
	H^2O	Colch.	H^2O	Colch.	H^2O	Colch.	Primary	Secondary
1.	*		*			*	††	0
2.	*			*		*	††	0
3.		*	*		*		†††	†††
4.		*		*	*		†††	†††
5.		*	*			*	†	0
6.		*		*		*	†	0

0, no infection; †, slight infection; ††, medium infection; †††, very good infection.

in colchicine and removed to water were heavily infected. This was true also with the secondary hosts (tests 3 and 4). The infection in the original hosts that were kept in colchicine was slight and could be observed only under the microscope. The secondary hosts in this case were not infected at all (tests 5 and 6).

This experiment answered the questions that were raised earlier: (1) The parasite did infect hosts that were grown in colchicine. This was illustrated in the primary hosts of tests 3–6 and the secondary hosts of test 4. (2) The parasite could infect while both members were in the chemical, as shown in the primary hosts of tests 3–6; however the parasitic zoospores had to be completely formed before put into colchicine. The parasite that was established in the primary host during the first 24-hour period was unable to complete its development during the second period and infect the secondary hosts (tests 1, 2, 5, and 6). This fact answered question 3: Once established in the host and then treated, the parasite will not continue to grow. The parasitic growth was inhibited, but it could resume growth when put into water (tests 3 and 4).

To determine the effect of the chemical on the postinfection development of *O. incrassata,* host cultures in water were infected with the parasite and incubated at 20° C. for 24 hours. After this period infection was rather extensive, but sporangia and zoospores had not yet formed. Half of the cultures were immersed in 0.1% colchicine, and the other half were left as controls in water. After six hours the controls had sporangia completely formed, and in many cases the zoospores were delimited and even discharged. Some gametangia had been differentiated. The treated parasites were disintegrating. In infected cultures put into colchicine, the parasite disintegrated but the host continued to grow. Eventually the parasite disappeared. In the control the parasite grew and the host protoplasm disappeared.

The tests with the parasite could possibly serve to illustrate that the physiology of the host was not greatly altered by the colchicine. This was shown earlier when the host grown in colchicine could be infected. As further evidence, additional tests were made. Earlier experiments using *S. delica* showed that if oogonial initials were put into colchicine solution at a certain time, some of them would revert to the mycelial stage (Slifkin, 1967). These could be easily recognized by the empty oogonial wall that interrupted the mycelium. This experiment was repeated. When the oogonium had reverted to the mycelial stage in colchicine, an excessively large number of parasite zoosporangia were introduced just as the zoospores were being released. At the end of two hours, a period long enough for the parasite to penetrate the host, the cultures were removed to water and the parasite allowed to continue development. Every converted hypha was infected.

Colchicine was applied to *Saprolegnia* sp. infected with *O. incrassata* at different stages in the life cycle of the parasite. As shown in the experiments above, the *Olpidiopsis* zoospores will swim and infect the host while they are in a colchicine solution. If host mycelium with established infection is put into the chemical, the protoplasm disintegrates, or growth is stopped. When the parasite was placed into the chemical just before zoospore delimitation, discharge tubes formed, but the protoplasm degenerated and did not produce zoospores.

Discussion

These studies present further evidence that the somatic nuclei of *Saprolegnia delica* are not affected by colchicine. Even though *Olpidiopsis incrassata* may have disturbed the wall of the host and allowed the chemical to enter and affect the parasite to the extent that it actually degenerated in some cases, the host continued to grow. In the absence of colchicine, the parasite developed at the expense of the host protoplasm within the area around it.

Walne (1966) showed that colchicine affects more than just the spindle fibers. Changes in cell size and shape, in nuclear volume and shape, in the number and arrangement of organelles, in motility, and at the cell surface occurred when she applied 0.005M colchicine to *Chlamydomonas eugametos*. Slifkin (1963) changed the carbon-nitrogen ratio of the substrate in which *Saprolegnia* sp. grew and induced immunity to *O. incrassata*. This indicated that the metabolism of the host could be changed and that infection by *Olpidiopsis* was affected by it. Colchicine did not affect the host in this way. These studies showed that the parasite could infect a culture of *Saprolegnia* that had grown in a colchicine solution. The chemical did not alter the host physiology in such a way as to make it immune to *O. incrassata*. The growth of the parasite, however, was stopped by colchicine.

Summary

Olpidiopsis incrassata infected *Saprolegnia delica* that had been grown in a 0.1% colchicine solution. The parasite could infect while both members were in the chemical, but further growth of *O. incrassata* was stopped. *S. delica* continued to grow in the colchicine during degeneration of the parasite.

LITERATURE CITED

McLARTY, D. A. 1941. Studies in the family Woroninaceae. I. Discussion of a new species including a consideration of the genera *Pseudolpidium* and *Olpidiopsis*. Bull. Torrey Bot. Club **68**: 49–66.

SLIFKIN, M. K. 1962. A new method for the purification and preservation of *Olpidiopsis incrassata*. Mycologia **54**: 105–106.

———. 1963. Parasitism of *Olpidiopsis incrassata* on members of the Saprolegniaceae. II. Effect of pH and host nutrition. Mycologia **55**: 172–182.

———. 1967. Nuclear division in *Saprolegnia* as revealed by colchicine. Mycologia **59**: 431–445.

WALNE, P. 1966. The effects of colchicine on cellular organization in *Chlamydomonas* I. Light microscopy and cytochemistry. Amer. Jour. Bot. **53**: 908–916.

Meiosis in the Antheridium of *Achlya ambisexualis* E 87

Alma Whiffen Barksdale
The New York Botanical Garden, Bronx, N.Y.

Early reports of meiosis during gametogenesis in certain members of the Saprolegniales (25, 26, 27) and Peronosporales (23, 24) have not been confirmed until recently. Before the work of Sansome (18, 19) it was unanimously concluded from cytological studies on various members of the Saprolegniales (2, 3, 4, 5, 6, 7, 8, 9, 10, 11, 12, 16, 17, 21, 29) that the single nuclear division observed in the sex organs was mitotic. On the other hand, investigation of germinating oospores led to the conclusion that the first two nuclear divisions in the oospore were meiotic (20, 30). Consequently, the oospore has come to be regarded as the probable site of meiosis (14, 15).

The purpose of this paper is to present photomicrographs of nuclear division in the antheridium of *Achlya ambisexualis* Raper, Strain E87, as evidence that meiosis occurs during gametogenesis in this fungus.

Materials and Methods

Achlya ambisexualis E87 was grown in 5 ml. of Medium A (1) in a petri dish, 60 mm. in diam. The dish was incubated at 25° C. for 72 hours, and then 0.5 ml. of an aqueous solution of hormone A, containing 10,000 dilution units per ml. was added to induce the formation of antheridia (1).

The mycelial mat bearing thousands of antheridia was fixed, 14 or 17–19 hours after the addition of hormone A, in either Newcomer's fixative (13) for 18–24 hours or in an alcohol-acetic and propionic acid mixture (22) for 15–20 minutes. For the study of antheridial initials, the mycelium was fixed 1, 1½, or 3 hours following the addition of hormone A. The fixed mycelium was hydrated by passage through a graded series of water-ethanol mixtures to water. It was then subjected to hydrolysis either for 5 minutes at room temperature and 4 minutes at 60° C. in 1N HCl or for 35 minutes at 37° C. in a solution containing 0.4 mg./ml. of ribonuclease Sigma (60 Kunitz units/mg.) and 0.1M THAM [tris (hydroxymethyl) amino methane], adjusted to pH 6.8. Following hydrolysis, the mycelium was washed, first in water and then in buffer (3 parts 0.2% w.v. KH_2PO_4 and 7 parts 0.4% w.v. $Na_2HPO_4.7H_2O$; pH 7.2). The material was stained for 7 minutes in a solution (28) prepared just before using by mixing 10 parts of above buffer with 1 part stock solution of Giemsa stain (methanol, 250 ml.; neutral glycerin, 250 ml.; and Giemsa stain Matheson Coleman & Bell, 3.5 g.). The stained mycelium was washed briefly in water and transferred to buffer; immediately thereafter, small bits of mycelium were cut from the mat and spread on a microscope slide in a drop of Valnor mounting medium (an aqueous solution of plastic prepared by the Valnor Corp., Brooklyn, N. Y.). The cover glass was pressed down firmly over the mount.

The stained preparations were examined under a Leitz Ortholux microscope fitted with 10X or 16X compensating eye pieces, an apochromatic 90X/1.4 oil-immersion objective and an aplanatic oil-immersed condenser, N.A. 1.25. A green filter was placed in the path of the light below the condenser. Photomicrographs were taken on 35 mm. Adox KB14 film, which was developed in FR X 22 developer. Negatives were enlarged 6 or 10 times when printed.

Observations

• *Antheridial branches.*—Examination of antheridial branches fixed at different stages after induction revealed no dividing nuclei. The nuclei from the main hypha appear to flow into the antheridial branch as it elongates (2–4 hours post-induction). These nuclei are elliptical at first (1.8–2.7 x 3.8–5.2 μ), but they become spherical (2.4–2.7 x 2.8–3.2 μ) as the tip expands (Figs. 1, 13). Only the nuclei at the distal end become enclosed in the antheridium.

• *First nuclear division in the antheridium.*—Following the delimitation of the antheridium (6–8 hours post-induction), the nuclei enlarge to ca. 3.2 x 3.6 μ and become uniformly distributed in the peripheral cytoplasm (Fig. 2). The number of nuclei varies with the size of the antheridium. In the pre-meiotic nucleus there is a prominent nucleolus, measuring 1.2–1.4 μ in diameter, and a network of chromonemata with 3–5 chromocenters (Fig. 14). The nuclei continue to enlarge until they attain a maximal size of 4.6–4.8 x 4.8–5.2 μ (Figs. 16, 17) at mid-prophase.

The nuclear membrane persists through diakinesis and may still be present after the spindle

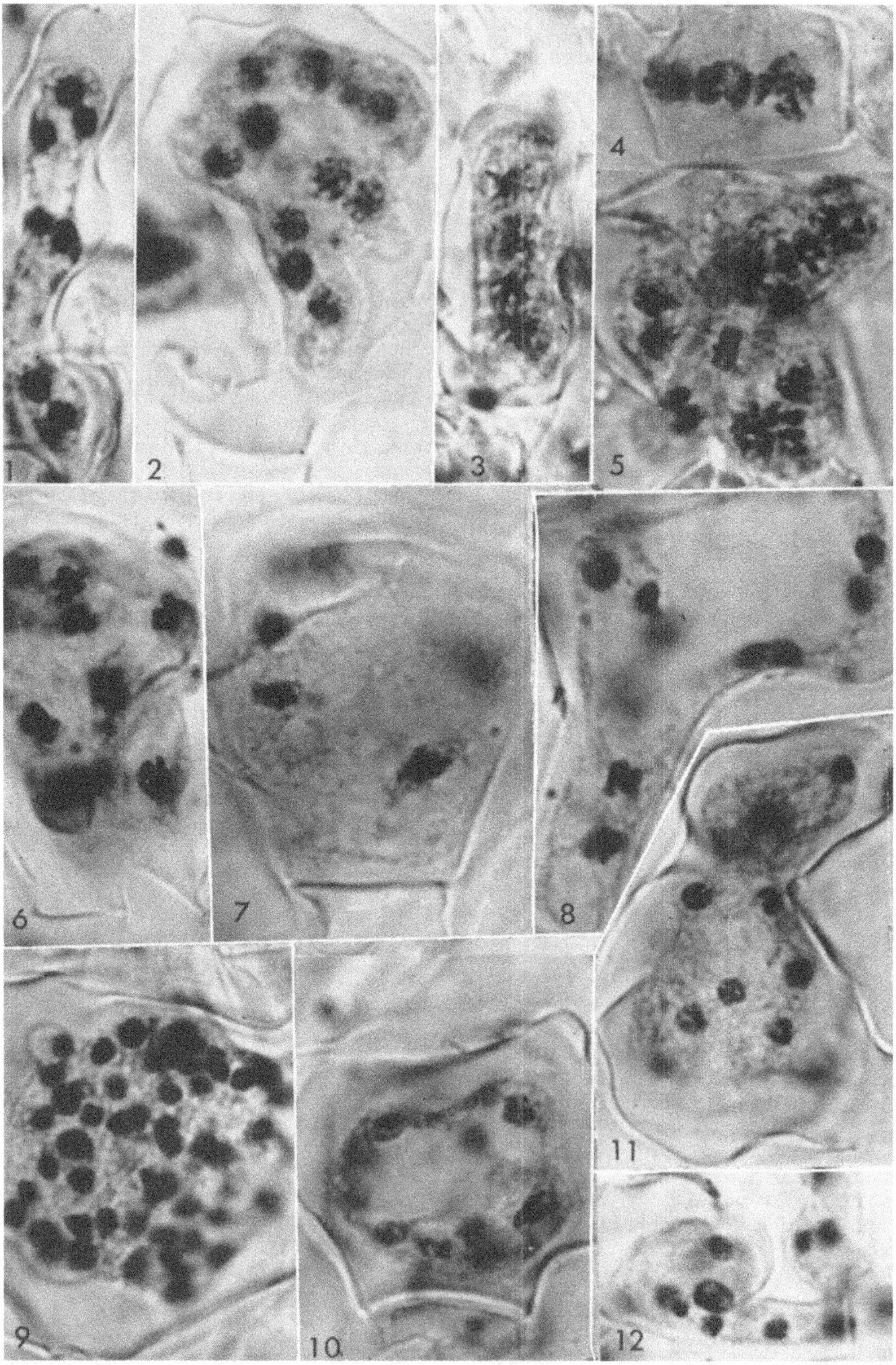

FIGS. 1–12. Antheridial branch and antheridia showing nuclei in various stages of division. X2800. Fig. 1. Tip of antheridial branch. Fig. 2. Pre-meiotic nuclei. Fig. 3. Pachytene nuclei. Fig. 4. Diplotene nuclei. Fig. 5. Nuclei at diakinesis. Fig. 6. Metaphase I. Fig. 7. Anaphase I. Fig. 8. Telophase I. Fig. 9. Prophase II and telophase II. Fig. 10. Prophase II and anaphase II. Fig. 11. Gametic nuclei. Fig. 12. Gametic nuclei, one distintegrated, and a nucleus that did not divide.

is formed. The latter point is difficult to ascertain since the spindle, which is intranuclear in origin, has the same dimensions as the nucleus, 4.4–4.6 μ in width and 4.8–5.2 μ in length, and appears to be co-extensive with the nucleus (Figs. 35, 40). Centrosomes were not seen, although they may be present.

We were unable to obtain well-spread leptotene chromosomes. Their slender threads were so closely entwined that it was not possible to count them. Pachynema proved to be the earliest stage at which a count could be made: three bivalent chromosomes were evident (Figs. 21, 29). The centromere is acentric in each of the three chromosomes. The nucleolar organizer is associated with the largest chromosome pair (chromosome 1). Traces of nucleolar material are visible until early diplonema. In Figure 29 the nucleolus can be seen attached to chromosome 1 at the left of the centromere. The other chromosomes are much shorter than chromosome 1. When contracted, chromosome 2 is seen to have a terminal satellite. Chromosome 3 reaches a state of maximal contraction earlier than 2 and 1, and in late prophase, the chromosomes of this third pair may be found widely separated. The estimated length of the chromosomes at early pachynema is as follows: chromosome 1, 17 μ; chromosome 2, 6.6 μ; and chromosome 3, 4.7 μ.

In the leptotene nucleus, the chromonemata appear as slender threads along which there are bead-like thickenings that stain deeply (Fig. 15). During the transition from leptotene to what appears to be the beginning of pachytene, the configuration of the chromonemata within the nucleus changes from that of a diffuse reticulum to a compact spiral in which the chromosomes lie in parallel coils (Figs. 16, 17). Zygonema probably occurs during the early part of this transition. The paired homologues can be distinguished readily at mid-pachytene (Figs. 18, 19), but before this time they can be resolved only with difficulty.

As the chromosomes shorten and thicken, they become folded into loops that form a complex bowknot (Figs. 22, 23, 24). In late pachytene and early diplonema super-coiling is very evident (Figs. 30, 31), and as the gyres of the super-coils come closer together, the bivalents become more compact and shorter (Figs. 32, 33).

In Figure 29, probably late pachytene, chromosome 1 measures approximately 9 μ in length; chromosome 2, 5 μ; and chromosome 3, 3 μ. By late diplotene (Fig. 32), the chromosomes measure, respectively, 4.5 μ, 2.1 μ, and 1.6 μ.

The bivalents begin moving onto the spindle before fully contracted (Figs. 34, 35, 36). Their orientation on the spindle at metaphase could not be determined with certainty. In Figure 38, the appearance of the middle bivalent is, perhaps, indicative of co-orientation.

No anaphase figures were found in which the chromosomes were well separated. Most had the appearance of Figures 41 and 42. In a few, it was fairly plain that there were three chromosomes destined for each pole (Fig. 43). Likewise, telophase figures in which the chromosomes could be counted were rare. Figure 46 was the best found. At one pole three chromosomes can be distinguished in different planes of focus.

At late telophase, the dyads elongate (Figs. 49, 50), and a nuclear membrane begins to form around them (Fig. 51). The nucleus enters a brief interphase before the beginning of the second division. During this interphase the nucleolus is reorganized, and the chromosomes, which at first are dense and darkly stained (Fig. 52), become thread-like and stain lightly (Fig. 53). These interphase nuclei measure 1.9–2.1 x 2.0–2.4 μ.

• *Second nuclear division in the antheridium.*— Within a single preparation of antheridia, almost all stages from pre-meiotic through post-meiotic may be found, but within any one antheridium the nuclei behave synchronously. The second nuclear division, however, is not as synchronous as the first. Some nuclei may be in prophase while others are in anaphase (Figs. 9, 10). The time required for the two meiotic divisions is estimated to be 4 to 6 hours. The second division is assumed to be of much shorter duration than the first because of the relative rarity of antheridia containing nuclei in the second stage of division.

The nucleolus disappears as the chromatin begins to condense in early prophase. Condensed chromatin appears first just under the nuclear membrane on one side of the nucleus (Fig. 54, and nucleus on right side of Fig. 58). In side view, strands of chromatin are seen projecting inward from this area (Fig. 55).

During the transition from prophase (Fig. 56) to metaphase, the chromosomes contract (Figs. 57, 58) and become arranged in what appears as a compact coil (Fig. 59). Nothing indicative of spindle fibers was seen in any of the preparations examined, but it is assumed that the closely associated dyads are oriented with their centromeres in a single plane corresponding to a metaphase plate. Figures 62, 63, and 64 are thought to represent polar views, and Figure 65 a lateral view of metaphase. In

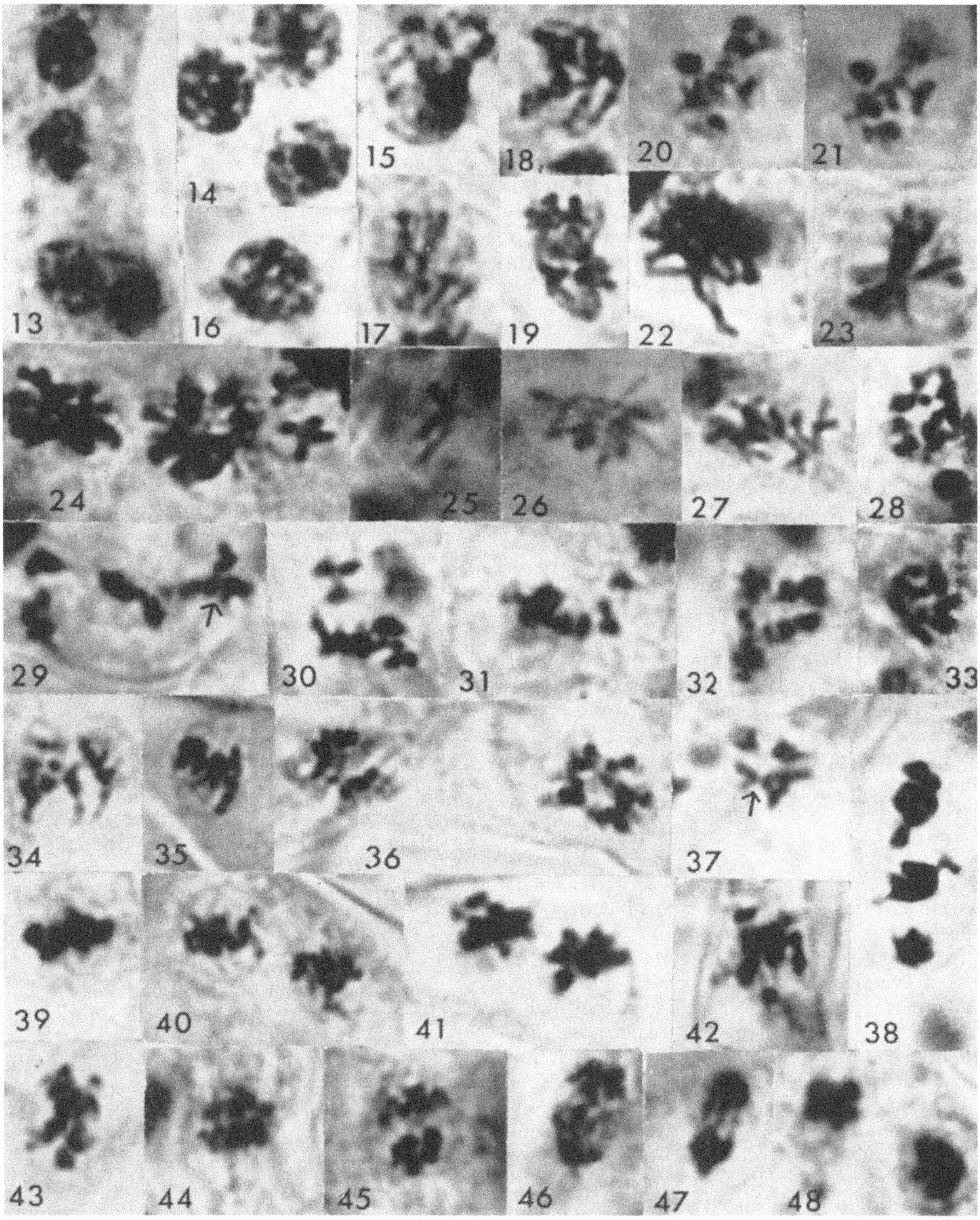

FIGS. 13–48. Stages in the first nuclear division. X3600. Fig. 13. Nuclei in antheridial branch. Fig. 14. Pre-meiotic nuclei in the antheridium. Fig. 15. Leptotene. Figs. 16 and 17. Post-zygotene, early pachytene. Figs. 18 and 19. Pachytene. Note double strand in lower half of 19. Figs. 20 and 21. Two focal planes in the same nucleus showing three pachytene bivalents. Chromosome 3, above to the left, and chromosome 2, below the large chromosome 1. Figs. 22 and 23. Mid-pachytene. The chromosome loops have been flattened and spread in 22. Fig. 24. Two nuclei at pachytene. Note x-shaped bivalent at right. Fig. 25. A pair of chromosomes, perhaps chromosome 2. Figs. 26 and 27. Late pachytene. In 27, paired chromosome 3 forms an X between and above chromosome 2 on the left and chromosome 1 on the right. Fig. 28. Nuclear configuration at pachytene. Fig. 29. Bivalents at late pachytene. Centromere is indicated by pointer. Figs. 30 and 31. Bivalents at early diplonema. In 30, the third bivalent is not in focus. Fig. 32.

the latter figure the small, stained body is actually connected to one of the dyads in another plane of focus. This body is thought to be nucleolar. The same structure and its attachment, lightly stained, appear in Figures 60 and 61.

The beginning of anaphase is indicated by the unilateral pulling away of the three dyads from the "metaphase clump" (Figs. 66, 67). In figures 68 and 69, the monads have begun to separate. The short arm of chromosome 1 seems to precede the other chromosomes to the pole (Figs. 70, 71). Anaphase figures measure 3.4–4.8 μ in length. At telophase the univalents show considerable contraction (Figs. 76, 77). In Figure 78, the longest chromosome in the telophase plate measures less than 2 μ in length.

As the chromosomes become more diffuse (Figs. 79, 80, 81), the nucleolus is reconstituted (Fig. 82) and membranes are formed around the haploid gametic nuclei (Fig. 84). The gametic nuclei measure 1.4–1.8 x 1.6–2.0 μ.

• *Disintegration of nuclei in the antheridium.*— A very few nuclei fail to divide again at the end of the first division (Figs. 12, 85). A much larger number of nuclei disintegrate at the end of meiosis. The disintegrating nuclei appear as solid, darkly stained bodies (Figs. 81, 83, and 85).

Nuclei were counted in each of 38 antheridia representing three nuclear stages: pre-meiotic, interphase I, and interphase II (post-meiotic). The number of nuclei per antheridium varied according to the stage and size. The antheridia ranged in size from 14 x 16 μ to 40 x 24 μ. The counts are recorded in Table I.

As expected, the mean number of nuclei at the end of the first division is about double the initial number. A mean of 32 is expected at the end of the second division, but the actual mean is 25. If this discrepancy is due to disintegration of gametic nuclei, then about 25 per cent of the reduction products are lost.

Table I

The size and number of nuclei per antheridium (a) before division, (b) at end of first division, and (c) at end of second division. Thirty-eight antheridia were counted.

Stage	Size	Number of Nuclei: Range	Number of Nuclei: Mean
(a) Pre-meiotic	3.2 x 3.6 μ	4–13	8
(b) Interphase I	2.0 x 2.2 μ	10–24	17
(c) Interphase II	1.6 x 1.8 μ	13–36	25

Discussion

While this study of nuclear division in the antheridium of *A. ambisexualis* E87 is far from complete, the following observations lead us to conclude that the nuclear divisions taking place in the antheridium are meiotic. First, the stages associated with the first division, particularly prophase, differ from those observed in the second division. Such a difference would not be expected were the divisions merely mitotic. Further, the number of chromosomes found at the poles at the end of each one of the divisions is three. It is most unlikely that the diploid number is three. Of course, the question crucial to this point, the number of chromosomes present at the beginning of the first division, has not been answered, nor has it been possible to demonstrate synapsis of the chromosomes.

If reduction in *A. ambisexualis* is indeed gametic rather than zygotic, it is of interest to know whether pairing takes place between chromosomes that are expanded, as they are in the Metazoa, or contracted, as they are at synapsis in *Neurospora* and other fungi having zygotic reduction (15). It is probable that the chromosomes of *A. ambisexualis* do not contract prior to synapsis. Were they to do so, it should be easy to distinguish and count the contracted chromosomes.

Further studies are needed, not only of the antheridia and oogonia of *A. ambisexualis*, but also of more representatives of the Saprolegniales and other members of the Oomycetes to learn whether meiosis during gametogenesis is the exception or the rule in this group.

Late diplonema. At the left, chromosome 3 is below chromosome 2; the L-shaped chromosome 1 is to the right. Fig. 33. Nuclear configuration at diakinesis. Fig. 34. Early diakinesis. Figs. 35 and 36. Bivalents becoming aligned on the spindle at pro-metaphase. Lateral view in 35 and on left side of 36. Polar view on right side of 36. Fig. 37. Polar view of bivalents at pro-metaphase. Chromosome 1 is across bottom, pointer indicates centromere; above, chromosome 3 at left and chromosome 2 at right. Fig. 38. Bivalents at metaphase. From top to bottom: Nos. 1, 2, and 3. Fig. 39. Lateral view of metaphase. Figs. 40 and 41. Two nuclear figures at early anaphase. In 41, the chromosomes are heavily stained. Fig. 42. Anaphase. Chromosome 3 preceding the others to the poles. Fig. 43. Disjunction of bivalents. Fig. 44 and 45. Late anaphase. Fig. 46 and 47. Telophase. Fig. 48. Daughter nuclei at end of first division.

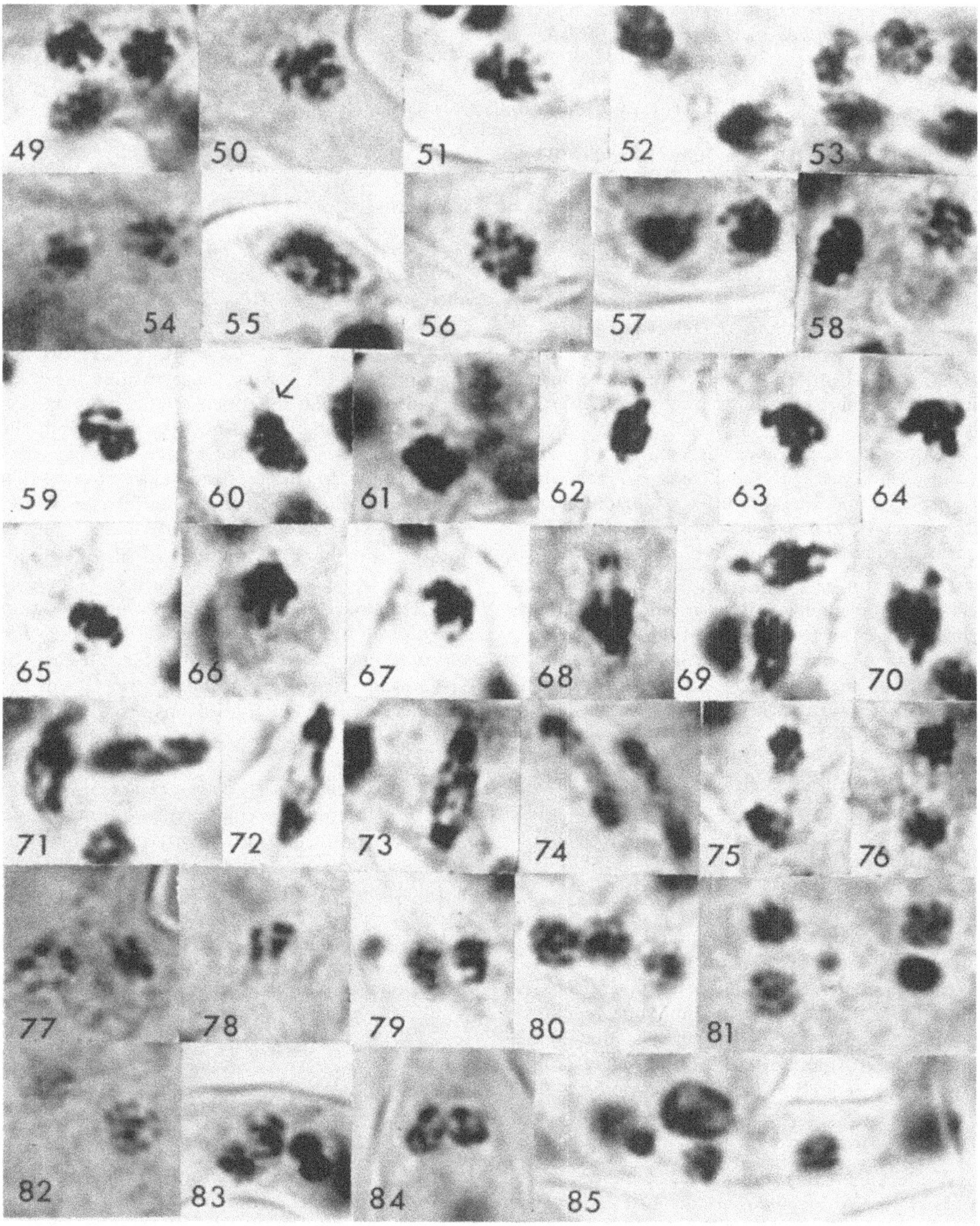

FIGS. 49–85. Interphases I and II and stages in second division. X3600. Figs. 49, 50, and 51. Transition from telophase I to interphase I. Figs. 52 and 53. Nuclei at interphase I. Fig. 54. Condensation of chromatin at beginning of second division. Fig. 55. Condensed chromatin projecting inward from nuclear membrane. Fig. 56. Prophase. Figs. 57 and 58. Contraction of dyads in late prophase. Nucleus at right at 58 is at the same stage as nuclei of 54. Fig. 59. Coiled configuration of dyads at metaphase. Figs. 60 and 61. Metaphase showing attached nucleolar organizer. Figs. 62, 63, and 64. Polar views of metaphase. Figs. 63 and 64 are two focal planes of the same metaphase figure. Fig. 65. Lateral view of metaphase. Figs. 66 and 67. Beginning of anaphase. Figs. 68 and 69. Early stage in separation of monads. Figs. 70 and 71. Monads moving to the poles. Figs. 72 and 73. Late anaphase. Figs. 74, 75, and 76. Telophase figures. Figs. 77 and 78. Telophase plates. The sister nucleus in 78 is not in focus. Figs. 79, 80, and 81. Transition

Summary

Dividing nuclei in antheridia of *Achlya ambisexualis* E87, fixed and stained by the Giemsa technique, were studied and photographed. There are two nuclear divisions, which differ from one another in ways indicating that the divisions are meiotic. The haploid number of chromosomes is believed to be three.

Until recently, the oospore has been considered to be the site of meiosis in the Saprolegniales and Peronosporales. Evidence is here given in support of the view that the sex organs are the site of meiosis in these fungi.

ACKNOWLEDGMENTS

I am especially indebted to several persons for encouragement and assistance during the four years in which this work has been in progress. Mr. J. S. Furtado adapted the Giemsa staining technique to *Achlya ambisexualis* and made the first of the antheridial preparations studied. Miss Diana Chen devoted much effort to achieving better fixation of the antheridia. Miss Barbara Bloswick carried on, patiently and skillfully making new preparations whenever called for. Dr. Elsa O'Donnell gave unstintingly of her time and expert knowledge of cytology.

The Leitz Ortholux microscope that made this study possible was purchased with funds from NSF Grant G–10838. This research has been supported by NIH Grant HD–00850 from the Institute for Child Health and Human Development.

LITERATURE CITED

1. BARKSDALE, ALMA W. 1963. The uptake of exogenous hormone A by certain strains of *Achlya*. Mycologia **55**: 164–171.
2. BHARGAVA, K. S. 1946. Oogenesis and fertilization in *Isoachlya anisospora* var. *indica*. Trans. Brit. Mycol. Soc. **29**: 101–107.
3. CARLSON, MARGERY C. 1929. Gametogenesis and fertilization in *Achlya racemosa*. Ann. Botany **43**: 111–117.
4. CLAUSSEN, P. 1908. Uber eintwicklung und befructung bei *Saprolegnia monoica*. Ber. d. Deut. Botan. Ges. **26**: 144–161.
5. COOPER, G. O. 1929. A cytological study of fertilization of *Achlya hypogyna* Coker and Pemberton. Trans. Wis. Acad. Sci., Arts and Letters **24**: 303–308.
6. COUCH, J. N. 1932. The development of the sexual organs in *Leptolegnia caudata*. Amer. J. Botany **19**: 584–599.
7. DAVIS, B. M. 1903. Oogenesis in *Saprolegnia*. Botan. Gaz. **35**: 233–249, 320–349.
8. HÖHNK, W. 1935. Zur cytologie de oogon und entwicklung bis *Saprolegnia ferax* (Gruith) Thuret. Naturwiss. ver Bremen. **29**: 308–323.
9. KASANOWSKY, V. 1911. *Aphanomyces laevis* de Bary I. Entwicklung der sexualorgane und befruchtung. Ber. d. Deut. Botan. Ges. **29**: 210–228.
10. MÄCKEL, H. G. 1928. Zur cytologie einiger Saprolegniaceen. Jahrb. wiss. Botan. **69**: 517–548.
11. MOREAU, F., ET MME. MOREAU. 1935. Les phénomènes cytologique du developement de l'oeuf et de la fécondation chez les champignons du groupe des Saprolégnieés. Compt. Rend. Acad. Sci. **201**: 1208–1210.
12. MÜCKE, M. 1908. Zur kenntnis der entwicklung und befruchtung von *Achlya polyandra* de Bary. Ber. d. Deut. Botan. Ges. **26**: 367–378.
13. NEWCOMER, EARL H. 1953. A new cytological and histological fixing fluid. Science **188**: 161.
14. OLIVE, L. S. 1953. The structure and behavior of fungus nuclei. Botan. Rev. **19**: 439–586.
15. ———. 1965. Nuclear behavior during meiosis. *In* G. C. Ainsworth and A. S. Sussman, eds., The Fungi. Vol. I, pp. 143–161. Academic Press, New York.
16. PATTERSON, P. 1927. Fertilization and oögenesis in *Achlya colorata*. Jour. Elisha Mitchell Sci. Soc. **43**: 119–136.
17. RAPER, J. R. 1936. Heterothallism and sterility in *Achlya* and observations on the cytology of *Achlya bisexualis*. Jour. Elisha Mitchell Sci. Soc. **52**: 274–293.
18. SANSOME, EVA. 1963. Meiosis in *Pythium debaryanum* Hesse and its significance in the life history of the biflagellatae. Trans. Brit. Mycol. Soc. **46**: 63–72.
19. ———. 1965. Meiosis in diploid and polyploid sex organs of *Phytophthora* and *Achlya*. Cytologia **30**: 103–117.
20. SCHRADER, E. 1938. Die Entwicklung von *Thraustotheca clavata*. Flora **132**: 125–150.
21. SHANOR, L. 1937. Observations on the development and cytology of the sexual organs of *Thraustotheca clavata* (de Bary) Humph. Jour. Elisha Mitchell Sci. Soc. **53**: 119–136.
22. SINGLETON, J. R. 1953. Chromosome morphology and the chromosome cycle in the ascus of *Neurospora crassa*. Amer. J. Botany **40**: 124–144.
23. STEVENS, F. L. 1899. The compound oosphere of *Albugo bliti*. Botan. Gaz. **28**: 149–176, 225–245.
24. ———. 1901. Gametogenesis and fertilization in *Albugo*. Botan. Gaz. **32**: 77–98, 157–169, 238–261.

---

from telophase to interphase II. Note disintegrated nucleus in 81. Fig. 82. Reappearance of nucleolus in interphase II. Figs. 83 and 84. Two pairs of gametic nuclei. In 83, there is also a pair of disintegrated nuclei. Fig. 85. Enlargement of nuclei in Fig. 12.

25. TROW, A. H. 1895. The karyology of *Saprolegnia*. Ann. Botany **9**: 609–652.
26. ———. 1899. Observations on the biology and cytology of a new variety of *Achlya americana*. Ann. Botany **13**: 131–179.
27. ———. 1904. On fertilization in the Saprolegnieae. Ann. Botany **18**: 541–569.
28. WARD, E. W. B., AND K. W. CJURYSEK. 1962. Somatic mitosis in *Neurospora crassa*. Amer. J. Bot. **49**: 393–399.
29. WOLF, F. T. 1938. Cytological observations on gametogenesis and fertilization in *Achlya flagellata*. Mycologia **30**: 456–467.
30. ZIEGLER, A. W. 1953. Meiosis in the Saprolegniaceae. Amer. J. Botany **40**: 60–66.

Genetic and Hormonal Regulation of Heterothallism in the Water Molds

J. Thomas Mullins

Department of Botany, University of Florida, Gainesville, Fla.

Heterothallism in the biflagellate water molds was first demonstrated by Couch (1926) in *Dictyuchus.* The pattern of sexuality was shown, however, to be quite different from that previously described in the heterothallic Mucorales, where strains of only two self-sterile, cross-fertile classes occurred in natural populations and among progeny of crosses (Blakeslee, 1906; Burgeff, 1912, 1914). In *Dictyuchus,* by contrast, the germination of oospores originating in crosses of male × female yielded: (1) only male; (2) male, female, mixed, and sterile; and (3) female, mixed, and sterile. In addition, oospores of a parthenogenetic isolate yielded progeny like the parent and, in a very few cases, progeny that were male. The progeny reported resulted from the germination of 16 oospores.

Additional examples of heterothallic species were subsequently found among the three major orders of biflagellate fungi: (1) Saprolegniales, *Achlya bisexualis* Coker and A. Couch (Coker, 1927), *A. regularis* (Coker and Leitner, 1938, reduced to synonymy with *A. bisexualis* by Johnson, 1956), and *A. ambisexualis* Raper (Raper, 1940); (2) Leptomitales, *Sapromyces reinschii* (Schroeter) K. Fritsch (Weston, 1938, Bishop, 1940); and (3) Peronosporales, *Phytophthora omnivora* (Leonion, 1931), *Peronospora parasitica* (Pers.) de Bary (de Bruyn, 1935, 1937). The germination of oospores was not reported in any of these papers.

The first extensive analysis of this pattern of sexuality was with *A. ambisexualis* (Raper, 1940, 1947, 1957). Ten isolates, collected in northern Illinois, were found to belong to six distinct classes when mated in all possible combinations and in matings with an additional isolate from England (E87) as follows (from left to right in decreasing maleness and increasing femaleness):

Male Male and/or Female *Female*

E87 → 11, 159 → 89 → 107, 190 → 88 → 155 → 78, 80, 184

Barksdale (1960) extended this to show that interthallic sexual reactions occur between members of homothallic as well as heterothallic species of *Achlya.* The strains studied could be assigned to one of twelve interacting types, on the basis of their sexual responses.

In *Phytophthora infestans* a slightly different system may occur (Smoot, et al., 1958; Galindo and Gallegly, 1960). Here all isolates collected from nature were bisexual and self-sterile, although relative degrees of maleness and femaleness were found.

The underlying genetic basis for this type of multiple sexual strains has remained obscure because of our inability to induce significant numbers of oospores to germinate, the only bright spot being Couch's analysis of *Dictyuchus.* Two glimmers of hope have recently appeared to brighten an otherwise dark picture. The first was the preliminary genetic analysis of progeny from crosses of male × female in *A. ambisexualis* and *D. monosporus* (Mullins and Raper, 1965). In both genera progeny included male, female, and intergrade strains of varied sexual potency. In addition, certain intergrades were self-fertile. The progeny tested were either initially derived from single zoospores produced by germ sporangia or directly from the germ mycelium. There was no case in which a single germinated oospore produced more than one type of progeny, although it could be an intergrade. It seems possible that what Couch interpreted as partial sexual segregation during early egg (oospore) germination or in early formed sporangia was only the marginal sexual differences between his progeny and either parent and that his strains were actually intergrades.

Second, it was found that in male × female matings of *A. bisexualis,* the progeny were either male or female (Barksdale, 1966). None of the progeny reacted sexually with both parents, and no apparent differences in sexual potency among males or females were found. However, one female strain was self-fertile and produced progeny that were largely female, but not self-fertile. In addition, the fact that two male progeny that resembled the F_1 male parent were produced by this self-fertile female was interpreted to mean that the nuclei were diploid and heterozygous and that meiosis did not occur in the oospore that gave rise to the self-fertile female.

These preliminary genetic analyses indicate that: (1) the life cycle is diploid, with meiosis occurring in the gametangia and the gametes comprising the only haploid phase; and (2) the control of sexual expression and mating competence resides in a more complex genetic system than paired alleles at a single locus.

Recent cytological evidence (Sansome, 1962, 1963, 1965; Barksdale, 1965) also tends to support the view that meiosis occurs in the gametangia of the water molds rather than in the germinating oospore as has been generally accepted (Ziegler, 1953). Apparently this was first suggested by Trow (1899) from work on a variety of *A. americana*. He reported four chromosomes in the nuclei of the oogonia and antheridia. After fertilization the oospore contained a single nucleus, which while undergoing successive divisions had eight chromosomes. Thus the nuclei present in the germ-tube of the oospore contained twice the number of chromosomes as did the nuclei in the gametangia.

The sexual reproductive process in *Achlya*, and probably in the other water molds, is initiated and coordinated throughout its entire course by a series of specific, diffusible hormones (Raper, 1951). Again it was Couch (1926) who, as a part of his study of heterothallism in *Dictyuchus*, first attempted experimentally to demonstrate the presence of hormones in the control of sexual organ development. Although his results were negative, they provided the basis for the subsequent demonstration of hormones in *Achlya* (Raper, 1952).

In *Achlya* four distinct morphological phases, each initiated and controlled by one or more hormones, have been recognized (Raper, 1951). The first phase is initiated by the secretion of hormones A and A_2 by the vegetative female thallus and results in the production of numerous short branches on each of the male hyphae. In a normal cross with a female, these branches are attracted to the oogonia, where upon contact they first develop septa that delimit them from the parent male hyphae and then function as antheridia. If hormone A is added to an isolated male culture, again the branches are induced and some will be delimited by septa in the usual way (Barksdale, 1963). We have verified the formation of these septa with time-lapse photography both in typical crosses of male × female and in hormonally induced males (Woodworth and and Mullins, 1966). It is interesting to note that hormone A apparently initiates a series of events in isolated male cultures that culminate in the induction of meiosis in the antheridial cells (Barksdale, 1965; Mullins, 1965). Hormone B is likely to have a similar role in female strains.

It seemed logical that a prerequisite for the induction of antheridial branching might be an enzymatic softening of the primary wall structure of the hyphae. The generally accepted view that in the Saprolegniaceae the hyphal wall contains cellulose has been recently supported by X-ray analysis and electron microscopy (Parker, et al., 1963). The analysis of cellulase activity from hormonally induced male mycelia and vegetative male mycelia was thus carried out. It was found that induced mycelia show a rapid rise in cellulase activity to a peak that coincides in time with the appearance of branch primordia (Thomas and Mullins, 1967). This is followed by a decline to the vegetative level after about 24 hours. The decline results from the release of the enzyme into the medium.

A similar rise in cellulase production occurs when branching is induced with metabolites such as casein hydrolysate or amino acid mixtures. However, the morphology of the induced branches is quite different, and they are not delimited from the parent hyphae by septa.

The rise in cellulase activity apparently depends upon protein synthesis. There is a 25 per cent increase in the incorporation of C^{14}-leucine into protein during the two-hour period following the addition of the inducer. Various inhibitors of protein synthesis, such as DL-para-fluorophenylalanine and puromycin, completely prevented branching and the rise in cellulase.

It was of interest to determine whether the induction of cellulase was limited to those strains of *Achlya* that respond to the hormone. Three strains were selected for testing on the basis of their varying ability to take up exogenous hormone A from the medium and to produce antheridial branches. They were: (1) 734, a strong female strain of *A. ambisexualis*, which secretes but does not take up hormone A; (2) E15, a homothallic strain of *A. con-*

spicua, which reacts as a male to 734 and takes up hormone A rapidly; and (3) 10, an intergrade of *A. ambisexualis*, which requires a two-hour incubation in hormone A before any measurable uptake occurs (Barksdale, 1963). It was found that E15, which branches in response to hormone A, shows a marked increase in cellulase production, whereas, strains 10 and 734 which do not branch in response to hormone A, also failed to show an increase in cellulase (Thomas, 1966).

The mechanism of response of female strains to hormone B, which is produced by sexually induced male strains, is similar to that described for male strains (Mullins and Thomas, 1967).

The basic information on the vegetative respiratory metabolism of *A. ambisexualis* has recently become available (Warren, 1966, 1967). Both manometric studies of intact mycelium and spectrophotometric studies of enzyme activity of mycelial extracts were done. Evidence for two glycolytic pathways, the hexose monophosphate and the Embden-Meyerhof-Parnas, was presented. In addition glycerolphosphate dehydrogenase and lactic dehydrogenase activity was observed. The standard Krebs cycle enzymes and cytochrome oxidase were demonstrated. Preliminary work showed that the respiratory rate of male strains was more than doubled in the presence of hormone A. The maximum increase was reached after two hours. The observation that the male filtrate was superior as a suspending medium tends to support Raper's (1950) description of hormone A^1, which is secreted by the vegetative male, as an enhancing agent of hormone A.

Today, more than 40 years after Professor Couch's initial discovery of heterothallism in the water molds, finds us just beginning to appreciate what an excellent system he described for asking fundamental questions about the relationship between genetic and hormonal control of sexual morphogenesis.

LITERATURE CITED

BARKSDALE, A. W. 1960. Inter-thallic sexual reactions in *Achlya*, a genus of the aquatic fungi. Amer. J. Bot. **47**: 14–23.

———. 1963. The uptake of exogenous hormone A by certain strains of *Achlya*. Mycologia **55**: 164–171.

———. 1965. Personal communication.

———. 1966. Segregation of sex in the progeny of a selfed heterozygote of *Achlya bisexualis*. Mycologia **58**: 802–804.

BISHOP, H. 1940. A study of sexuality in *Sapromyces reinschii*. Mycologia **32**: 505–529.

BLAKESLEE, A. F. 1906. Zygospore germinations in the Mucorineae. Ann. Mycol. **4**: 1–28.

BURGEFF, H. 1912. Über sexualität, variabilität, und vererbung bei *Phycomyces nitens* Kunze. Ber. deuts. bot. Ges. **30**: 679–685.

———. 1914. Utersuchungen über variabilität, sexualität, und erblichkeit bei *Phycomyces nitens* Kunze. I. Flora **107**: 259–316.

COKER, W. C. 1927. Other water molds from the soil. Jour. Elisha Mitchell Sci. Soc. **42**: 207–226.

COKER, W. C., AND J. LEITNER. 1938. New species of *Achlya* and *Apodachlya*. Jour. Elisha Mitchell Sci. Soc. **54**: 311–318.

COUCH, J. N. 1926. Heterothallism in *Dictyuchus*, a genus of the water moulds. Ann. Botany (London) **40**: 849–881.

DE BRUYN, H. L. D. 1935. Heterothallism in *Peronospora parasitica*. Phytopath. **25**: 8.

———. 1937. Heterothallism in *Peronospora parasitica*. Genetics **19**: 553–558.

GALINDO, J. A., AND M. E. GALLEGLY. 1960. The nature of sexuality in *Phytophthora infestans*. Phytopath. **50**: 123–128.

JOHNSON, T. W. 1956. The Genus *Achlya*: Morphology and Taxonomy. Univ. Mich. Press, Ann Arbor.

LEONIAN, L. H. 1931. Heterothallism in *Phytophthora*. Phytopath. **21**: 941–955.

MULLINS, J. T. 1965. Unpublished.

MULLINS, J. T., AND J. R. RAPER. 1965. Heterothallism in biflagellate aquatic fungi: preliminary genetic analysis. Science **150**: 1174–1175.

MULLINS, J. T., AND D. DES S. THOMAS. 1967. Biochemical mechanisms in the hormonal control of sexual morphogenesis in *Achlya*. ASB Bulletin **14**: 36.

PARKER, B. C., R. D. PRESTON, AND G. E. FOGG. 1963. Studies of the structure and chemical composition of the cell walls of Vaucheriaceae and Saprolegniaceae. Proc. Roy. Soc. B, London **158**: 435–445.

RAPER, J. R. 1940. Sexuality in *Achlya ambisexualis*. Mycologia **32**: 710–727.

———. 1947. On the distribution and sexuality of *Achlya ambisexualis*. Amer. J. Bot. **34**: 609.

———. 1950. Sexual hormones in *Achlya*. VI. The hormones of the A-complex. Proc. Nat. Acad. Sci. US **36**: 524–533.

———. 1951. Sexual hormones in *Achlya*. Amer. Sci. **39**: 110–120.

———. 1952. Chemical regulation of sexual processes in the Thallophytes. Bot. Rev. **18**: 447–545.

———. 1957. Hormones and sexuality in lower plants. Sym. Soc. Exp. Biol. **11**: 143–165.

SANSOME, E. R., AND B. J. HARRIS. 1962. Use of camphor-induced polyploidy to determine the place of meiosis in fungi. Nature **196**: 291–292.

———. 1963. Meiosis in *Pythium debaryanum* Hesse and its significance in the life-history of the biflagellate. Trans. Brit. Mycol. Soc. **46**: 63–72.

———. 1965. Meiosis in diploid and polyploid sex organs of *Phytophthora* and *Achlya*. Cytologia **30**: 103–117.

SMOOT, J. J., et al. 1958. Production and germination of oospores of *Phytophthora infestans*. Phytopath. **48**: 165.

THOMAS, D. DES S. 1966. Hormone-enzyme regulation of morphogenesis in the water mold *Achlya ambisexualis* Raper. Unpublished Ph.D. thesis. University of Florida.

THOMAS, D. DES S., AND J. T. MULLINS. 1967. Role of enzymatic wall-softening in plant morphogenesis: hormonal induction in *Achlya*. Science **156**: 84–85.

TROW, A. H. 1899. Observations on the biology and cytology of a new variety of *Achlya americana*. Ann. Botany (London) **13**: 131–179.

WARREN, C. O., JR. 1966. Respiratory metabolism in the heterothallic water mold *Achlya ambisexualis* Raper. Unpublished Ph.D. thesis. University of Florida.

———. 1967. Respiratory metabolism in the heterothallic water mold *Achlya ambisexualis* Raper. ASB Bulletin **14**: 44.

WESTON, W. H., JR. 1938. Heterothallism in *Sapromyces reinschii*. Preliminary note. Mycologia **30**: 245–253.

WOODWORTH, R., AND J. T. MULLINS. 1966. Unpublished.

ZIEGLER, A. W. 1953. Meiosis in the Saprolegniaceae. Amer. J. Bot. **40**: 60–66.

Factors Affecting Oogenesis and Oospore Germination in *Achlya hypogyna*[1]

JOHN C. CLAUSZ
Department of Botany, University of North Carolina at Chapel Hill, N. C.

A variety of laboratory conditions have been applied to oospores of species of the Oomycetes in order to induce germination. Successful oospore germination techniques can be grouped into methods involving (1) transfer of oospores to fresh water or water containing a nutritional source, (2) light requirements, and (3) temperature requirements. Age of oospores has been shown to be an important factor leading to oospore germination.

This work was designed to study the effect of temperature on oogenesis and the effect of age of oospores, light, temperature, nutritional sources, and desiccation on germination of oospores of *Achlya hypogyna*.

Methods and Materials

A single-spore, bacteria-free culture of *Achlya hypogyna* was obtained by using the techniques described by Szaniszlo (1965). His methods were employed also for inoculation of sterile hempseed halves that were subsequently placed in 150 ml. of a water extract of soil. The water extract of soil was prepared by placing about 250 grams of soil in a 2-liter flask with 1500 ml. of distilled water. After vigorous shaking, the suspension was permitted to stand overnight before filtering and autoclaving.

After cultures were grown for various lengths of time at 20° C., an oogonial suspension was prepared by homogenizing the mycelial colony (minus the hempseed) in a Waring blendor. The resulting suspension had oogonia intact, but the mycelial fragments were devoid of protoplasm. About 150 xg was sufficient to sediment the oogonia and mycelial fragments into a loose pellet. After the supernatant was discarded, the pellet was diluted up to 25 ml.

Since oogenesis is not synchronously initiated, oospores may be in various stages of maturation, germination, or disintegration. In order to describe this heterogenous group of oospores, 500 oospores were classified into 4 groups:

Group I. Mature oospores are characterized by oil globules arranged at the periphery of the oospore (Fig. 1, M.O.).

Group II. Oospores exhibiting initial stages of germination have small oil globules dispersed through the dark and granular cytoplasm (Fig. 1, O.I.S.G.).

Group III. Disintegrating oospores have oil globules coalesced into one or several large oil globules (Fig. 3, D.O.), or the oospores are nearly empty except for particles moving about by Brownian movement.

Group IV. Germinating oospores have germ tubes growing through the hypogynous antheridium (Fig. 2, G.O.).

Prior to treatment, oogonial suspensions were examined to determine how many of 500 oospores were in each group. These 500 oospores were counted in groups of 100; each group consistently repeated the first.

Results

Effects of temperature on oogenesis.—Hempseed and cornmeal agar (CMA) plate cultures were grown at 15° C. and 25° C. for 10 days prior to examination.

The number of oospores per oogonium ranged from 2–6 when cultures were grown at 15° C. (Figs. 5 and 7). There were many large papillae on oogonia pictured in Figures 5 and 7. The size of the papillae varied, but they were larger than those on oogonia from cultures grown at 25° C., in which the oogonia were almost smooth (Figs. 6, 8). The number of oospores per oogonium was usually one or two at the higher temperature.

Effect of age on untreated oospores.—Oogonial suspensions were prepared from cultures that were 4, 6, 7, 8, and 10 weeks old. The percentage of oospores in Groups I–IV was determined for each suspension.

The results (Table I) reveal that the percentage of oospores in the four classes varied with age of the untreated culture. The percentage of mature oospores increased up to culture age of 8 weeks, then dropped with increased age of culture. No germination occurred in untreated cultures until they were 8 weeks old. Only those oospores represented by Groups I and II were able to germinate after treatment.

---

[1] This paper is based on a thesis submitted to the faculty of the University of North Carolina at Chapel Hill in partial fulfillment of the requirements for the degree of Master of Arts in the Department of Botany.

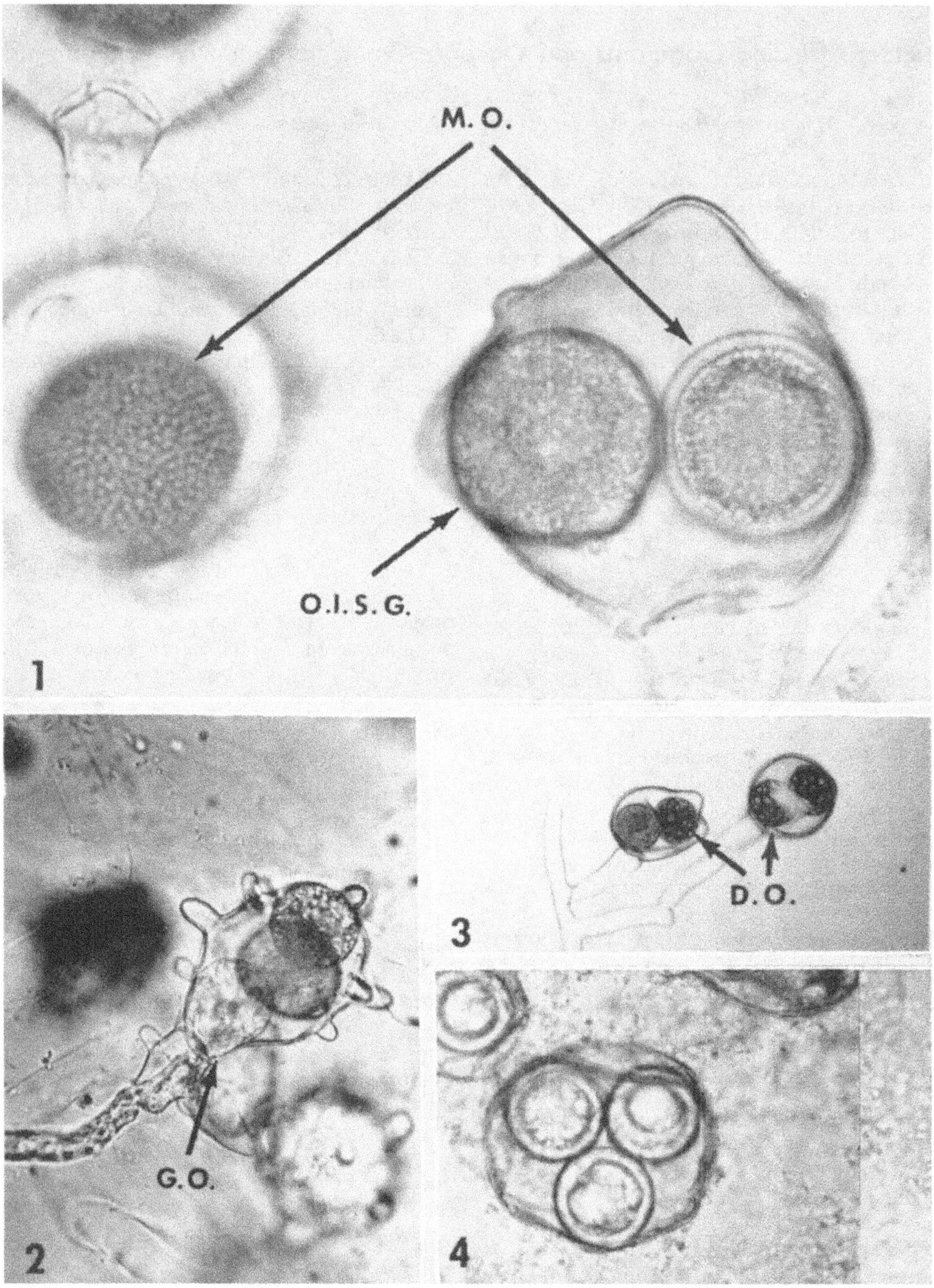

FIG. 1. Oogonia with mature oospores (M.O.) and an oospore in the initial stage of germination (O.I.S.G.). X1000. FIG. 2. Oogonium containing a germinating oospore. X400. FIG. 3. Oogonia with disintegrating oospores. X250. FIG. 4. Oospores after drying and rewetting. X400.

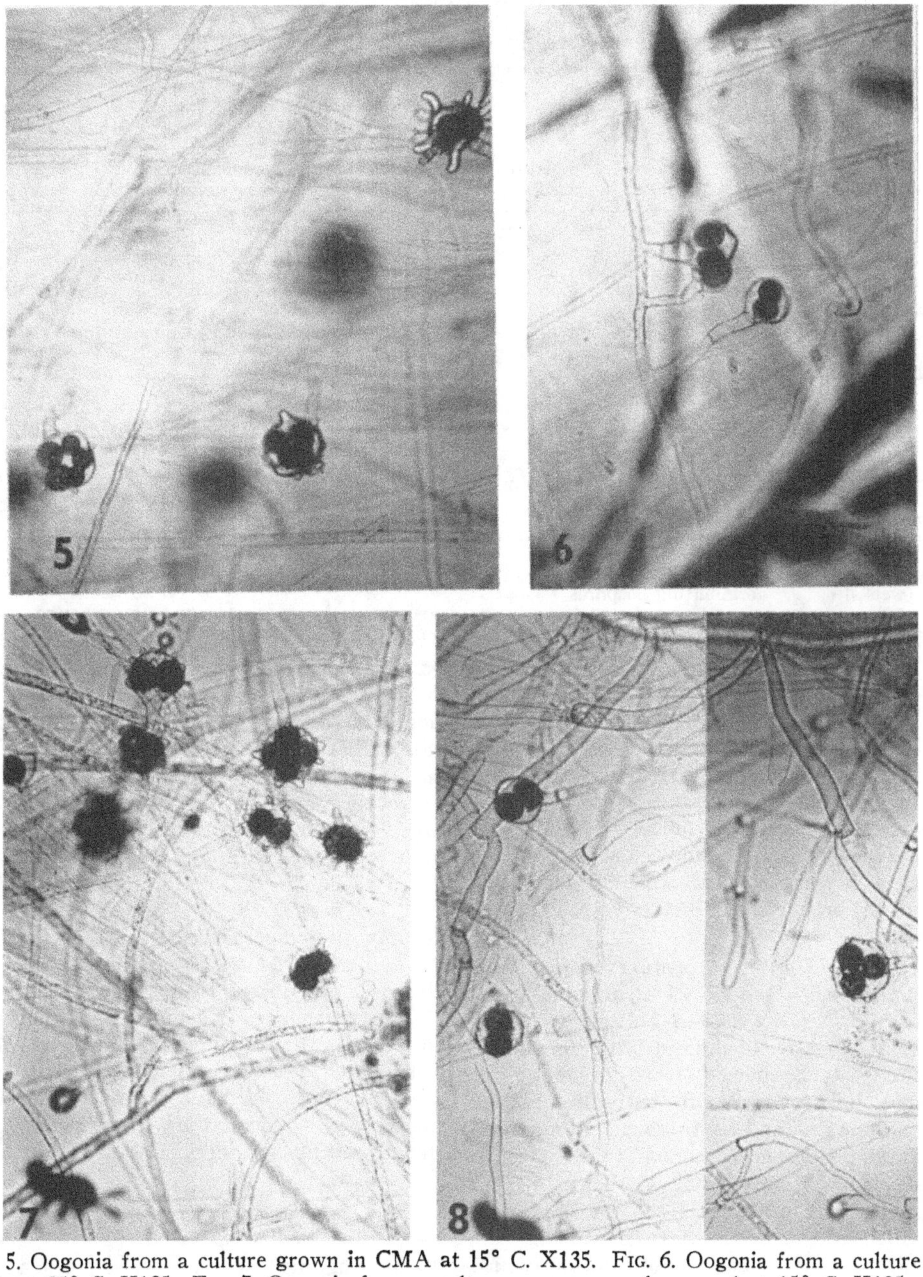

FIG. 5. Oogonia from a culture grown in CMA at 15° C. X135. FIG. 6. Oogonia from a culture grown in CMA at 25° C. X125. FIG. 7. Oogonia from a culture grown on a hempseed at 15° C. X125. FIG. 8. Oogonia from a culture grown on a hempseed at 25° C. X125.

Effect of age on germination percentage.— One ml. of each of the suspensions described above was dispensed into sterile humidifying chambers (a small, 55-mm. diameter petri dish in a large, 90-mm. diameter petri dish that had water in its bottom). The chambers were placed in an incubator at 20° C. with 140–160 foot-candles of fluorescent light for 48 hours. At the end of the experimental period, 500 oospores were classified. The percentage change in the four groups was calculated.

After treatment, oospores germinated if cultures from which they were obtained were 6 weeks of age or older (Table II). As the age of the culture increased, the percentage of germinating oospores increased also. Germina-

Table I

Percentage of oospores in groups I–IV as a function of age

Age of culture	Percentage of oospores in groups: I Mature oospores	II Oosp. initial stage germ.	III Disintegrating oospores	IV Germinating oospores
4 weeks	51.0%	32.2%	16.8%	0.0%
6 weeks	55.4	21.0	23.6	0.0
7 weeks	58.8	10.2	31.0	0.0
8 weeks	65.3	12.2	22.1	0.4
10 weeks	39.8	12.2	46.6	1.4

Table II

The effect of age on germination percentage after exposure to 20° C. in continuous illumination

Age of culture (weeks)	Increase in percentage of germinating oospores
4	0.0%
6	0.2
7	0.4
8	0.8
10	4.9

tion was not observed from 4-week-old cultures. The 8-weeks requirement could be reduced to 6 weeks by the washing technique involved in preparation of the oogonial suspension.

Effect of light and temperature on germination of oospores.—One ml. of an oogonial suspension from an 8-week-old culture was dispensed into a sterile humidifying chamber. Some of the chambers were wrapped in aluminum foil to make them unilluminated. The unilluminated and continuously illuminated chambers were placed in incubators at 5-degree intervals from 5° C. to 35° C. with 140–160 foot-candles of fluorescent light for 48 hours, after which 500 oospores were classified.

Germination occurred in both the presence and absence of light (Group IV in Table III). At 20° C. a 1.8 per cent and 0.8 per cent increase in percentage of germinating oospores over the control was observed in the light and dark, respectively.

The indication that germination occurs only in the dark at 10° C. and 15° C. is unlikely since data from experiments on a 7-week-old oogonial suspension revealed that the opposite might be true. Light seems to promote germination to some degree. At all temperatures at which germination occurred, except 10° C. and 15° C., there was a greater increase in percentage of germinating oospores when the oogonial suspensions were exposed to continuous illumination (Group IV in Table III). Perhaps this effect is better reflected by the decrease in percentage of mature oospores (Group I in Table III). Upon treatment at each temperature, a greater decrease in percentage of mature oospores resulted when the oogonial suspension was in continous illumination. This decrease was also reflected by an increase in disintegrating oospores. The disintegrating oospores likewise showed a greater change in percentage when the oogonial suspension was in continuous illumination (Group III in Table III). Both these results are depicted graphically in Figures 9 and 10. Note that at all temperatures the greater difference always occurred in the oogonial suspensions that were subjected to continuous illumination.

The percentage of germinations increased up to 20° C., then decreased until no oospores germinated at 35° C. It appears that temperatures above 25° C. have a detrimental effect on oospores as indicated by large percentages of disintegrating oospores (Group III in Table III).

Effects of nutritional sources on germination.—An oogonial suspension (1 ml. containing 10,800 oogonia) was prepared from a 9-week-old culture and dispensed onto agar plates containing different nutritional sources. These cultures were incubated at 20° C. for 12 hours, after which the number of oogonia with germinating oospores in an area 4 cm. square was determined. The number of oogonia with germinating oospores on the whole agar surface and percentage of germinations per ml. were calculated.

Oogonia placed on CMA plates had a greater

Table III

Effect of light and temperature on the germination of oospores. Percentage of 500 oospores in one of four groups before and after experimental conditions and percentage of deviation from the control count

	Group I		Group II		Group III		Group IV	
	Mature oospores		Initial stages of germ.		Disintegrating oospores		Germinating oospores	
Condition	%	Dev.	%	Dev.	%	Dev.	%	Dev.
Control—Before Treat.	65.3		12.2		22.1		0.4	
5° C. Cont. Illum.	57.0	−8.3	11.6	−0.6	32.8	+10.7	0.6	+0.2
Unillum.	58.7	−6.6	14.1	+1.9	27.2	+5.1	0.0	−0.4
10° C. Cont. Illum.	57.6	−7.7	14.2	+2.0	27.8	+5.7	0.4	0.0
Unillum.	59.0	−6.3	14.8	+2.6	25.6	+3.5	0.6	+0.2
15° C. Cont. Illum.	60.0	−5.3	15.3	+3.1	24.3	+2.2	0.4	0.0
Unillum.	61.0	−4.3	14.2	+2.0	23.4	+1.3	1.4	+1.0
20° C. Cont. Illum.	54.6	−10.7	14.2	+2.0	29.0	+6.9	2.2	+1.8
Unillum.	57.8	−7.5	14.2	+2.0	26.8	+4.7	1.2	+0.8
25° C. Cont. Illum.	51.0	−14.3	16.0	+3.8	31.8	+9.7	1.2	+0.8
Unillum.	56.6	−8.7	14.0	+1.8	28.8	+6.7	0.6	+0.2
30° C.* Cont. Illum.	51.4	−7.4	4.0	−6.2	44.4	+13.3	0.2	+0.2
Unillum.	53.6	−5.3	3.6	−6.6	42.8	+11.7	0.0	0.0
35° C. Cont Illum.	33.5	−31.8	0.0	−12.2	66.5	+44.4	0.0	−0.4
Unillum	39.0	−26.3	0.0	−12.2	61.0	+38.9	0.0	−0.4

*Data from experiments on a 7-week-old culture

number of germinating oospores than oogonia placed on other agar media (Table IV). Determining whether the increase was due to mineral, carbon, or nitrogen sources was not possible. Even though there was a marked increase in the number of germinations on CMA, the increase yielded germinations in only 2 per cent of the total number of oogonia.

Effect of desiccation on germination.—When hempseed cultures were dried to room humidity at 25° C., the oil globules in oospores coalesced into one or several large oil globules. When rewetted with water extract of soil, no germination occurred, and the oospores did not return to their mature appearance (Fig. 4).

Discussion and Conclusions

The effect of temperature on oogenesis has been reported previously by Reischer (1949). High temperatures (25° C.) tend to suppress papillation, whereas low temperatures (15° C.) permit growth of normal-appearing papillae for *Achlya colorata*. The morphogenetic effect of temperature on oogenesis in *Achlya hypogyna* is very similar.

Sussman (1966) reviewed dormancy and spore germination in fungi and established terminology that will be followed here:

Dormancy—Any rest period or reversible interruption of the phenotypic development of an organism.

Maturation—The complex of changes associated with the development of the resting stage of dormant organisms or of the germinable stage in those without a dormant period.

Activation—The application of environmental stimuli which induce germination.

Afterripening—The treatments undergone in nature that lead to germination; activation under natural circumstances.

Germination—A process leading to the first irreversible stage that is recognizably different

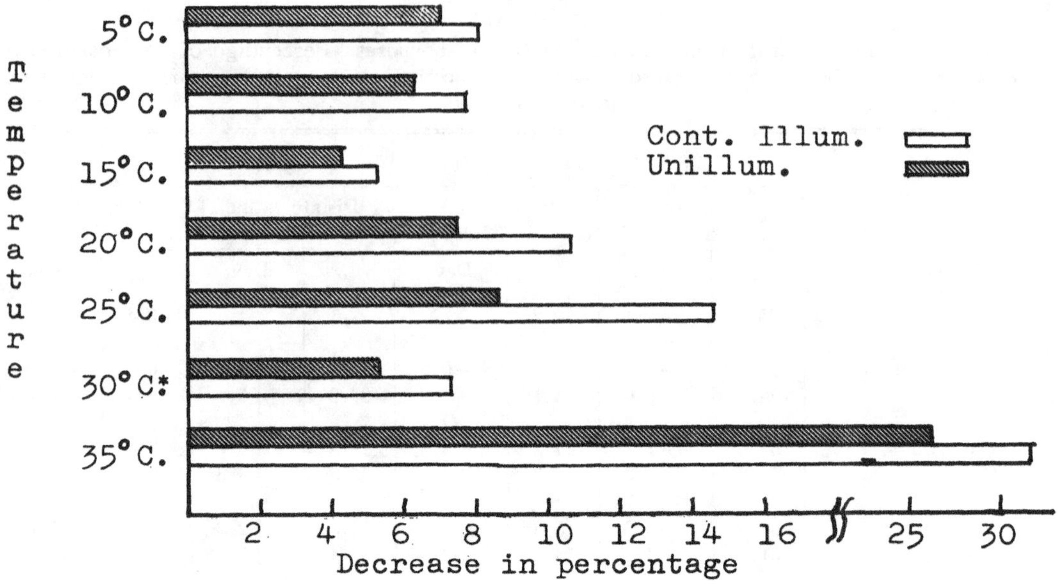

FIG. 9. Effect of light and temperature on the decrease in percentage of mature oospores (Table III, group I). * Data from experiments on 7-week-old culture.

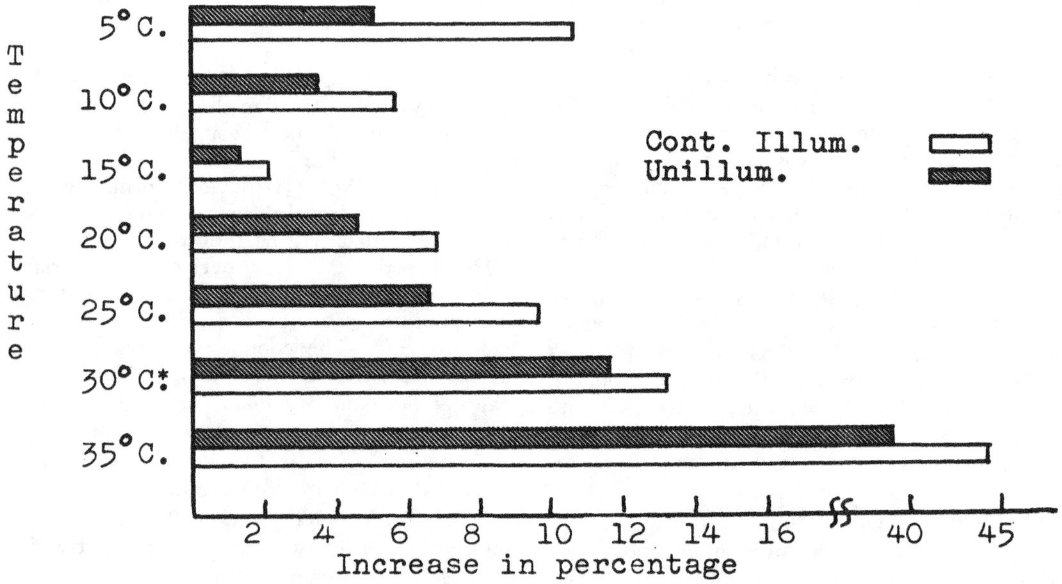

FIG. 10. Effect of light and temperature on the increase in percentage of disintegrating oospores (Table III, group III). * Data from experiments on 7-week-old culture.

Table IV
Effect of nutritional sources on germination

Nutritional source	Number of oogonia with germinations per 4 sq. cm.	Number of oogonia with germinations per 63.7 sq. cm.	Percentage of total number of oogonia with germinations
Agar	1	15.9	0.147%
Agar - Medium A*	1	15.9	0.147
Agar - Soil Water	1	15.9	0.147
Agar - Salts**	1	15.9	0.147
Agar - Glucose (0.5%)	1	15.9	0.147
Agar - Starch (0.5%)	2	31.8	0.294
Agar - Peptone (0.05%)	2	31.8	0.294
Cornmeal Agar	13	207.0	1.91

*Barksdale, A.W. 1963. Mycologia 55: 164.
**As in Medium A

from the dormant organism, as judged by physiological or morphological criteria.

When these terms are applied to the oospores and oospore germination of *Achlya hypogna*, one notes the following sequence. Mature oospores first appear when a culture is 8 days old; thus, maturation of oospores appears to be complete in 8 days. An afterripening period of 7 weeks then follows: this can be abbreviated by 2 weeks through activation by the washing technique involved on preparation of the oogonial suspension. Increasing age of culture increases the percentage of germination, probably because the number of oospores that have gone through the 7-week dormancy period is greater. Activation by the described treatment thus yields a greater percentage of germination.

In contrast, oospores of *Achlya recurva*, applying the same terminology, appear mature in 3 days and are capable of germination immediately (Latham, 1935).

Sorokine (1876), Zeigler (1948), and Scott (1961) found that light is necessary for germination of oospores of species that they worked with. Light does not appear to be necessary for oospore germination in *Achlya hypogyna*, but it may promote the germination process. Zeigler did state that occasionally he found oospores of species he studied germinating in the dark.

For *Achlya hypogyna*, 20° C. is the optimum temperature for germination. Schössler (1929), Blackwell (1943a, 1943b), and Barksdale (personal communication) have used low and freezing temperatures to reduce the afterripening period. These temperatures were not successful with *Achlya hypogyna*.

The only treatment that substantially increased germination is a nutritional one. The experiment designed to test whether carbon source, nitrogen source, or salts was the nutritional source enhancing germination was not fruitful. Only when a nutritional source was added did oospores germinate for Trow (1901), Coker (1923), Couch (1926), Smoot, et al. (1958), and Scharen (1960).

Results presented here agree with those of Blackwell and Waterhouse (1931), Salvin (1942), and Blackwell (1943b) as to the effect of desiccation on germination. Indications are that in *Achlya hypogyna* desiccation kills oospores.

Summary

Temperature affects oogenesis in *Achlya hypogyna*. Cultures grown at high temperatures have oogonia with small and few papillae and few oospores per oogonium. Cultures grown at low temperatures have oogonia with many large papillae and several to many oospores per oogonium.

The oospores of *Achlya hypogyna* appear mature after 8 days of culture. An afterripening period of 7 weeks follows, which can be reduced to 5 weeks through activation by the washing involved in preparation of the oogonial suspension. In contrast to the reports of previous workers, light was not found to be necessary for oospore germination. The optimum temperature for germination was 20° C. The treatment that enhanced germination was a nutritional one. The number of oogonia on CMA with germinating oospores is 13 times the number of the control. Desiccation kills oospores.

ACKNOWLEDGMENTS

I wish to thank Dr. W. J. Koch, who directed this study, and Drs. P. J. Szaniszlo and C. J. Umphlett for their assistance in preparing the manuscript.

LITERATURE CITED

BLACKWELL, E. M., 1943a. The life history of *Phytophthora cactorum*. Trans. Brit. Mycol. Soc. **26**: 71–89

———. 1943b. Presidential address on the germination of the oöspores of *Phytophthora cactorum*. Trans. Brit. Mycol. Soc. **26**: 93–103.

BLACKWELL, E. M., AND G. M. WATERHOUSE. 1931. Spores and spore germination in the genus *Phytophthora*. Trans. Brit. Mycol. Soc. **15**: 294–310.

COKER, W. C. 1923 The Saprolegniaceae. The University of North Carolina Press, Chapel Hill, N.C. 201 pp.

COUCH, J. N. 1926. Heterothallism in *Dictyuchus*, a genus of the water moulds. Ann. Botany **40**: 849–882.

LATHAM, D. H. 1935 *Achlya recurva* Cornu from North Carolina. Jour. Elisha Mitchell Sci. Soc. **51**: 183–188.

REISCHER, H. S. 1949. The effect of temperature on the papillation of oögonia of *Achlya colorata*. Mycologia **41**: 398–402.

SALVIN, S. B. 1942. Preliminary report on the intergeneric mating of *Thraustotheca clavata* and *Achlya flagellata*. Amer. J. Bot. **29**: 674–676.

SCHAREN, A. L. 1960. Germination of oöspores of *Aphanomyces euteiches* imbedded in plant tissue. Phytopathology **50**: 274–277.

SCHÖSSLER, L. A. 1929. Geschlechterverteilung und fakultatuve parthenogenesese bei Saprolegniaceen. Planta Archiv. für Wissenschaftiche Botanik **8**: 629–670.

SCOTT, W. W. 1961. A monograph of the genus *Aphanomyces*. Va. Ag. Exp. Sta. **151**: 22.

SMOOT, J. J. et al. 1958. Production and germination of oospores of *Phytophthora infestans*. Phytopathology **48**: 165–171.

SOROKINE, N. 1876. Quelque mots sur le development de L'*Aphanomyces stellatus*. Ann. Sci. Nat. 6 Sér. **3**: 46–52.

SUSSMAN, A. S. 1966. Dormancy and spore germination. *In* G. C. Ainsworth and A. S. Sussman, eds., The Fungi. Vol. II, Ch. 23. Academic Press, New York.

SZANISZLO, PAUL J. 1965. A study on the effect of light and temperature on the formation of oögonia and oöspheres in *Saprolegnia diclina*. Jour. Elisha Mitchell Sci. Soc. **81**: 10–15.

TROW, A. H. 1901. Observations on the biology and cytology of *Pythium ultimum*. Ann. Botany **15**: 267–312.

ZEIGLER, A. W. 1948. A comparative study of zygote germination in the Saprolegniaceae. Jour. Elisha Mitchell Sci. Soc. **64**: 13–40.

Is *Blakeslea* a Valid Genus?

B. S. Mehrotra and Usha Baijal
Department of Botany, University of Allahabad, Allahabad, India

The genus *Blakeslea* was created by Thaxter in 1914 to distinguish it from the genus *Choanephora*. The main basis of separation was the presence of sporangiola in the former and conidia in the latter, the sporangia being common in both.

Since the time the genus *Blakeslea* was established, workers have doubted the validity of this difference between the two genera. Even Thaxter himself had suspected that the conidia in the genus *Choanephora* are monosporous sporangiola.

Sinha (1940) for the first time observed conidia and even monosporous sporangiola in *Blakeslea trispora* which he considered to be identical to conidia of *Choanephora*. From this he concluded that there is no difference between the two genera.

Later Poitras (1955) took up this problem. He subjected the conidia of *Choanephora* to various treatments to prove that the wall of the conidia is double layered. He also observed that the outer thin sporangiolic membrane was drawn out for a short distance with the germ tube, indicating that the inner spore wall is separate from the outer sporangiolic membrane. Poitras believed that these observations confirm that the conidia in *Choanephora* are monosporous sporangiola. After having concluded that in both the genera, *Blakeslea* and *Choanephora*, sporangiola are present, he concurred with the proposal of Sinha (1940) to merge *Blakeslea* with the earlier known genus *Choanephora*.

In spite of all this, a feeling existed among workers (Hesseltine, 1962; Mehrotra and Mehrotra, 1964) that from a practical point of view, at least, the retention of the genus *Blakeslea* is amply justifiable. This is because in *Blakeslea trispora* conidia like that of *Choanephora* are very rare (only Sinha [1940] has observed them), and in *Choanephora* the conidia, though double walled, cannot be interpreted as true (monosporous) sporangiola since under normal and usual conditions they behave as conidia.

The new species of *Blakeslea* described below confirms the difference between true monosporous sporangiola and the ones found in Choanephoras. Besides the typical sporangia, borne on circinate sporangiophores, this species produces (in all normal laboratory media tried) true monosporous sporangiola, each having a punctate wall that always breaks on maturity, liberating a single sporangiospore with appendages often clearly visible at both the ends (neither Sinha [1940] nor Poitras [1955] saw appendages over the "conidia" of *Choanephora*). It seems therefore proper to retain the genus *Blakeslea* to include the species in which sporangiola (trisporic in *B. trispora*; monosporic in *B. monospora*) are usually produced under normal conditions, and to confine the generic limits of *Choanephora* to include the species in which conidia (of course, double walled and phylogenetically or by homology monosporous sporangiola) are present.

A description of the new species follows:

Blakeslea monospora sp. nov.

Coloniae in agaris culturae avenaceo, SMA et solanaceo albae, floccosae; hyphae eseptatae, irregulariter ramosae, usque 6 μ latae; sporangiophora e mycelio aereo et substrato oriunda, plerumque ad basim attenuata, ad apicem latioria; sporangiophora sporangia ferentia circinata, eramosa, punctata, plerumque ad apicem brunneola, ad basim hyalina, sporangiophoro quoque sporangium unum apicaliter ferenti; sporangia e globosa dorsiventricaliter compressa, atra, majoria 50–200 μ, minoria 10–40 μ diam., tunica fragili rufo-brunnea punctata praedita, in dimidia aequalia dua dirumpentia; columellae ex ovales pyriformes, hyalinae usque pallide brunneolae, 7–129.5 x 5.2–94.5 μ, saepe in maturitate e sporangiophoro disjunctae; sporangiosporae numerosae, ellipsoideae, rufo-brunneae, 15–21 (–25.5) x 7.5–10.5 μ, plerumque 16.5–18 x 9 μ, longitudinaliter striatae, utrinque appendicibus numerosis hyalinis ornatae, usque 37.5 μ longae; sporangiophora sporangia ferentia erecta hyalina eramosa vel dichotome vel subdichotome ramosa, plerumque in ambitu undulata, ramo quoque in vesiculam globosam 28–50.7 μ diam. terminato; sporangiola dehiscentia, singulatim supra papillas breves enata, plurimum 4.5–6 μ longa, per superficiem totam vesiculae terminali vel rarius in processibus infra vesiculam terminalem oriunda, tunica punctata fragili, in maturitate

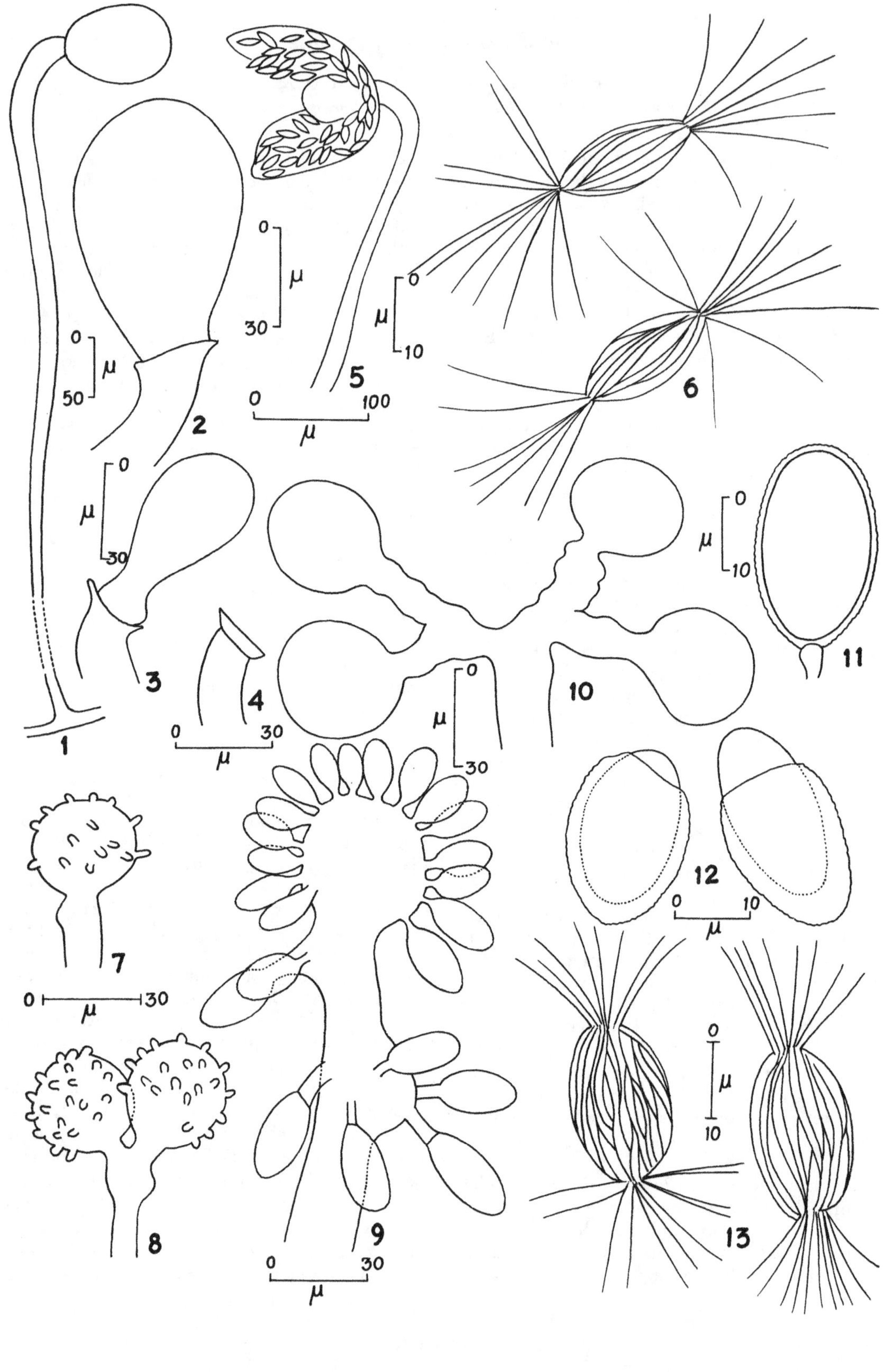

0
μ
50
1
2
0
μ
30
3
4
0
μ
30
0
μ
30
5
0
100
μ
0
μ
10
6
7
0
μ
30
8
9
0
30
μ
10
0
μ
30
0
μ
10
11
12
0
10
μ
0
μ
10
13

sporam unicem liberante; sporangiosporae e sporangiolo ellipsoideae, eis e sporangio enatis latiores, rufo-brunneae, 18–24 x (9.7–) 10.5–13.5 µ, plerumque 19.5–21 x 12–13.5 µ, longitudinaliter striatae, utrinque appendiculatae, appendicibus eis e sporangio enatis tenuiores.

Hab. e solo isolata, Vindhyachal.

Colonies on oatmeal agar,[1] SMA,[2] and potato-dextrose agar[3] white, floccose; hyphae nonseptate, irregularly branched, up to 6 µ in width; sporangiophores arising from aerial as well as substrate mycelium, usually narrow toward base and broader toward the tip, sporangiophores bearing sporangia circinate, unbranched, punctate, usually brownish toward the tip and hyaline at base, each bearing a sporangium at the tip; sporangia globose to dorsiventrally compressed, black, larger ones 50–200 µ, smaller ones 10–40 µ in diameter wall fragile, reddish brown, punctate, breaking into two equal halves; columellae oval to pyriform, hyaline to light brown in color, 7–129.5 x 5.2–94.5 µ, often disjointed from the sporangiophore on maturity; sporangiospores numerous, ellipsoidal, reddish brown, 15–21 (25.5) x 7.5–10.5 µ, mostly 16.5–18 x 9 µ, longitudinally striate, provided with numerous hyaline appendages at both the ends, up to 37.5 µ in length; sporangiophores bearing sporangiola erect, hyaline, unbranched or branched dichotomously or subdichotomously, usually wavy in outline, each branch terminating into a globose vesicle, 28–50.7 µ in diameter; sporangiola dehiscent, borne singly over short papillae, mostly 4.5–6 µ in length, arising from the entire surface of a terminal vesicle or rarely from lateral vesicular outgrowths below the terminal vesicle, wall punctate, fragile, liberating a single spore on maturity; sporangiospore from a sporangiolum ellipsoidal, more broad than the sporangiospore originating from a sporangium, reddish brown, 18–24 x (9.7) 10.5–13.5 µ, mostly 19.5–21 x 12–13.5 µ, longitudinally striate, with appendages at both ends; appendages more delicate than those of the sporangiospores originating from a sporangium.

Type: M-26, isolated from a soil sample collected at Vindhyachal. Culture deposited in BSM Culture Collection, Botany Department, University of Allahabad. It will also be deposited at NRRL, Peoria, Illinois, and Centraal Bureau voor Schimmelcultures, Baarn, Holland.

[1] Oatmeal agar: oatmeal. 20 gm.; yeast extract, 0.5 gm.; agar, 20 gm.; distilled water, 1,000 ml.

[2] Synthetic mucor agar (SMA): dextrose, 40 gm.; asparagine, 2 gm.; KH_2PO_4, 0.5 gm.; $MgSO_4$, 0.25 gm.; thiamine chloride, 0.5 mg.; agar, 20 gm.; distilled water, 1,000 ml.

[3] Potato dextrose agar (PDA): dextrose, 20 gm.; potato (peeled and sliced), 200 gm.; agar, 20 gm.; distilled water, 1,000 ml.

LITERATURE CITED

Hesseltine, C. W. 1962. Personal communication.

Mehrotra, B. S., and M. D. Mehrotra. 1964. Taxonomic studies of Choanephoraceae in India. Mycopath. Mycol. Appl. **22**: 21–39.

Poitras, A. W. 1955. Observations on asexual reproductive structures of the Choanephoraceae. Mycologia **47**: 702–713.

Sinha, A. 1940. A wet rot of leaves of *Colocasia antiquorum* due to secondary infection by *Choanephora cucurbitarum* Thaxter and *Choanephora trispora* Thaxter. (*Blakeslea trispora* Thaxter). Proc. Indian Acad. Sci. **11**: 167–176.

Thaxter, R. 1914. New or peculiar Zygomycetes. *Blakeslea, Dissophora, Haplosporangium*, nova genera. Bot. Gaz. **58**: 353–366.

Figs. 1–13. *Blakeslea monospora* sp. nov. Camera lucida drawings. Fig. 1. A sporangiophore. Figs. 2–3. Columellae. Fig. 4. Tip of the mature sporangiophore after the columella has disjointed. Fig. 5. Upper portion of a sporangiophore showing manner of opening of sporangium. Fig. 6. Two sporangiospores. Figs. 7–8. Upper portions of sporangiophores with vesicles at their tips. Fig. 9. An abnormal sporangiophore with sporangiola borne on lateral vesicular outgrowths. Fig. 10. Upper portion of the branched sporangiophore with vesicles at their tips. Fig. 11. A monosporus sporangiolum. Fig. 12. Two sporangiola liberating their respective sporangiospore. Fig. 13. Two sporangiospores from two sporangiola.

Overleaf

Figs. 14–21. *Blakeslea monospora* sp. nov. Photomicrographs. Fig. 14. Upper portion of a sporangiophore with sporangiola at their tips (as seen in petri dish). X180. Fig. 15. An enlarged sporangiospore from a sporangium (note the presence of appendages at both the ends). X1400. Fig. 16. Upper portion of a sporangiophore with columella at its tip. X600. Fig. 17. Sporangiospores from a sporangium (note the presence of longitudinal striations). X1100. Fig. 18. Upper portion of a sporangiophore with vesicles at its tip. X600. Figs. 19–20. Three sporangiola each liberating a single sporangiospore. X1400. Fig. 21. A sporangiospore from a sporangiolum (note the presence of appendages). X1400.

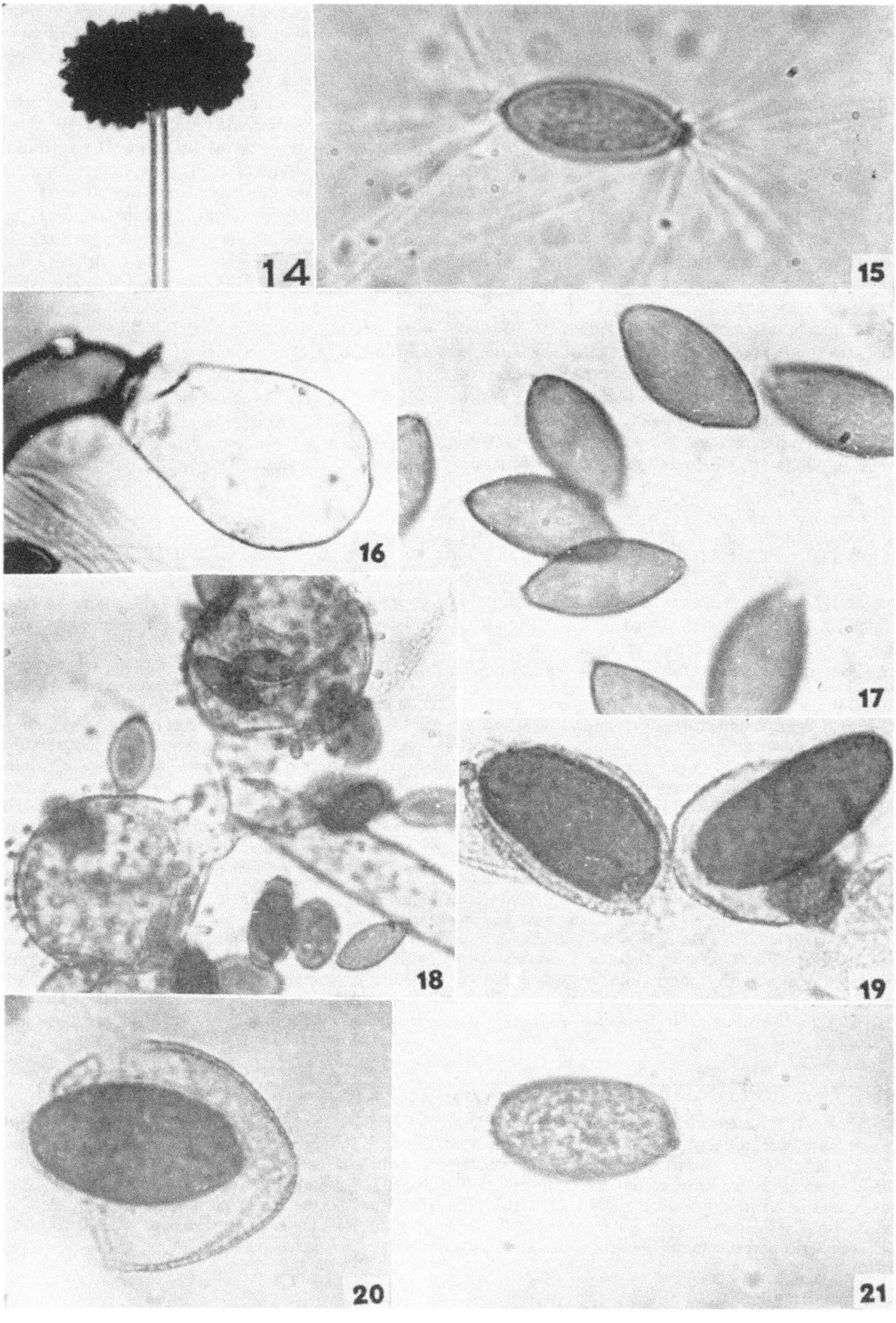
14
15
16
17
18
19
20
21

Two New Species of *Conidiobolus* from India

M. C. SRINIVASAN AND M. J. THIRUMALACHAR
Hindustani Antibiotics, Ltd., Pimpri, Poona, India

Summary

Studies on the morphology of spore forms of two new species of *Conidiobolus* are presented. *C. couchii* Srin. & Thirum. forms smooth zygospores and elongated secondary conidia which are borne and discharged forcibly in a manner similar to the secondary globose conidia—a feature hitherto known only in *C. eurymitus* Drechsl. *C. lobatus* Srin. & Thirum. is characterized by a highly lobulate to lichenoid mycelial colony, and is non-zygosporic.

In the present paper an account is given of two new species of *Conidiobolus* isolated from plant detritus samples collected in India. One of them forming smooth zygospores has been named in honor of Dr. J. N. Couch, who was the first to isolate and study in detail a smooth-zygospore-forming species of *Conidiobolus*—*C. brefeldianus* (Couch, 1939). Type cultures of the new species have been deposited with the American Type Culture Collection, Rockville, U.S.A.; Centraalbureau voor Schimmelcultures, Baarn; Commonwealth Mycological Institute, England; and the Indian Agricultural Research Institute, New Delhi.

Conidiobolus couchii Srinivasan & Thirum. sp. nov.

Mycelium hyalinum, incoloratae, ramosae, plerumque 3.5–6 μ latae. Conidiophoris simplices, nan-ramosis, 30–50 μ altis, 4–7 μ latis, apice unum primiforme conidium 10–16 μ latis, nunc hypham germinationis vel conidium formae elongatae ferens.

Conidia formae elongatae violenter absilientia incolorata nonnihil fusiformis vel elongato ellipsoidis, 12–17 μ longa et 8–12 μ lata, zygosporis globosis vel subglobosis, 12–16 μ diam. parietibus laevibus 0.8–1.2 μ.

Hab. in foliis putrescentibus, Poona, April 7, 1967.

Mycelium thin-walled, hyaline with vacuolate contents, branched, 3.5 to 6 μ wide, becoming disjointed into hyphal segments; conidiophores undifferentiated, unbranched, phototrophic, 30–50 μ high, 4–7 μ wide, developing conidia apically, primary conidia 10–16 μ in diameter with a small conical basal papilla, germinating directly or forming secondary conidia which are

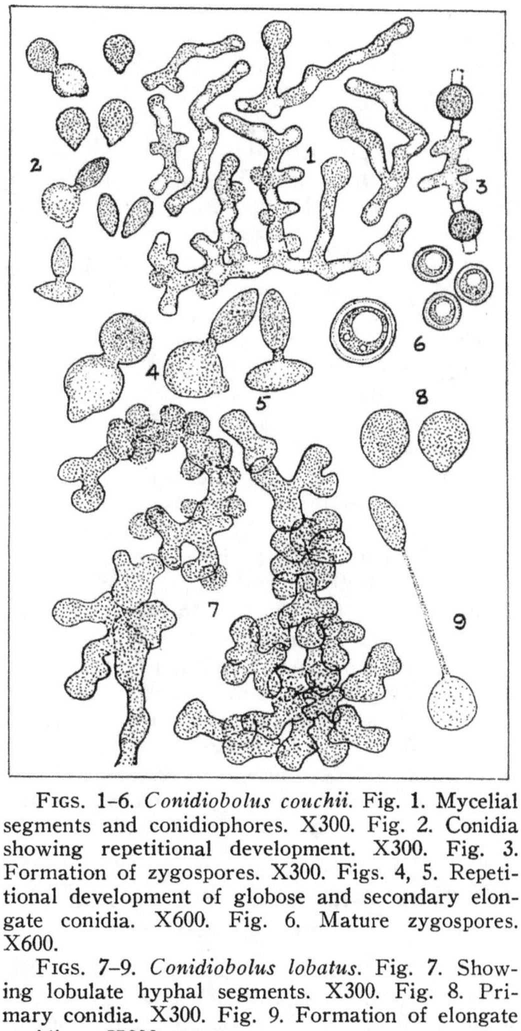

FIGS. 1–6. *Conidiobolus couchii*. Fig. 1. Mycelial segments and conidiophores. X300. Fig. 2. Conidia showing repetitional development. X300. Fig. 3. Formation of zygospores. X300. Figs. 4, 5. Repetitional development of globose and secondary elongate conidia. X600. Fig. 6. Mature zygospores. X600.

FIGS. 7–9. *Conidiobolus lobatus*. Fig. 7. Showing lobulate hyphal segments. X300. Fig. 8. Primary conidia. X300. Fig. 9. Formation of elongate conidium. X300.

globose or ovate to almond-shaped and developing on short tubular conidiophore. Both globose and almond-shaped secondary conidia forcibly discharged, the latter 12–17 μ long and 8–12 μ wide; zygospores globose to subglobose, 12–16 μ in diameter, smooth, with wall 0.8–1.2 μ thick (Figs. 1–6).

Hab. Isolated from plant detritus, Poona University Campus, Poona, April 7, 1967.

The fungus is slow-growing and forms a flat, somwhat mucoid colony on yeast-glucose-agar medium. The mycelium is composed of thin, delicate hyphae with vacuolate cytoplasm, which give rise to numerous, closely set, short, lateral branches. Hyphal segments are delimited through the withdrawl of contents and septation. The conidiophores are formed in large numbers in young cultures and represent undifferentiated, unbranched lateral branches which are phototrophic and developing globose conidia terminally. In young cultures (3–4 days old) large numbers of conidia are discharged which form a white deposit on the upper lid of the petri dish above the colony. Repetitional development of conidia takes place by the formation of globose secondary conidia or, more characteristically, by the development of an elongate-ovate or almond-shaped secondary conidium in a manner similar to the secondary conidia described in *C. eurymitus* by Drechsler (1965). These secondary conidia developed on short extensions of the globose conidia are also discharged with force, and as pointed out by Drechsler (1965), these are different from the elongated secondary conidia in other members of the genus which are borne aloft on tall, slender, filamentous conidiophores from which they later become detached passively. In cultures 4–5 days old, zygospore formation is initiated and takes place by conjugation between hyphal segments from the same or different mycelial branch. Mature zygospores were smooth walled, 12–16 μ in diameter, with a wall 0.8–1.2 μ thick.

The fungus represents a second species of *Conidiobolus*, besides *C. eurymitus*, in which elongate secondary conidia are formed and discharged with force. In its spore measurements, *C. couchii* is very small compared with *C. eurymitus.*

Conidiobolus lobatus Srinivasan & Thirum. sp. nov.

Mycelium hyalinum, septatum, 6–8 μ latis, ramosis lobulatus, colony lichenoidis; conidiophoris non-ramosis, 40–60 μ altis, apice unum primiforme conidium ferentes; primiforme conidiis globosis 20–24 μ diam. ad basin papilla 1–2 μ altis 2–4 μ latis. Conidiis elongate- ellipsoidis 18–25 μ longis 8–10 μ latis. Zygosporis ignota.

Hab. in foliis putrescentibus, New Delhi, January 1, 1967.

Mycelium thin-walled, 6–8 μ wide, septate and developing numerous lobed to irregularly dilated branches which grow together to form a lichenoid colony; conidiophores simple, 40–60 μ high, unbranched, and bear conidia terminally; primary conidia globose, 20–24 μ in diameter, with an obtuse or somewhat tapering basal papilla 1–2 μ high and 2–4 μ wide. Secondary ellipsoid conidium on filamentous conidiophores sometimes formed, 18–25 μ long and 8–10 μ wide. Zygospores not observed (Figs. 7–9).

Hab. Isolated from plant debris, New Delhi, India. January 1967.

The fungus is slow-growing and on yeast-glucose agar forms a flat, dull-white, somewhat mucoid colony with a metallic luster. Microscopic examination shows the colony to be composed of a highly lobulate mycelium which imparts a lichenoid appearance. The development of conidiophores and conidia are relatively sparse. Growth on cornmeal agar is very poor.

The conidia germinate by simple germ tubes, commonly 1–3 in number, and the mycelium often becomes disjointed following septation. Early in its development the mycelium gives rise to numerous knob-like lateral branches which expand into irregularly lobed or dilated structures. A large number of such branches growing together give the characteristic lobulated appearance to the colony. This pattern of mycelial growth has been observed on all agar media on which the fungus makes good growth. Primary conidia are commonly 20–24 μ in diameter, and in repetitional development globose secondary conidia and occasionally elongate conidia in filamentous conidiophore are formed. Zygospore formation was not observed on any of the media.

In forming a lichenoid thallus, the fungus is similar to *C. lichenicolus* (Srinivasan and Thirumalachar, 1967). It differs, however, in being nonzygosporic and possessing smaller primary conidia (20–24 μ) compared with 25–35 μ in *C. lichenicolus.*

LITERATURE CITED

COUCH, J. N. 1939. A new *Conidiobolus* with sexual reproduction. Amer. Jour. Bot. **26**: 119–130.

DRECHSLER, C. 1965. A robust *Conidiobolus* with zygospores containing parietal protoplasm. Mycologia **57**: 913–926.

SRINIVASAN, M. C., AND M. J. THIRUMALACHAR. In press. Studies on species of *Conidiobolus* from India—V. Mycopath. Mycol. Appl.

Studies in the Genus *Prototheca*.
I. Literature Review

WM. BRIDGE COOKE[1]
U. S. Department of the Interior
Federal Water Pollution Control Administration, Cincinnati, Ohio

Introduction

Prototheca is a genus of colorless organisms usually assigned to the green algae as a nonpigmented derivative of such a genus as *Chlorella*. Krüger (1894) considered it funguslike when he first recognized it, but no one had been able to relate it to a well-characterized fungal group. In 1895 Saccardo placed it in the *Endomycetaceae* of the *Ascomycetes*. West (1916) placed it in the *Chlorophyceae* and included it in the subfamily *Oocysteae* as a genus of colorless organisms. To the family *Oocystaceae* Fritsch (1948) assigned ". . . the genus *Prototheca* of Krüger which appears to be no more than a colorless *Chlorella*. . . ."

Redaelli and Ciferri (1935, 1936) laid more stress on the colorless nature of the cells of this genus and followed Printz (1927) in including it in the *Protothecaceae*, a family of nonpigmented organisms related by loss of chlorophyll to the *Chlorellaceae*. Ciferri et al. (1957) merely refer to the *Prototheceae*.

Strains of species of *Prototheca* have been isolated from slime fluxes of trees, feces and fingernails of man, potato skin, acid stream water, diseased tissues associated with a case of bovine mastitis, sludges in waste stabilization ponds, and other habitats. It appears to be one of the easier genera that can be obtained experimentally in the laboratory. Beijerinck (1904), Granick (1948), and Butler (1954) have obtained it as a laboratory mutant of *Chlorella* by cultural manipulations and the use of radiation.

Apparently Beijerinck (1904) was the first to note that colorless strains of *Chlorella* occur naturally or can be induced in the laboratory. The names *Chlorella protothecoides* and *Prototheca chlorelloides* have been attributed to him. Meyer (1933) discussed some of the problems arising after finding hyaline strains of *Chlorella variegata* Beijer.

Fritsch (1948) derived the nonmotile algae from the motile algae, an indication that the coccoid habit arose from prolongation of the sedentary phase of a flagellated individual. In addition, he stated that "loss of green colour is a frequent phenomenon in cultures provided with organic nutrient, and the saprophytic tendency of many Chlorococcales no doubt affords scope for the origin of permanently colourless forms."

Granick (1948) showed that X-radiated cells produce green, yellow, and white mutants when grown in darkness on 3% agar that included 0.7% glucose and inorganic salts. Mutant strain W_5, isolated as a colorless colony, gave rise to pale green colonies, and these produced colorless colonies. On solid media both types of colonies produced a brown pigment that was identified as protoporphyrin, a precursor of chlorophyll.

Butler (1954) produced a series of white colonies from a parent (C–1) culture of *Chlorella*. The range in sizes of cells was relatively uniform; typical "autospores" (aplanospores) were produced by all three strains; plastids were produced by two of these strains; and starch was synthesized, although weakly, in one. The colorless nature of these cells was stable in culture. From this work it was inferred that *Prototheca* was derived from the *Chlorophyceae*.

From the time of the first report of a species of *Prototheca*, interest has been shown in its physiology. Krüger (1894) studied assimilation of carbon and nitrogen compounds by various strains and included a summary of his studies with his description of the two species he found. *P. moriformis* grew in glucose, galactose, and glycerin, but not in lactose, maltose, or dextrin. It grew on peptone, asparagin, ammonium tartrate, and inorganic ammonium salts, but not on nitrate. No fermentation was observed. Moist heat at 52° to 53° C. was lethal, as was dry heat at 70° to 75° C. The organism survived a 4-hour exposure to –83° C. Its cardinal temperature points were minimum 6° to 10° C., optimum 29° to 31° C., and maximum about 38° C. For *P. zopfii*, Krüger (1894) noted that assimilation and fermentation reactions are essentially the same as those for *P. moriformis*.

The summary of Barker's work (1935, 1936),

[1] In charge, Fungus Studies, Biological Treatment Research Activities, Cincinnati Water Research Laboratory.

listed by Albritton (1954) in his Table 21 on algal nutrition, shows that the strain of *P. zopfii* used requires thiamine or else pyrimidine and thiazole. This organism utilized glucose, but not arabinose, maltose, sucrose, or xylose. Although it did not grow on citric, fumaric, and succinic acids, at pH 3.0 to 5.5 it could use lactic and pyruvic acids. Of the alcohols (nomenclature and order of Albritton, 1954) tested, i-butanol, n-butanol, ethanol, glycerol, n-pentanol, and n-propanol were utilized, but i-pentanol only partially, and methanol and i-propanol not at all. Among the fatty acids tested, growth occurred on i-butyric, n-butyric, i-caproic, n-caproic, n-descylic, n-heptylic, n-nonylic, n-octylic, propionic, and n-valeric, but only weakly on i-valeric. Ammonium ions and peptone were available as nitrogen sources.

Ashford et al. (1930), in describing the new species *P. portoricensis* (with an average of two autospores per cell) and its variety *trispora*, (with an average of three autospores per cell), presented some correlative physiological information. Under the conditions of their experiments, *P. portoricensis* assimilated glucose, galatose, and levulose preferentially; it used maltose, saccharose, lactose, dextrin, starch, and glycerin weakly, and did not utilize trisaccharides or starches. The strain of this species that they isolated assimilated peptone well; used asparagin, acid ammonium tartrate, and ammonium chloride weakly; and failed to assimilate nitrates, nitrites, or glycine. There was little variation from an initial pH of 6.4 in culture without an indicator, but when brom-thymol-blue was used, the pH of the culture was stated to drop to as low as 4.45. This pH change may have occurred when the indicator was added to the nutrient solution, according to the authors. Optimum temperature for growth was found to be 30° to 32° C. and maximum temperature 36 to 38° C. The variety *trispora* of this species had similar assimilation patterns and growth temperatures. However, the pH of the growth medium was stated to reach only 5.1 in the presence of the indicator brom-thymol-blue. Neither the species nor the variety could ferment glucose.

In describing *P. ciferrii*, based on larger cells and certain physiological properties, Negroni and Blaisten (1941) found its optimum temperature to be 30° C.; no fermentation occurred, and there was no modification of the pH in culture media with carbohydrates. Glucose and levulose were used strongly, but utilization of mannose and galactose was weak. In nitrogen auxanograms using glucose as the carbon source, these writers found growth with peptone positive, but negative with other compounds tested.

In comparing the vitamin requirements of strains of *Prototheca*, Ciferri (1956) found that *P. portoricensis* var. *trispora* required thiamine unless it was grown in a peptone-containing medium, presumably with vitamin contaminants; that none of the strains of *Prototheca* could grow on media containing only glucose, salts, and amino acids; and that achloric mutant strains of *Chlorella pyreniodosa* did not require vitamins or other growth factors. All previous work had shown that nitrate could not be used as a nitrogen source.

In a study of carbon and nitrogen sources, Ciferri (1957) and Ciferri, Montemartini, and Ciferri (1957) used a liquid medium of the following basic composition per liter of distilled water; carbon source, 34 g.; KH_2PO_4, 5 g.; nitrogen source, 0.25 mg. N per ml.; ferric chloride, trace. For carbon sources the pH was adjusted to 6, and KNO_3 was used as the source of nitrogen; (presumably nitrate-nitrogen was adequate for growth of all strains tested); for nitrogen sources the pH was adjusted to 5, and glucose was used as the carbon source. Growth on carbon sources was measured by oven dry weights of cells produced, and on nitrogen sources, by photoelectric turbidometry. Those species tested for carbon sources included *P. portoricensis* and its variety *trispora*, *P. zopfii*, *P. ciferrii*, and *P. chlorelloides*; in addition, *P. moriformis* was tested for nitrogen sources. All strains tested grew weakly on xylose. Good growth was obtained on glycerin, glucose, levulose, and galactose; poor growth, on mannose and saccharose; and little or no growth on lactose, raffinose, and inulin. The nitrogen sources tested included peptone, arginine, glutamic acid, aspartic acid, proline, histidine, leucine, norleucine, phenylalanine, alanine, urea, glycine, $NaNO_3$ and NH_4Cl. *P. ciferrii* grew best and used all these compounds; *P. moriformis* demonstrated poorest growth. In general, growth was better on the compounds mentioned early in the list, but some growth was obtained on all.

Hartsell (1956) published a table in which Hedrick reported that one strain of *P. zopfii* had survived preservation under paraffin oil during the five-year period prior to the date of the report. The medium on which the culture was kept at room temperature was composed of 3 g. each of malt extract and yeast extract and 5 g. each of peptone and glucose per liter of distilled water.

In 1959, Tubaki and Soneda reported results of studies made on a number of cultures that were obtained from Ciferri, Wickerham, and others and included two strains isolated in Japan. They found no definite physiological characters upon which species could be differentiated, although two groups could be separated on the basis of optimum temperature requirements. Their specific differentiation was based on morphological criteria, such as size and shape of parent and daughter cells. All cultures tested assimilated glucose, galactose (although sometimes weakly), ethanol, (one weak, one negative), and glycerol (one weak); all except *P. moriformis* grew weakly to moderately on lactic acid. One strain showed weak growth on methanol, and three on acetone; eleven grew well on acetic acid, and four on propionic acid. No growth was obtained on maltose, saccharose, lactose, cellobiose, butanol, butyric acid, or formic acid. All nitrogen sources were utilized, including peptone, urea, asparagin, KNO_3, $(NH_4)_2SO_4$, and sodium glutamate. All species demonstrated a definite thiamine deficiency, and riboflavin promoted growth in most instances. No visible amylolytic, cellulolytic, or pectinolytic activities were indicated when type strains were grown on starch, cellulose, and pectin agars.

In the study of populations of yeast-like organisms in slime fluxes of *Ulmus carpinifolia* on the campus of the University of California at Davis, Phaff, Yoneyama, and Carmo-Sousa (1964) found that *Prototheca moriformis, P. zopfii*, and *P. ciferrii* were present. *P. moriformis* was present from April to November with some degree of regularity. *P. zopfii* appeared in only three samplings between September and January, and *P. ciferrii* appeared sporadically between July and November. None of these species assumed a dominant role in the population, although at times *P. moriformis* was more abundant than one of the two common yeasts.

An unidentified strain of *Prototheca* has been found by Christina M. Richards in work with frogs (*Rana pipiens*) at the University of Illinois (1958) and later at Idaho State College. Organisms found in the intestines of tadpoles whose growth was reduced by crowding included cells that appeared to belong to a species of this genus. Photographs and dried preparations were submitted to the writer, and intestinal contents streaked on agar slants were submitted to Dr. Phaff. The growth was not obtained in pure culture, and no specific determination has been attempted.

In his conclusion, Butler (1954) remarked that Beijerinck thought that *Prototheca* formed one transition from algae to fungi. Ciferri, Montemartini, and Ciferri (1957) state that since the *Protothecea*e "... are better adapted to the heterotrophic life than the colorless *Chlorella* strains, it may be inferred that the *Protothecea*e could not have arisen from the green, allied forms, or that the separation occurred in the remote past." Tubaki and Soneda (1959) suggest a relationship with the fungi on the basis of pigmentation. They rule out, however, relationship between this genus and two fungus genera of uncertain position that also produce endospores: *Protomyces* and *Coccidioides*.

Because of the loss of pigmentation, the naturally occurring strains of achloric chlorelloid organisms are grouped together in the genus *Prototheca*. It seems to this writer that at some time during the existence of the genus *Chlorella*, or an ancestral alga similar to it, one or more strains developed the ability to grow as saprobes in environments into which stray cells of a parent organism migrated. These saprobes have reportedly lost the ability to produce thiamine under certain conditions. They never developed an ability to utilize more compounds than simple sugars, alcohols, organic acids, ammonium compounds, and amino acids that may be thought to be present in many habitats in nature as contents of exudates, degradation products, or end products of the metabolism of other organisms. The genus *Prototheca* is thus considered to be an alga rather than a fungus, occupying the same ecological niche as many yeasts or yeast-like fungi, but always showing the algal relationship by the method of reproduction.

The nuclear condition of the members of this genus has never been investigated. No one has ever reported conjugation of any type; thus it is uncertain to what extent hybrid vigor is maintained by chromosomal or gene exchange between adjoining populations or strains of one or more species. Experiments reported above on achloric strains of *Chlorella* seem to indicate that the dual phenomenon and heterokaryosis are not operable here. However, insufficient experimental and cytological work is available in published form to determine whether this is true or what type of parasexual cycle might be operative.

BIBLIOGRAPHY

ALBRITTON, E. C. 1954. Ed., Standard Values in Nutrition and Metabolism. W. B. Saunders Co., Philadelphia, xiii. 380 pp.

ANDERSON, E. H. 1945. Nature of the growth factor for the colorless alga *Prototheca zopfii*. Jour. Gen. Physiol. **28**: 287–296.

———. 1945. Studies in the metabolism of the colorless alga *Prototheca zopfii*. Jour. Gen. Physiol. **28**: 297–327.

ASHFORD, B. K., R. CIFERRI, AND L. M. DALMAU. 1930. A new species of *Prototheca* and a variety of the same isolated from the human intestine. Arch. f. Protistenk. **70**: 619–638.

BARKER, H. A. 1935. The metabolism of the colorless alga, *Prototheca zopfii* Krüger. Jour. Cell. Comp. Phys. **7**: 73–93.

BEIJERINCK, M. W., 1904. *Chlorella variegata*, ein bunter Mikrobe. Rec. Trav. Bot. Neerl. **1**: 14–27.

BOIDIN, J., M. C. PIGNAL, F. MERMIER, AND M. ARPIN. 1963. Quelques levures camerounaises. Cahiers de la Maboké. **1** (2) : 85–100.

BUTLER, E. E. 1954. Radiation-induced chlorophyll-less mutants of *Chlorella*. Science **120**: 274–275.

CHODAT, R. 1913. Monographies d'algues en culture pure. Mat. Flore Crypt. Suisse. **4** (2) : 1–266 (121–123).

CIFERRI, O. 1956. Thiamine-deficiency of *Prototheca*, a yeast-like achloric alga. Nature **178**: 1475–1476.

———. 1957. Metabolismo comparativo delle Protothecae e delle mutanti achloriche di Chlorella. Giorn. di Microbiologia **3** (2) : 97–108.

CIFERRI, R., A. MONTEMARTINI, AND O. CIFERRI. 1957. Caratteristiche morfologiche e assimilative e speciologia delle Protothecae. Nuovi Annali d'Igene e Microbiologia **8**: 554–563.

FRITSCH, F. E. 1948. The Structure and Reproduction of the Algae. Cambridge University Press, Cambridge. Vol. I. xvii, 791 pp.

GRANICK, S. 1948. Protoporphyrin 9 as a precursor of chlorophyll. Jour. Biol. Chem. **172**: 717–727.

HARTSELL, S. E. 1956. Microbiological process report. Maintenance of cultures under paraffin oil. Appl. Microbiol. **4**: 350–355.

HENNEBERG, W. 1936. Handbuch dr Gärungsbakteriologie. **2**: 307–308.

ITERSON, G. VAN, JR., L. E. DEN DOOREN DE JONG, AND A. J. KLUYVER. 1940. Martinus William Beijerinck. Martinus Nijhoff, The Hague.

JOHANSEN, D. A. 1940. Plant Microtechnique. McGraw-Hill Book Co., New York. xi, 523 pp.

KRÜGER, W. 1894. Kurz Characteristik einiger niederen Organismen in Saftflüsse der Laubbäume. 1. Ueber einen neuen Pilz-typus, repräsentiert durch die Gattung *Prototheca* (*P. moriformis* (sic) et *P. zopfii*). 2. Ueber zwei aus Saftflussen rein gezüchtet Algen. Hedwigia **33**: 241–251, 251–266.

MEYER, H. 1933. Das Chlorose- und Panaschüreproblem bei Chlorellen, II. Teil. Beih. Bot. Centr. Abt. 1. **51**: 170.

NEGRONI, P., AND R. BLAISTEN. 1941. Estudio morfologico y fisiologico de una neuva especie de *Prototheca*: *P. ciferrii* n. sp. aislada de epidermis de papa. Mycopathologia **3**: 94–104.

PHAFF, H. J., M. YONEYAMA, AND L. DOCARMO-SOUSA. 1964. A one year quantitative study of the yeast flora in a single slime flux of *Ulmus carpinifolia* Gled. Riv. Patol. Veg. Ser. III. **4**: 485–497.

PRINGSHEIM, E. G. 1946. Pure Cultures of Algae, Their Preparation and Maintenance. Cambridge Univ. Press, Cambridge. Vol. I. xii, 119 pp.

PRINTZ, A. 1927. In Engler and Prantl, Die Naturlichen Pflanzenfamilien. 2d ed. Chlorophyceae **3**: 1–463. Pp. 131–132.

REDAELLI, P., AND R. CIFERRI. 1935. Pouvoir pathogène pour les animaux des algues coprophytes achloriques du genre *Prototheca*. Observations sur les Protothecaceae. Bull. Sex. Ital. Sox. Intern. Microbiol. fasc. **8–9**: 1–8.

——— AND ———. 1936. Argomenti a favore di una sistemazione del genre *Blastocystis* nelle Algae. Boll. Ist. Sieroter. Milanese **15**: 154–160.

RICHARDS, C. M. 1958. The inhibition of growth in crowded *Rana pipiens* tadpoles. Physiological Zoology **31**: 138–151.

SACCARDO, P. A. 1895. Sylloge fungorum. Pavia. Vol. XI.

TUBAKI, K., AND M. SONEDA. 1959. Cultural and taxonomical studies on *Prototheca*. Nagaoa **1959**: 25–34.

WEST, G. S. 1916. Algae. Cambridge Univ. Press, Cambridge. Vol. I. x, 475 pp.

Studies in the Genus *Prototheca*.
II. Taxonomy

WM. BRIDGE COOKE[1]
U. S. Department of the Interior
Federal Water Pollution Control Administration, Cincinnati, Ohio

In the last ten years, two comprehensive studies of the genus *Prototheca* have been published. Ciferri, Montemartini, and Ciferri (1957) presented a scheme of classification, based on cultures available to them, that included four species and one variety. Tubaki and Soneda (1959) described five species known to them. They recognized *P. wickerhamii* from the United States as a new species, but their own recently isolated strains from Japan proved to be cospecific with established species.

The life cycle of *Prototheca,* insofar as it has been described by direct observation or inference and as observed by the writer, is as follows: Upon release from the sporangium, the sporangiospore increases in size, assumes the shape of the parent cell, and goes through an assimilative phase that may last for a long or a short period. By the time the cell has reached one-half to two-thirds the size of the parent cell, cleavage lines appear within it. In some cases the cell merely develops a heavier wall around itself. Usually one, two, four, or more cleavage lines develop; the units of cytoplasm develop walls around themselves within the areas of the cleavage planes; and two to twenty or more spores develop, which at first are triangular or irregular in shape. This shape may remain irregular until the sporangial wall is burst, or the spores may become rounded or ovate in the image of the parent cell before the wall breaks. The sporangium wall apparently breaks by pressure from the enlarging spores, and the latter are released in a passive way. They may remain in a cluster or separate after release. The cycle is then repeated when conditions of nutrition, temperature, and moisture are favorable.

The structures with which we are concerned, then, are assimilative cells, sporangial initials, sporangia, and sporangiospores. Unfortunately, within one life history these four terms must be applied to one and the same cell. In the following descriptions of species and varieties, all cells will be considered assimilative cells, and the measurements will be given for "cells." Cells were measured before cleavage lines develop and during early stages of spore formation. When one, two, or three thick-walled spores are produced, they are referred to as hypnospores. If cleavage occurs, the cell is referred to as a sporangium and the daughter cells as aplanospores, although they could also be termed autospores.

The large cyst-like structures that are produced one, two, or three per cell are considered gemmae, cysts, hypnospores, or resting spores (West, 1916), since these are the terms used for other members of the *Protococcales.* Development is quite similar to that illustrated by West for *Chlorella* in his text figure 120, A–I, p. 194, and rupture of the wall of the cells containing cysts or other spores is like that of *Oocystis* illustrated in his text figure 122, A–F, page 196. West referred to the type of reproduction demonstrated in this genus as "multiplication" rather than "asexual reproduction," since there is no flagellation. The hypnospores seen in two- and three-celled sporangia in all strains observed, however, are very similar to spores illustrated for other species, although, when thin-walled, they are true nonmotile aplanospores.

Measurements in the following description are based on growth in liquid wort (diamalt broth) culture.

Concerning the average numbers of aplanospores produced per sporangium in cultures growing on potato dextrose agar slides and in liquid wort, all cultures examined produced from 1 to 7 or as many as 10 to 20 aplanospores per sporangium. Because of this variability in numbers, counting the aplanospores and then deriving an equation on which to base some sort of average for each species or variety did not yield satisfactory information on which to base species or varieties in this genus.

In spite of the facts that records of *Prototheca* are scattered in the literature, that isolates of the genus are derived from various types of habitats in many parts of the world, and that usually these are obtained by students

[1] In Charge, Fungus Studies, Biological Treatment Research Activities, Cincinnati Water Research Laboratory.

of yeasts searching for naturally occurring species of their own type of organisms, to date, only one name for this genus has been found in the literature on colorless algae. My expanded description of the genus *Prototheca* is as follows:

***Prototheca* Krüger**
Hedwigia **33:** 263. 1894

Colorless, hyaline cells producing a white-to-cream-colored yeast-like growth on agar slants; cells separate, no pseudomycelium or chain-like development, not multiplying by budding, but by irregular cleavage of the protoplasm into 2 to 20 or more units around which walls develop; some cells becoming hypnospores by encystment of contents developing a thick, smooth wall around the protoplasm within the cell; by thickening of walls some cells may produce 2 or 3 hypnospores; walls of aplanospores thin, becoming visible upon disintegration of sporangium wall; aplanospores at first polyhedral from crowding, then, in face or side view triangular-ovate to trapezoidal-ovate in the sporangium, becoming globose or ovate before rupture of the sporangium wall or shortly thereafter; aplanospores remaining in a clustei for a short time after rupture of the sporangium wall, then separating and appearing like assimilative cells; cleavage of protoplasm initiated in cells of any size, usually starting regularly in 1, 2, and 3 planes, then becoming irregularly arranged; sporangia and aplanospores globose, broad-ovate, subcylindric or cylindric, varying in size between and within species, no plastid-like structures observed; no sexual reproduction observed.

Lectotype: *P. zopfii* Krüger

Krüger did not state which of the two species (*P. zopfii* and *P. moriformis*) he proposed should be considered the type of the genus. Negroni and Blaisten (1941), however, selected *P. zopfii,* which has been retained because, while it is not the first of the two species described by Krüger, it is the one on which most of his and numerous later studies were based. In selection of the type, it should be noted that the epithet "moriformis" adequately describes any of the species considered valid here, at least while the aplanospores lie within the sporanium, but the epithet "zopfii" memorializes one of the more important nineteenth-century German microbiologists.

Two courses are open for characterizing species in this genus. One is based on the morphology of the cells as they appear in culture. Because there is no special configuration of the cell, no special markings on the wall, and no special structure within the wall, this course is limited to the description of the shape and the measurement of the size of the cells. The second is based on the physiology of the strains studied. Here one could look for special patterns in carbon or nitrogen assimilation, vitamin requirements, pH and temperature ranges, and other growth characteristics. Work done in my laboratory and in others, however, indicates that the pattern of growth is greatly similar in most media, except for the utilization of certain carbon sources.

The most variable characteristic in those strains of the genus *Prototheca* studied to date is the size of the cells, which fluctuates more obviously within the species with larger cells. This fluctuation may result from the fact that cells of all sizes have been measured rather than cells that have reached or passed one developmental stage or another. On the basis of cell size, four groups may be suggested: *P. moriformis* Krüger, with small cells; *P. zopfii* Krüger, with medium-sized cells; *P. ciferrii* Negroni and Blaisten, *P. wickerhamii* Tubaki and Soneda, and *P. stagnora* W. B. Cooke, with medium-large cells; and *P. portoricensis* Ashford, Ciferri, and Dalmau, with large-sized spores. In the several tests used, cells maintain a relatively consistent pattern of size within strains assigned to one or another of these species. As cell size increases, the spread between width and length becomes greater, indicating a tendency of development within the genus from spherical to short-ovate cells through long-ovate to short-cylindric cells.

Until examined microscopically, isolates assignable to this genus appear like yeast cultures. Since they can be handled readily in the same way that yeasts are handled, it is convenient to use the techniques of yeast identification in studying individual isolates. This is done by using the techniques described by Lodder and Kreger-van Rij (1952) for individual cultures. In screening the ability of the isolated strains to use additional carbon sources in the Wickerham series (Wickerham and Burton, 1948), strains can be inserted with other yeast and yeast-like strains in replicator (Fig. 1) series, where as many as 25 strains can be inoculated simultaneously on agar plates.

On the basis of these tests, it was found that in addition to size, certain strains could be grouped according to their ability to use certain sugars. All strains grew on glucose, galactose, and glycerol; in addition, strains of two species utilized trehalose, and strains of one species utilized sucrose. The species that utilized trehalose could be separated easily on

FIG. 1. Equipment used for replication of yeast cultures on a variety of media. Left, rear: a block of oak wood, relatively resistant to repeated autoclaving, with spaces for thirty 9 x 30 mm.-shell vials. Center, rear: the replicator, made of stainless steel, with a strap handle of the same material, on the top side (against the table), with 25 wire "needles" 6 cm. long, soldered into holes. Holes in the side pieces coincide with the poles of the stands in the front row. Left, front: A wood block used for storage of the replicator; holes in the wood under the petri dish receive the replicator needles. Center, front: wooden block mounted on stainless steel plate in which guide posts are welded; wooden block has recessed area to hold petri dish in position during inoculation. Right, front: wooden block mounted as in center piece; 25 holes are bored at equal depths in block to accommodate 9 x 30 mm. shell vials, 16 in outer row, 8 in inner row and 1 in center, to match positions of needles in replicator. Notches in front edge of all instruments indicate starting point of numbering of each culture and match red pencil mark on bottom of petri dish in which agar with various test compounds has solidified and surface-dried before inoculation. The four petri dishes show different colony types obtained from 100 strains of unknown species of yeasts.

Modified from Beech, Carr, and Codner (1955) by Miller and Phaff (unpublished), then by Sanitary Engineering Center model shop.

the basis of cell size; *P. moriformis* has the smallest cells of the species studied, and *P. wickerhamii* has medium-large cells. The three species that utilized neither trehalose nor sucrose could also be separated on the basis of cell size: *P. zopfii* has medium-sized cells, *P. ciferri* has medium-large cells, and *P. portoricensis* has the largest cells of the strains that were studied. This leaves the species that utilizes sucrose to be identified, and no name could be found for it.

Prototheca stagnora W. B. Cooke, sp. nov. (Figs. 2, 3)

Cellulae ellipsoideae, 10.8–18.0 x 14.4–21.6 μ, vel 14.4–21.6 x 18.0–27.0 μ; aplanosporae ellipsoideae, 1.8–5.4 x 3.6–7.2 μ vel 3.6–7.2 x 7.2–14.4 μ. Cultura in agarico farinacea, hyalina; glucosum, sucrosum, galactosum, ethanolo et glycerolo assimilantur, sed non trehalosum.

Streak cultures white, smooth to minutely reticulate, dull to shiny, pasty, with a slightly raised center and an even to scalloped margin; cells of two overlapping sizes, 10.8–18.0 x 14.4–2.16 μ, or 14.4–21.6 x 18.0–27.0 μ; aplanospores usually 3.6–7.2 x 7.2–14.4 μ, sometimes, or in unopened sporangia, 1.8–5.4 x 3.6–7.2 μ. Excellent growth on glucose, ethanol, and glycerol; good to excellent but latent growth on sucrose; fair to good growth on galactose, DL-lactic acid, and succinic acid.

Cells of cultures assigned to this species show a relatively uniform pattern of size, although in some isolates the cells become slightly larger than in others. Utilization patterns on glucose, ethanol, and glycerol are regular, while on other carbon sources patterns of utilization vary with the strains tested. Results reported above were obtained on solid agar media. Dr. H. J. Phaff at Davis, California, confirmed the tests in liquid culture and found that sucrose was utilized latently, slowly coming up to the full 3+ level considered positive.

Culture SEC–L–1690 (ATCC 16528) has been chosen as the type strain and is deposited

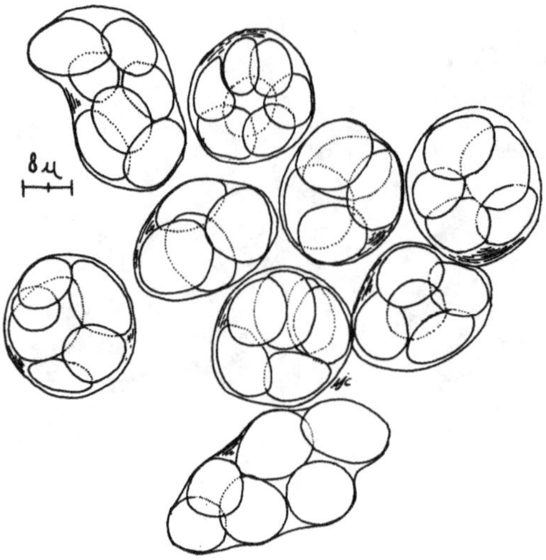

FIG. 2. 1690—*Prototheca stagnora.*

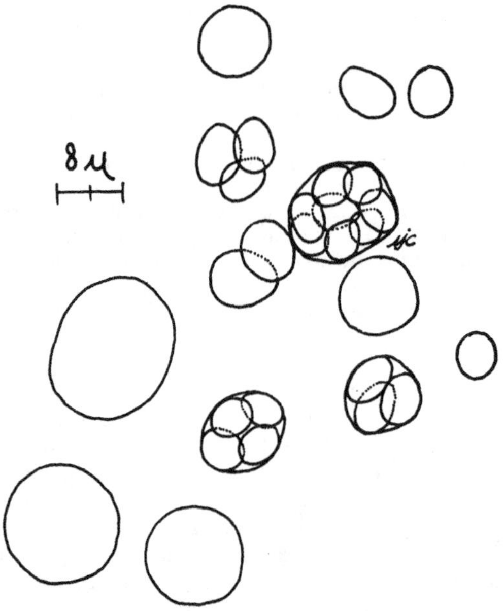

FIG. 3. 62–335—*Prototheca stagnora.*

in the American Type Culture Collection, the Northern Utilization and Development Laboratory (NRRL), the culture collection of the Department of Food Technology, University of California, Davis, and the Centraalbureau voor Schimmelcultures (CBS), Baarn, Holland.

Sources of cultures examined and habitats: Waste stabilization ponds (the Latin word for pond is *stagnum*), near Lebanon, Warren Co., Ohio. L–893 and L–894: Oct. 24, 1960, from sludge near effluent pipe in bottom of Pond #1, isolated in YNB–1% glucose broth; L–1146, Nov. 7, 1960, from sludge near effluent pipe in bottom of Pond #1, isolated in YNB–1% glucose broth; L–1146, Nov. 7, 1960, from sludge near effluent pipe in bottom of Pond #1, isolated in YNB–1% glucose broth; L–1177, Nov. 7, 1960, from pond water in middle of Pond #3, isolated in YCB-KNO_3 broth; L–1690 (ATCC 16528) type, isolated March 13, 1961, from bottom sludge near effluent pipe, Pond #1, on neopeptone-dextrose-rose bengal-Aureomycin agar. Sewage Treatment Plant, Dayton, Montgomery Co., Ohio. 62–265: June 14, 1962, from raw sludge, in YNB–1% glucose broth; 62–320, 62–321: June 14, 1962, from raw sludge in YNB–20% glucose broth; 62–335, 62–339, 62–344 (ATCC 16527): June 14, 1962, from digested sludge in YNB–20% glucose broth: 63–558: April 26, 1963, from digested sludge in YNB–1% glucose broth. Sewage treatment plant, Hamilton, Butler Co., Ohio 63–462: April 5, 1963, from digested sludge, on neopeptone-dextrose-rose bengal-Aureomycin agar.

To date, on the basis of cell size and carbon assimilation patterns, six species of *Prototheca* have been recognized. As additional isolates are found and demonstrated to have different assimilation and size patterns, or habitat requirements, other species may be recognized and added to this list. As the strains upon which this study were based become better known, an approach to a monographic treatment of this genus can be made.

LITERATURE CITED

CIFERRI, R., A. MONTEMARTINI, AND O. CIFERRI. 1957. Caratteristiche morfologiche e assimilative e speciologia dell Protothecae. Nuovi Annali d'Igene e Microbiologia **8**: 554–563.

BEECH, F. W., J. G. CARR, AND R. C. CODNER. 1955. A multipoint inoculator for plating bacteria and yeasts. J. Gen. Microbiol. **13**: 408–410.

LODDER, J., AND N. J. W. KREGER VAN RIJ. 1952. The Yeasts, a Taxonomic Study. North Holland Publishing Co., Amsterdam. xi, 713 pp.

NEGRONI, P., AND R. BLAISTEN. 1941. Estudio morfologico y fisologico de una neuva especie de *Prototheca: P. ciferrii*, n. sp. aislada de epidermis de papa. Mycopathologia **3**: 94–104.

TUBAKI, K., AND M. SONEDA. 1959. Cultural and taxonomical studies on *Prototheca*. Nagaoa **1959**: 25–34.

WEST, G. S. 1916. Algae. Cambridge University Press, Cambridge. Vol. I. x, 475 pp.

WICKERSHAM, L. J., AND K. A. BURTON. 1948. Carbon assimilation tests for the classification of yeasts. J. Bact. **56**: 363–371.

Invasion of Some Tropical Timbers by Fungi in Brackish Waters

DONALD D. RITCHIE[1]
Naval Research Laboratory, Washington, D. C.

Many factors affect the ability of fungi to attack wood. The obvious environmental factors are temperature, pH, the relative amount of water present, oxygen supply, and sometimes light. The morphology of the fungus and its enzymatic composition, which is to say its genetic capabilities, are also critical. Equally important are the characteristics of the wood: its physical hardness, its possession of required nutrients, and the possible presence of inhibitory compounds. All wood, being mostly cellulose, is potentially available to many fungi—mainly, of course, the cellulolytic ones.

Animals, too, attack wood, even though some—e.g., termites—obtain nourishment from it indirectly through symbionts. The requirements of animals being generally similar to but not identical with those of fungi, a series of trials has been made in an effort to compare the activities of some animals, notably marine borers, with those of some salt-tolerant fungi.

Some woods are known to be more resistant than others to infestations by marine borers, and this resistance is known to be related to the chemical composition rather than to the hardness of the tissues, although the compounds responsible for the resistance are not known. In an extensive series of tests, Southwell et al. (1965) exposed 114 species of tropical woods for fourteen months in a variety of sites, including waters of different salinities, and tropical soils. They established the resistance or susceptibility of a number of species to shipworms (Teredinidae) and other wood-boring invertebrates (Limnoria and pholads), and to some extent to termites and terrestrial fungi.

The possibility of combined attack by marine borers and salt-tolerant fungi has been considered. Some observations indicate that both fungi and animals help destroy wood in marine or estuarine habitats, with fungi facilitating the activity of borers (Meyers and Reynolds, 1957) . Some experiments seem to discount the likelihood of this sequence of events (Ray and Stuntz, 1959), but the matter is not unequivocally settled.

This combination of facts, suppositions, and knowledge gaps, plus the availability of selected wood samples and a fortunate conjunction of time and exposure sites, prompted the inception of the experiments to be reported in the present paper. The intent of these trials was to determine whether those species of wood which are susceptible to borer attack are also susceptible to fungus attack, and to find what effect environmental conditions, especially salinity, may have on the invasion of wood by salt-water-inhabiting fungi.

Materials and Methods

From the species tested by the Southwell group, eight were chosen as representing a range of susceptibility or resistance to borers. None was completely resistant to borers in the brackish waters of Miraflores Lake in the Canal Zone, Panama (average salinity, 0.74 parts per thousand), but several were resistant in the saltier waters of the nearby oceans, bays, and estuaries. The species of woods, selected for exposure to marine fungi, listed with their resistance to borer attack, are these:

Chrysophyllum cainito L. (caimito, star apple) —resistant

Pouteria chiricana Standley (nispero de monte) —resistant

Ocotea rodiaei (Schomb.) Mez. (greenheart)— —resistant

Tabebuia pentaphylla (L.) Hemsl. (roble de sabana)—moderately resistant

Pinus caribaea Mor. (Nicaraguan pine)—susceptible

Prioria copaifera Gris. (cativo)—susceptible

Rhizophora mangle L. (red mangrove)—susceptible

Symphonia globulifera L.f. (cerillo)—susceptible

The samples, which had been dried for three years in air-conditioned storage, were cut into strips measuring about 100 x 6 x 6 mm. These were suspended in the water by nichrome wires so that even at low tide they were covered. The exposure sites were chosen partly for variety and partly by expedience, because many desirable sites were either inaccessible or subject to vandalism. All the sites were in the Canal Zone near the entrances to the Panama Canal. The sites, with their arbitrarily assigned numbers, representative temperature readings, and exposure periods are given in Table I.

[1] On leave from Barnard College, Columbia University, New York, N. Y.

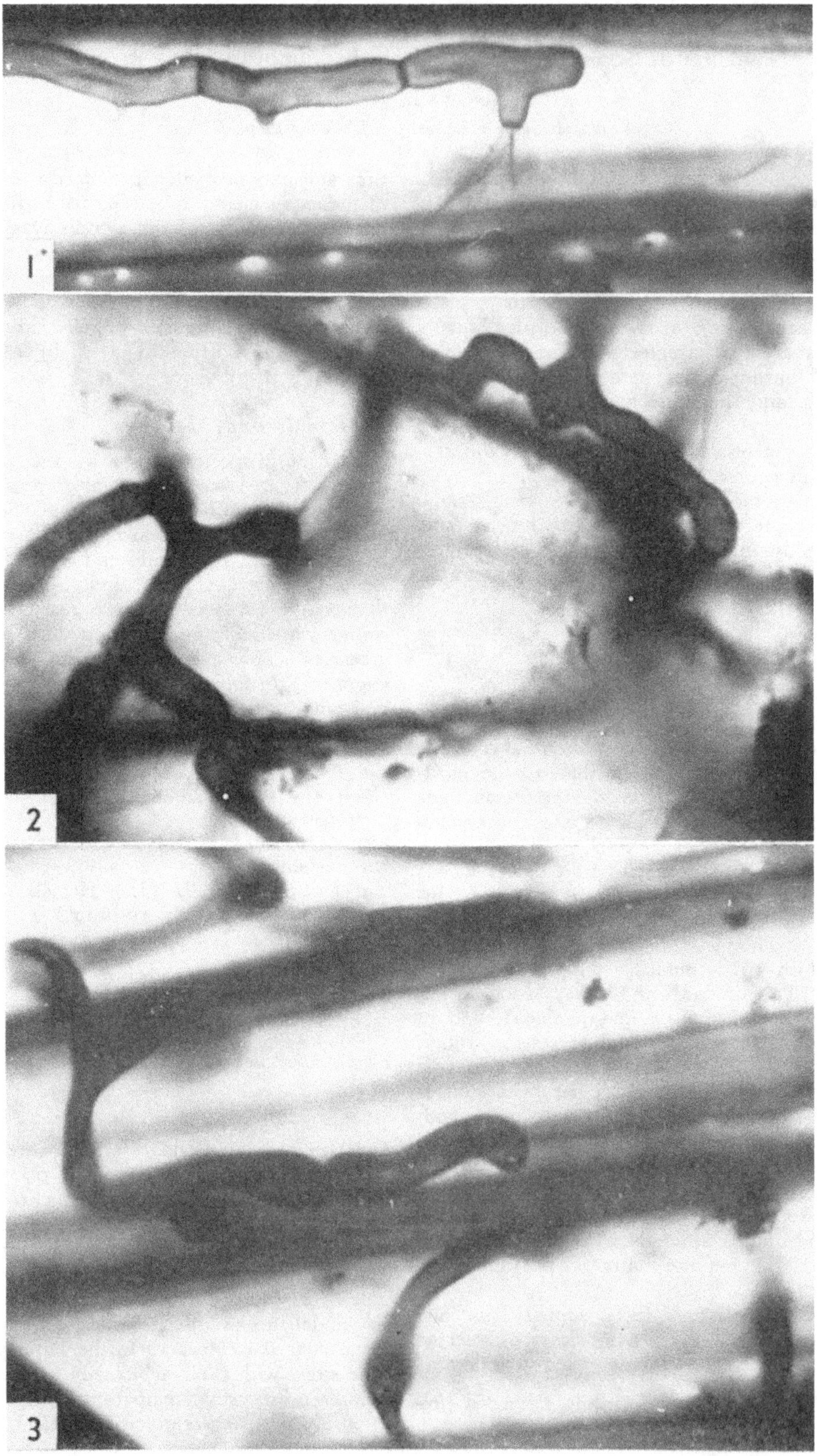
1
2
3

Table I

Sites of exposure of wood samples

No.	Station	Exposure Period	Water Temperature, C°.
0	Coco Solo Pier 1	7/23–9/3	27.2
1	Galeta Point	7/23–9/5	27.8
2	Coco Solo Pier 1	7/23–9/3	27.2
3	Coco Solo Ramp	7/23–8/30	27.8
4	Fort Sherman Ferry Slip	7/24–8/27	29.1
5	Four-hundred foot Hill	7/23–8/26	30.0
6	Four-hundred foot Hill	7/28–8/28	30.0
7	Galeta Point	7/23–9/5	27.8
8	Fort Sherman Light	7/24–8/27	32.9
10	Galeta Radio Station	7/23–8/23	28.3
11	Coco Solo Ramp	7/23–8/30	27.8
13	Fort Amador Pier	7/25–8/23	26.7

Salinities were determined by the chromate-silver-nitrate method of Harvey (1957). Temperatures were relatively constant, with a minimum of 26.7° C. at Fort Amador (Pacific side) and a high of 32.9° C. at the Fort Sherman Light (Atlantic side). Salinity varied because of run-off from rains, which, on the Atlantic side where most of the exposure sites were, measured 15.37 inches during July and 22.18 inches during August, 1963.

On being lifted from the water, wood samples were dropped individually into separate, sterile containers. In the laboratory, each sample was divided aseptically into three pieces. One portion was air dried and stored; one portion fixed in formaldehyde, ethanol, and acetic acid; and one portion used for isolation of fungi.

The fixed material was used for recognition of hyphal invasion. Soaking in glycerol-ethanol softened the wood before it was cut longitudinally on a sliding microtome at 25 μ. The method of Hubert (1922) uses methyl violet and Bismarck brown to differentiate lignified walls, which stain orange to red; cellulose walls, which stain light purple; and fungus mycelium, which stains dark purple.

Evaluation of the amount of fungus infestation was made by inspecting five fields of view in each of five different sections. Subjective bias in choice of fields was avoided by evaluating the first field to come into focus, then throwing the section out of focus while the slide was being moved to a new position, re-focusing, and evaluating the new field where the slide stopped. Each field of view, as seen with a 4-mm. microscope objective, was scored on a five-point scale from zero to 4 (field crowded with hyphae). Twenty-five readings were made on each wood sample at each exposure site. The evaluations, expressed on the 0–4 scale, were averaged for each species at any one site; then the averages were averaged for the species at all sites.

Fungi grew from teased slivers of wood which were placed on the surface of agar made up with sea water, 0.5% glucose, and 0.1% yeast extract.

Results

On a basis of the amount of visible mycelial infestation in the sections, the eight species of woods could be grouped into three classes, with clear gaps in the susceptibility spectrum. With a score of zero representing total resistance and 4.0 representing massed hyphae in every field of view examined, the species were rated as shown in Table II. Obviously *Rhizophora mangle* (red mangrove) and *Ocotea rodiaei* (greenheart) are in a class by themselves with respect to freedom from fungus invasion. *Rhizophora* was practically fungus-free in all the exposure sites, but one *Ocotea* sample (from Galeta Point) did have some hyphae in a few wood cells.

Most of the other species showed some scattered mycelium in most fields and in many sections. Their ratings were bunched in a close

Figs. 1–3. Hyphae in sections of wood of *Prioria copaifera*, X300. Methyl violet and Bismarck brown. (1) Thin extrusion of hypha extending toward vessel wall. (2) Hyphae passing (lower left) or about to pass (upper right) through pits in parenchyma cell. (3) Branching mycelium in vessel. Pore at end of hypha (right of center) from which penetrating hypha will extend.

Table II

Rating of wood species with respect to invasion of fungi in estuarine waters, with score from 0 (no invasion) to 4 (complete infestation)

Species	Rating
Rhizophora mangle	0.05±0.12
Ocotea rodiaei	0.11±0.24
Chrysophyllum cainito	0.44±0.39
Pouteria chiricana	0.45±0.30
Pinus caribaea	0.57±0.35
Tabebuia pentaphylla	0.59±0.69
Symphonia globulifera	0.84±0.49
Prioria copaifera	1.78±0.57

Table III

Genera of fungi isolated from wood samples, listed in order of frequency of isolation

Number of Isolations	Genus
29	*Fusarium*
18	*Tritirachium*
15	*Pestalotia*
13	*Phoma, Trichoderma*
12	*Sterile mycelia*
9	*Paecilomyces*
8	*Curvularia*
6	*Aspergillus*
4	*Pencillium*
2	*Geotrichum, Gliocladium, Phomopsis*
1	*Ascomycete (indet.), Candida, Cephalosporium, Coniothyrium, Epicoccum, Haplosporangium, Hormodendrum, Lulworthia, Mucor, Nigrospora, Oedocephalum, Sphaeropsis, Spicaria, Zygodesmus*

sequence from 0.44 to 0.83, with no conspicuous gaps.

Prioria copaifera (cativo) was alone in its susceptibility, achieving a rating of 1.78 after submersion of only four to six weeks. Some of the individual samples, which contributed to the average rating, scored as high as 2.7 (at the Fort Sherman ferry slip) and 3.0 (at the Coco Solo Pier #1). The numerical ratings indicate only that fungus infestation was common; they do not show that the vigor of the invading hyphae, as evidenced by their large size and rich branching, was greater than that in other samples. One *Pestalotia*, present in several samples, was unusually active, and had a unique method of growing from cell to cell. (Ritchie, 1967).

The fungi isolated from the immersed samples are shown in Table III. Many of them are obviously terrestrial molds which have sufficient tolerance of estuarine salinities to inhabit salty or brackish waters. This is verified by the fact that although many will grow in sea water, they can be made to grow better on dilute media.

Infestation was in general neither better nor worse in the less saline exposure sites where outflow from fresh creeks diluted the inshore water, especially after heavy rains. The amounts of fungus growth in the nine sites which yielded the fullest data (some samples were lost from some sites) are compared in Table IV.

Table IV

Average ratings (relative fungus infestation) of all wood species, listed by exposure site in decreasing order of salinities

Site #	Salinity range, parts per thousand	Rating
8	30.01–30.32	0.4
13	29.06–32.30	0.4
1	29.51–31.67	0.4
7	29.21–31.47	0.4
10	26.57–29.61	0.3
2	23.32–24.70	1.0
11	14.20–24.30	0.5
6	11.90–22.22	1.1
5	11.89–21.20	0.5

Discussion

The way in which fungi invaded these particular species of woods, under the environmental conditions provided, brings up several notable points.

Fungus succession, as an ecological phenomenon, is poorly understood (Johnson and Sparrow, 1961). Among marine and estuarine fungi, however, the pioneers in tropical as in temperate climates are usually the imperfects. Although ascomycetes are common in saline habitats, the current tests, being limited to six weeks or less, allowed time for only two to establish themselves. One of them, on *Chrysophyllum* in Panama Bay, was never identified, and one was the ubiquitous *Lulworthia medusa* (Ell. & Ev.) Cribb & Cribb, which occurred on *Prioria copaifera,* also in Panama Bay. Ascomycetes are typically slower in the succession of fungi appearing in substrates exposed in salt water.

The growth of essentially terrestrial molds in saline habitats, where they find osmotic and ionic conditions less than optimal, is not surprising. The decision as to what makes the best conditions for growth of a given strain of organism is based on laboratory experiments, and does not prove either that the conditions which seem best in the laboratory are necessarily the best for the organism in nature, or that the organism is restricted to the habitat which in the opinion of an experimenter is optimal for the organism (Ritchie, 1959).

The wood species under observation here were rated for resistance to decay by bacteria and terrestrial fungi as a part of the marine borer screening program (Southwell et al., 1965). In that project, sample stakes were driven partially into the soil, many of them in tropical rain forest conditions where warmth, moisture, and abundant nourishment could encourage fungal growth. After 18 months, the samples were rated by macroscopic inspection for evidence of decay. When fungal activity in soil and in salt water is compared, only one species, *Ocotea rodiaei,* proved resistant in both habitats. *Rhizophora mangle,* which had the least mycelium of any species in the present series, showed considerable terrestrial decay. As to susceptibility, again only one species was consistent in soil and in salt water: *Prioria copaifera.* This species was also heavily damaged by marine and brackish-water borers. Cativo must be delicious. No clear relation between resistance to terrestrial and to salt-water fungi appears, although some parallels can be found in *Ocotea* and *Prioria.*

The uncertainty of a relation between fungus and borer destruction of wood in salt water still remains. The present tests, however, indicate little cooperation between plants and animals in this matter. A comparison of the data in Table II with those of Southwell et al. (1965) shows that one species may be free from salt-water fungi, riddled by borers, and rotted by terrestrial fungi (*Rhizophora*). Another (*Ocotea*) may have general resistance to all three kinds of organisms; another (*Pouteria*) may have moderate resistance to salt-water fungi, high resistance to borers, and low resistance to terrestrial decay. Only *Prioria* proved to be generally susceptible. Inasmuch as no pattern is discernible, a relation between fungi and food-eating animals seems unlikely.

Although the relatively short exposure time of these trials limits the practicality of generalization, the similarity of infestation in all the sites indicates that salinity was a weak factor in determining the rate and the amount of fungus growth. An apparently considerable difference separates the ratings at sites five and six (ratings of 0.5 and 1.1 respectively), yet their salinities were almost the same, and the sites were in fact only a few feet apart. Then sites five and thirteen, having similar amounts of infestation (ratings 0.5 and 0.4 respectively) were at the extremes of the salinity range, and were indeed in different oceans.

Summary

Eight species of tropical timber woods (*Chrysophyllum cainito, Ocotea rodiaei, Pinus caribaea, Pouteria chiricana, Prioria copaifera, Rhizophora mangle, Symphonia globulifera,* and *Tabebuia pentaphylla*) were submerged in estuaries near the entrances of the Panama Canal. These species, chosen for their known susceptibility or resistance to attack by marine borers, were invaded by salt-tolerant fungi, mostly imperfect species. The amount of fungus infestation varied from almost none to very heavy in the space of less than six weeks, but the amount of mycelium in the woods bore no apparent relation to either animal attack or concentration of solutes in the water. *Ocotea rodiaei* (greenheart) and *Rhizophora mangle* (red mangrove) were practically fungus-free at the end of the test. *Prioria copaifera* (cativo) was filled with hyphae, and the others were invaded by scattered mycelium.

ACKNOWLEDGMENTS

The wood samples used in the present tests were provided by the Corrosion Laboratory, Canal Zone, Panama, of the Naval Research Laboratory, Washington, D. C. The laboratory also furnished logistic support and physical facilities for exposure procedures, isolation, and cultivation of the fungi. Most of the preparation of sections for evaluation was done by Misses Malke Engel and Joan Gardner, at Barnard College, Columbia University, through a training grant from the National Science Foundation to Yeshiva University.

LITERATURE CITED

HARVEY, H. W. 1957. The Chemistry and Fertility of Sea Waters. Cambridge University Press, Cambridge.

HUBERT, E. E. 1922. A staining method for hyphae of wood-inhabiting fungi. Phytopathology **12**: 440–441.

JOHNSON, T. W., JR., AND F. K. SPARROW. 1961. Fungi in Oceans and Estuaries. J. Cramer, Weinheim.

MEYERS, S. P., AND E. S. REYNOLDS. 1957. Incidence of marine fungi in relation to wood-borer attack. Science **126**: 969.

RAY, D. L., AND D. E. STUNTZ. 1959. Possible relation between marine fungi and Limnoria attack on submerged wood. Science **129**: 93–94.

RITCHIE, D. 1959. The effect of salinity and temperature on marine and other fungi from various climates. Bull. Torrey Bot. Club. **86**: 367–373.

———. 1967. Penetration of wood cells by special extensions of *Pestalotia* hyphae. Mycologia **59**: 417–422.

SOUTHWELL, C. R., C. W. HUMMER, JR., B. W. FORGESON, T. R. PRICE, T. R. SWEENEY, AND A. L. ALEXANDER. 1965. Natural resistance of woods to marine borer and other biological deterioration in tropical environments. Travaux du Centre de Recherches et d'Études Océanographiques, Paris. **6** (new series): 419–432.

Fungus Spores in Lake Singletary Sediment

FREDERICK A. WOLF

Department of Botany, Duke University, Durham, N. C.

Introduction

Lake Singletary is among the several bay lakes situated in the Coastal Plain of North Carolina. The origin of these lakes has been the subject of considerable study, partly because of peculiarity of their contour, since each is elongated in a northwest-southeasterly direction. The studies by Frey (1949, 1950, 1951) contain summaries, to date, of the results of these studies and of their interpretation.

Lake Singletary is one of the smaller of these lakes, having a surface area of 569 acres and a maximum depth of about 12 feet. It lacks both inlet and outlet streams so that its water supply is limited to that from direct rainfall and from drainage from the immediately surrounding area. The water has a dark color due to finely suspended products of fires, and has an acidity of pH 4.5. These water factors very probably are responsible for the sparse population of aquatic plants along the lake border and also for the sparse algal growth throughout the lake.

The rate of sedimentation in this lake, based on radiocarbon dating, has been determined to be approximately one inch per 750 years. A series of samples from different profile levels was recently carbon-dated by Whitehead (1966). His results provide the following data. At the 1.04–1.07 meter depth, the age is 5,750 ± 155 years (I–1752); at the 1.41–1.49 meter depth, 11,000 ± 200 years (I–1751); at the 1.69–1.74 meter depth, 16,200 ± 290 years (I–1750); at the 2.02–2.07 meter depth, 35,800 $\pm^{3200}_{2000}$ years (I–1748); and at the 2.30–2.34 meter depth, > 40,000 years (I–1749).

From pollen analyses, Frey (1951) identified the following kinds of tree pollen in the sediment of Lake Singletary: *Abies, Acer, Betula, Carpinus, Carya, Castanea, Chamaecyparis, Fagus, Fraxinus, Juglans, Liquidambar, Nyssa, Ostrya, Quercus, Picea, Pinus, Salix, Taxodium, Tilia, Tsuga, and Ulmus.* He reported that the stratum containing pollen of *Tsuga, Betula, Carya,* and *Quercus,* found immediately above a layer containing pollen of *Pinus* and *Abies,* has a radiocarbon age of 10,540 ± 510 years. He also noted that spores of *Asplenium, Isoetes, Lycopodium,* and *Sphagnum* are interspersed among the tree pollen.

The results of these foregoing studies of pollen analyses and of radiocarbon age of Lake Singletary sediment indicate that a "made-to-order" opportunity exists for determining whether spore analysis can be employed to complement these results. Moreover, special consideration appears never to have been given to the fungus spore content of the sediment of this lake or of any other of the bay lakes. The present study was undertaken, therefore, to determine whether the findings from identification of the kinds of spores in Lake Singletary sediment and of their stratigraphic distribution can be correlated with previous palynological studies.

Materials and Methods

Two cores of sediment, extending to a depth of 2.6 meters, were taken from a point near the center of the lake where the water depth approximated 9 feet. The sediment at the water-sediment interface is a soft, coal-black ooze that has a slimy feel. With increase in depth to about a meter, the black sediment becomes more firm and paste-like. There is a thin layer of brown clay with a sparse content of sand beyond the one-meter depth. The remainder of the core beyond this clay layer is sand that is infiltrated with a small amount of black char.

Samples for microscopic examination were taken at 0.10-meter intervals. Those from the upper meter consisted of one-eighth cc. of material, and each of those from the remainder of the core consisted of one cc. All preparations from the water-sediment interface consisted of dilute suspensions in water, and the kinds of spores found by this technique are shown in Plate I. Attempts were also made with this material to concentrate the spores (a) by flotation in salt solution, (b) by centrifugation at different rates, and (c) by flotation in dilute hydrogen peroxide to which potassium hydroxide had been added. Partly because of paucity of spores, these three procedures proved to be of little value.

The larger sand grains from samples of the sand-containing material were first removed by decantation, using water that was made alkaline to aid in freeing spores that might otherwise have remained attached to the grains of sand.

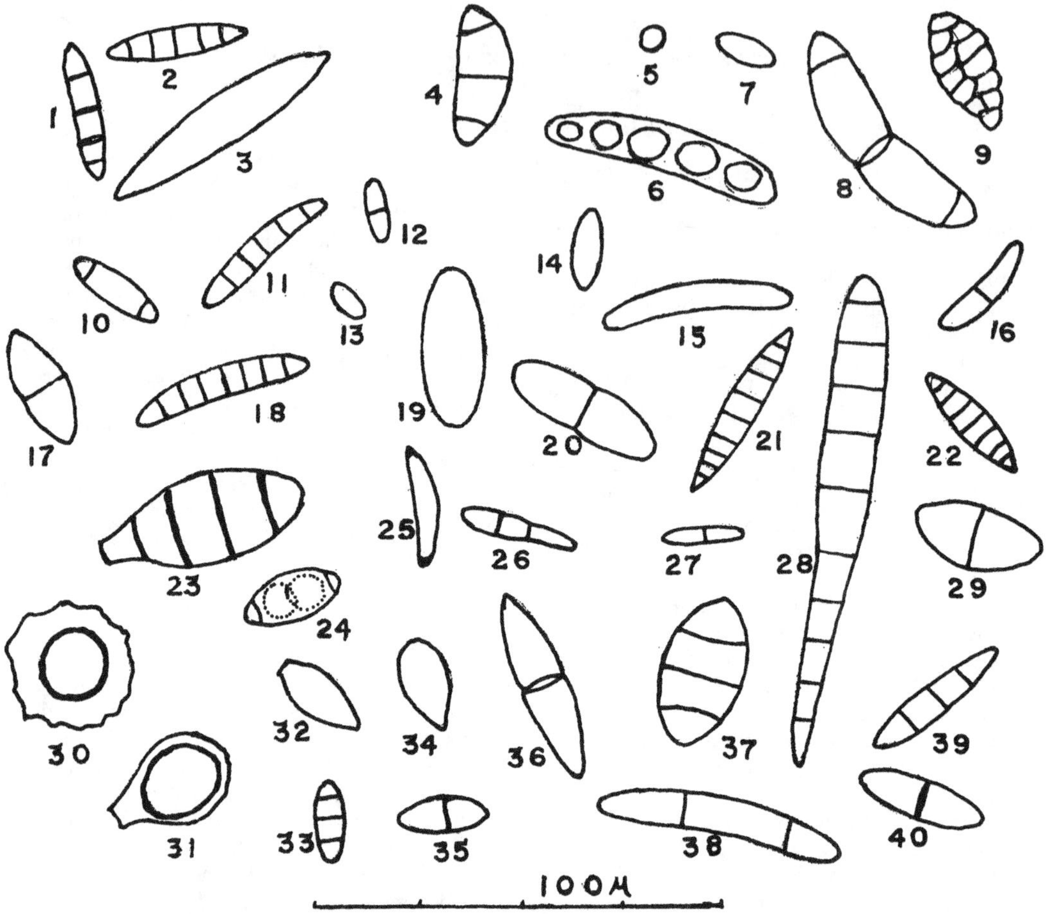

PLATE I. Representative kinds of fungus spores deposited at the water-sediment interface of Lake Singletary. 1, 11, 18, *Camposporium*; 2, 21, 22, and 39, *Leptosphaeria*; 4, *Curvularia*; 6, *Helminthosporium*; 7 and 14, *Peziza*; 9, *Pleospora*; 10 and 24, *Hendersonula*; 12 and 27, *Didymosphaeria*; 15, *Aschersonia*; 17, *Glonium*; 23, *Bactridium*; 26, *Piricularia*; 28, *Clasterosporium*; 29, *Dermatella*; 30 and 31, *Pythium*; 32 and 34, *Fusicladium*; 33, *Septogloeum*; 35 and 40, *Diplodina*; 37, *Trichocladium*. Others not identified.

All the samples from all profile depths were desilicified for 48 hours or more with hydrofluoric acid. They were then subjected to acetolysis by the method of Erdtman (1960).

Aliquot parts of the residue remaining after acetolysis were mounted in glycerine to which a trace of eosin had been added. Drawings of the kinds of spores found at the surface of the core and of those at the different profile depths, shown in Plates II and III, were made with the aid of a camera lucida. Their magnification is shown by comparison with the scale. The total number of spores at different depths was derived by computation based upon the area of the cover glass and that of the microscopic field surveyed. These results are shown graphically in Plate IV.

Results and Interpretations

The kinds of spores found by direct examination of dilute suspensions of material taken from the water-sediment interface, Plate I, do not accord with expectations. It was anticipated that the spores of such genera would be found to predominate as are at present known to be associated with pines and turkey oak, for the reason that these kinds of trees presently are most abundant in the immediate environs of the lake. According to check lists, the kinds of fungi on pines include many Basidiomycetes —i.e., Thelephoraceae, Polyporaceae, and Agaricaceae and also leaf and stem rusts and needle cast fungi. Those on oaks include members of the Sphaeriales on the branches and imperfect fungi on the foliage. But spores of none of

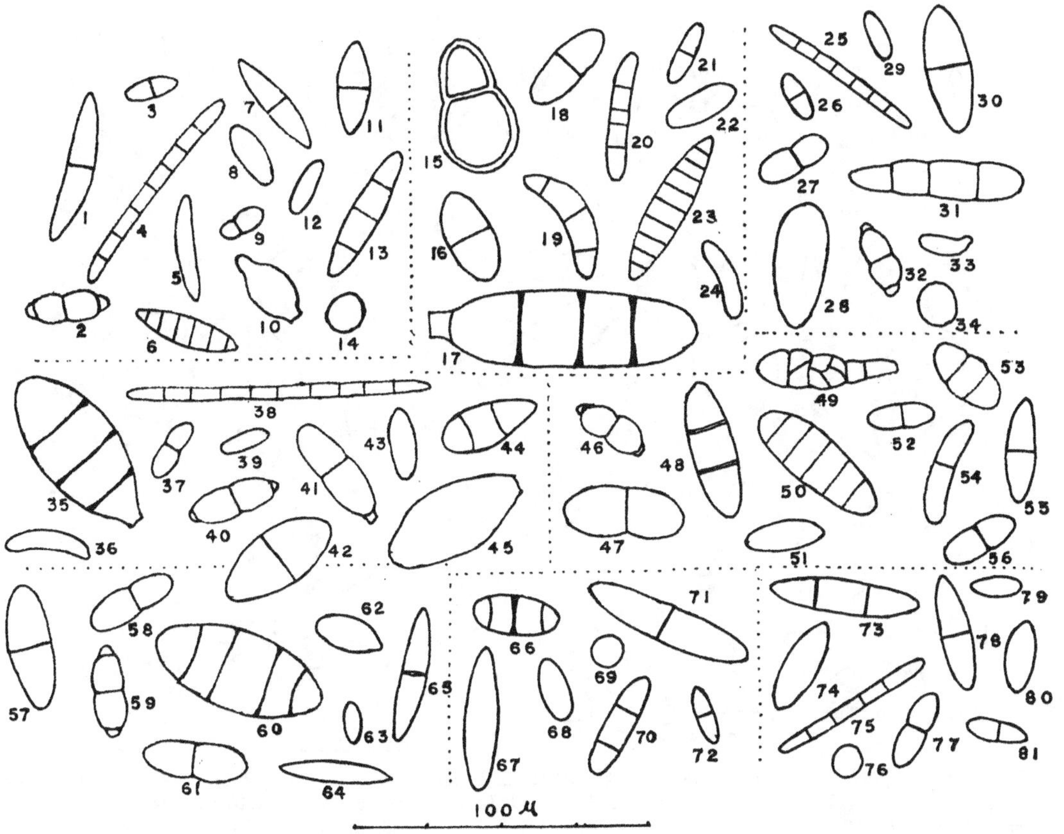

PLATE II. Spores from profile depths of 0.10 to 0.80 meters. 1–14, at 0.10 meters; 15–24, at 0.20 meters; 25–34, at 0.30 meters; 35–45, at 0.40 meters; 46–56, at 0.50 meters; 57–65, at 0.60 meters; 66–72, at 0.70 meters; and 73–81, at 0.80 meters. 2, 32, 40, and 46, *Hendersonula*; 3, 26, 72, and 81, *Didymosphaeria*; 4, 20, 25, 38, and 75, *Camposporium*; 6 and 23, *Calonectria*; 7, 30, 65, and 78, *Scolecotrichum*; 8, 22, 43, 68, and 80, *Peziza*; 9, *Ohleria*; 11 and 55, *Glonium*; 17 and 35, *Bactridium*; 19, *Curvularia*; 27, 56, 58, 61, and 77, *Nectria*; 44 and 60, *Trichocladium*; 48 and 73, *Dentryphiopsis*; 49, *Alternaria*; 50, *Brachysporium*; others not identified.

these kinds of fungi were among those identified. Consideration of the following facts may aid in accounting for lack of occurrence of the kinds of spores that were expected: (1) generic identification of isolated spores is attended with great uncertainty, especially of those that are one-celled or two-celled, since members of many taxonomic groups have such kinds; (2) spores belonging to different families or other taxonomic groups may have the same appearance, and such kinds can be identified only after examination of their spore-producing structures; (3) acetolysis treatment may cause hyaline spores to become dark-walled; and (4) the identification of spores that are thousands of years old must be based on resemblance with contemporary genera, with the result that the assignment of names remains tentative except for kinds of fungi that possess spores having distinctive morphologic features.

Comparison of the kinds of fungus spores found at different profile depths, Plates II and III with those in Plate I fails to reveal relationship between spore analyses and the palynological findings published by Frey (1951). This outcome is counter to that which was anticipated, and presently adequate explanations are not available. It may be suggested, however, that a better understanding of the results of spore analyses might be attained were experiments performed in which spores of different species of fungi were tested (1) to determine their tolerance to treatment by acetolysis, and (2) to determine their ability to withstand decomposition in sediment. Such tests might reveal that certain kinds are destroyed by acetolysis, and also that some can retain their

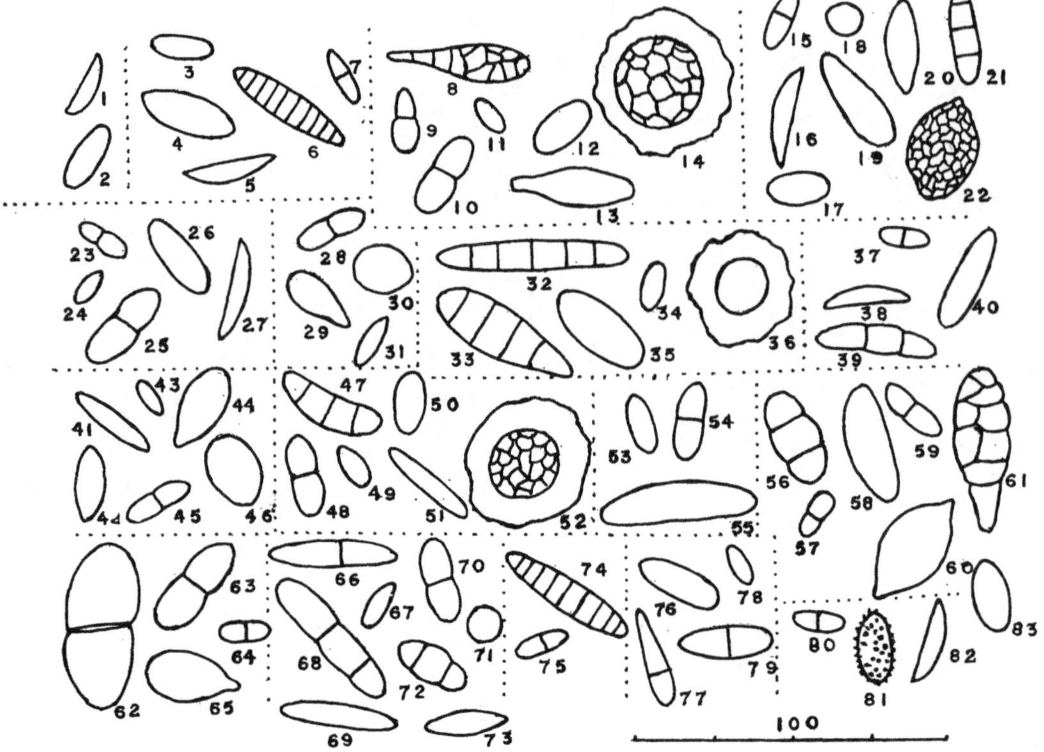

PLATE III. Spores from profile depths of 0.90 to 2.50 meters. 1–2, at 0.90 meters; 3–7, at 1.00 meter; 8–14, at 1.10 meters; 15–22, at 1.20 meters; 23–27, at 1.30 meters; 28–31, at 1.40 meters; 32–36, at 1.50 meters; 37–40, at 1.60 meters; 41–46, at 1.70 meters; 47–52, at 1.80 meters; 53–55, at 1.90 meters; 56–61, at 2.00 meters; 62–65, at 2.10 meters; 66–73, at 2.20 meters; 74–75, at 2.30 meters; 76–79, at 2.40 meters; and 80–83, at 2.50 meters.

1, 5, 16, 27, 38, and 82, *Volutella*; 6 and 74, *Calostilbella*; 7, 75, and 80, *Didymosphaeria*; 8 and 61, *Alternaria*; 9, 23, 37, 45, and 64, *Mycosphaerella*; 11, 24, 34, 67, and 78, *Peziza*; 14, 36, and 52, *Pythium*; 15 and 45, *Nectria*; 21 and 32, *Trichocladium*; 22, *Coniosporium*; 39, 56, 68, and 72, *Pleospora*; 47, *Curvularia*; 81, *Ascobolus*. Others not identified.

gross form indefinitely, whereas others become completely decomposed, even within a brief period. Such an outcome is predicated upon the results of recent studies that deal with the structure and chemical composition of the cell walls of fungi. It has been shown that the cell walls of fungi are laminated as, for example, in the studies of *Fusarium culmorum* by Marchant (1966). He found, by microchemical tests, that the outer wall layer of this species of *Fusarium* is constituted of xylans and that the inner wall is chitinous. Aronson et al. (1967) determined that the cell walls of certain Oomycetes, widely believed to be constituted of cellulose, are largely composed of acid-soluble glucans (glucose, mannose, xylose, and fucose) with only a small proportion of cellulose. The Oomycetes studied included *Phytophthora cinnamomi, Phytophthora parasitica,* and *Pythium debaryanum.* O'Brien and Ralph (1966) fractionated the cell-wall substances of twenty Basidiomycetes and two Pyrenomycetes. They determined that the walls of the Basidiomycetes are composed of glucans. Their analyses of the cell-wall constituents of the two Pyrenomycetes, *Daldinia concentrica* and *Xylaria hypoxylon,* showed that both contain glucose and mannose. However, *D. concentrica* cell walls also contain galactose, whereas this aldose sugar is lacking in *X. hypoxylon.*

The population of spores is small throughout the entire profile of Lake Singletary (Plate IV) whereas pollen is abundant at all depths. Because of paucity of spores, the differences in number found at the different levels (Plate IV) may not be significant. Evidence indicates that no marked differences existed at any time within the sediment in conditions favorable or unfavorable for the growth and reproduction of fungi. For this reason it may be assumed that

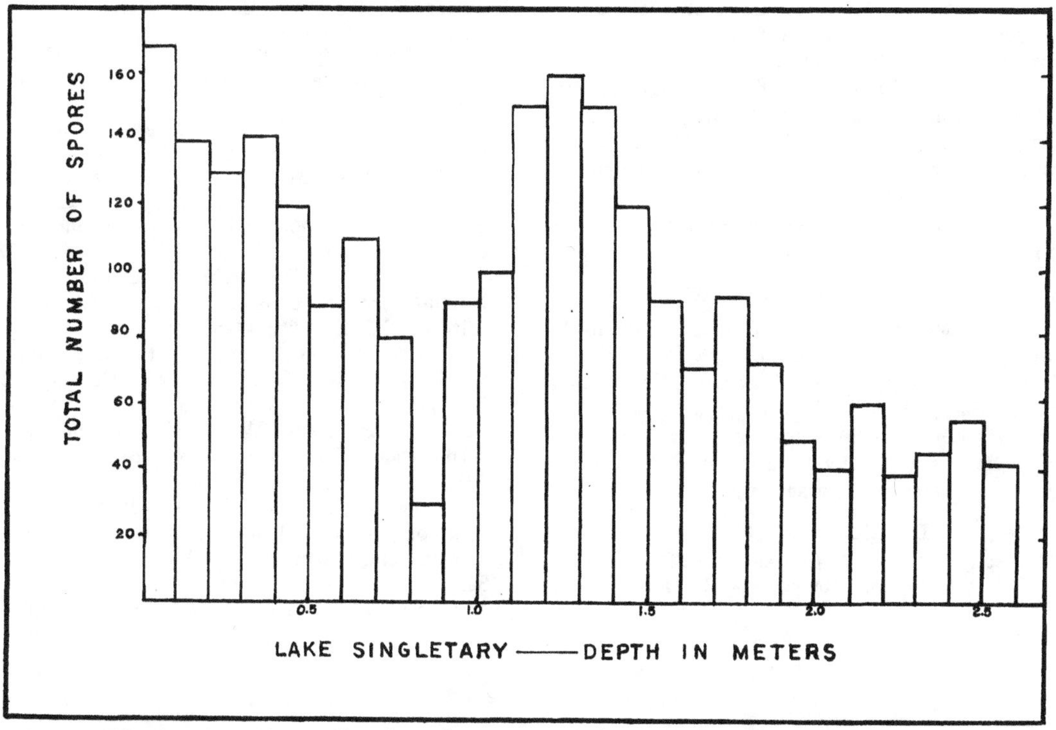

PLATE IV. Stratigraphic profile of total spore population in sediment of Lake Singletary, at depth intervals of 0.10 meters.

paucity of fungus spores and contrasting abundance of pollen in the sediment may not be attributed to marked environmental changes. A more probable explanation for spore paucity would be either that few spores were deposited or that the kinds deposited were nonresistant to decomposition.

In recent years, nonpetrified fungus spores of certain kinds have been found in sediment whose deposition dates from the late Pleistocene. It is understandable that the existence of nonpetrified, apparently intact spores in alluvium of this age has been regarded with disbelief. It should be borne in mind, however, that the environmental conditions in sediment —i.e., lack of oxygen, lack of light, and uniformity of temperature and moisture—probably have remained constant for many years. If, under these conditions, the complete decomposition of buried spores did not take place during a period of a few days or weeks, it becomes credible that they might be preserved, without petrification, for indefinite periods, even for hundreds or thousands of years. In fact, evidence that fungi are preserved in a nonpetrified state in sediments whose deposition long outdates the Pleistocene has been supplied (Dilcher, 1963, 1965). He found in Henry County, Tennessee, well-preserved leaves and leaf fragments of *Sapindus* and *Chrysobalanus* in a layer of gray clay. On these leaves were the fruiting elements of members of the families Meliolaceae, Micropeltaceae, Microthyriaceae, Tuberculariaceae, and Dematiaceae. He identified and described these fungi on the basis of resemblance with contemporary genera. The clay layer in which the leaves and associated fungi had been buried occurred at a depth of 10 to 20 feet beneath the soil surface. The deposition of this clay layer took place during the Eocene Epoch, which occurred, as dated by the potassium argon method, approximately 60 million years ago.

Summary

This report involves the results of examination of sediment of Lake Singletary to determine the kinds and distribution of fungus spores at different depths.

The kinds of spores found are not identical with those known to be associated with the dominant kinds of plants—i.e., the kinds of trees that once grew or are now growing in the environs of this lake.

Spores were sparse in sediment at the water-sediment interface and also throughout the entire profile. The reasons for paucity of spores in this lake remain unknown, but the probable controlling factors are discussed.

It was anticipated that the regional vegetational changes that have taken place, as revealed by results of previously published pollen analyses, would be found to supplement or complement the findings from spore analyses. Evidence to support this anticipated relationship, however, was not found.

ACKNOWLEDGMENTS

Assistance in obtaining cores of Lake Singletary sediment was provided by Dr. D. A. Livingstone and Mark Mantuani, Department of Zoology, Duke University, and I am grateful for their help and counsel. Financial support was provided by Grant GB5517 from the National Science Foundation.

LITERATURE CITED

ARONSON, J. M., B. A. COOPER, AND M. S. FULLER. 1967. Glucans of Oomycete cell walls. Science **155**: 322–325.

DILCHER, D. L. 1963. Eocene epiphyllous fungi. Science **142**: 667–669.

———. 1965. Epiphyllous fungi from Eocene deposits in Western Tennessee. Paleontographica Abt. B. **116**: 2–54.

ERDTMAN, G. 1960. The acetolysis method. Svensk. Tidskr. **54**: 561–564.

FREY, D. G. 1949. Morphometry and hydrography of some natural lakes in the North Carolina Coastal Plain: The bay lake as a morphometric type. Jour. Elisha Mitchell Sci. Soc. **65**: 1–37.

———. 1950. Carolina bays in relation to the North Carolina Coastal Plain. Jour. Elisha Mitchell Sci. Soc. **66**: 44–52.

———. 1951. Pollen succession in the sediments of Singletary Lake, North Carolina. Ecology **32**: 518–533.

MARCHANT, R. 1966. Wall structure and spore germination in *Fusarium culmorum*. Ann. Bot. **30**: 820–830.

O'BRIEN, R. W., AND B. J. RALPH. 1966. The cell wall composition and taxonomy of some Basidiomycetes and Ascomycetes. Ann. Bot. **30**: 831–843.

WHITEHEAD, D. R. 1966. Personal communication.

Isolation of the Perfect State of *Microsporum gypseum, Nannizzia incurvata* Stockdale, 1961, from Soil in North Carolina

BEULAH M. ASHBROOK[1]

Institute of Pathology, University of Tennessee, Memphis, Tenn.

The perfect state of *Sabouraudites* (Achorion) *gypseum* was described first by Nannizzi (1927). He obtained cleistothecia by growing an isolate from a skin lesion on forest soil mixed with feathers and old skin. Nannizzi classified the perfect state of *Microsporum gypseum* in the family Gymnoascaceae and named it *Gymnoascus gypseum.* Most mycologists discredited his findings, as he had not worked with pure cultures.

Griffin (1960), in Australia, isolated the perfect state of *Microsporum gypseum* from the soil. The characteristics of this isolate were compared with Nannizzi's description, and the two organisms were concluded to be the same. The name *Gymnoascus gypseum* was retained.

Szathmáry and Herpay (1960) in Hungary also described perithecial formation of *M. gypseum* isolated directly from soil. However, they classified it as *Achorion gypseum.*

In 1961, Stockdale in England reported her observations on cleistothecial formation in a strain of *M. gypseum.* This strain was isolated from a skin lesion on a worker in a plant nursery. Stockdale decided that this fungus had characteristics that distinguished it from all known genera of the Gymnoascaceae and proposed the name *Nannizzia incurvata,* gen. nov., sp. nov.

This paper will report the isolation of *Nannizzia incurvata* directly from soil near Chapel Hill, North Carolina.

Materials and Methods

Eight soil samples from around and inside an old, abandoned chicken house were collected on two different occasions. The fungus was grown from the soil by Vanbreuseghem's (1952) technique. Soil samples were put into sterile petri dishes, and enough sterile distilled water was added to each dish to moisten the soil. Small bits of sterilized horse hair (mane) were sprinkled over the top of the moistened soil, and the cultures were incubated at room temperature.

[1] This paper was submitted for this special issue by Norman F. Conant, School of Medicine, Duke University, Durham, N. C.

Observations

Cleistothecia were observed after 7–8 days as small white clumps of hyphae in the midst of fluffy white to buff mycelium, growing on the bits of hair on the surface of the soil (Pl. I, Figs. 1, 2).

Microscopic examination of a cleistothecium shows a thick weft of hyphae with numerous thick-walled echinulate, spindle-shaped macroconidia with 4–6 septations. The peridial hyphae usually end in a blunt projection but may extend as smooth-walled, septate hyphae that are either straight, loosely coiled, or tightly coiled (Pl. I, Fig. 3). The cells of the peridial hyphae are moderately thick-walled, densely asperulate, and more or less constricted. Branches of the peridial hyphae curve inward toward their axis, with the outermost ones curving over the cleistothecium.

By crushing or dissecting the cleistothecium, the asci and ascospores are released (Pl. I, Fig. 4). The asci are globose to ovate and measure 5.1–7.0 µ in diameter; they contain eight ascospores. The ascospores are yellow in mass, smooth-walled, and lenticular and measure 1.2–2.5 x 2–3 µ.

Sabouraud's agar cultures, at 14–16 days, revealed the asexual stage of *M. gypseum.* A saline suspension was made of the typical macroconidia and sprinkled over the surface of sterilized soil and hair, moistened with sterile distilled water, and left at room temperature. In 8 days, cleistothecial formation was again noted on hairs in the midst of fluffy white mycelium. Observations made during the following month again revealed the formation of asci and ascospores.

Discussion

Two separate isolations from soil collected from in and around an old chicken house yielded the perfect state of *Microsporum gypseum, Nannizzi incurvata* Stockdale, 1961.

Observations and microscopic examinations at 2–3-day intervals show these isolates to be compatible with Stockdale's description. Stockdale's single ascospore culture revealed her strain of *Nannizzia incurvata,* isolated from a skin lesion, to be heterothallic. She also re-

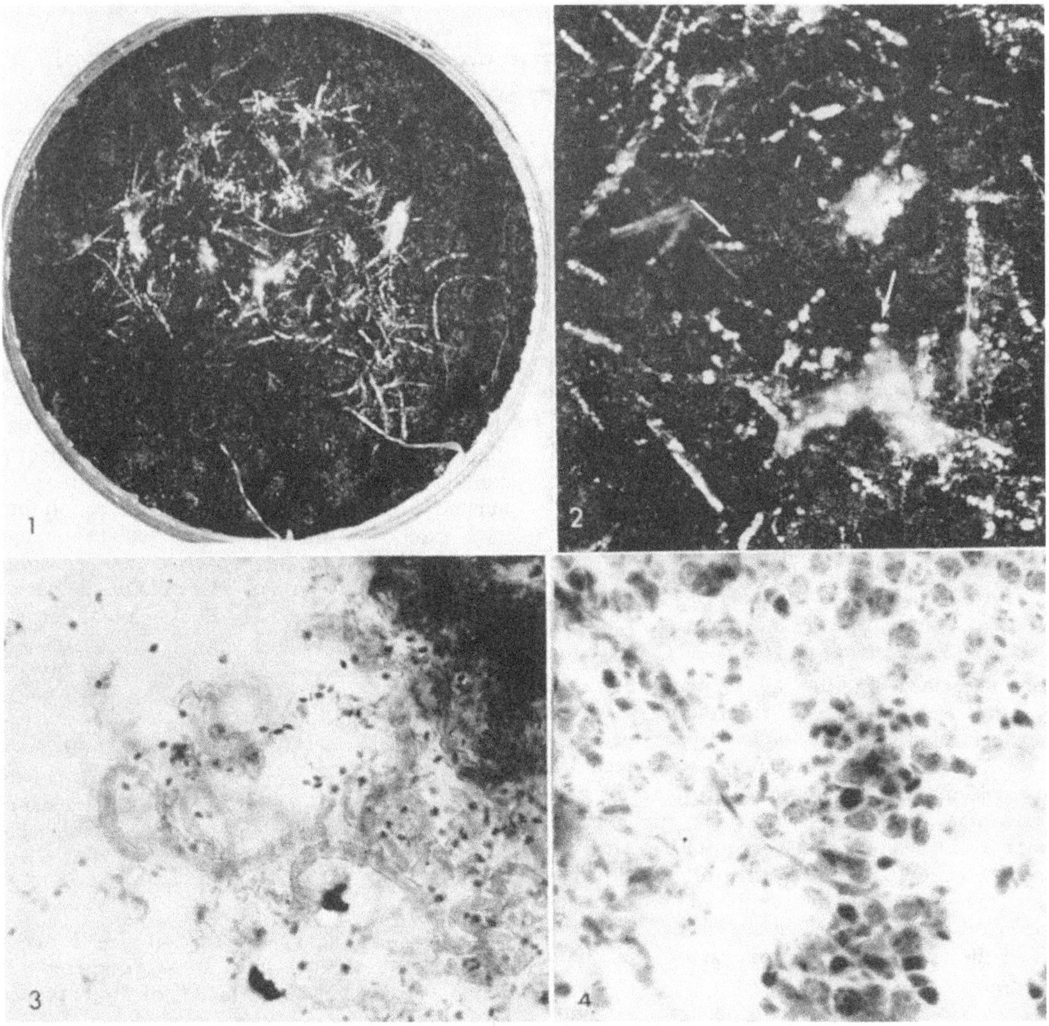

PLATE I. Fig. 1. Cleistothecia in soil culture with horse-hair bait. Fig. 2. Enlargement to show cleistothecia. X4. Fig. 3. Peridial hyphae. X375. Fig. 4. Asci and ascospores. X1000.

ported other studies in which she used various strains of *M. gypseum* to support her findings.

Following this hypothesis, the production of the sexual stage in these isolations would indicate that both + and — strains of *Microsporum gypseum* were growing together in the soil. To substantiate this, single ascospore cultures should be done.

Summary

1. A report is made of the isolation of *Nannizzia incurvata* Stockdale 1961, directly from soil taken from inside and from around an old chicken house near Chapel Hill, N. C.

2. The soil and horsehair-bait technique was used for isolation.

3. The description of this fungus is compatible with Stockdale's isolation and description.

ACKNOWLEDGMENTS

The author wishes to acknowledge gratefully the advice and encouragement given her by Dr. John N. Couch, Department of Botany, University of North Carolina, and Dr. Norman F. Conant, Department of Microbiology and Immunology, Duke University. Thanks also to Miss Marion Seiler, Department of Botany, University of North Carolina, for photographs of soil cultures; and to C. M. Bishop of Department of Pathology, Duke University, for photographs of microscope slides.

REFERENCES

GRIFFIN, D. M. 1960. The re-discovery of *Gymnoascus gypseum*, the perfect stage of *Microsporum gypseum*, and a note on *Trichophyton terrestre*. Trans. Brit. Mycol. Soc. **43**: 637.

———. 1960b. Perfect stage of *Microsporum gypseum*. Nature **186**: 94–95.

NANNIZZI, A. 1927. Ricerche sull'origine saprofitica dei funghi delle tinge. II. *Gymnoascus gypseum* sp. n. forma ascofora del Sabouradites (Achorion) gypseum (Bodin) Ota et Langeron. Atti. Accad. Fisiocr. Siena **10**: 89–97.

STOCKDALE, PHYLLIS M. 1961. *Nannizzia incurvata* gen. nov., sp. nov., a perfect state of *Microsporum gypseum* (Bodin) Guiart et Grigorakis. Sabouraudia **1**: 41–48.

SZATHMÁRY, S., AND HERPAY, Z. 1960. Perithecium-formation of *Microsporum gypseum* and its cognate, *Epidermophyton radiosulcatum* var. flavum Szathmáry 1940 on soil. Mycopathologia, **13**: 1–14.

VANBREUSEGHEM, R. 1952. Technique biologique pour l'isolement des dermatophytes du sol. Ann. Soc. Belge de Med. Trop. **32**: 173–178.

Powdery Mildew of Oak Caused by a Species of *Typhulochaeta*[1]

W. G. Solheim, Dan O. Eboh, and Jerry McHenry
Department of Plant Pathology, University of Arizona, Tucson, Ariz.

Several species of the Erysiphaceae are known to parasitize oaks throughout the United States. Among the more common ones are species in the genera *Erysiphe, Microsphaera, Cystotheca,* and *Phyllactinia. Saccardia quercina* Cooke has been reported from Georgia. While this fungus was originally placed in the Erysipheae, it has been transferred by Theissen and Sydow (1917) to the Saccardiaceae of the Myriangiales.

During work with materials to be filed in the Phytopathological Herbarium of the University of Arizona, a collection of mildew on Arizona oak was found. This collection was dated February 28, 1965, and was from the Santa Rita Mountains, Arizona. The specimen proved to be most interesting. Mr. McHenry had studied it at an earlier date. His notes indicated collections of the fungus were reported from the Santa Catalina Mountains near Tucson in December, 1947, and from Rucker Canyon, Cochise County, in 1957. Neither of these two collections have been found to date. The authors recollected mildewed oak in abundance in the Santa Rita Mountains on March 15 and April 5, 1967.

The fungus causing this powdery mildew is a species of the genus *Typhulochaeta.* This genus was described by Ito and Hara in a paper by Ito (1915) and is based on *T. japonica* Ito et Hara on *Quercus glandulifera.* Hirata (1966) gives the known range of the fungus as Japan and China.

The Arizona material has many points of similarity with *T. japonica* but differs from it in several respects. *T. japonica* is described as having 90–160 appendages; the Arizona fungus has 5–18. The appendages are described by Ito (1915) as follows: "They are simple and clavate in shape, with a close resemblance to the fructification of a species belonging to the genus *Typhula* of *Basidiomycetes* in their outlines, and also to the immature stage of the penicillate cells of *Phyllactinia.* They are colorless and measure 45–65 µ in length and 10–15 µ in width. Their walls are thick with a thin central granular protoplasmic thread. The wall, especially in the apical portion undergoes a mucilaginous change. It swells up in water to a slight degree but considerably in potassium hydroxide and mineral acids, rupturing the wall at apex and extruding in a form of round mucilaginous mass." In the Arizona material the appendages at maturity consist of a stalk extending to the tip with a wall about 1 µ thick and capped with an amber-colored, elliptical mass of waxy material (Figs. 3, 4, 5). The waxy material may be formed as suggested by Ito. The surface cells of the cleistothecia as shown in the drawings in Ito's paper appear to be more or less isodiametric and polygonal in shape. In the Arizona specimens these cells have a very irregular contour, as shown in Fig. 8. Because of these differences, varietal status is assigned the Arizona fungus.

Typhulochaeta japonica Ito et Hara var. *couchii* var. nov.

Hypophylla; mycelio pannos albidos ad marginem foliorum formante, vel late effuso; cleistotheciis sparsis vel numerosis, gregariis, globosis, 75–190 µ diam., brunneis, pellucidis, ascos et ascosporas divulgantibus, cellulis irregularibus, 7.5–22 x 4.5–15.5 µ, cum hyphis numerosis et irregularibus ad basem egredentibus; appendicibus 5–18 supra medium cleistothecii egredentibus, nonseptatis, hyalinis, sursum leniter attenuatis et crasse pileatis, 37–70 x 3.5–8 µ, pileo electrino, 11–20 µ diam.; ascis 7–16, obovoideis, saccatis vel elongatis, breviter pedicellatis, 65–95 x 32–44 µ, muris 2–2.5 µ crassis; sporis 8, ellipsoideis, hyalinis, 24–35 x 11–18 µ.

Hab. in foliis *Querci arizonicae* Sarg. (Fagaceae), Madera Canyon, Santa Rita Mountains, Santa Cruz County, Arizona, Amer. bor. W. G. Solheim, Dan O. Eboh, and Jerry McHenry, 6581.

Hypophyllous, forming white patches along margins of leaves and eventually covering much or all of the leaf surface, becoming peppered with the light brown cleistothecia; cleistothecia sparce to abundant, gregarious, globose, 75–190 µ diam., wall 8–15 µ diam., light brown, transparent, revealing asci and ascospores, cells very irregular in outline, 7.5–22 x 4.5–15.5 µ, with numerous irregular hyphae at base which attach the cleistothecia to the surface mycelium; appendages 5–18, arising from the upper part

[1] University of Arizona, Agricultural Experiment Station, Department of Plant Pathology, Journal Article #1233.

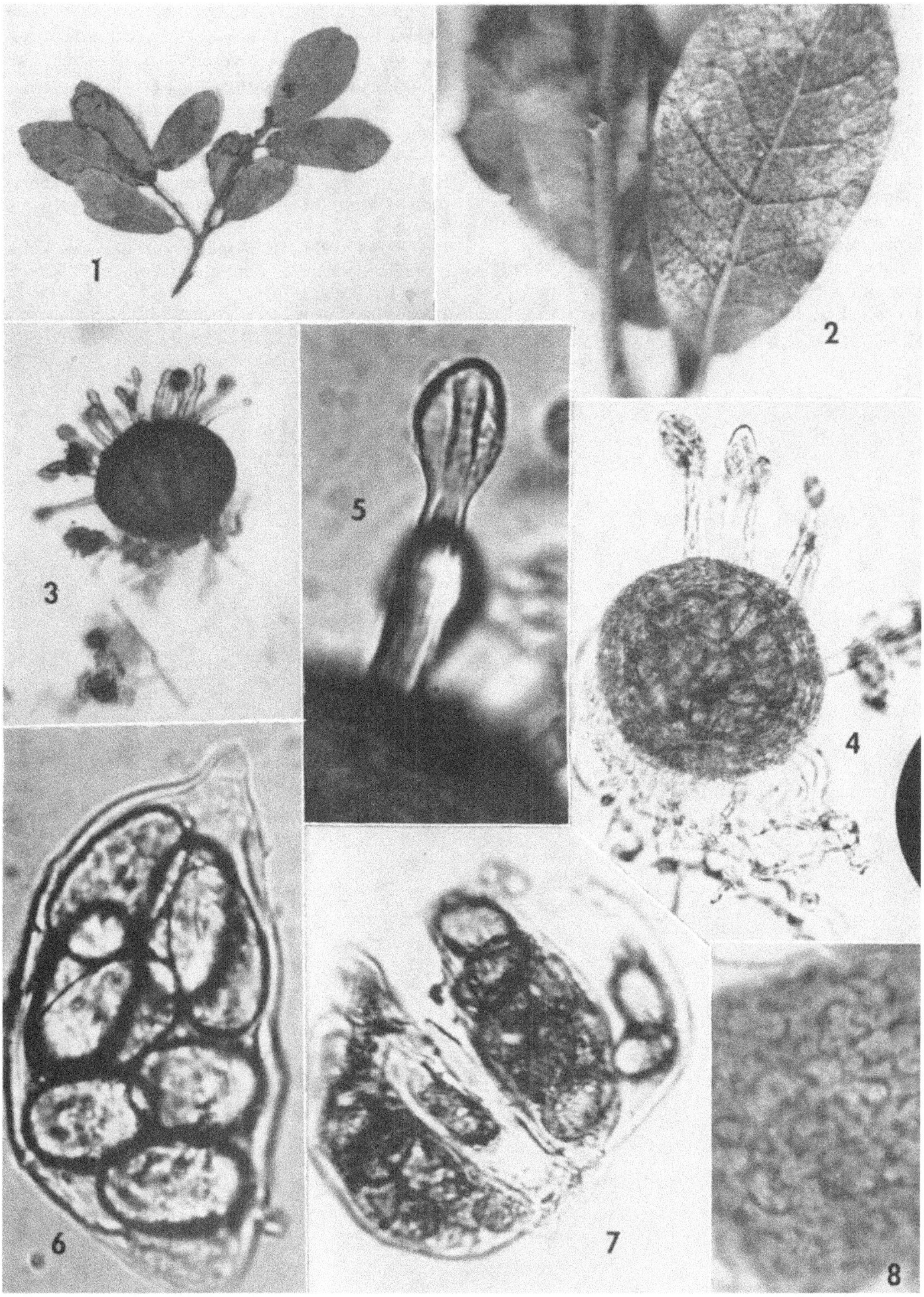

FIGS. 1–8. Figs. 1, 2. Infected leaves. Figs. 3, 4. Cleistothecia showing appendages, translucent character of wall and basal hyphae. Fig. 5. Tip of appendage with enlarged amber colored cap. Fig. 6. Ascus with 8 ascospores. Fig. 7. Fascicled asci. Fig. 8. Outer cells of cleistothecium.

of the cleistothecium, hyaline, nonseptate, tapering slightly upward and capped with a bulbous, amber-colored head, 37–70 x 3.5–8 μ, bulbous head 11–20 μ diam.; asci 7–16, obovate, saccate to elongate, fascicled, short pedicellate, 65–95 x 32–44 μ, wall 2–2.5 μ diam.; ascospores 8, elliptical, hyaline, 24–35 x 11–18 μ.

On leaves of *Quercus arizonica* Sarg. Madera Canyon, Santa Rita Mountains, Santa Cruz County, Arizona, April 5, 1967, W. G. Solheim, Dan O. Eboh, and Jerry McHenry, 6581.

Type material will be distributed through Mycoflora Saximontanensis Exsiccata and also deposited at the University of North Carolina at Chapel Hill and the University of Arizona.

The authors gratefully acknowledge their indebtedness to Dr. Donald M. Ayers, Professor of Classics, University of Arizona, for checking and correcting the Latin diagnosis.

LITERATURE CITED

HIRATA, KOJI. 1966. Host range and geographical distribution of the powdery mildews. Niigata, Japan. 472 pp.

ITO, SEIYA. 1915. On *Typhulochaeta*, a new genus of Erysiphaceae. Bot. Mag. Tokyo, **29**: 15–22. Pl. II, Figs. 1–16.

THEISSEN, F., AND H. SYDOW. 1917. Synoptische Tafeln. Ann. Myc. **15**: 389–491. Pp. 433–447.

The First Ascomycete from the Deep Sea

JAN KOHLMEYER
University of North Carolina Institute of Marine Sciences, Morehead City, N. C.

While many species of higher marine fungi have been collected close to the seashore, knowledge on occurrence of these organisms in deep oceans is rare. No marine Ascomycetes have yet been found to live in considerable depths of the sea. With baiting methods, Höhnk (1959) isolated in the laboratory Fungi Imperfecti from depths of 4,610 meters in the North Atlantic. Other papers dealing with fungal species recovered from benthonic sediments and waters have been compiled by Johnson and Sparrow (1961). These species were also isolated by baiting or incubating sediments or water samples, methods which cannot determine whether the fungi were active in that particular habitat or merely present in form of spores or hyphal fragments. The following note gives a description of the first marine Ascomycete ever found growing in deep ocean waters.

The Environment

In recent years much interest has been shown in biodeterioration of materials in the deep ocean (Muraoka 1964, 1965, 1966a, b; Turner, 1965). Among samples of engineering materials, wood panels were submerged in depths of 2,340–6,800 feet (713–2,073 m.) for periods ranging from 4 months to 3 years. Test panels of these experiments have kindly been put at my disposal by the U. S. Naval Civil Engineering Laboratory (Port Hueneme, California) for examination for marine fungi.

Table I

Factors	Surface Water	Test Site I
Depth, m.	—	1,615
Temperature, °C.	13.0	2.53
Dissolved oxygen conc., ml./l.	5.6	1.26
Salinity, °/°°	33.6	34.56
Hydrostatic pressure, at	—	176
Current, knots	—	0.033
Sediment	—	green mud, containing Foraminifera glauconite, quartz, etc.

A Submersible Test Unit (STU I–1) was placed on the ocean floor in March, 1962, at Test Site I (81 nautical miles southwest of Port Hueneme, California) and recovered on 25 February, 1965. The conditions at this test site are summarized in Table I (after Muraoka, 1966a; current corrected according to latest data, J. S. Muraoka, personal communication).

The wood panel containing the fungus described below was attached to the test unit about 6 feet above the sediment.

Description of the Fungus

The dried wood sample examined for fungi was a ¾ x 3 x 24-inch pine panel, riddled by marine borers (*Xylophaga washingtona* Bartsch and *X. duplicata* Knudsen). The surface of the wood was softened up to 400 μ deep by fungal and bacterial action. Thin cross-sections through the wood revealed typical tunnels in the secondary wall layers, a type of degradation known as "soft rot" caused by Ascomycetes and Fungi Imperfecti (Kohlmeyer, 1958). Several fruiting bodies of an Ascomycete were found embedded in the surface layers. Most of the ascocarps were empty; the content of the remainder was collapsed by drying.

Fruiting bodies: 150–230 μ high (including papilla), 215–300 μ wide, subglobose or ellipsoidal, partly or completely immersed, solitary, blackish brown above, light brown to hyaline below, coriaceous, papillate (Figs. 1, 2). **Walls** 40 μ thick in the papillae, laterally 15–20 μ thick (Fig. 3). **Papillae** short, obtusely conical, ostiolate; no periphyses present. **Paraphyses:** not seen; centrum of immature ascocarps probably filled with early deliquescing, hyaline pseudoparenchyma. **Asci:** not seen, early deliquescing; ascospores frequently in bundles. **Ascospores:** 50–71 x 4–5.5 μ, filiform or elongate fusoid, straight or slightly curved, apically somewhat attenuate and thick-walled (Fig. 5), one-celled, hyaline (Fig. 4); no end chambers or appendages present.

Taxonomical Considerations

At this time, it is not possible to assign the deep sea fungus to any described species. Even assignment to genus is difficult. Among marine

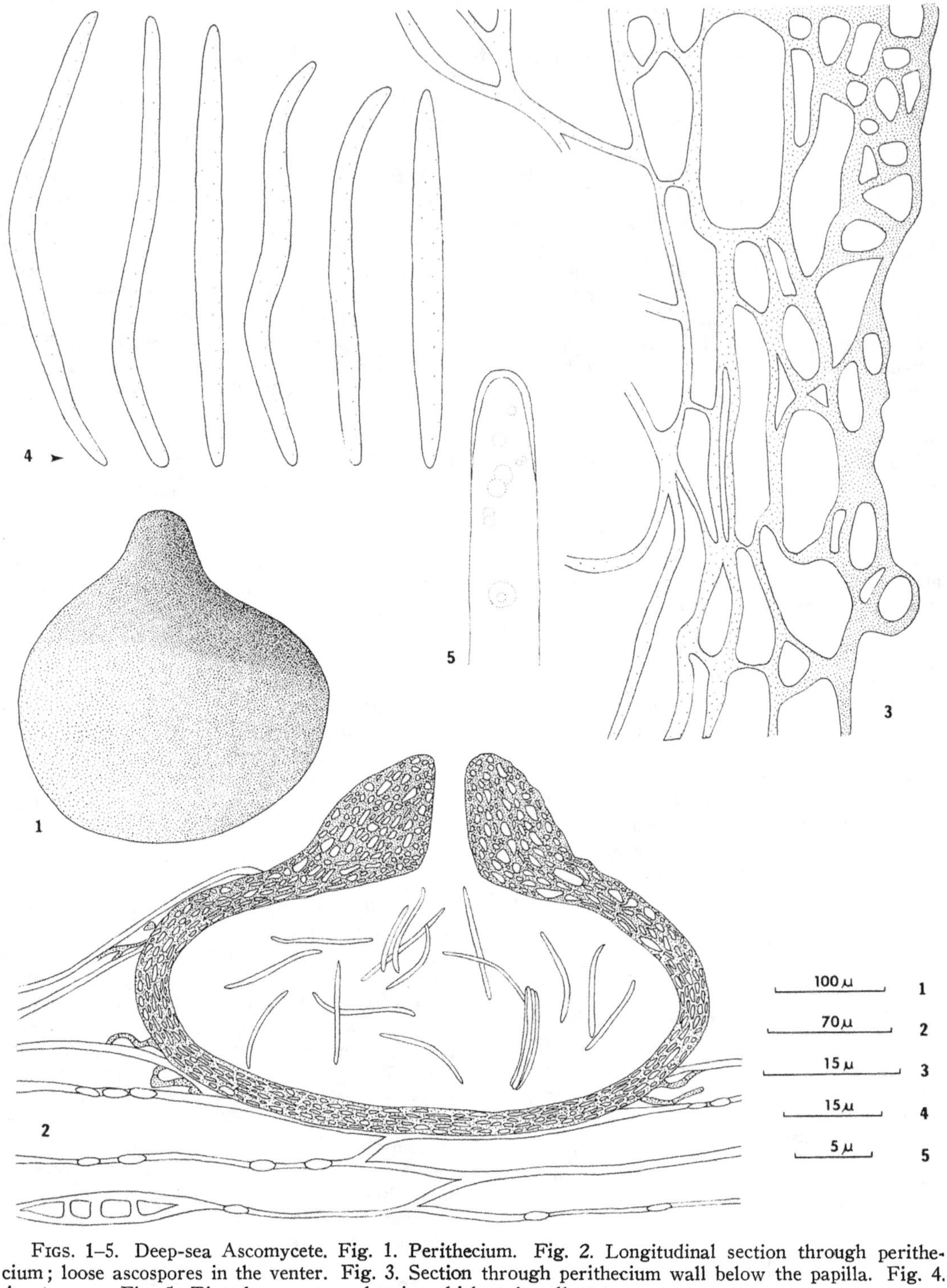

FIGS. 1–5. Deep-sea Ascomycete. Fig. 1. Perithecium. Fig. 2. Longitudinal section through perithecium; loose ascospores in the venter. Fig. 3. Section through perithecium wall below the papilla. Fig. 4. Ascospores. Fig. 5. Tip of ascospore, showing thickened wall at apex.

Ascomycetes are the following genera with scolecosporous representatives (J. and E. Kohlmeyer, 1964): *Lindra* Wilson, *Lulworthia* Sutherland, *Maireomyces* Feldmann, *Ophiobolus* Riess, *Trailia* Sutherland. In *Lindra, Ophiobolus,* and *Trailia* the ascospores are septate, and in *Lulworthia* the ascospores have typical end chambers. Thus, *Maireomyces* is the closest genus. However, this monotypic genus is incompletely known, and no type material of *M. peyssoneliae* Feldm. is available (v. Arx and Müller, 1954). The Ascomycete from the deep ocean has ascospores similar to those of *M. peyssoneliae* and *Ophiobolus australiensis* Johnson et Sparrow, which are both considerably longer. It probably belongs to the Halosphaeriaceae.

Although the species described above is new to science and may even represent a new genus, it is not named here because important features of the fungus, like the structure of centrum and asci, are not recognizable in the material at hand. However, since fungal samples of the deep sea will probably continue to be scarce in the years to come, it seemed appropriate at least to report on the existence of Ascomycetes in deep ocean habitats.

General Conclusions

Besides the wood panel described before, I examined samples of three other test units, exposed off the Californian coast for 1 year, 13 months, and 3 years, respectively. While hyphae were present, none of the specimen bore any fungal fruiting structures. Apparently, hyphal growth and development of ascocarps are very slow in the deep sea. As cellulosic material is rarer in the deep ocean than along the shore, it may take a long time until freshly deposited wood becomes infested by spores deriving from old material on the bottom of the sea.

In all probability only species of fungi adapted to high hydrostatic pressures, low temperatures, and low oxygen concentrations can grow in the deep sea. The constant darkness of this environment might also be one of the factors that restrict development of fungal species.

Summary

Fruiting bodies of an Ascomycete developed on a pine panel submerged off the California coast at 1,615 meters for 3 years. The temperature at the test site was 2.53° C., the dissolved oxygen concentration 1.26 ml./l., and the hydrostatic pressure 176 at. The fungus is illustrated and its taxonomic position and some ecological questions are discussed.

ACKNOWLEDGMENTS

My sincere thanks are due to Mr. J. S. Muraoka (Port Hueneme, California) for sending wood samples of the U. S. Naval Civil Engineering Laboratory and to Dr. R. J. Menzies (Beaufort) for the loan of one panel. The illustrations were kindly prepared by Mrs. Erika Kohlmeyer.

LITERATURE CITED

VON ARX, J. A., AND MÜLLER, E. 1954. Die Gattungen der amerosporen Pyrenomyceten. Beitr. Kryptogamenflora Schweiz **11**: 1–434.

HÖHNK, W. 1959. Ein Beitrag zur ozeanischen Mykologie. D. Hydrographische Z. Reihe B, Nr. **3**: 81–87.

JOHNSON, T. W., JR., AND SPARROW, F. K., JR. 1961. Fungi in oceans and estuaries. J. Cramer, Weinheim, West Germany. Pp. i–xxiv, 1–668.

KOHLMEYER, J. 1958. Holzzerstörende Pilze im Meerwasser. Holz als Roh—und Werkstoff **16**: 215–220.

KOHLMEYER, JAN, AND KOHLMEYER, ERIKA. 1964. Synoptic Plates of Higher Marine Fungi. J. Cramer, Weinheim, West Germany. Pp. 1–64, pl. 1–12.

MURAOKA, J. S. 1964. Deep-ocean biodeterioration of materials. Part I. Four months at 5,640 feet. U. S. Naval Civil Engineering Lab., Port Hueneme, Calif., Tech. Rept. R**329**: 1–35.

———. 1965. Deep-ocean biodeterioration of materials. Part II. Six months at 2,340 feet. U. S. Naval Civil Engineering Lab., Port Hueneme, Calif., Tech. Rept. R**393**: 1–42.

———. 1966a. Deep-ocean biodeterioration of materials. Part III. Three years at 5,300 feet. U. S. Naval Civil Engineering Lab., Port Hueneme, Calif., Tech. Rept. TR**428**: 1–47.

———. 1966b. Deep-ocean biodeterioration of materials. Part IV. One year at 6,800 feet. U. S. Naval Civil Engineering Lab., Port Hueneme, Calif., Tech. Rept. R**456**: 1–45.

TURNER, RUTH D. 1965. Some results of deep water testing. Ann. Rep. American Malacological Union: 9–11.

The Case of *Lambertella brunneola*: an Object Lesson in Taxonomy of the Higher Fungi

Richard P. Korf and K. P. Dumont
Department of Plant Pathology, Cornell University, Ithaca, N. Y.

Introduction

Microscopic features of both somatic and reproductive structures have long been regularly stressed in descriptions of micromycetes, since gross characters are for the most part lacking. With the macromycetes, however, a different tradition in fungal description has evolved. The *gross morphology* of the sporocarp of Ascomycetes and Basidiomycetes played the major role in their classification and in the delimitation of taxa at ordinal through infraspecific ranks. This is not to imply that the microscope was not used in the study of the higher fungi, for in fact some microscopic features were regularly described, and taxa were not infrequently distinguished on these alone.

What have been, traditionally, the microscopic elements which mycologists have used in their descriptions of macromycetes? Until relatively recently they have been almost exclusively those of the *hymenium* (or other sporogenous zones). With the Basidiomycetes, sizes, shapes, and pigmentation of the basidiospores, basidia, basidioles, and hymenial cystidia were usually detailed in descriptions. Almost wholly lacking were descriptions of the microscopic features of the trama, context, and covering sterile layers. In Ascomycetes the same tradition prevailed: ascospores, asci, and paraphyses were described, while the remainder of the ascocarp was normally discussed in gross morphological terms only.

The assumption that the reproductive and accessory structures of the hymenium are logical candidates for the most valid characters upon which to base a classification is presumably inherited from phanerogamic botanists, whose thinking was dominated by floral structures. Even today many mycologists assign far greater taxonomic weight to characters of ascospores and basidiospores than to other microscopic or even macroscopic elements.

Basidiomycete workers no longer look merely at hymenial elements under the microscope. Today, even the beginning mycology student is faced with keys to genera and higher groups (Smith and Shaffer, 1964) which include such terms as bilateral and inverse trama, epicutis, dichophyses, trichodermium, and acanthophyses for structural elements or tissues. All of these terms are lacking in the "best" mycological glossary available 30 years ago (Snell, 1936). Other terms describing microchemical reactions, such as the amyloid or dextrinoid reaction of spores or of trama, or the presence of carminophilous granules, are "recent" innovations in fungal taxonomy.

In Ascomycetes, the examination of the sterile tissues of the ascocarp is also being emphasized. For example, no modern taxonomist can give a diagnosis of a species of *Leptosphaeria* without providing details of the pseudothecial structure (Müller, 1950). In Discomycetes, the first worker to emphasize microanatomy of the apothecium as a basis for constructing a classification was probably Rehm (1887–96), followed by Durand (1900) and Lagarde (1906). A precise codification of tissue types found in apothecia was proposed by Starbäck (1895), but was nearly forgotten until Nannfeldt (1932) revived the use of the terms *textura globulosa, textura oblita,* etc. The use of tissue types and other microanatomical characters in distinguishing taxa at various ranks may be noted in many recent Discomycete monographs (van Brummelen, 1967; Denison, 1965; Dennis, 1949, 1956, 1960; Dissing, 1966; Hennebert and Groves, 1963; Kimbrough and Korf, 1967; Le Gal, 1953; Sánchez and Korf, 1966; White, 1941).

The Case of **Lambertella brunneola**

Lambertella brunneola (Pat.) Le Gal is presented here merely as one example of the kind of problem that besets even the most careful worker. It is chosen because a complete analysis is available through monographic studies on *Lambertella* by the junior author, and in part because it allows the senior author to rectify in print one of his errors. It also offers present-day pertinence, since all of the observations cited here have been made in the past 40 years, the period of "modern" fungal taxonomy.

The five collections that have been referred to *L. brunneola* in the literature, one of them the type of *L. viburni* Whetzel, have all been examined microscopically in the course of our studies. Differences among the collections in

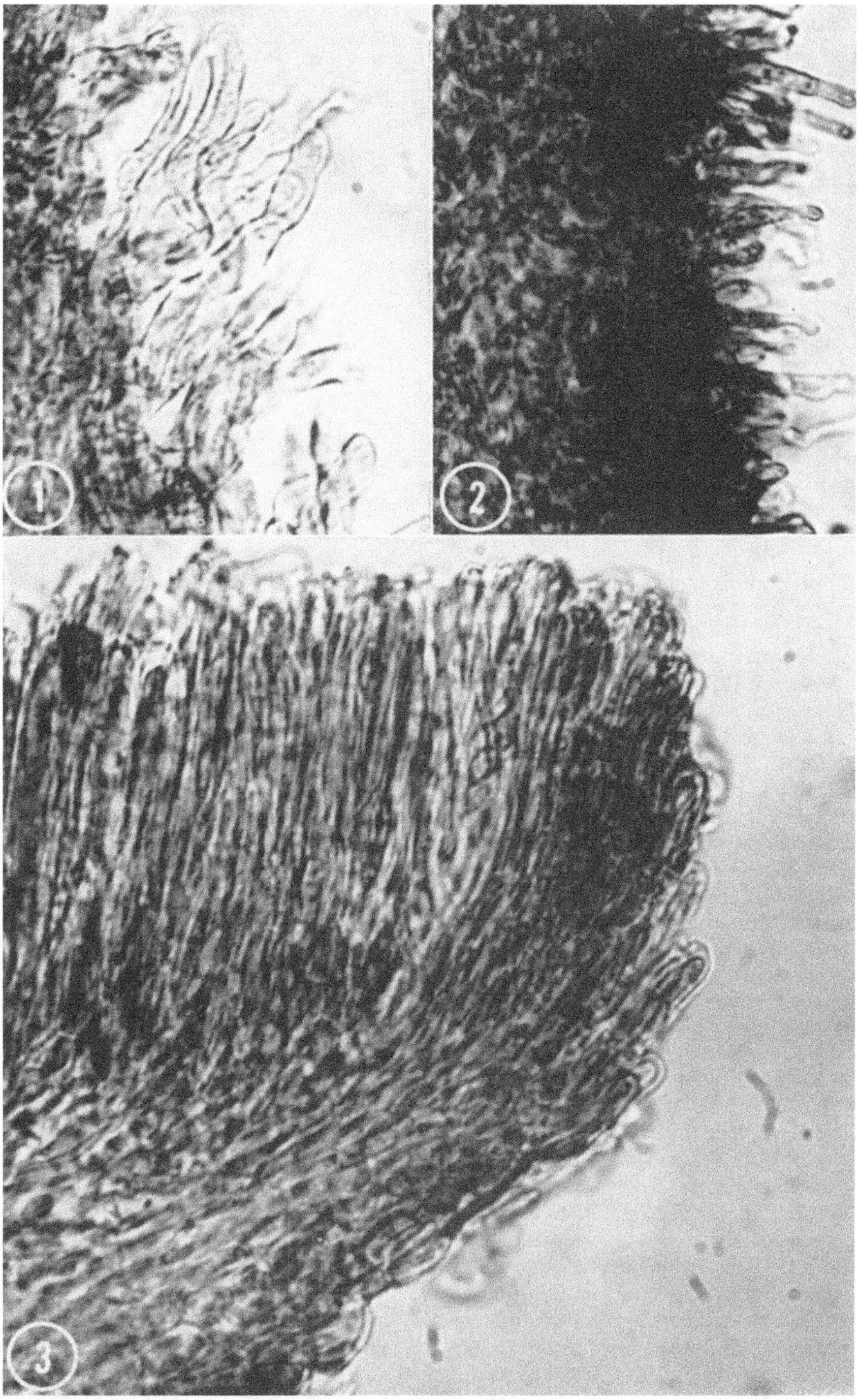

FIGS. 1–3. Photomicrographs of portions of vertical sections of apothecia from collections referred to *Lambertella brunneola*. X880. Fig. 1. Broad, granularly roughened tomentum hyphae arising from the ectal excipulum of an Indian specimen (CUP 49369). Fig. 2. Tapering, smooth tomentum hyphae arising from the ectal excipulum of a Japanese specimen (CUP 49370). Fig. 3. Short, blunt, marginal cells and absence of tomentum hyphae below in the type specimen from Madagascar (CUP 49371). Photographs by the authors.

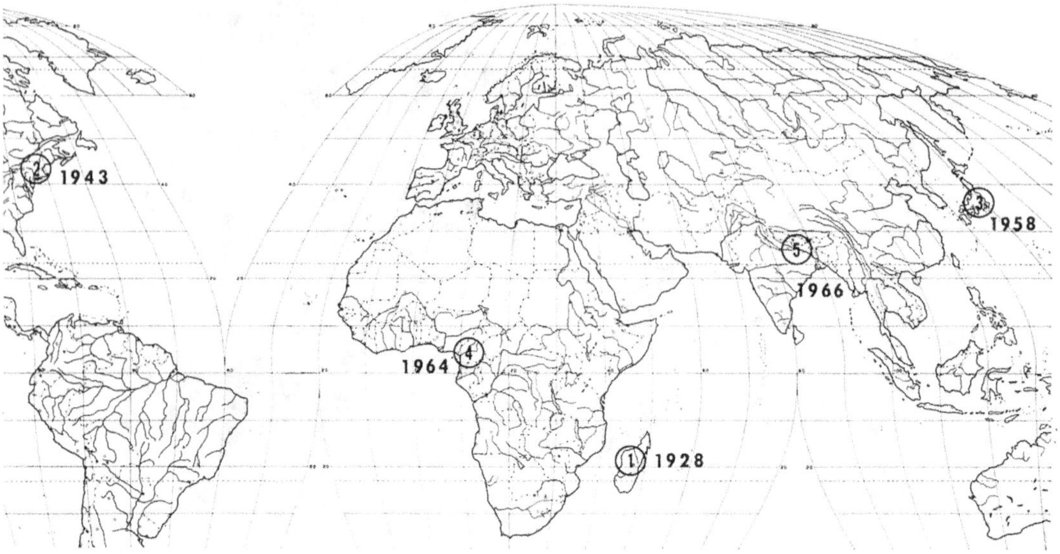

FIG. 4. The presumed distribution of *Lambertella brunneola* based on the five collections recorded in the literature prior to 1967.

macroscopic features are not striking. Microscopically, however, differences are pronounced, particularly in sterile structures and tissues of the ascocarp. As one example, "tomentum hyphae" clothe the exterior of the apothecia in certain collections, but differ markedly among the collections (Figs. 1 and 2). They may be lacking or merely represented by short, hyphal ends near the margin in other collections (Fig. 3). Our studies do not confirm the presumed *world-wide* distribution of *L. brunneola* (Fig. 4), since it is our opinion that each of the five collections represents a distinct taxon (some of which, at the conclusion of our studies, may not even fall within the revised generic limits of *Lambertella*).

L. brunneola now appears to be known from a single collection from Madagascar. For the other four taxa we do not yet have valid names. *L. viburni* needs only a Latin diagnosis to make it conform to the Code of Nomenclature (Professor Whetzel, an outspoken critic of the Code's requirement for a Latin diagnosis, purposefully did not provide one). The other three may unfortunately require new names, though older names for them may be hidden in such unwieldy genera as *Helotium* or *Peziza.*

A History of Five Collections and Four Descriptions

The initial collection from Madagascar was described by Patouillard (1928) as *Helotium brunneolum* Pat. with a brief, five-line diagnosis, divided almost equally between gross morphological features and a microscopic analysis of asci, ascospores, and paraphyses (Fig. 5 A). On re-examination of this material, Le Gal (1953) found that Patouillard had apparently studied immature apothecia and had overlooked the fact that the ascospores turn brown at maturity, a major generic feature of *Lambertella.* She transferred the species to *Lambertella* and provided excellent, lengthy, detailed descriptions and illustrations of the hymenial structures (Fig. 5 B). Unfortunately, neither she nor Patouillard discussed microanatomy of the apothecium.

The second collection was made in New Hampshire (U.S.A.) and described by Whetzel (1943) as *L. viburni.* He presented detailed descriptions and illustrations of the hymenial elements. He stressed the characteristic brown ascospores, the peculiar deliquescence of the asci (which occurs in only a few other species of the genus), the presence of black-line stromata in the affected leaves, and the presence of spermatia in spermagonia. Whetzel did provide some microscopic information on nonhymenial structures, but regrettably failed to discuss the diagnostic microanatomy of the apothecium (Fig. 5 C). Though *she did not examine* any of Whetzel's ample material, Le Gal (1953) synonymized Whetzel's species with Patouillard's *H. brunneolum* on the basis of Whetzel's diagnosis.

The third known collection was found in Japan and reported by Korf (1958), who pro-

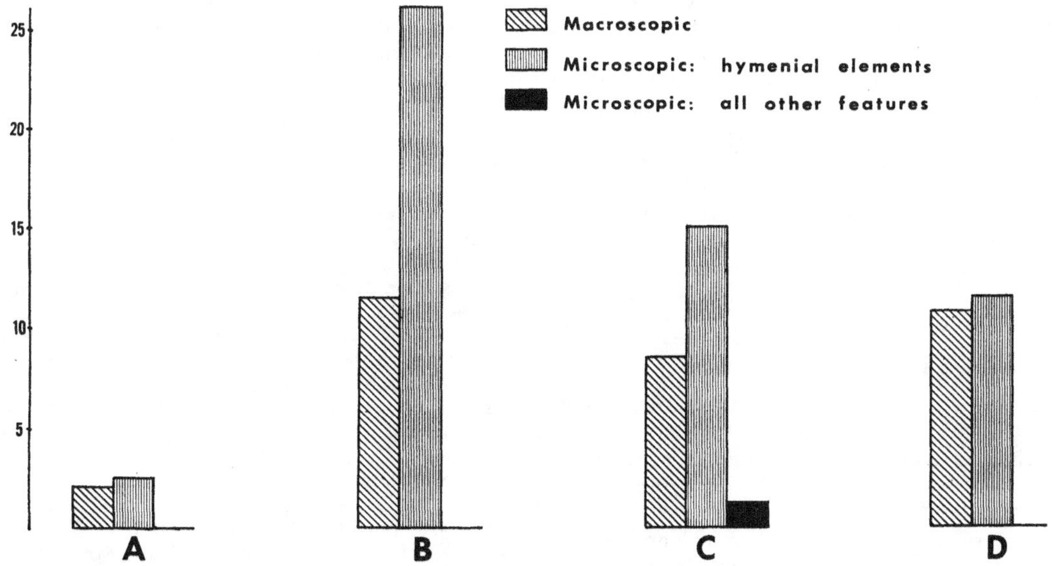

Fig. 5. Number of lines devoted to description of gross (macroscopic) features, of the hymenial elements, and of all other microanatomical features (see key) in the four known descriptions of "*Lambertella brunneola.*" Data are taken from: A (Patouillard, 1928), B (Le Gal, 1953), C (Whetzel, 1943), D (Tewari and Pant, 1966). Note that only in the Whetzel description is there any microanatomy presented other than that of the hymenial elements (i.e., asci, ascospores, and paraphyses).

vided no description. He commented on *L. viburni* and accepted Le Gal's decision on *L. brunneola,* noting that "From her excellent description, there seems to be no doubt of the correctness of this synonymy." It should be noted that *he did not examine* either the Patouillard or the Whetzel collection and based his identification solely on the literature.

The fourth report was based on material from the Cameroons (Berthet, 1964). Hitherto unknown and informative cytological details were presented, as well as macroscopic and microscopic features of cultures. No description of the apothecia was given. Though apparently *he did not examine* any of the earlier collections, he also placed *L. viburni* in synonymy with *L. brunneola.*

The Tewari and Pant (1966) report of *L. brunneola* from India represents the most recent record. They presented a formal description of their fungus, which gave details of the hymenial elements but failed to mention the microstructure of the apothecium (Fig. 5 D). Though *they did not examine* any of the earlier specimens, they also placed *L. viburni* in synonymy with *L. brunneola.*

Lesson One

When a taxon is placed in synonymy by a recognized authority, all too frequently later authors tend to accept such synonymy without dissent. Mme. Le Gal is without doubt one of the foremost authorities on Discomycetes living today. Her conclusion that *L. viburni* and *L. brunneola* are identical carried so much weight that none of the succeeding authors questioned the synonymy. Without examining any of the material on which the synonymy was based, and each in fact only examining a single collection in hand, the later authors all earnestly (or perhaps pedantically) listed *L. viburni* as a synonym. Such is, however, an inherent problem in dealing with identifications, for which there is no easy solution. We are certainly not proposing casual use of type or authentic specimens, since these represent voucher material upon which names are based, and should be reserved for the use of the monographer. Even the most ample type specimen can only have a limited number of mounts made from it before it is exhausted, and unfortunately most types are fragmentary.

Lesson Two

Existing literature descriptions may easily lead to misidentifications, since the only microscopic features usually provided are those of the hymenial elements. Le Gal, Korf, Berthet, and Tewari and Pant displayed common faults in our example, in part in not examining additional material and in part in relying on the descriptions at their disposal, all of which failed to provide adequate data on apothecial structure upon which to base a correct identification. A stronger charge can be laid upon them: *not once* in the six literature reports did any author provide microanatomical details of the medullary and excipular structures of the apothecium. Far too much emphasis was placed by all of these workers on comparison of written descriptions of only three hymenial structures (asci, ascospores, and paraphyses) with the material they had in hand.

Conclusion

If we continue to ignore the exceptionally complicated and often diagnostic sterile tissues of the apothecium, and continue to provide extensive descriptions of only the hymenial elements, literature descriptions in the Discomycetes are doomed to yield misidentifications. Hymenial analysis at the expense of microanatomical description of the sporocarp will yield similar misidentifications in all groups of higher fungi. Mycologists must change their habits. Workers must now provide for *known* macrofungi new and ample descriptions in which microanatomy of the sterile elements is stressed. Eventually, by reference to such expanded literature descriptions alone, we should be able to identify with much greater precision. It is imperative, of course, for workers who provide the new microscopic details to deposit their specimens in a permanent herbarium for future reference by monographers.

Summary

Lambertella brunneola has been thought to have a world-wide distribution on the basis of five widely separated collections, which on microscopic analysis prove to represent instead five distinct taxa. The example is selected merely as an object lesson in the dangers of identification based on the literature alone. The failure of existing descriptions to give microanatomical details of the ascocarp and excessive reliance on characters of the hymenial elements are held to be the cause of the accumulation of such misinformation.

ACKNOWLEDGMENTS

This work has been supported by National Science Foundation grant GB–2339, "Monographic Studies of the Discomycetes." We wish to express our sincere thanks to Dr. Roger Heim, Muséum National d'Histoire Naturelle, Paris, for the loan of the type specimen of *Helotium brunneolum* Pat., and to Dr. P. Berthet, Université de Lyon, and Mr. V. J. Tewari, Banares Hindu University, for loaning the specimens upon which their reports of *Lambertella brunneola* were based.

LITERATURE CITED

BERTHET, P. 1964. Essai biotaxinomique sur les discomycètes. Thèse. Faculté des Sciences de l'Université de Lyon. 157 pp. Published by the author.

BRUMMELEN, J. VAN. 1967. A world-monograph of the genera *Ascobolus* and *Saccobolus* (Ascomycetes, Pezizales). Persoonia Suppl. **1**: 1–260.

DENISON, W. C. 1965. Central American Pezizales. I. A new genus of the Sarcoscyphaceae. Mycologia **57**: 649–656.

DENNIS, R. W. G. 1949. A revision of the British Hyaloscyphaceae with notes on related European species. Commonwealth Mycol. Inst. Mycol. Papers **32**: 1–97.

———. 1956. A revision of the British Helotiaceae in the herbarium of the Royal Botanic Gardens, Kew, with notes on related European species. Commonwealth Mycol. Inst. Mycol. Papers **62**: 1–216.

———. 1960. British Cup Fungi and their Allies. An Introduction to the Ascomycetes. Ray Society, London. 280 pp.

DISSING, H. 1966. The genus *Helvella* in Europe. Dansk Bot. Arkiv **25** (1): 1–172.

DURAND, E. J. 1900. The classification of the fleshy Pezizineae with reference to the structural characters illustrating the bases of their division into families. Bull. Torrey Bot. Club **27**: 463–495.

HENNEBERT, G. L., AND J. W. GROVES. 1963. Three new species of *Botryotinia* on Ranunculaceae. Canad. J. Botany **41**: 341–370.

KIMBROUGH, J. W., AND R. P. KORF. 1967. A synopsis of the genera and species of the tribe Theleboleae (= Pseudoascoboleae). Am. J. Bot. **54**: 9–23.

KORF, R. P. 1958. Japanese discomycete notes I-VIII. Sci. Rept. Yokohama Nat. Univ. II. **7**: 7–35.

LAGARDE, J. 1906. Contribution à l'étude des discomycètes charnus. Ann. Mycol. **4**: 125–256.

LE GAL, M. 1953. Les discomycètes de Madagascar. Prodr. Flore Mycol. Madagascar **4**: 1–465.

MÜLLER, E. 1950. Die schweizerischen Arten der Gattung *Leptosphaeria* und ihrer Verwandten. Sydowia, Ann. Mycol. **4**: 185–319.

NANNFELDT, J. A. 1932. Studien über die Morphologie und Systematik der nicht-lichenisierten inoperculaten Discomyceten. Nova Acta Reg. Soc. Sci. Upsal. IV. **8** (2): 1–368.

PATOUILLARD, N. T. 1928. Contribution à l'étude des champignons de Madagascar. Mém. Acad. Malgache **6**: 1–49. [1927.]

REHM, H. 1887-96. Ascomyceten: Hysteriaceen und Discomyceten. *In* RABENHORST, L., Kryptogamen-Flora von Deutschland, Oesterreich und der Schweiz II. **1** (3): viii + 1275 + 169 pp.

SÁNCHEZ, A., AND R. P. KORF. 1966. The genus

Vibrissea, and the generic names *Leptosporium*, *Apostemium*, *Apostemidium*, *Gorgoniceps* and *Ophiogloea*. Mycologia **58**: 722–737.

SMITH, A. H., AND R. L. SHAFFER. 1964. Keys to Genera of Higher Fungi. Univ. Michigan Biol. Station, Ann Arbor, Michigan. 120 pp.

SNELL, W. H. 1936. Three Thousand Mycological Terms. Rhode Island Bot. Club Publ. 2. 151 pp.

STARBÄCK, K. 1895. Discomyceten-Studien. Bihang Kongl. Svenska Vet.-Akad. Handl. XXI. **3** (5): 1–42.

TEWARI, V. P., AND D. C. PANT. 1966. Ascomycetes of India I. Mycologia **58**: 57–66.

WHETZEL, H. H. 1943. A monograph of *Lambertella*, a genus of brown-spored inoperculate discomycetes. Lloydia **6**: 18–52.

WHITE, W. L. 1941. A monograph of the genus *Rutstroemia* (Discomycetes). Lloydia **4**: 153–240.

Hypogeous Ascomycetes from Idaho

LILIAN E. HAWKER
Department of Botany, University of Bristol, Bristol, England

During a collecting excursion in the summer of 1962, organized by Dr. A. H. Smith of the University of Michigan and based at McCall, Idaho, five species of hypogeous Ascomycetes were collected. Of these, a species of *Elaphomyces* was found in quantity on a number of occasions and in several localities. Several collections were made of three members of the Tuberales, *Barssia oregonensis, Hydnotrya cerebriformis,* and *Geopora magnata.*

Single specimens of *Piersonia alveolata* and *Tuber californicum* were also collected.

Elaphomyces granulatus *var.* asperulus (*Vitt.*) *Hawker*

Numerous collections, many of them very large, were made in coniferous woods in the McCall district of Idaho. These collections were made at Payette Lake, Upper Payette Lake, Toller's Ditch, Burgdorf, Squaw Meadow, Brundage Mountain, Brundage Reservoir, Blacklea Creek, Granite Mountain, and Granite Lake Trail.

With the possible exception of two small collections made by Dr. A. H. Smith at Brundage Reservoir on August 13, these are all believed to be the same species and to be the same as two earlier collections made by him (SM65426, mature specimens; SM65421, immature specimens).

General Macroscopic Characters of Fruit Body

Young fruit bodies irregularly globose or dented, coarsely granulose-squamulose, squamules breaking off in powdery mass, embedded in irregular crust of soil and mycelium, often with associated yellow mycelial strands, peridium bright ochre yellow, becoming ochraceous buff when dry. Mature fruit bodies globose or commonly flattened to almost disc-shaped, squamulose, "crust" well developed, separating readily from fruit body, no mycelial strands seen, peridium dingy ochraceous buff, drying to potato-colored (buckthorn brown, Ridgway), size up to 4.5 cm. diam.

Peridium up to 3 to 4 mm., thinning to ca. 2 mm. when mature; outer layer, whitish in section; inner layer, thick, vinaceous brown in section, not mottled, darkening on exposure, drying dingy ochre; consisting of interwoven hyphae.

Gleba at first vinaceous buff masses, separated by whitish sterile veins (one collection from Granite Mountain with blue veins), with water drops in some specimens, later whole contents powdery brown-black.

Asci club-shaped to globose, 35–50 μ diam.; 6-spored, or less commonly 8-spored; evanescent.

Ascopores, at first pale brown, smooth-walled when released from ascus and ca. 15 μ diam., then enlarging, darkening (brown-black in mass), becoming irregularly sculptured 28–33–35 μ diam. (See Hawker [1968] for Stereoscan electron microscopy of spore sculpturing.)

The species *E. granulatus* Fr. has been the subject of recent consideration. Lange (1956) and Eckblad (1962) restricted this species to forms in which the mature spores are covered with irregular spines, and transferred forms with warty spores to *E. asperulus* (Vitt.). Hawker et al. (1967) pointed out that Vittadini (1831, 1842) gave no clear description of spore ornamentation but separated *E. asperulus* from *E. granulatus* on peridial characters, the peridium of the former being reddish in section and that of the latter pallid. Hawker et al. showed that the final form of the ascospore ornamentation depended largely on rate of drying of the fruit body and its age when collected. They suggested that forms with pallid peridia (as seen in section) placed in *E. granulatus* by Berkeley and Broome (1837–85) should be included in *E. granulatus* var. *granulatus* Hawker, and those that show mainly reddish coloration of the peridium in section should be included in *E. granulatus* var. *asperulus* (Vitt.) Hawker.

The Idaho material differs from *E. granulatus* (sensu Berkeley and Broome) in a number of characters: (1) paler color of young fruit-bodies, (2) redder color of mature specimens, (3) prevalence of characteristically flattened mature specimens, (4) reddish tinge of inner layer of peridium, (5) spores brown-black instead of blue-black in mass, (6) larger size of spores av. 33 μ diam. as against 28 μ for British material—which, however, is rather variable in spore size. Size of spores also alters during maturation. It is, therefore, appropriately placed in *E. granulatus* var. *asperulus.*

The two small collections made by Dr. A. H. Smith at Brundage Reservoir on August 13, 1962, differed in having a thinner peridium, spores blue-black in mass, ca. 27 µ diam. In all but the thin peridium, these specimens more nearly approach the British material attributed to *E. granulatus.* A third collection made in the same locality on the same day by the writer is intermediate between these two collections and the larger number of collections described above. There is insufficient material to determine whether the collections of August 13 belong to the same or a different variety or species.

Barssia oregonensis *Gilkey*

Previous records are from Oregon and California only (Gilkey, 1925). Collections were made at Boulder Creek (a single small, immature specimen in July, 1962, and an overripe specimen in August), Upper Payette Lake (a large collection including immature and fully mature specimens, August), and Brundage Mountain (a single doubtful immature specimen in early July). The Idaho material was closely similar to authentic material kindly supplied by Dr. Gilkey (Gilkey collection No. 918), and the identification has been confirmed by her.

Macroscopic Characters of Fruit Bodies

Irregularly globose, or flattened, hollow (or, in very young specimens, stuffed), up to 2 cm. longest diam., pinkish buff to cinnamon buff, surface slightly scabrous (less so than in Gilkey's material), waxy, internal cavity much convoluted (Fig. 1(A) ab), single opening showing whitish gleba. Odor of mature speci-

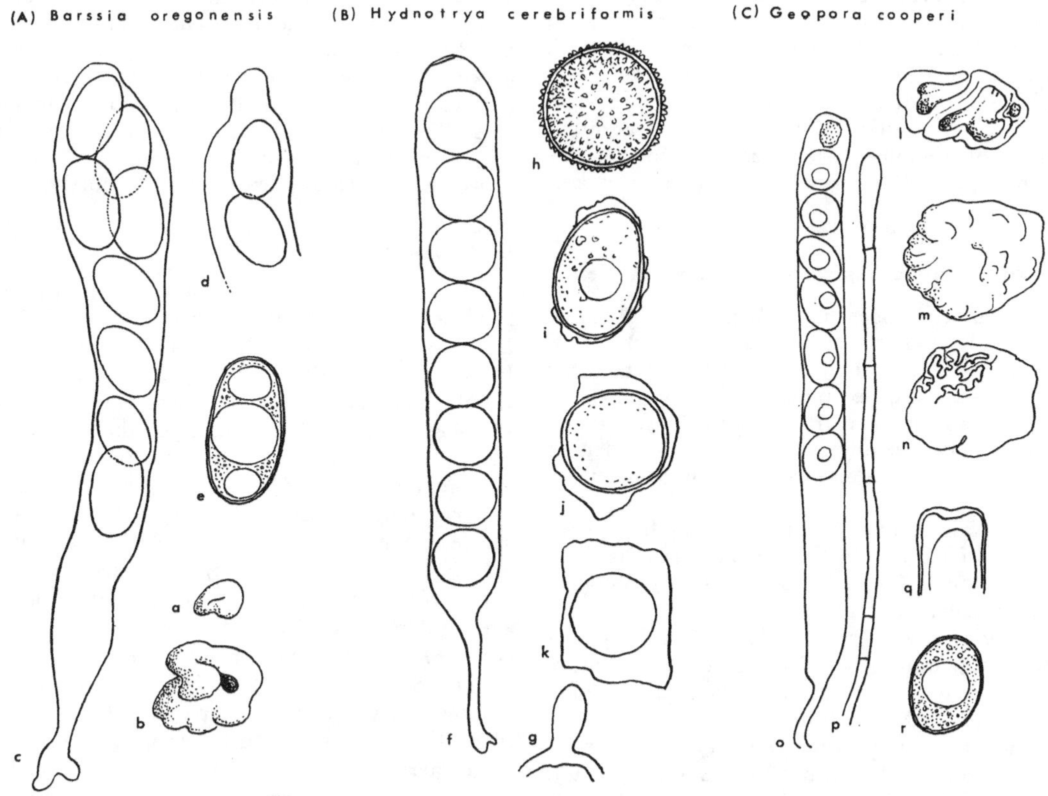

FIG. 1. (A) *Barssia oregonensis:* a and b, young mature fruit bodies; c, ascus; d, tip of ascus with a pronounced apical papilla; e, mature spore showing smooth wall and prominent oil drops.

(B) *Hydnotrya cerebriformis:* f, mature ascus showing apical pore; g, young ascus growing out from ascogenous hypha; h, i, j, k, various types of ascospore—h, globose spiny; i, ellipsoidal and warty; j and k, globose with prominent flanges. All these types were found in the same fruit body.

(C) *Geopora cooperi:* l, section through immature specimen, gleba only silghtly convoluted; m, mature fruit-body; n, same in section showing much convoluted gleba; o, mature ascus; p, paraphyse; q, ascus tip (enlarged); r, spore showing single prominent oil drop.

Fruit bodies X1 (*Barssia*) and ½ nat. size (*Geopora*); whole asci X400, spores X500.

mens unpleasant (reminiscent of *Hydnotrya tulasnei*), absent in young fruit bodies.

Asic banana-shaped up to 220 x 35 μ, tapering to base, often with papillate apex, curved, 8-spored, uniseriate or partially biseriate (Fig. 1(A) cd).

Paraphyses, slender, filamentous, septate, lacking swollen apex, projecting beyond asci.

Ascospores, smooth, ellipsoid, hyaline, 22–26 x 10–13 μ in single immature specimen of Boulder Creek collection, average 33 x 17 μ in upper Payette Lake collection (Fig. 1 (A) e). Gilkey gives 26–32 x 12–17 μ.

Hydnotrya cerebriformis *Hark*

This species, was previously known only from the type locality, Nevada County, California (Gilkey, 1953). Collections were made in July and August, 1962, from the following localities; Brundage Mountain, Brundage Reservoir, Upper Payette Lake, Granite Twin lakes, Duck Lake-Payette Lake Trail, N. Fork Lake Fork, Vulcan Hot Springs S. Fork.

Mature spores were not seen until late August. A similar delay in the maturation of spores of a number of species of Tuberales has been described for British material (Hawker, 1954).

Macroscopic Character of Fruit Bodies

Irregularly globose up to 4 cm. diam. or, more commonly, somewhat flattened, closely convoluted, young fruit bodies creamy white when freshly exposed, becoming reddish on exposure to air, fully mature fruit bodies dingy salmon pink. Gleba opening on surface, main opening on top, others at additional points in some specimens.

Peridium minutely scurfy or papillate, consisting of parallel hyphae of bulbous cells (19–20 μ diam.,) outer cells irregularly inflated, giving scurfy effect.

Gleba waxy cream to blush pink in young specimen, dingy purplish salmon in mature ones; cavities, formed by numerous convolutions, crowded, not stuffed.

Asci in regular hymenium, ca. 120–140 x 30 μ, tapered at extreme base (Fig. 1 (B) f, g), rounded apex with plug-like thickening, resembling a dehiscence pore, 8 ascospores, usually uniseriate, sometimes incompletely biseriate.

Paraphyses slender filamentous (6 μ diam.) septate, slightly inflated at tip, irregular length, some projecting above asci for distance of up to 8 μ.

Ascospores, spherical or slightly ellipsoidal, at first smooth and hyaline (17–22 μ diam.) containing a single large central oil drop, often with large wing-like hyaline flanges, later becoming usually red brown and minutely papillose but some becoming flanged or warty (cf. *H. tulasnei*). Both types may be found in same specimen or even in same ascus.

Flanged spores tend to collapse in dried material, suggesting that these are abnormal and that the flanges normally disappear during maturation (Fig. 1 (B) h–k).

Agreement with Gilkey's descriptions (1939, 1954) of the type material is fairly close, and the identification has been confirmed by her. The chief differences are a wider range of peridial form and the presence of some flanged spores. Gilkey (1954) has pointed out the extreme variability in fruit body form in other species of *Hydnotrya* and *Gyrocratera.* Observations (Hawker, unpublished) on British material of *H. tulasnei* and *G. ploettneriana* show a similar variability. The Idaho material can be placed in a series of increasing complexity roughly corresponding to degree of maturity, the most mature specimens being most like the type as described by Gilkey. It has already been pointed out that the flanged spores are probably abnormal and abortive.

Geopora cooperi *Hark*

Numerous collections from the following localities from June to August; Payette Lake, Brundage Mountain, N. Fork Lake Fork, Hazard Creek, Warm Lake, and the Seven Devils National Park. The material was very variable. It was at first thought that at least two species were present, but further material exhibited a complete range of form, and the material must therefore be accepted as a single species. A mature specimen from Payette Lake submitted to Dr. H. Gilkey was identified by her as *Geopora magnata.* Some specimens, however, fitted more closely to her description of *Geopora cooperi,* and further collections made by Dr. Smith and his associates in 1964 were assigned to *G. cooperi* (Burdsall, 1965). It is likely that the distinction between these species is unwarranted.

Burdsall noticed that when fresh fruit bodies were broken open, puffing of ascospores occurred. Dehiscence of the ascus was by a hinged operculum, but this was not always discernible in old specimens. A vestigial operculum or pore was seen by the writer in *Hydnotrya cerebriformis* but was not observed to open.

Macroscopic Characters of Fruit Body (Fig. 1 (C) l, m, n)

Shape and size irregular (up to 10 cm. diam.) at first waxy, like tallow, brittle, dingy white, readily breaking up into flakes, finely tomentose with felt of white hairs in which soil is entangled, later clay-colored, surface drier, lobing and folding of fruit body becoming more complex and less liable to flaking. Interior dingy white in young specimens. In mature specimens, gleba moist (but not milky) when cut, translucent, mainly dingy cream, hymenium white opaque line, venae externae salmon to coffee-colored. Odor absent or pleasant, of citrus fruit.

Peridium up to 700 μ thick. Inner layer of interwoven hyphae, outer layer of inflated cells, dingy, yellow to reddish brown according to maturity, terminating in slender hyaline septate hairs (5–7 μ diam.) projecting at least 200 μ beyond surface.

Gleba loosely folded, variable, chambers hollow, venae externae seen as cinnamon or even salmon-colored line, due to continuation of colored, hairy peridium around folds. Hymenium pure white.

Asci, ca. 220 x 22 μ, tapering to base, square apex with thickened pad later opening as dehiscence pore (Fig. 1 (C) o, q), usually 8-spored, uniseriate, but some spores may abort, spores may be arranged with long axis at right angles to axis of ascus and may then appear to be spherical.

Paraphyses slender (diam. 4–5 μ) septate, hyaline, slightly inflated apex, not projecting above level of mature asci (Fig. 1(C) p).

Ascospores ellipsoidal, but owing to position sometimes taken up in ascus, may appear to be spherical, 22–25 x 13–15 μ, large central oil drop, smooth thin wall (0.5 μ) (Fig. 1 (C) r).

Piersonia alveolata *Hark*

A single specimen collected on Brundage Mountain July 5, 1962, was attributed to this species although it did not agree exactly with Gilkey's description (1954).

Tuber californicum *Hark*

A single nearly mature specimen was collected at Payette Lake, August 21, 1962. This corresponded fairly closely with the descriptions given by Gilkey (1939, 1954), but the peridium had a pink tinge and the spores were rather smaller, possibly owing to immaturity. It resembled *T. puberulum* Berk. et Br. in many respects but differed in the form and relative scarcity of the peridial hairs and in fruit body color. It is doubtful whether the differences between these two "species" are sufficient to justify separating them.

Macroscopic Characters of Fruit Body

Fruit body not lobed, 9 x 8 x 5 mm., pinkish buff when fresh, soon pale ochraceous. *Gleba* pinkish, veins white when fresh, soon becoming clay-colored.

Peridium 200 μ thick, outer layers cellular, inner of interwoven hyphae, extreme outer layer pigmented and cells occasionally prolonged into long, septate, external hairs.

Asci, pyriform or subglobose, av. diameter 77 μ at widest part (i.e., rather smaller than sizes given by Gilkey but specimen not fully mature), not stalked, 1–2 spores, less commonly 3–4, usually all but one or two abortive (Fig. 2).

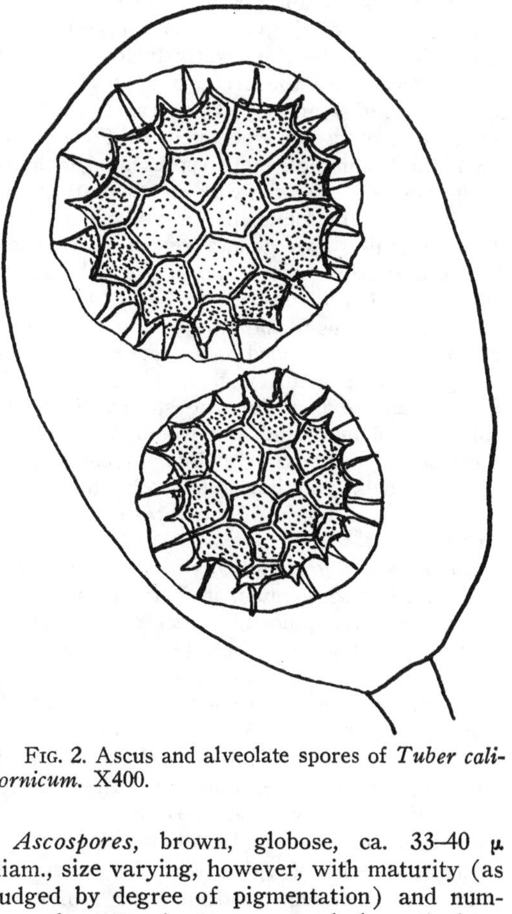

FIG. 2. Ascus and alveolate spores of *Tuber californicum.* X400.

Ascospores, brown, globose, ca. 33–40 μ diam., size varying, however, with maturity (as judged by degree of pigmentation) and number of spores in ascus; regularly reticulate; usually about six alveoli across diameter; epispore ca. 4 μ thick (Fig. 2).

Discussion of the Incidence of Hypogeous Ascomycetes in the McCall District of Idaho

Since the species of *Elaphomyces* and a number of hypogeous Basidiomycetes were found in quantity during the whole period of the excursion (June to September) and in a number of localities, the infrequency of representatives of the Tuberales is worth noting.

In a study of British hypogeous fungi (Hawker, 1954) based on over 1,000 collections, the writer pointed out differences in the incidence of fruit bodies of *Elaphomyces* spp., members of the Tuberales and hypogeous Gasteromycetes. The majority of British species of Tuberales—with the exception of species of *Genea, Gyrocratera,* and *Hydnotrya* but including most of the numerous recorded species of the genus *Tuber*—were favored by a neutral or slightly alkaline reaction of the soil. No estimates of soil pH were made in Idaho but the nature of the vegetation suggested that it was likely to be comparatively low. This in itself would limit the frequency of almost all species of *Tuber* and allied genera. It is significant that only one specimen of *Tuber* was found, and that the genera represented included a species of *Hydnotrya,* representatives of which genus are found in acid soils in Britain. The remaining two genera represented in the Idaho collection show morphological resemblances to *Hydnotrya.*

The hypogeous Basidiomycetes found during this excursion included many species of *Rhizopogon.* This genus is poorly represented in Britain, and, except for one species, is found only in acid moorland and woodland soils. In contrast, *Hymenogaster,* which is common in alkaline chalk and limestone soils in Britain, was not found in the Idaho forests during the writer's visit.

Elaphomyces granulatus, which was common in Idaho, is also common in Britain and grows well in acid woodland soils, such as the tertiary plateau gravel south of the Thames. This fungus, however, is also found in less acid soils and is able to tolerate a wider range of soil pH than are most Gasteromycetes and Tuberales.

Hydnotrya tulasnei and the related *Gyrocratera ploettneriana* are found in Britain under coniferous trees, whereas most of the British Tuberales are associated with deciduous trees such as beech. *Elaphomyces granulatus* in Britain is commonly associated with pine but has also been recorded under deciduous trees, such as oak; other British species of *Elaphomyces* are found most frequently under beech. Among British hypogeous Gasteromycetes there is considerable variation in tree associations, but *Rhizopogon* is seldom found except in association with pine. Thus the prevalence of conifers and the comparative absence of deciduous trees in Idaho may also be a factor in limiting the incidence of members of the Tuberales to those genera, such as *Hydnotrya,* normally found in coniferous woods.

A further factor may well be the climate. In South Britain, members of the Tuberales can usually be found at all times of the year, the partially decayed fruit bodies of the preceding year still being found in January and February alongside new fruit-body initials.

The young fruit bodies are not readily identified, except by someone familiar with the genus, until July or even later, when the spores develop and mature. In unusually severe winters the fruit-body initials either abort with the onset of cold weather in January or are delayed in development and may not mature until late autumn. Many species are, however, common in mountainous regions on the continent of Europe, where the winter is normally severe, with heavy snowfalls. Thus it is unlikely that the Idaho winter is a limiting factor, although the late thaw may delay maturation of the fruit bodies. Mature spores were seen only in collections made after mid-August. It is more likely that the hot, dry summers are limiting. Fruit bodies of the common British species of Tuberales are scarce in unusually dry summers. Conditions in Idaho, therefore, are likely to be unsuitable for any species except those of unusually tough, leathery texture, or those growing particularly rapidly. Observations in Britain suggest that the tough texture of fruit bodies of *Elaphomyces* enables them to survive periods of drought which would destroy the more fragile or fleshy fruit bodies of most members of the Tuberales, and that the rapid growth of most hypogeous Basidiomycetes enables them to develop and reach maturity during periods of moist conditions, following a storm, which are not long enough to permit development of the slower-growing fruit bodies of *Tuber* and allied species.

Thus the hypogeous flora of Idaho is influenced by climate, soil pH, and the predominantly coniferous tree cover. There may well be other factors such as aeration and mineral content of the soil. Further ecological study of this community would be rewarding. Since hypogeous fungi are notoriously difficult to grow in artificial culture, the conditions favoring a particular species are at present deduced only from field notes of large numbers of collections. The collections made in Idaho were too few to yield such information except in comparison with studies elsewhere.

ACKNOWLEDGMENTS

Thanks are due to Dr. A. H. Smith, University of Michigan, who invited me to join the expedition to Idaho, for permission to publish these results; to other members of the party who collected much of the material; to Professor H. M. Gilkey, State University of Oregon, Corvallis, who confirmed the identification of some of the material of Tuberales; and to Dr. J. Fraymouth of this department who sorted and checked much of the dried material.

REFERENCES

Berkeley, M. J., and C. E. Broome. 1837–1885. Notices of British Fungi. Ann. Mag. Nat. Hist. (1) **6**: 430–439.

Burdsall, H. H. 1965. Operculate asci and puffing of ascospores in *Geopora* (Tuberales). Mycologia **57**: 485–488.

Eckblad, F. E. 1962. Studies in the hypogeous fungi of Norway II. Revision of the genus *Elaphomyces*. Nytt. Mag. Bot. **9**: 199–209.

Gilkey, H. M. 1939. Tuberales of North America. Oregon State Monograph 1. 63 pp.

———. 1954. Tuberales. North American Flora. Vol. II, Part 1. New York Bot. Garden. 34 pp.

———. 1954a. Taxonomic notes on Tuberales. Mycologia **46**: 783–793.

Hawker, L. E. 1954. British hypogeous Fungi. Phil. Trans. Roy. Soc. Lond. B. **237**: 429–546.

———. 1968. Wall ornamentation of species of *Elaphomyces* as shown by the Stereoscan electron microscope. Trans. British. Mycol. Soc. In press.

Hawker, L. E., J. Fraymouth, and M. de la Torre. 1967. The identity of *Elaphomyces granulatus*. Trans. Brit. Mycol. Soc. **50**: 129–136.

Lange, M. 1956. Danish hypogeous Macromycetes. Dansk. Bot. Ark. **16**: 5–84.

Vittadini, C. 1831. *Monographia Tuberacearum*. Milan.

———. 1842. *Monographia Lycoperdineum*. Milan.

Ascocarp Development in *Leptosphaerulina argentinensis*

WILLIAM C. DENISON AND ROBERT C. CARLSTROM
Department of Botany and Plant Pathology, Oregon State University, Corvallis, Ore.

Leptosphaerulina McAlpine is a genus of Loculoascomycetes with unilocular pseudothecia resembling those of *Pleospora*, but differing in their lack of pseudoparaphyses, their smaller size, and their saccate asci. Müller (1951) and Wehmeyer (1955) have described ascocarp development in *Leptosphaerulina australis*, a species with consistently 4-septate ascospores. This paper describes ascocarp development in a closely related species with 5-septate ascospores.

Materials and Methods

The cultures used in this study were isolated in October, 1966, from specimens of diseased bent grass (*Agrostis palustris*) from Hermiston, Oregon. Fragments of grass with attached pseudothecia were surface sterilized in 10% chlorox and planted in water agar supplemented with streptomycin. Ascospores discharged from pseudothecia germinated on the agar surface and were transferred, singly, to sterile slants.

The medium employed routinely was a modified cornmeal agar consisting of a commercial preparation of cornmeal agar supplemented with 1 gram each of yeast extract and malt extract, plus 10 grams of plain agar per liter of medium (Alexopoulos, pers. comm.).

Portions of agar cultures containing developing pseudothecia were fixed in FAA, embedded in paraffin through a tertiary butyl alcohol series, sectioned at 15 μ, and stained with safranin and fast green (Johansen, 1940).

Nuclei in hyphae and in developing asci were stained in acetocarmine using the techniques of Austin (1959) and Tandler (1959).

Observations

Cultures derived from single ascospores form pseudothecia readily in culture. Young colonies are white, becoming brown, then black, as pseudothecia develop. At room temperature, pigmentation is present on the fourth day, meiotic asci on the sixth, and mature ascospores on the eighth. Mature colonies are flat and black, with numerous pseudothecia and almost no aerial hyphae. No conidial stages have been seen. The formation of sectors with reduced pigmentation and fewer pseudothecia is common in some isolates.

Freshly discharged ascospores germinate immediately on a suitable substrate, producing one or more germ tubes within an hour. Germ tubes from the same or adjacent ascospores anastomose freely, as do actively growing hyphae in general. Hyphae in young colonies and at the margin of actively growing older ones are hyaline, thin-walled, 4–7 μ wide, and conspicuously vacuolated. The first sign of fruiting in a maturing culture is the development of pigmented fertile hyphae. These are wider, 7–12 μ, have thicker, brownish walls, are septate at shorter intervals, and are frequently swollen just below the septum. They grow just below the surface of the agar. Hyphae in young colonies have scattered, solitary nuclei, but by the time that fertile hyphae appear, most of the nuclei in both types of hyphae are paired (Fig. 1).

Pseudothecial initials begin as swollen terminal or intercalary cells of the fertile hyphae. The protoplast of such a cell cleaves into several cells. Branches, from the same and adjacent hyphae, grow closely about it, forming a knot of cells (Fig. 5). Possibly one or more of the hyphae connected to the pseudothecial initial at this stage are trichogynes or antheridial hyphae, but differentiated gametangia were not observed.

Asci first appear as larger cells with larger nuclei and more densely staining cytoplasm (Fig. 6). The system of ascogenous cells from which they arise is located originally near the center of the pseudothecial initial, but with continuing growth of both asci and pseudothecium it comes to lie at the bottom of the locule. In stained sections ascogenous cells are difficult to distinguish from adjacent sterile ones, but in squashed preparations they remain connected to each other and to the young asci and may be freed from the remainder of the ascocarp (Fig. 2). Ascogenous cells continue to produce new asci after the first ones have matured and discharged their ascospores, so that several generations of asci may occupy the same locule.

Karyogamy occurs while the asci are very small, 6–10 μ. Thus it is difficult, if not impossible, to distinguish young, dikaryotic asci from the ascogenous cells. After karyogamy, the diploid nucleus enlarges to roughly three times the diameter of haploid nuclei and the ascus

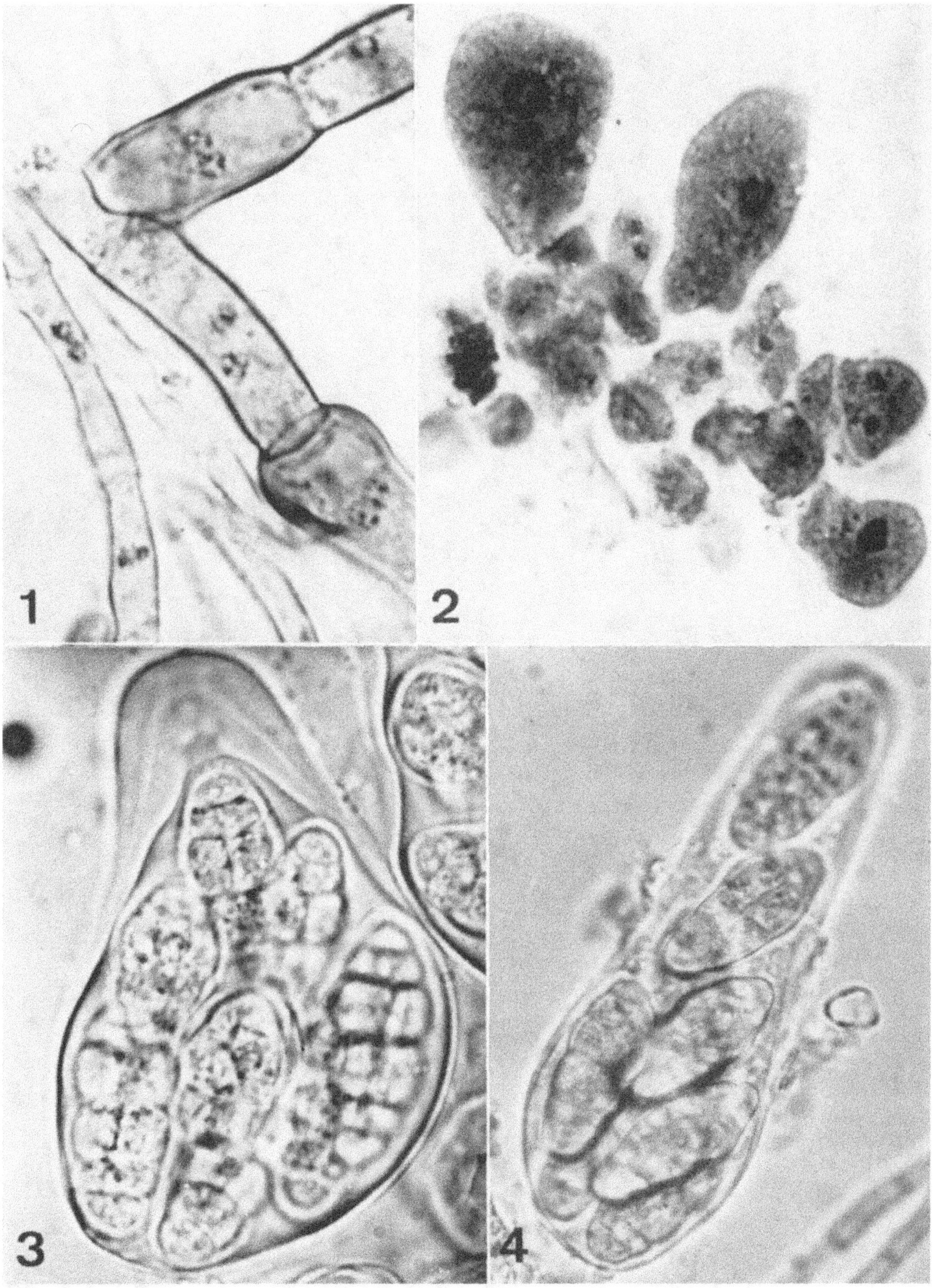

PLATE I. Fig. 1. Fertile and vegetative hyphae showing paired nuclei X1600. Fig. 2. Ascogenous cells with three young diploid asci. X1600. Fig. 3. Mature ascus and ascospores. X1300. Fig. 4. Ascus with partially extended inner wall. X1100.

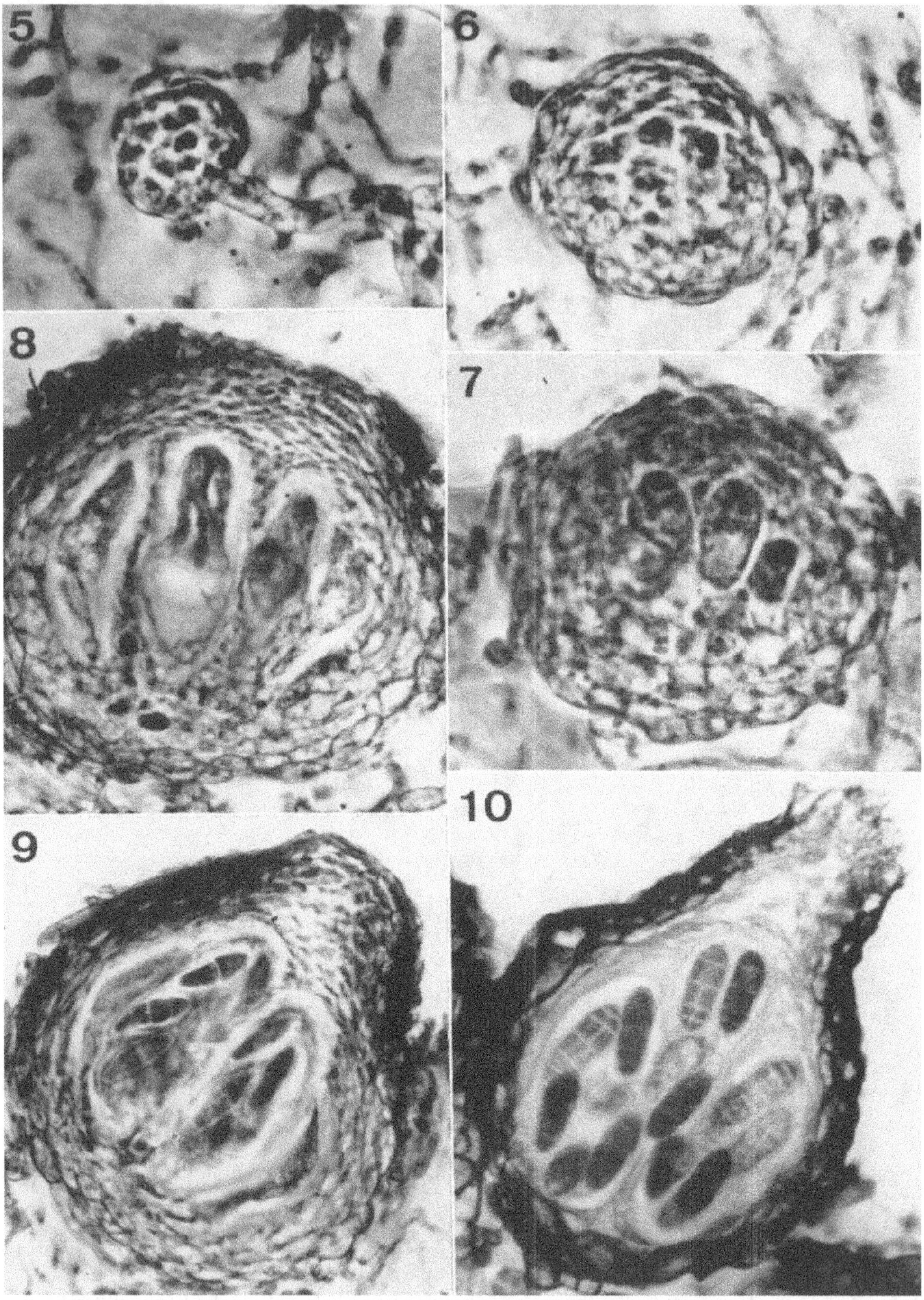

PLATE II. Figs. 5–10. Successive stages in ascocarp development, from pseudothecial initial to mature pseudothecium. X460.

elongates, reaching a length of 60–70 μ at meiosis. Three synchronous divisions, the first two presumably meiotic, are followed by the delimitation of eight ascospore initials. Further divisions within each initial result in the formation of multinucleate ascospores with five transverse septa and one or more longitudinal ones (Figs. 3, 10). Young asci are thin-walled, but about the time that meiosis occurs the walls in the upper half of the ascus become greatly thickened (Fig. 8).

Differentiation of the pseudothecial wall begins about the time that meiosis occurs in the asci. Starting in a collar-like zone near the top of the pseudothecium, the cells in the outermost layers stop growing and their walls become thickened and black (Fig. 8). This process spreads until all but the apex of the pseudothecium is enclosed in a black rind. The cells at the apex continue to grow, forming a papilla-like ostiolar protrusion on the previously globose pseudothecium (Fig. 9). At maturity the cells in the interior of the ostiole disintegrate, forming a passage to the outside (Fig. 10). Pseudoparaphyses are not formed. The asci and pseudothecia grow at comparable rates so that there is relatively little intrusive growth of asci into sterile tissue. However, strands of sterile tissue are sometimes seen between developing asci (Fig. 8) and their remains may be found in mature pseudothecia.

Mature pseudothecia are pear-shaped, (95) 120–150 (175) μ high by (80) 95–125 (160) μ broad, black, and glabrous. Mature asci are saccate, (50) 60–85 (100) by (25) 30–40 (50) μ, with a thickened wall at the apex and a slender, claw-like attachment at the base (Fig. 3). Mature ascospores are yellowish, (20) 24–32 (40) by (8) 10–12 (14) μ, with five transverse septa and one or more longitudinal ones. They taper gradually toward each rounded pole, and the third cell from the top is usually the broadest. The asci are bitunicate: at discharge, or when mounted in water, the outer wall ruptures and the inner one emerges as a cylindrical extension (Fig. 4). Normally this extension projects through the ostiole so that spores are forcibly ejected into the air.

Discussion

Our fungus is very like a small *Pleospora*; for some time we mistakenly classified it in that genus. It differs in its lack of pseudoparaphyses, its smaller pseudothecia, and its saccate asci. Graham and Luttrell (1961) have described several species of *Leptosphaerulina* occurring in North America. Of these species, our fungus most nearly resembles *L. argentinensis.*

The development of *L. argentinensis* closely parallels that described for *L. australis* (= *Pleospora gaeumannii* = *Pseudoplea gaeumannii*) by Müller (1951) and Wehmeyer (1955). In *L. argentinensis* we found no evidence of differentiated gametangia. This is in agreement with Wehmeyer's findings in *L. australis,* but not with those of Müller, who described ascogonia. Since *L. argentinensis* is homothallic (fruiting cultures develop from single, uninucleate ascospore initials) and since a majority of nuclei are paired by the time pseudothecial initials appear, gametangia, if formed, might be superfluous. Wehmeyer found that in *L. australis* the asci arise from binucleate cells scattered throughout the center of the primordium, whereas in *L. argentinensis* we find that asci arise successively from a differentiated layer of ascogenous cells which occupy a basal position in mature pseudothecia.

Because of its short life cycle and its prolific production of pseudothecia in culture, *L. argentinensis* is an excellent example of the Loculoascomycetes to use in introductory mycology laboratories.

ACKNOWLEDGMENTS

We want to thank Mrs. Carol A. Franklin for her assistance in preparing paraffin sections. We are also grateful to Mr. H. Bierman, County Agricultural Agent, Hermiston, Oregon, who supplied the grass specimens from which our cultures were isolated.

LITERATURE CITED

Austin, A. P. 1959. Iron-alum aceto-carmine staining for chromosomes and other anatomical features of Rhodophyceae. Stain Tech. **34**: 69–76.

Graham, J. H., and E. S. Luttrell. 1961. Species of *Leptosphaerulina* on forage plants. Phytopath. **51**: 680–693.

Johansen, D. A. 1940. Plant microtechnique. McGraw-Hill, New York.

Müller, E. 1951. Über die Entwicklung von *Pleospora gaeumannii* nov. spec. Ber. Schweiz. Bot. Ges. **61**: 165–174.

Tandler, C. J. 1959. An alkali-formaldehyde squash technic for plant cytology and cytochemistry. Stain Tech. **34**: 234–236.

Wehmeyer, L. E. 1955. The development of the ascocarp in *Pseudoplea gaeumannii.* Mycologia **47**: 163–176.

Heteroecism in *Puccinia duthiae*

M. J. Narasimhan[1] and M. J. Thirumalachar
Hindustani Antibiotics, Ltd., Pimpri, Poona, India

The grass rust *Puccinia duthiae* Ell. & Tracy on *Andropogon pertusus* (*Bothriochloa pertusa*) was first described from India based on the collections made by Duthie in 1889 near Sahranpur, India. It is known to parasitize other species of *Bothriochloa* and *Dichanthium*, which are important forage grasses in the tropics. *Bothriochloa glabra*, *B. insculpta*, *B. intermedia*, *B. kuntziana*, and *Dichanthium annulatum* are some of the other hosts. Cummins (1953) records the distribution of this rust in Mauritius, South Africa, Australia, Zanzibar, and China. He also cites the collections received by him from Thirumalachar on *B. pertusa* and *D. annulatum* from Mysore State, India. The rust was later described by Doidge under the name *Puccinia amphilophidis* Doidge, which is a synonym of *P. duthiae*.

The rust is well distributed all over the plains in India and causes heavy infection of the leaves of *B. pertusa* and *D. annulatum*. The infection first appears in the month of August soon after the rains. The grass being perennial, rusted leaves can be collected during most part of the year, but in summer months only telial stages can be collected.

Uredia are hypophyllous and rarely epiphyllous, on yellowish to reddish-brown leaf spots, erumpent and pulverulent. The sori are subepidermal, and paraphysate, the paraphyses being clavate and capitate, yellowish-brown, 40 to 90 μ long and 10 to 25 μ wide at the apex. Urediospores are ovate-ellipsoid, cinnamon-brown, deeper in color at the apex, apically up to 2.5 μ thick, with 4 to 5 equatorial germ pores, wall finely echinulate, spores measuring mostly 18–25 x 25–35 μ. Telia follow uredia in development; black, pulvinate, spores compactly grouped. Teliospores broadly ellipsoid, usually rounded at the top and narrowing at the base, slightly constricted at the septum, 22–30 x 35–45 μ; wall chestnut-brown, smooth, 2.5 to 3.5 μ thick on the sides, and apically thickened up to 8 μ. Pedicel yellowish-brown, fragile, persistent, up to 12.0 μ long.

Freshly collected urediospores germinated readily within 8 hours in moist chambers at 25° C. When the spore material was exposed to 37° C. for 15 days, their viability was reduced to 20 per cent. Inoculation experiments were carried out on healthy seedings of *Dichanthium annulatum* in the greenhouse. The inoculated plants were enclosed in moist chambers for 24 hours, after which they were transferred outside for observation. Heavy infection took place on the inoculated leaves, and uredo sori erupted out after 12 to 15 days.

The teliospores were germinated by the method described by Thirumalachar and Narasimhan (1953). Teliospores collected in the previous season and stored at room temperature germinated more readily than those that were collected fresh in the same season. Under favorable conditions of moisture and temperature of 25° C., the teliospore germinations were abundant—up to 80 per cent. Both the cells of the teliospore germinate simultaneously (Fig. 1) and develop 4-celled basidia. The basidiospores are developed on short sterigmata, and at maturity are discharged forcibly. Under excessive moisture conditions the basidiospores

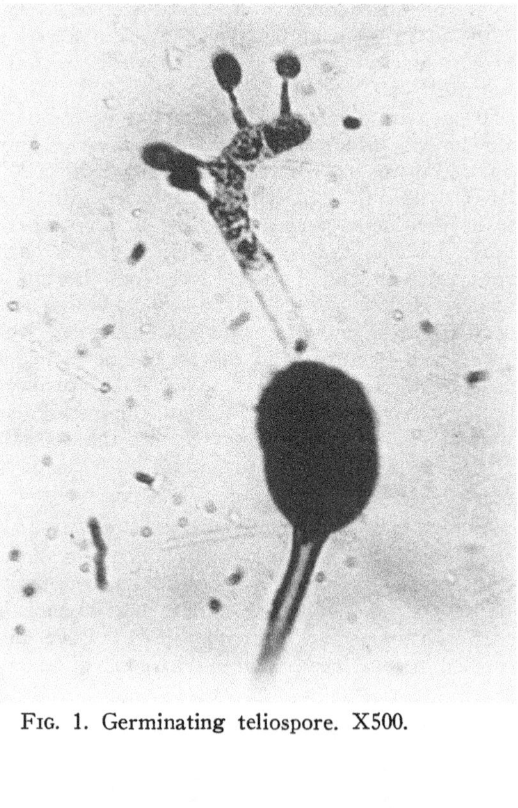

Fig. 1. Germinating teliospore. X500.

[1] The senior author is grateful to the Council of Scientific and Industrial Research for the grant-in-aid, which enabled undertaking this work.

often develop secondary sporidia, or long germ tubes.

Intensive field studies on the first outbreak of the rust on *Dichanthium annulatum* and its possible association with any aecial stage on other plants were carried out by the senior author. Detailed studies indicated that the rust was possibly connected genetically with an aecial stage on *Barleria cuspidata* Heyne, a wild shrub in the dry land forests. Cross inoculation experiments were done with basidiospores of *Puccinia duthiae* on leaves of *B. cuspidata* and aeciospores from *B. cuspidata* on *Dichanthium annulatum*; these gave positive results showing their genetic connection. An account of these studies is given below.

The aecial stage on *Barleria cuspidata* is borne on perennial hyphae, which incites the formation of systemic infection, with resultant large witches' brooms (Fig. 2). The systemic growth is a result of the infection of meristematic tissues in the axillary bud, so that the unfolding shoots are pale yellow, with shortened internodes and thick, narrow leaves that are all covered with pycnia and aecia (Fig. 3). When the young plant is infected artifically with basidiospores, the infection spots are circular, slightly raised, yellow, bearing pycnia and aecia. The systemic infection usually takes place in mature plants, where the infection of the axillary shoot brings about the malformation of the shoots.

FIG. 2. Showing the systemically infected and healthy plants of *Barleria cuspidata*.

FIG. 3. Systemic infection showing aecial pustules produced by artificial inoculation of axillary shoot. Natural size.

For inoculation experiments, young seedlings of *Barleria cuspidata* were raised in the greenhouse. Germinating teliospores with basidiospores were used for inoculating young, unfolding leaves, which were incubated in moist chambers. Success of inoculation was evident a week after, from the development of yellow infection spots, which expanded and showed numerous pycnia after 10 days. Aeciospore development followed 15 days after the development of the first infection spot. In the cross inoculation studies, aeciospores from *Barleria cuspidata* were used for inoculating *Dichanthium annulatum,* the technique being similar to that for urediospore inoculations. Successful inoculation with aeciospores was evident by the development of uredia on *Dichanthium annulatum* after 12 days. The heteroecious nature of the rust was thus established by cross-inoculation experiments.

The systemic aecial infection on *Barleria cuspidata* closely resembles that of *Puccinia cacao* McAlp. on *Hygrophila spinosa,* another member of the Acanthaceae. The systemic mycelium bears numerous pycnia during summer months (when the temperature is above 35° C.). These are all sterile without spermatia or paraphyses. During the cooler part of the year after the rains, when the temperature is about 25° C., the pycnia are fertile, and the development of aecia follows. No aecia are formed in association with sterile pycnia. This phenomenon was first observed by Arthur and Cummins (1936) in the Philippine rust *Aecidium manilense* Arth. & Cumm., which was shown by Thirumalachar and Narasimhan (1954) to be the aecial stage of *Puccinia cacao.* The latter authors gave a detailed morphological study of the rust. *Puccinia duthiae* is therefore another rust which perennates by the formation of systemic aecial mycelium.

The aecial stage must clearly be placed under *P. duthiae*. The unconnected aecial stage commonly occurring in India on *Barleria cuspidata* was previously described as *Aecidium barleriae* by Salam and Ramachar (1956). This was changed to *A. Salamii* by Laundon (1963), since the specific name *A. barleriae* had been pre-empted by *A. barleriae* Doidge (1948) from South Africa. The difference between *A. barleriae* Doidge and *A. salamii* Laundon appears to be only in the external symptoms. Doidge's material has circular infection spots, while those of *A. salamii* collected in India have witches' brooms also. We have already indicated that all these symptoms are the stages of one and the same rust, depending upon the place of infection. The development of aecia depends upon the temperature conditions. Only sterile pycnia are formed under hot weather, when the witches' brooms appear smooth and glistening with tiny specks of pycnia only. Hence any diagnoses differentiating *A. barleriae* Doidge and *A. salamii* on profuse or scattered nature of aecia is incorrect. The measurements of sori and spores in both *A. barleriae* and *A. salamii* are identical, and the two rusts are one and the same. The uredial and telial stages also occur in South Africa, which Doidge (Bothalia 3: 496, 1939) had named *Puccinia amphilophidis*. This was shown to be identical with *P. duthiae* by Cummins (1953). It is therefore concluded that *Puccinia duthiae* is a heteroecious rust, the aecial, uredial, and telial stages of which occur in both India and South Africa and possibly other countries.

LITERATURE CITED

ARTHUR, J. C., AND G. B. CUMMINS. 1936. Philippine rusts in Clemens Collection 1923–1926. Philippine Jour. Sci. **61**: 463–488.

CUMMINS, G. B., 1953. The species of *Puccinia* parasitic on the Andropogoneae. Uredineana, Encyclopediae Mycologique, Vol. XXIV. Paris. Pp. 1–89.

DOIDGE, E. M. 1948. South African rust fungi. Bothalia **4**: 895.

LAUNDON, G. F. 1963. Rust fungi. I. On Acanthaceae. Mycol. Paper No. 89, Commonwealth Mycological Institute, England. Pp. 1–89.

SALAM, M. A., AND P. RAMACHAR. 1956. Jour. Indian Bot. Soc. **35**: 152.

THIRUMALACHAR, M. J., AND M. J. NARASIMHAN, 1953. Notes on some mycological methods. Mycologia **45**: 461–468.

———. AND ———. 1954. Morphology of spore forms and heteroecism in *Puccinia cacao*. Mycologia **46**: 222–228.

An Unusual New Heterobasidiomycete with Tilletia-Like Basidia[1]

LINDSAY S. OLIVE

Department of Botany, University of North Carolina at Chapel Hill, N. C.

Introduction

During the course of plating out dead plant materials from South Carolina in a search for mycetozoans, a remarkable new heterobasidiomycete was found on dead florets of a single collection of the large plume grass *Erianthus giganteus*. In view of the fact that the basidia, basidiospores, and life cycle resemble in several significant ways those of the smut fungi, the organism, described here as a new genus and species, is being placed in a newly erected family of the Ustilaginales.

The isolation procedures and formulas for agar media used in this study have been described in a recent publication (Olive, 1967).

Results

Prior to a discussion of the life cycle, descriptions of the new taxa are presented.[2]

Filobasidiaceae L. Olive, familia nova

Basidia in circulis laxis v. densis e mycelio zygodesmatibus ornato oriunda, gracilia, haud septata, in apice basidiosporas sessiles producentia, probasidia crasse tunicata carentia.

Basidia arising in loose or dense groups from a mycelium with clamp connections, slender, nonseptate, bearing sessile thin-walled basidiospores terminally; thick-walled probasidia lacking, blastospores present.

Filobasidium L. Olive, gen. nov.

Characters iidem ac familiae.

With the characteristics of the family. Type species: *F. floriforme*.

Filobasidium floriforme L. Olive, sp. nov.

Basidia in floribus emortuis graminum circulis albis 140–210 x 350–1400 μ, vel sparsa, occurentia, 3.2–7 x 57–212 (–265) μ; basidiosporae ovatae v. ellipticae, 6–9 x 10.5–16.5 μ, plerumque 6–8 in basidii apice sublatae, germinatione blastosporas producentia; mycelium quoque blastosporas gignens. Fungus heterothallicus, saprogenus v. aegre parasiticus.

Basidia appearing on dead grass florets in small white patches (in agar plates), 140–210 x 350–1400 μ, or loosely scattered, measuring 3.2–7 x 57–212 (–265) μ; basidiospores ovate to elliptical, thin-walled, mostly 6–8, borne terminally, 6–9 x 10.5–16.5 μ, germinating to produce blastospores; mycelium also producing blastospores. Heterothallic and saprophytic or partially mycoparasitic.

A single collection on dead florets of *Erianthus giganteus* (Walt.) Muhl., Bluffton, South Carolina, November 19, 1966 (type).

The new fungus was first observed on dead florets (collected from standing inflorescences) placed on the surface of two weakly nutrient agar media—lactose-yeast extract (LY) and hay infusion (HI) agars—in petri dishes, but it has not been identified on unplated florets. The organism has been maintained in culture and studied on these same media.

Mature basidia and basidiospores were present when the fungus was first observed. Basidiospores, which readily become detached onto the tip of a needle, were transferred to fresh plates of LY agar, where they very quickly began to bud out oval to elliptical blastospores smaller than the basidiospores (Fig. 15). In a few days, conspicuous yeast-like colonies had developed. These were at first whitish to cream-colored, but after about two weeks they developed a light pinkish tinge.[3] Only yeast-like colonies lacking hyphae developed in single-spore or mixed colonies on LY agar.

The mycelial stage was first obtained in culture by isolating hyphal fragments from a small colony found growing on the agar near a grass floret in one of the original isolation plates. This was a simple task with a deFonbrune micromanipulator, as the hyphae are surprisingly resistant to damage during micromanipulation. Colonies developing from hyphae on LY and HI agars tend to have a compact growth pattern and to remain rather limited in size (Fig. 3). The hyphae have typical clamp connections (Figs. 9, 14), with branching at the clamps (Fig. 13). Although cytological

---

1 This study was supported by Grant GB-5508 from the National Science Foundation.

2 The author is grateful to Dr. D. P. Rogers for preparation of the Latin descriptions.

3 The blastospores are utilizable as a food source by the cellular slime mold *Acrasis rosea* Olive & Stoianovitch in laboratory culture.

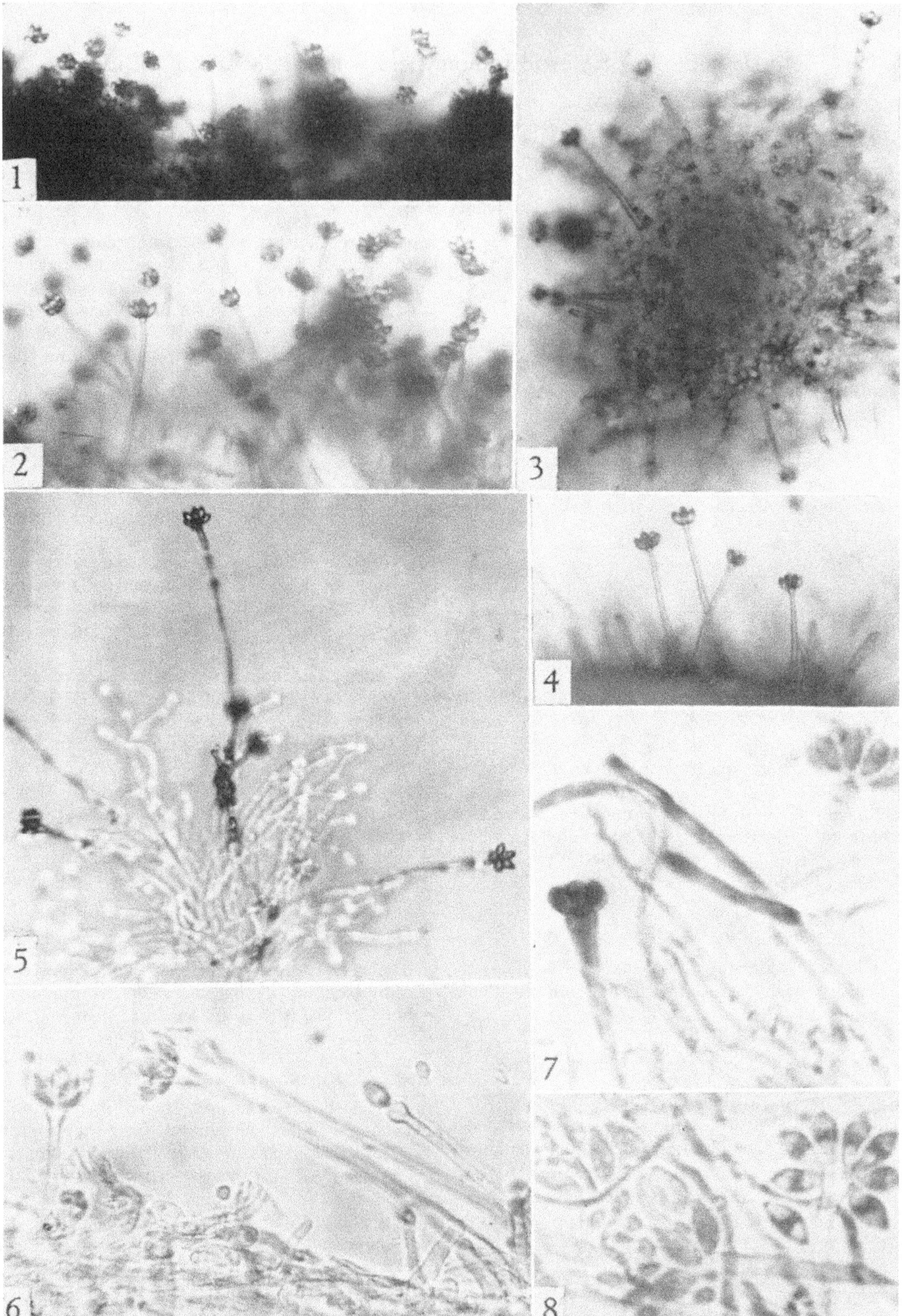
1
2
3
4
5
6
7
8

studies have not yet been made, it has been determined with cotton blue staining and a phase microscope that the hyphal cells are binucleate.

Within a few days, occasionally in two or three days, mature basidia appear in cultures and may eventually become very abundant. On the grass florets the sporulating colonies are whitish. On agar they are also whitish at first, but after a few weeks on LY agar they develop a light pinkish tinge. Basidia on grass florets and those on agar cultures are similar in appearance (Figs. 1–5). They are elongate and slender, usually slightly broader near the base, tapering apically and then expanding rather abruptly at the apex (5–7 μ), as shown in many of the figures (Figs. 1–8).

The basidiospores are budded out directly from the expanded apex (Fig. 7) and are most often 6–8 in number, though 5-spored basidia are occasionally found. The most common number of spores is 7. They typically occur in a petal-like whorl, giving the basidium a flower-like appearance in apical view (hence the specific name).

Sometimes small isolated groups of hyphae on agar give rise to only a few basidia (Fig. 5), and under certain conditions single hyphae that have grown out from the main bounds of a colony produce basidia singly. Such basidia usually arise at or near the surface of the agar, and frequently they grow out horizontally on the agar surface before turning away from it and producing the basidiospores (Fig. 5). Some of the longest basidia (up to 265 μ) are produced in this manner. These probably exceed in length any that are produced under natural conditions. If the humidity in petri dish cultures is reduced with the aid of a porous clay cover, the average length of the basidia is significantly reduced.

Under the humid conditions that often prevail in petri dish cultures, it is not uncommon to find attached basidiospores budding out blastospores apically. Basidiospores that become detached and fall back onto the mycelium or nearby agar soon bud out blastospores that, in turn, continue to bud profusely. Also, blastospores are budded out both terminally and laterally from hyphae of the dikaryotic mycelium (Fig. 12). The yeast-like development is favored by media such as LY and cornmeal-dextrose agars that contain sugars, but it is much less profuse on HI agar. The mycelium can be prevented from producing blastospores by growing it on HI agar and covering it with a block of agar so that it does not become exposed to air. This is somewhat surprising in view of the fact that much of the blastospore production in uncovered colonies occurs on hyphae sunken in the agar medium.

In agar colonies basidia arise mostly from branching clusters of hyphae with clamp connections (Fig. 9). The thin-walled probasidia are typically slightly broader than the hyphae bearing them and contain a denser cytoplasm (Figs. 9–11). By elongating, they develop directly into mature basidia, with or without a slight demarcation between the original probasidial boundary and the remainder of the basidium (Fig. 10). During spore formation the basidium becomes empty of protoplasm (Figs. 6, 11). The basidiospores are not forcibly discharged but are easily detached. Ballistospores are not found in *F. floriforme.*

The seven basidiospores of a single basidium were isolated onto agar, and seven yeast-like colonies resulted, none of which have produced hyphae on various agar media after about three months of culture. Transfers from these colonies were mixed together in a small amount of water, after which sterilized florets of *Erianthus giganteus* were dipped into the suspension and transferred to plates of HI medium. On the following day dikaryotic hyphae were observed growing on the agar surface adjacent to the florets. Within two weeks basidia had developed on some of the florets and on the agar nearby.

In the original platings of the *Erianthus* inflorescence, it was noticed that a species of *Alternaria* was quite abundant and even grew intermixed with *F. floriforme* on the florets. When the above described experiment was carried out with the *Alternaria* inoculated in the vicinity of the florets, the development of dikaryotic hyphae and basidia on the florets and adjacent agar surfaces was notably increased. Microscopic examination of these cultures revealed that irregular branches of limited development, but often elaborate in appearance (Fig. 18), were arising from clamp connections on the dikaryotic mycelium, and some were applying themselves to the *Alternaria* hyphae (Figs. 16, 17). Since many of these *Alternaria* hyphae were either dead or in a dying condition, it seems clear that the basidiomycete was parasitizing its associate. Similar evidence of myco-

FIGS. 1–8. *Filobasidium floriforme.* Fig. 1. Basidia on dead grass floret. X63. Fig. 2. Basidia in agar culture. X195. Fig. 3. Colony on agar. X63. Fig. 4. Basidia in agar culture. X195. Fig. 5. Small colony on agar. X195. Figs. 6–8. Basidia and basidiospores from agar cultures. Eight spores are clearly shown in Figure 8. X643.

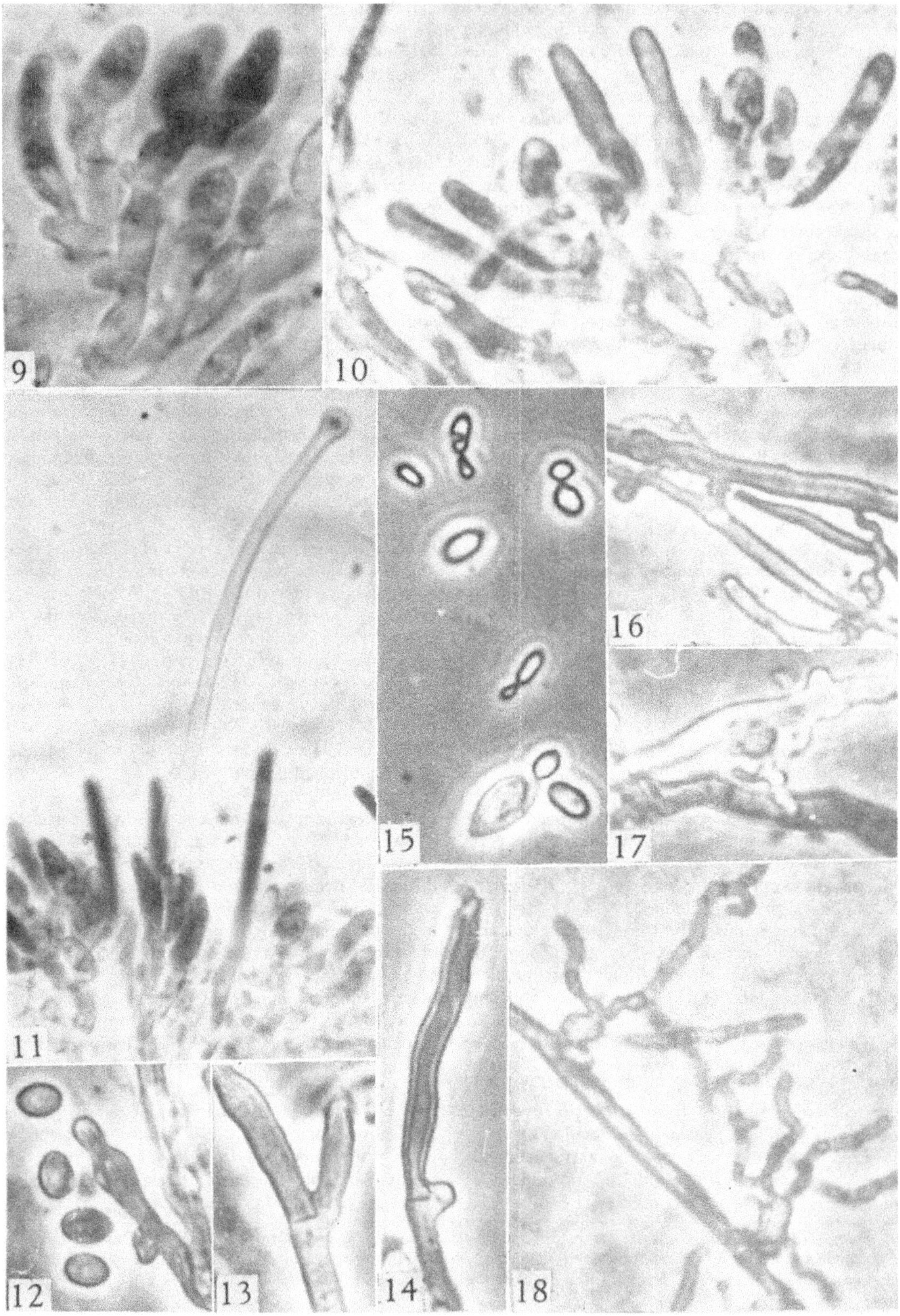
9
10
11
15
16
17
12
13
14
18

parasitism was reported by the writer (1946) for two tremellaceous fungi, and striking examples of it have been described recently for several of the higher basidiomycetes by Griffith and Barnett (1967). In all these species, special branches arising from the basidiomycete hyphae apply themselves to the hyphae of the host fungus. These branches were earlier referred to by the writer (1946) as "haustorial branches," but since they do not appear to extend themselves into the host cells in the manner of typical haustoria, it is now proposed that they be called *haustorioid branches*. It should be noted that *F. floriforme* develops and sporulates quite well on ordinary agar media in the absence of the *Alternaria*, and its parasitism must therefore be considered facultative. However, without the host fungus, the haustorioid branches of the basidiomycete are simpler and less abundant.

The seven previously mentioned single-spore cultures derived from a single basidium were mixed in various combinations on HI, LY, cornmeal, and water agars, as well as on a weak malt agar recommended for mating studies in the smut fungi. Only on certain batches of HI agar have reasonably satisfactory results been obtained in dikaryon initiation, and experiments are now under way to discover what factors may favorably influence the occurrence of dikaryotization. The meager information available clearly demonstrates that *F. floriforme* is heterothallic. When any member of a group of four cultures was mixed on HI medium with any member of the remaining group of three, the dikaryophase was initiated. This occurred by means of a short connecting tube between two compatible blastospores, a hypha with clamp connections then growing out from the fusion product. The first dikaryotic hyphae always start out by growing down into the agar medium beneath or at the edge of the colony of budding blastospores. These hyphae proliferate sparsely, and generally after a period of 1–2 weeks some of the hyphae grow back up to the agar surface and produce basidia singly or in small groups, sometimes within the bounds of the yeast-like colony but more often outside it. All dikaryons had true clamp connections and sporulated abundantly in colonies developing from transfers to fresh plates. Preliminary observations of single-spore cultures derived from several basidia indicate segregation in the basidium for factors affecting growth rate and viability. Some spores carried lethal factors, and after a short period of budding, all growth ceased.

Discussion

The possibility that the new fungus might belong to the family Cryptobasidiaceae described by Malençon (1953) has been considered and rejected. Members of this family are parasites of flowering plants (in Africa and South America), on which they cause hypertrophy. The mycelium is endophytic, and the fruiting pustules break through the overlying host tissue as the basidia develop. The basidia are elongate and nonseptate and bear up to eight sessile basidiospores. Each basidiospore is surrounded by a thick wall which encloses one or more separable chlamydospores at maturity. At germination each chamydospore gives rise to a filament that bears elongate budding sporidia. In *Coniodyctium chevalieri* Har. & Pat., the only species studied in any detail, amictic basidia arise from a mycelium with uninucleate cells. Sexuality, therefore, appears to be absent. Malençon suggests that the group is intermediate between the Exobasidiaceae and the Ustilaginales. Probably an equally convincing case could be made for allying the family more closely with the Ustilaginales, but its final dispensation must await further studies. In any event, the group seems too highly modified to permit the inclusion of *Filobasidium*.

From the time of the Tulasnes (1847), most students of the Ustilaginales have recognized two families in the order—the Ustilaginaceae, with basidiospores of indefinite number budded out laterally from a septate basidium, and the Tilletiaceae, with basidiospores formed terminally and simultaneously on a nonseptate basidium. In both families basidia commonly arise from persistent probasidia with more or less thickened walls. Recently, Fischer and Holton (1957), considering the existence of intermediate types of development, proposed combining both families into one, the Ustilagi-

Figs. 9–18. *Filobasidium floriforme*. Fig. 9. Probasidia at ends of branching hyphae with clamp connections. X1500. Fig. 10. Probasidia and young basidia. X1250. Fig. 11. Developing basidia and empty mature basidium. X563. Fig. 12. Blastospores budding out from tip of hyphal branch with clamp connection. X1700. Figs. 13, 14. Hyphae showing clamp connections. X1700. Fig. 15. Budding basidiospore and blastospores. X1700. Figs. 16, 17. Haustorioid branches of *F. floriforme* attached to hyphae of *Alternaria*. X1700. Fig. 18. Haustorioid branches with clamp connections at their bases, from cultures with *Alternaria*. X1700. All photographs except Figures 9 and 11 were taken with a Zeiss phase-contrast microscope. Most of the material was mounted in lactophenol with cotton blue.

naceae. However, since the two types of basidial development appear to be distinct for most smut fungi, the conventional segregation of the order into Ustilaginaceae and Tilletiaceae is adopted here.

The decision to include the Filobasidiaceae as a third family of the Ustilaginales is based on the writer's belief that *Filobasidium floriforme* has many features in common with the smut fungi and that it fits less well in any other known order. Under the circumstances, the alternative course of erecting a new order to accommodate this single species seems inappropriate. Affinities with the smut fungi are clearly indicated by the basidial morphology, sessile basidiospores, blastospore production, and method of dikaryotization. It is not suggested that the new fungus be called a smut, but that it be considered a form sufficiently related to be placed in the same order with the true smuts.

Although smuts are generally considered as basidiomycetes with thick-walled probasidia of mostly intercalary origin appearing in black powdery masses, there are several genera in which the probasidia commonly arise terminally on the hyphae, and in a number of species of these genera they are thin-walled and pale to colorless. In *Entyloma* these thin-walled probasidia may, with little or no resting period, give rise directly to basidia within the host tissue. As in *Filobasidium floriforme,* these basidia are slender and non-septate and bear the sessile basidiospores terminally. In addition, a dikaryotic mycelium with clamp connections has been described in *Entyloma.* Finally, the existence of mycoparasitism in *F. floriforme* may be a further indication of relationship with the smut fungi, which, though parasitic on higher plants, may also be readily cultured in the laboratory. A few have even been able to complete their life cycle in artificial culture. For citations and further information on these subjects, the reader is referred to the comprehensive treatments of Fischer (1953) and Fischer and Holton (1957).

The mating reactions between single-spore cultures of a single basidium of *F. floriforme* reveal only that the fungus is heterothallic. Further analyses are required to determine whether one or two loci for mating type exist, and additional collections are needed to test for the possible existence of multiple alleles for mating type. In view of the conflicting information on this subject in the true smuts, an intensive study of the compatibility mechanism in *F. floriforme* would seem advisable, with the hope that it might help clarify this difficult problem for the Ustilaginales as a whole. Such studies are now in progress.

The phylogenetic position of the new fungus is uncertain. If, as is generally believed, the septate basidium is more primitive than the nonseptate, then *F. floriforme* must be considered a derived rather than primitive form within the Ustilaginales. If so, then it may very well have had a tilletiaceous ancestor with many of the charasteristics of *Entyloma.*

Summary

A new heterobasidiomycete, *Filobasidium floriforme* nov. gen. et sp., with slender, nonseptate basidia that bear 6–8 sessile basidiospores terminally, has been described. Persistent probasidia are absent. The fungus is placed in a newly erected family, the Filobasidiaceae, of the Ustilaginales. Yeast-like colonies of budding cells develop from single-basidiospore isolations. The origin of dikaryotic hyphae with clamp connections from anastomosing blastospores in mixed compatible platings demonstrates that the fungus is heterothallic. It is capable of saprophytic growth as well as mycoparasitism on *Alternaria.*

Type material of *Filobasidium floriforme* has been deposited in the mycological collections of the New York Botanical Garden, and cultures of the organism are being maintained in our laboratory.

LITERATURE CITED

GRIFFITH, N. T., AND H. L. BARNETT. 1967. Mycoparasitism by basidiomycetes in culture. Mycologia **59**: 149–154.

FISCHER, G. W. 1953. Manual of the North American Smut Fungi. Ronald Press, N.Y.

FISCHER, G. W., AND C. S. HOLTON. 1957. Biology and Control of the Smut Fungi. Ronald Press, N.Y.

MELANÇON, G. 1953. Le *Coniodyctium chevalieri* Har. & Pat., sa nature et ses affinités. Bull. Soc. Myc. France **69**: 77–100.

OLIVE, L. S. 1946. New or rare heterobasidiomycetes from North Carolina. II. Jour. Elisha Mitchell Sci. Soc. **62**: 65–71.

———. 1967. The Protostelida—a new order of the Mycetozoa. Mycologia **59**: 1–29.

TULASNE, L. R., AND C. TULASNE. 1847. Mémoire sur les Ustilaginées comparées aux Urédinées. Ann. Sci. Nat. Bot., Sér. 3. **7**: 12–127.

Genetic Regulation of Sexual Morphogenesis in *Schizophyllum commune*

JOHN R. RAPER AND CARLENE A. RAPER
The Biological Laboratories, Harvard University, Cambridge, Mass.

The mating system in the Homobasidiomycete *Schizophyllum commune* stands in stark contrast to that of most other organisms. The worldwide population of this species consists of almost 42,000 distinct mating types, and any individual is continuously competent to mate with more than 98 per cent of all other individuals.

The earlier work on the mating system in *S. commune* and related higher Basidiomycetes sought primarily an understanding of its genetic basis. Each stage of genetic clarification, however, has revealed new opportunities for a study of the sexual cycle, and the entire system has now been exposed to the point that it constitutes an exceptional example of genetic regulation of a morphogenetic progression in an eukaryotic organism.

This brief paper is a synthesis of recent studies too numerous to cite individually. A comprehensive treatment of the various topics summarized here as well as a complete bibliography are available in Raper (1966), and only more recent relevant work will be cited.

The regulatory system consists of two clearly separable parts: the *regulating component* and the *regulated component*. The regulating component comprises four primary loci, each with an extensive series of multiple alleles, grouped in two interacting factors, the *A* and *B* incompatibility factors. The *A* and *B* factors serve to initiate or to inhibit two interrelated morphogenetic progressions. The regulated component consists of a large but unknown number of genes, scattered throughout the genome, that function in the transformation of two compatible haploid plants into a dikaryon. The dikaryotic state is genetically and physiologically equivalent to the diploid, with the critical difference that the two genomes in each dikaryotic cell are indefinitely maintained in two separate nuclei.

To understand the nature of the regulatory system, it is first necessary to examine what, in fact, is regulated. The haploid, homokaryotic mycelium is composed of uninucleate cells with simple septa (Fig. 2) and contains nuclei of only a single genotype, which includes one *A* factor and one *B* factor. When two such mycelia meet, contiguous cells of the two fuse as a matter of course to establish binucleate fusion cells. When the nuclei brought together in the fusion cells carry different *A* factors and different *B* factors, there ensues a sequence of events that converts both mates into a fertile heterokaryon, the dikaryon. This sequence comprises six distinct stages (traced by the central vertical line in Figure 1 and illustrated at the right), and these may be briefly described as follows:

1. *Nuclear migration.*—The nuclei of the fusion cells pass into the hyphae of both mates and migrate (presumably through progressively disrupted septa) throughout both pre-established mycelia.
2. *Nuclear pairing.*—When the migrating nuclei reach the apical cells of either mycelium, a pairing of the migrating nucleus with the nucleus already present is established in each apical cell.
3. *Conjugate division.*—This connubial arrangement persists throughout the growth of the dikaryon and is maintained through all subsequent cell divisions by the synchronous division of the two nuclei of the dikaryotic pair.
4. *Hook-cell formation.*—Early in the process of conjugate division, a short, lateral outgrowth, directed basipetally, develops on the dividing cell, and into this hook-cell moves a daughter of one of the dividing nuclei.
5. *Septation of apical cell and hook-cell.*—Shortly thereafter, a septum is laid down across the apical cell immediately proximal to the hook-cell, and, simultaneously, a septum is laid down across the base of the hook-cell temporarily to entrap the nucleus.
6. *Fusion of hook-cell.*—The tip of the hook-cell then fuses with the subapical cell, and the entrapped nucleus is released into the subapical cell to join the daughter of the other divided nucleus of the original pair. The fused hook-cell apparatus is known as the clamp connection.

A dikaryon is thus established in each cell of the growing mycelium, and the nuclei of the two parental types are indefinitely maintained at an exact ratio of 1:1. The dikaryon is thus a highly specialized heterokaryon constituted of binucleate cells and embellished with clamp connections at all septa (Fig. 2).

All of these events occur only when the two mates have neither *A* nor *B* factor in common,

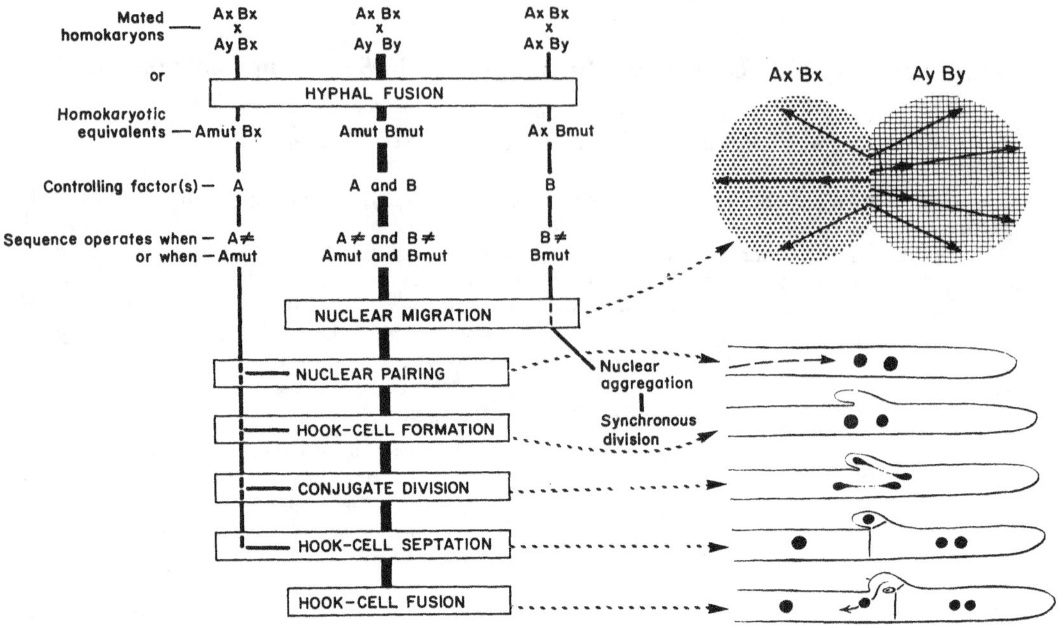

FIG. 1. Incompatibility-factor control of the stages in sexual morphogenesis in *Schizophyllum commune*. The sequence of events leading to the establishment of the dikaryon in a compatible mating ($A{\neq}B{\neq}$) is traced by the heavy central vertical line. The partial sequences evoked by the A and B factors separately in the $A{\neq}B{=}$ and $A{=}B{\neq}$ heterokaryons are traced by the thinner lines to left and right. The events comprising the morphogenetic progression are diagrammed on the right (from Raper, in press).

$A \neq B \neq$. When both A and B factors are common to the two mates, $A = B =$, none of the sequence takes place. The mycelia of the mated homokaryons remain essentially unchanged—i.e., their hyphae remain simply septate with uninucleate cells and no clamp connections (Fig. 2a). When either A or B factor is common, however, a heterokaryon forms that differs significantly from the dikaryon. In each case, only certain of the events of the morphogenetic sequence outlined above ensue: when the A factor is common, only nuclear migration occurs; when the B factor is common, there is no extensive nuclear migration, but all subsequent stages except the final one take place (Fig. 1). These facts signify that most of the morphogenetic process is constituted of two partial but interlocking sequences that are separately regulated by the A and B factors. The final stage, the fusion of the hook-cell, requires heterozygosity for both factors and is thus regulated by both.

The partial sequences operating in these common-factor interactions determine the characteristics of the resulting heterokaryons. The $A = B \neq$ interaction, in which only the B sequence operates, involves not only rapid heterokaryosis by nuclear migration but a continuing alteration in septal structure or stability. This and the lack of control over nuclear pairing result in the formation of a heterokaryon with cells of varying length and of variable nuclear content. The number of nuclei per cell ranges from 0 to 25 or more, and the aggregated nuclei appear to divide synchronously. The ratio of the two constituent nuclear types also varies greatly between $A = B \neq$ heterokaryons, from about 1:1 to thousands: 1. The $A = B \neq$ heterokaryon appears sickly and has short, irregularly shaped branches and relatively few aerial hyphae (Fig. 2). The $A \neq B =$ interaction, in which only the A sequence operates, involves little or no nuclear migration, and heterokaryosis is normally restricted to the immediate vicinity of the initial fusion cells. $A \neq B =$ heterokaryons synthesized from homokaryons carrying complementary biochemical mutations can be grown and maintained under the selective pressures imposed by a medium that lacks the specific nutrilites required by each mate. In such heterokaryons, nuclear pairing and conjugate division provide each apical cell with a nucleus of each of the two constituent types, but the failure of the last stage of the morphogenetic sequence leaves a nucleus permanently trapped in each hook-cell.

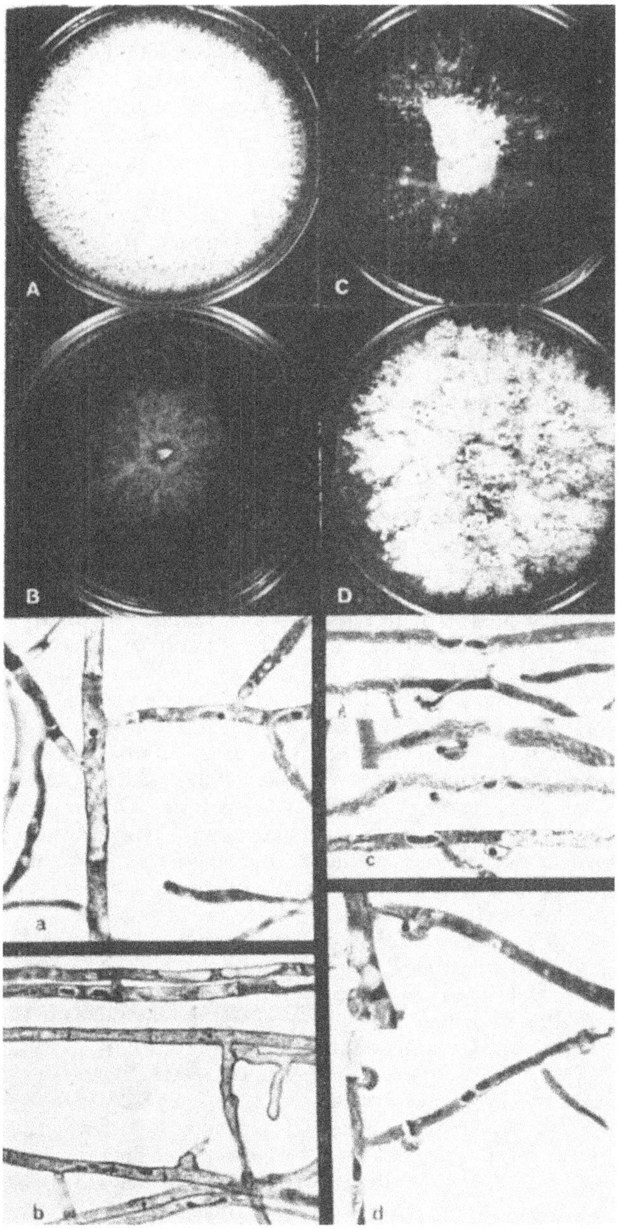

FIG. 2. Mycelial and hyphal characteristics of different mycelial types of *Schizophyllum commune*. Aa. Homokaryon: uninucleate cells and simple septa. Bb. $A{=}B{\neq}$ heterokaryon: sparse, depressed mycelium; cells of varying length and nuclear content (0–25); simple septa. Cc. $A{\neq}B{=}$ heterokaryon: nutritionally forced heterokaryotic growth from two biochemically deficient strains on minimal medium; growth sparse; binucleate apical cells with conjugate division; failure of fusion of hook-cell results in nucleate pseudoclamps; uninucleate subapical cells. Dd. $A{\neq}B{\neq}$ heterokaryon (dikaryon) with young fruiting bodies; cells regularly binucleate; clamp connections at all septa (from Raper and Raper, 1966).

The $A \neq B =$ heterokaryon is thus composed of uninucleate cells, save the binucleate terminal cells, and the over-all nuclear ratio is approximately 1:1 (Fig. 2). $A = B \neq$ and $A \neq B =$ heterokaryons typically produce no fruiting bodies and are thus sexually infertile.

The actual expression of the several stages of sexual morphogenesis, which are determined by the proper combination of unlike incompatibility factors, is under the further control of numerous genes distributed throughout the genome. Mutations in several of these genes have been found to disrupt the morphogenetic progression in various ways. Seventeen such "modifier mutations" have been characterized to date. They occupy a number of distinct loci, segregate normally, and assort independently of both *A* and *B* incompatibility factors. They belong to six different types, I–VI, each with its own pattern of dominance or recessiveness to the corresponding wild-type allele, interaction with other modifying genes, and specific effects on the developmental progression either in the A sequence, in the B sequence, or in both sequences. Whereas none of the modifiers is expressed in the homokaryons, their effects in heterokaryons indicate an intricate relationship among the events comprising dikaryosis (Raper and Raper, 1966; Koltin, 1967). Their specific and sometimes mutliple effects include the prevention of: nuclear migration (types IV and VI), nuclear pairing (I, II, and V), fusion of hook-cells (II), and nuclear aggregation in the B sequence (I, II, VI, and V), as well as the abnormal induction of hook-cells in the B sequence (I, II, III, and V). These effects of the modifier mutations permit several conclusions about the nature and interdependence of the specific stages in the progression: (1) Nuclear migration is not a necessary antecedent to subsequent events. (2) Nuclear aggregation is blocked in the B sequence by the operation of the A sequence and is independent of nuclear migration. (3) Hook-cell formation does not depend on nuclear pairing. (4) Conjugate division is probably a consequence of nuclear pairing. (5) Hook-cell septation is probably a consequence of conjugate division of paired nuclei, but it can also occur independently of conjugate nuclear division. (6) The control of hook-cell fusion, an integral component of both the A and B sequences, is independent of the controls of the preceding events. Continued study of the genetic control of these events, with particular emphasis on physiological and biochemical approaches (see below), should provide a better understanding of the course of sexual development.

The regulating component of this system, the *A* and *B* incompatibility factors, has a less complex genetic basis. The two incompatibility factors are located on different chromosomes, and each is constituted of two linked loci, α and β, each with multiple alleles. Insofar as is now known, both a single basic structure and a common pattern of intrafactor interactions apply to both factors. Linkage between $B\alpha$ and $B\beta$ is generally closer than that between $A\alpha$ and $A\beta$, but linkage of the loci within both factors is extremely variable: <.03–8 crossover units for B α-β, and 1–40 crossover units for A α-β. The *B* factors comprise three distinct classes with respect to linkage between α and β, and these distinctions are interpreted as being due to differences in the relative positions of the two loci (Koltin, 1967). Otherwise, the high degree of variation in recombination between the α's and β's within each factor, *A* and *B*, is controlled by genes that are external to the factors and specific for each factor (Simchen, 1967; Judith A. Stamberg, unpublished).

The number of alleles identified at $A\alpha$ and $A\beta$ is 9 and 26 respectively, and the projected number of alleles in the total worldwide population is 9 $A\alpha$'s and ca. 50 $A\beta$'s. A unique allelic combination of an $A\alpha$ and $A\beta$ determines a unique *A*-factor mating response, or *A*-factor phenotype, which can be identified only by mating tests. Differently constituted *A* factors are completely interactive—i.e., compatible, when they have different alleles either at $A\alpha$, at $A\beta$, or at both $A\alpha$ and $A\beta$. The *A* factor thus has about 450 alternate and equivalent phenotypes. The allelic series in the component loci of the *B* factor appear to be somewhat less extensive. The number of alleles identified at $B\alpha$ and $B\beta$ in the class of *B* factors with the highest frequency of recombination, Class I, is 7 and 7 respectively, and these numbers project to only 7 or 8 α's and 7 β's for the worldwide population. There are, therefore, about 50 Class I *B* factors. The numbers of α and β alleles in the other two classes of *B* factors have not yet been determined, but on the basis of the frequency and pattern of observed repeats in the sample available, it is estimated that the worldwide population comprises about 43 *B* factors of Classes II and III (Koltin, 1967). The total number of *B* factors in all three classes, therefore, is about 93. Since each combination of a specific *A* factor and a specific *B* factor constitutes a unique mating type, 450 *A* factors x 93 *B* factors yield a total of about 41,850 mating types in the worldwide population of *S. commune*.

The means by which the extensive series of

alleles at each of the incompatibility loci could have evolved is a persisting puzzle. It seems reasonable to assume that new alleles can originate from existing alleles either by recombination or by mutation. Attempts to demonstrate the former have failed: the evidence to date suggests that intralocus recombination does not occur at a frequency greater than 10^{-5}. Attempts to demonstrate the latter have *almost* succeeded: several new alleles have been derived by mutation from known alleles at the β locus of the *B* factor, but these, in their interactions with their progenitor alleles, are not quite equivalent to wild-type alleles. Even this partial success appears to depend upon at least two successive mutations in the same locus (Raper and Raudaskoski, in press). All of the known and specifically located primary mutations of the incompatibility alleles in both *S. commune* and the related species, *Coprinus lagopus* (some 22 to date) have occurred in only two of the four incompatibility loci, *A*β and *B*β, and these, without exception, have resulted in the loss of the normal discriminatory functions of self-recognition and self-incompatibility of the affected factor. Thus, in a homokaryon carrying a mutant *B*β allele, the B sequence functions, and the homokaryon is a phenotypic mimic of an $A = B \neq$ heterokaryon; a mutant *A*β permits the operation of the A sequence and determines a mimic of an $A \neq B =$ heterokaryon; a homokaryon carrying both a mutant *A*β and a mutant *B*β is phenotypically a dikaryon. Some secondary mutations of the *B*β locus have been shown to restore normal morphology and self-sterility in the homokaryon and to be compatible with the original, progenitor allele. These are the mutations that, except for certain pecularities in their incompatibility relationship with the progenitor—e.g., unilateral mating behavior, closely resemble new functional alleles.

Some recent, although preliminary, information about the biochemistry of the morphogenetic progression is pertinent to any interpretation of the regulating and regulated components as an integrated system. A few relevant facts have emerged, for instance, from a comparison of the proteins of the homokaryon with those of the dikaryon by serological and electrophoretic techniques. No constant or significant differences have been found between the protein spectra of two isogenic, compatible homokaryons, i.e., homokaryons with identical genotypes except for the alleles at all four incompatibility loci. When these spectra are compared with the spectrum obtained from the dikaryon synthesized from these homokaryons, however, numerous and significant differences are evident (C. S. Wang, unpublished). Specific proteins are present in the homokaryons that are absent in the dikaryon and vice versa, and these differences appear to relate, not to the differences in the specificity of the particular incompatibility factors present, but to the operation or inactivity of the morphogenetic sequence. Mutations that affect the system, both in the incompatibility factors and in the morphogenetic sequence, are currently being introduced into isogenic strains, and the protein spectra of these will be compared in order to test a number of specific predictions about the system. Another recent study has correlated the activity of a specific enzyme, the substrate of which is a component of the cell wall, with the operation of the B sequence (Wessels and Niederpruem, 1967). Here also, it is expected than an analysis of the effect of certain mutations, such as modifiers of nuclear migration, upon the activity of the enzyme will help to elucidate the part of the sequence involving septal instability and nuclear migration.

The mode of operation of the incompatibility factors in regulating sexual morphogenesis in the higher Basidiomycetes has long been the subject of speculation. Enough information has now become available, particularly in recent years, to permit the formulation of a model of "least implausibility" that is consistent with all the known features of the system. The proposed model is based on three postulates: (1) the genes of the *A* and *B* factors serve as master regulators for two parallel and interrelated biochemical pathways that underlie the events comprising dikaryosis; (2) the *A* and *B* factors inhibit the operation of these pathways in the homokaryon, with the result that the homokaryotic status is indefinitely maintained in isolation; and (3) in the interaction of paired, different *A* or *B* factors, the inhibition is released to permit $A = B \neq$ or $A \neq B =$ heterokaryosis when only one factor is different and dikaryosis when both are different.

The model as applied specifically to the *A* factor, for instance, would cast the *A*α and *A*β loci in the roles of regulator genes, the products of which control the A sequence. Each locus is assumed to elaborate a gene product, and the specific products of the two loci may be designated *Pa1, Pa2,* . . . , and *Pb1, Pb2,* . . . Cytoplasmic interaction between these primary gene products must also be assumed to account for the action of the *A* factor. The interactions involving *Pa* and *Pb* are of two types: (1) between the products of *A*α and *A*β of the *same* genome, e.g., *Pa1–Pb1, Pa1–Pb2,* etc., and (2)

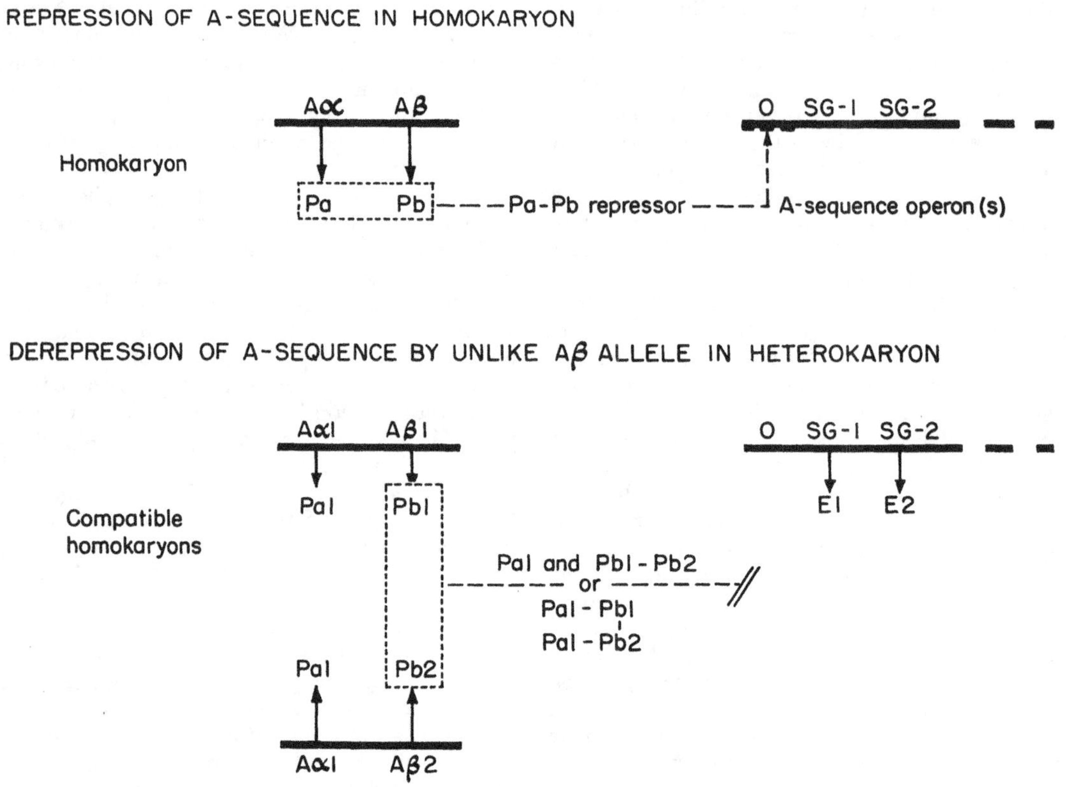

FIG. 3. Model for the regulatory role of the *A*-incompatibility factor. The two loci of the *A* factor, *Aα* and *Aβ*, behave as dual regulator genes, the products of which interact to form a repressor. In the interaction of different *A* factors, the products of unlike alleles at either locus interact to prevent the formation of the repressor with the consequent derepression of the *A* sequence (from Raper, 1966).

between products of the *Aα*'s or the *Aβ*'s of *different* genomes, e.g., *Pa1–Pa2* and *Pb1–Pb2*. The first interaction occurs continuously in the homokaryon, and its product represses the A sequence. It is assumed that all *Pa*'s are equally interactive with all *Pb*'s. The second interaction occurs continuously in heterokaryons with compatible *A* factors and disrupts the formation of the A-sequence repressor. It is assumed that: (1) this interaction prevails over the *Pa–Pb* interaction; (2) only different *Pa*'s or different *Pb*'s can interact; and (3) interaction between either different *Pa*'s or different *Pb*'s is sufficient for derepression of the A sequence (Fig. 3).

The primary mutations of *Aβ* are interpreted as alterations that result in the loss of the repressive activity of *Pa–Pb*. Their characteristic universal compatibility, however, suggests that they have retained the competence to derepress by interaction with the products of other wild *Aβ* alleles and have assumed, in addition, a new competence to interact with the product of the progenitor allele. The secondary mutations, which have been found to date only in the *Bβ* locus, are interpreted as alterations that (1) restore the repressive activity of *Pa–Pb*; (2) retain the competence to interact with other *Pb*'s; and (3) retain the competence of the primary mutation to interact with the progenitor *Pb*. They may be designated as new specificities of the *Bβ* allele.

Within this concept, the modifier mutations represent alterations in the structural genes that are responsible for the specific steps of the A and B sequences. The normal activity of these genes is regulated by the products of the appropriate combination of incompatibility alleles. The simplest explanation for repression and derepression of a whole array of genes would invoke at least one operon (probably several operons) for each of the incompatibility factors. The model is thus a modification of that proposed by Jacob and Monod for the regulation of β-galoctosidase synthesis in the bacterium *Escherichia coli*.

To rationalize these features in the context of a regulator-gene model as applied to the incompatibility factors of the higher fungi with their extensive series of multiple alleles requires versatile gene products. Each product must be capable of reacting in two specific manners, and, in addition, the product of one interaction, *Pa–Pb*, must be competent to interact with a repressible site(s) to regulate the appropriate sequence. Allosteric proteins could presumably accommodate these different activities, but further speculation serves no useful purpose.

Other models have been proposed and considered. For example, *Pa* and *Pb* may be precursors of an activator that induces the reactions comprising the A or B sequences. These precursors would be inactive in the homokaryon. Different *Pa*'s or different *Pb*'s in heterokaryons would interact to form an activator, which, in turn, would induce the relevant sequence.

No clear choice can now be made between these two types of regulation, as either may be involved on the basis of present evidence. The *repressor*-model is currently preferred because it accommodates the simpler interpretation of the numerous primary mutations of the incompatibility loci as "loss" mutations with respect to repressive activity. Within the confines of the *inducer*-model, all such mutations would have to be considered alterations that result in the formation of a specific activating substance.

In justification of such speculation, certain lines of investigation, primarily genetic, may be applied to test the validity of the proposed model. The fitness of the repressor-model over an inducer-model would be upheld if, for example: (1) $A\alpha$ and $B\alpha$ mutations have identical effects to those of $A\beta$ and $B\beta$; (2) deletion of any incompatibility locus has the same effects as the present primary mutations but are "recessive," i.e., will not interact with other alleles of the same series; and (3) recombination between mutant alleles and between mutant and wild alleles reveals separable functions. Both models predict that major distinguishable features in protein spectra should not be associated with specific incompatibility factors but rather with the events of sexual morphogenesis, e.g., the protein spectrum of an $A \neq B \neq$ heterokaryon should be essentially like that of a homokaryotic strain carrying primary mutations of both *A* and *B* factors.

Sexual morphogenesis in the higher Basidiomycetes is an admittedly complex business, and it would seem that a complete understanding of the system at the biochemical level is a very distant if not hopeless goal. Yet, as a morphogenetic process in an eukaryotic organism, it is ideally simple, as it involves only the transition from a uninucleate to a binucleate state. In other cases, less complex genetic systems may underlie grander morphological developments. This, however, appears unlikely, as it is only reasonable to expect a direct relationship between the magnitude of morphological change and the complexity of its genetic control.

ACKNOWLEDGMENTS

Much of the work summarized here has been done at Harvard University and has been supported by grants from the U. S. Public Health Service and the National Science Foundation.

REFERENCES

KOLTIN, Y. 1967. Structure and function of the *B* factor of *Schizophyllum commune*. Unpublished thesis, Harvard University, Cambridge, Mass.

KOLTIN, Y., J. R. RAPER, AND G. SIMCHEN. 1967. Genetics of the incompatibility factors of *Schizophyllum commune*: the *B* factor. Proc. Nat. Acad. Sci. U. S. **57**: 55–62.

RAPER, C. A., AND J. R. RAPER. 1966. Mutations modifying sexual morphogenesis in *Schizophyllum*. Genetics **54**: 1151–1168.

RAPER, J. R. 1966. Genetics of Sexuality in Higher Fungi. Ronald Press, New York.

RAPER, J. R. In press. Growth and reproduction of fungi. *In* F. C. Steward, ed., Plant Physiology, a Treatise. Vol. VI. Academic Press, New York.

RAPER, J. R., AND M. RAUDASKOSKI. In press. Secondary mutations at the $B\beta$ locus of *Schizophyllum*. Heredity.

SIMCHEN, G. 1967. Genetic control of recombination and the incompatibility system in *Schizophyllum commune*. Genet. Res. Camb. **9**: 195–210.

WESSELS, J. G. H., AND D. J. NIEDERPRUEN. 1967. Role of a cell-wall glucan-degrading enzyme in mating of *Schizophyllum commune*. Jour. Bacteriol. **94**: 1594–1602.

Further Studies on *Rhizopogon*. I[1]

ALEXANDER H. SMITH

Herbarium and the Department of Botany, University of Michigan, Ann Arbor, Mich.

It is now known from the work of Smith and Zeller (1966) that *Rhizopogon* is a large genus of around 140 species, and that its species show considerable diversity in color reactions with chemicals and with the air when the surface of the basidiocarp is injured. Such features in recent years have been given great prominence in the classification of both Hymenomycetes and Gastromycetes. In *Rhizopogon* Smith established the subgenus *Amylopogon* for species with amyloid spores. Certain species in which the spores were never amyloid at any developmental stage were also included, especially if a particular pattern of pigmentation and correlated chemical reactions were present. *R. chamaleontanus,* described herein, is of great interest in this respect. Sections of the gleba mounted in Melzer's showed individual spores that were dark violet in color in the cavities near the peridium while spores in the interior cavities were merely dingy ochraceous in the same mount. To my knowledge, such a situation has not been previously reported. This situation becomes of considerable theoretical interest relative to recognizing genera on the basis of the iodine reaction of the spores, as has been suggested for *Aleurodiscus.* Although the situation described for the spores of *R. chamaleontinus* appears to be constant for the species, to me it lends further weight to not recognizing *Amylopogon* as a genus—as some of my collegues have suggested (personal communication). However, it does necessitate a re-evaluation of the *R. atroviolaceus* group. Some of these species show a limited number of dark violet spores in a mount. It has been assumed previously that these were the most mature spores.

The specimens cited are deposited in the University of Michigan Herbarium. Color terms within quotation marks are taken from Color Standards and Color Nomenclature, R. Ridgway, 1912. Melzer's solution, the standard reagent for demonstrating a blue reaction with iodine for various hyphal components is simply referred to as Melzer's.

[1] I wish to acknowledge financial aid from Grant GB–2902 from the National Science Foundation for support of this project. The work on *Rhizopogon* was incidental to collecting species of *Pluteus.* I wish also to thank the Inter-Mountain Forest and Range Experiment Station of the U. S. Forest Service, U. S. Department of Agriculture, Mr. Joseph Pechanec, director, for its cooperation in furnishing accommodations and facilities for my party at the Priest River Experimental Forest for the season of 1966.

Rhizopogon alkalivirens Smith

This species was collected in woods at Cusick, Washington, on October 8, 1966, by K. A. Harrison (Smith 74030)—the first report of the species since it was described from material collected near New Meadows, Idaho in 1964. The large versiform spores in revived material are distinctly and persistently amyloid. When fresh, the gleba was violet where touched with a drop of Melzer's. The peridial surface, where touched with KOH, changed to pinkish and then olive; where touched with $FeSO_4$, it changed slowly to olive. As revived in KOH, sections of the peridium are brownish vinaceous, and there is considerable brownish vinaceous debris in the layer. In Melzer's dingy brown to orange-brown pigment balls are numerous in the peridium and in the mount generally. The gleba is fragile as dried and relatively soft, whereas the gleba in *R. griseogleba,* a rather similar species, is very hard. When fresh, the above-cited collection was mistaken for *R. idahoensis.* The interesting problem here is whether these species with the giant spores are producing basidiospores or conidia.

Rhizopogon avellaneitectus sp. nov.

Fructificationes 1–3.8 cm. latae, globosae, siccae, copiose fibrillosae, fibrillulus avellaneus; peridium demum incarnatum; sporae 6–7 x 2–2.3 µ, oblongae, inamyloideae. Specimen typicum in Herb. Univ. Mich. conservatum est; legit prope Cusick, Wash., 29 Sept. 1966, Smith *73737.*

Basidiocarp 1–3.8 cm. in diam., globose to flattened and irregular in outline, surface dry and matted loosely with a thin epicutis of grayish avellaneous fibrils, causing the basidiocarps to be very inconspicuous as they protrude through the litter of dead plant remains; beneath the fibrils is present and undertone of pinkish, with a few coarse rhizomorphs variously distributed over the surface; peridium in section pallid, becoming distinctly pinkish when cut and redder with the application of KOH; $FeSO_4$ on surface and on cut surface negative; C_2H_5OH and Guaic both dingy pink on surface; KOH on surface dingy brown; odor mild to slightly fragrant, taste mild. Gleba white to grayish to olive-buff, soft in white stage, cartilaginous when older, chambers labrynthiform. Columella none.

Spores 6–7 x 2–2.3 μ, smooth, with a false septum, oblong to narrowly subfusoid, yellowish hyaline in KOH and more yellowish in Melzer's. Basidia versiform, 7–10 μ broad, 4- or 6-spored, globose to clavate or subcylindric, hyaline in KOH; basidioles thin-walled but in oldest specimen thickened up to 1.5 μ and wall not gelatinous. Subhymenium cellular, the cells up to 12 μ diam., the hyphal walls not gelatinous. Tramal plates of hyaline interwoven gelatinous-refractive hyphae 3–6 μ broad (variously inflated and constricted) as revived in KOH. Peridium with an epicutis of appressed hyphae pale tan in KOH and with pale fulvous incrustations, hyphae 3–10 μ diam., and many with more inflated cells; beneath this a region continuing to cavities and with many inflated cells singly or in pockets or as cells in a filament (hence in a row), some cells with walls up to 1 μ thick, with orange-brown pigment balls in Melzer's in the peridium and mount, some hyphal fragments seen with partly amyloid walls (assumed to be of foreign origin). Clamp connections absent.

Scattered to solitary along old roads under *Pinus contorta* (some Douglas fir present in the area), near Cusick, Wash., Sept. 29, 1966, Smith 73737, type.

When collected, the species was thought to belong in section Villosuli, but the color of the epicuticular hyphae in KOH in both fresh and dried specimens rules this out. It is actually close to the *R. rubescens* group, but distinct because of the gray felt-like epicutis that develops early in the life of the basidiocarp. According to Smith and Zeller (1966), it would key out near stirps *Maculatus* but is neither of the two species placed there. It belongs in series *Versicolores* as a separate stirps.

Rhizopogon brunneo-fibrillosus sp. nov.

Fructificationes circa 5 cm. latae, globosae, valde brunneo-fibrillose reticulatae, in "KOH" rubrae demum badiae; sporae 5–6.5 x 2–2.2 μ, inamyloideae. Specimen typicum in Herb. Univ. Mich. conservatum est; legit prope Portland, Oregon, Oct. 1966, Oreg. Myc. Society (Smith, 74055).

Basidiocarp 5 cm. diam. globose or nearly so, with a conspicuous coating of brown (near cinnamon brown) rhizomorphs over a dingy ochraceous ground color that slowly changes to cinnamon buff by late maturity, not changing color when handled; KOH intensely brick red, going quickly to dark red-brown (dark bay brown); $FeSO_4$ slowly dark olive-brown; Guaic slowly reddish brown; C_2H_5OH brownish. Gleba olive-buff and retaining an olive tone in drying; chambers large, 2–4 per mm., consistency friable as dried. Columella lacking.

Spores 5–6.5 x 2–2.2 μ, oblong to narrowly ellipsoid, smooth, ochraceous buff in groups but nearly hyaline individually; in Melzer's pale dingy ochraceous and most with a false septum, thin-walled.

Basidia 12–25 x 7–12 μ, number of spores attached not ascertained. Basidioles resembling basidia and thin-walled (wall up to 0.5 μ thick); hymenium with the candelabra type of branching and hence the subhymenium not sharply distinct from the hymenium, the hyphae nongelatinous and 5–9 μ diam.; tramal plates of nongelatinous hyphae 4–12 μ diam., the cells often enlarged but the structure of the plate obviously filamentous; hyphal walls smooth, hyaline and 0.2–0.5 μ thick. Peridium with imbedded zhizomorphs but these not as numerous as in *R. ochoraceorubens,* of appressed hyphae 5–12 μ diam., hyphal cells often somewhat inflated but no sphaerocysts seen either singly or in groups, encrusting material present on walls and dingy vinaceous brown as revived in KOH, pigment pockets (probably massive incrustations) similarly colored (no red or olive coloration in material revived in KOH); in Melzer's with scattered dull orange-brown pigment bodies (globules or variously shaped bodies). Clamp connections none.

The type was sorted out from a mixed collection brought into the Annual Mushroom Show of the Oregon Mycological Society by an unidentified member. The type locality is somewhere in the Willamette River Valley.

This species differs from *R. libocedri* in that the dried basidiocarp did not darken in $FeSO_4$ to inky black; the spores had a false septum in Melzer's; the spores did not become at all dextrinoid; the subhymenium was nongelatinous (with branching of the candelabra type); and in the details of the peridium. No oleiferous hyphae were seen in *R. brunneo-fibrillosus.* From *R. fuscorubens* it differs in narrower spores, and in not having the rhizomorphs distinctively colored (to fuscous-red or magenta) as revived in KOH.

Its distinctive features as a species are the very small, narrow spores; the KOH reaction of fresh material (brick red, then bay) as contrasted with relatively little coloration as revived in KOH; the candelabra-type branching of the subhymenial hyphae, as well as their nongelatinous walls; and the dark, dull-brown rhizomorphs over a very dingy ochraceous ground color when fresh. The rhizomorphs do not show conspicuously on dried material. The crumbly consistency of the gleba and peridium (as dried) is such that sections made by free-hand technique are mostly fragments.

R. brunneo-fibrillosis is related to the *R. ochraceorubens* group.

Rhizopogon chamaleontinus sp. nov.

Fructificationes 1–2 cm. latae, globosae, vel irregulares, fibrilloso-subreticulatae, albidae tactu vinaceo-fuscae, demum atrofuscae, in "KOH" olivaceae demum atrae; sporae 6–9 x 3–4.5 (5) μ, leves, valde amyloideae demum inamyloideae. Specimen typicum in Herb. Univ. Mich. conservatum est; legit prope Hill's Resort, Priest Lake, Idaho, 24 Sept. 1966, Smith 73638.

Basidiocarps 1–2 cm. diam., globose to irregular; when young with numerous appressed fibrils and rhizomorphs but in age or as dried nearly glabrous; white when young, darkening to fuscous black in drying; $FeSO_4$ on surface quickly black, KOH olive then black; when handled staining fuscous to vinaceous fuscous. Gleba pallid, becoming wood brown and avellaneous as dried, chambers minute, consistency cartilaginous fresh, very hard when dry. Columella absent.

Spores variable, 6–9 x 3–4.5 (5) μ, elongate-drop-shaped to subelliptic or at times somewhat irregular; wall slightly thickened; when fresh the spores in the outer cavities (near peridium) violet black in Melzer's but those in the interior dingy ochraceous to hyaline (inamyloid). In revived material the same is true except that the spores in the outer cavities are dark violet and those in the interior are cinnamon buff in Melzer's (when strongly amyloid, this reaction may include the entire spore or part of it; some spores were seen with an inamyloid "plage"); as revived in KOH many spores are seen adhering to each other or with adhering debris.

Basidia 4–, 6-spored (not many clearly seen), 7–9 broad at apex, clavate, length variable. Basidioles apparently not becoming thick-walled. Subhymenium of subgelatinous hyphae, branching candelabra-like (not filamentose interwoven or cellular); hyphae hyaline in KOH and Melzer's, content in places rarely showing amyloid strands; hyphae of tramal plates gelatinous and interwoven, 4–7 μ diam. Hyphae of peridium 5–12 μ diam., the cells mostly elongate, some near the cavities vinaceous red in KOH as revived, this color at times extending to hyphae of the outer tramal plates; dark, dull orange-brown pigment balls and bodies numerous in Melzer's mounts but none violet; much debris in the layer (as seen in both KOH and Melzer's). Clamp connection none.

Priest Lake Idaho, Sept. 24, 1966. near Hill's Resort, Smith 73638.

This collection is worth placing on record because of the very clear pattern of amyloid reaction shown by the spores. Those in the outer cavities are dark violet viewed individually; the older ones in the interior cavities were inamyloid. This discovery was a distinct shock, because it had been assumed from studies of other species of *Rhizopogon* having dark violet spores in iodine that these were the older spores. The change is apparently one of maturation and appears to be rather sharply defined. The peculiar combination of features of this species is: coloration (white staining vinaceous to fuscous, as in *R. idahoensis*); the olive-to-black KOH reaction; the quick $FeSO_4$ change to black; lack of yellow tones at any time in the basidiocarp; candlebra-type branching of the subhymenium; and the spore features as discussed.

R. chamaleontinus appears distinct from *R. anomalus* in that few pale blue spores are present, and little material was seen in the interior of the tramal-plate hyphae.

Rhizopogon griseovinaceus sp. nov.

Fructificationes 2–2.5 cm. latae, subglobosae, siccae, minute areolatae, griseo-vinaceae; sporae 6–7.5 x 2.2–2.5 μ, inamyloideae, leves. Specimen typicum in Herb. Univ. Mich. conservatum est; legit prope Clear Lake, Lane Co. Oregon, 18 Oct. 1966, F. P. Sipe 1470.

Basidiocarps 2–3.5 cm. broad, subglobose to flattened, surface dry, minutely areolate, lacking obvious rhizomorphs, color tan when fresh (Sipe), as dried "ecru drab" (a medium vinaceous gray). Gleba olive buff as dried and very soft and friable (crumbly in attempts at sectioning), cavities labrynthiform. Columella lacking.

Spores 6–7.5 x 2.2–2.5 μ, narrowly elliptic to oblong, smooth, thin-walled, not truncate at place of attachment, hyaline in KOH, yellowish in Melzer's (mature dried specimens available).

Basidia 7–9 μ wide, of various lengths up to 25 μ long, hyaline, thin-walled, 6–7 spores seen attached (possibly 8-spored), merely yellowish in Melzer's solution. Basidioles thin-walled, hyaline. Subhymenium a nongelatinous cellular layer 1–2 cells deep, the walls thin and hyaline and content not distinctive. Tramal plates of hyaline thin-walled, nongelatinous hyphae 3–8 μ broad (or some cells inflated and broader), relatively few hyphae present, thus causing the plates to be very weak. Peridium with a trichodermial type of epicutis with the elements aggregating to form the small areolae or becoming appressed to the surface; the hyphal cells 9–15 (20) μ diam, often short (under 50 μ), the terminal cells elliptic to cystidioid, the walls greenish in KOH but fading to yellowish on standing, with dark olive rodlike particles in and between the cells in places; the subcutis of appressed hyphae at first, and also greenish in KOH as revived; tramal body of thin-walled hyaline hyphae closely interwoven, with some red pigment in the layer as revived in KOH, but this fading in a short time (15 minutes), inflated cells seen but no large pockets of sphaerocysts present; in Melzer's the epicuticular hyphae nearly hyaline but with dingy incrustations and a few broken pieces of hyphae partly amyloid, dark violet rods also present between the hyphae in places but these possibly the same as the olive particles seen in KOH mounts; tramal body with numerous orange-brown pigment balls in the layer as revived in Melzer's; no amyloid hyphae seen in the interior of the peridium. Clamp connections none.

The type was collected at Clear Lake, Lane County, Oregon, October 18, 1966, by Frank P. Sipe (1470). Dr. Sipe had gone to Clear Lake to search for another collection of the species named in his honor and found, among others, this one. It is obviously related to *R. semireticulatus* Smith but differs in the trichodermial type of epicutis, which is strongly olive as revived in KOH. The similarity is that the body of the peridium is red when first revived in KOH. The spores are shorter and narrower than for *R. semireticulatus* Smith.

In most species showing both the green and red reaction with KOH, the peridium is nearly fuscous as dried; this is not true of *R. grisiovinaceus*. The very fragile gleba as dried, the vinaceous gray color of the dried peridium, and the small but obvious areolae make a set of features for macroscopic recognition of dried specimens. $FeSO_4$ slowly blackens the dried peridium. Although the species must be placed in *Amylopogon* by all features save the reaction of the spores to iodine, it does show a slight similarity to members of the section *Villosuli* in the structure of the peridial epicutis.

***Rhizopogon hymenogastrosporus* Soehner**

Smith and Zeller (1966) overlooked an article by Soehner (1956) in which the above species was described. Since *R. hymenogastrosporus* appears to be close to *R. ventricisporus* Smith, the following translation of Soenner's description is given to enable the two species to be readily compared.

Fruit body in youth milk-white, as taken from the soil spotted pale olive yellow on a white ground color, when touched becoming spotted with dingy carmine-red, when dried dull brown, with the feel of a kid glove, more or less globose, hardly furrowed or warty, with a short rooting tuft of hairs at base, with a few appressed dark-colored rhizomorphs (pale reddish brown) toward the apex.

Peridium very thin, fresh 250–300 μ thick, of narrow, floccose, parallel hyphae.

Gleba white when young, then dingy yellow and toned greenish yellow, when dried gray-greenish brown; cavities irregular. Trama of large cells or broad-lumened (inflated ?) hyphae, in fresh material 60–100 μ thick.

Basidia 20–22 x 7–10 μ, sterigmata 3–5 μ long (but not shown in drawings) spores more or less cup-like at base.

Spores hyaline singly, in groups pale yellowish, a large number 8–10 x 5–6 μ, numerous Hymenogaster-like spores 10–13 x 7–9 μ also present, these versiform (with bumps, etc.), long ones often 16 μ long, 1–3 oil drops present; inverted cup-shaped at base.

Odor none.

In summer in spruce woods. Europe

R. ventricisporus dries ochraceous rather than brown, the subhymenium is of filaments 3–6 μ diam.; and Soehner does not mention rusty-ochraceous stains around worm holes in the gleba of his species. Probably other differences exist as well; they are therefore not considered synonymous on the basis of present information. *R. hymenogastrosporus* is of great interest mycologically, however. It and possibly *R. virens* are known from Europe and represent there the group in North America consisting of *R. hysterangioides, R. variabilisporus, R. griseogleba, R. clavitisporus,* and *R. subclavitisporus,* which all show the same type of variation in spore size and shape. A cytological study of this group would be most helpful to determine which of the spores are conidia and which are basidiospores.

Rhizopogon idahoensis Smith

During the 1966 collecting season, fifty-two collections of this species were made by Harrison and Smith in a woods of mixed conifers near Cusick, Washington. At the time this species was described Smith and Zeller (1966), Smith had certain reservations about it. Consequently the Cusick collections were a welcome find.

First, the species appeared to be associated with Douglas fir, but because of the mixture of conifers present, further observations are needed to establish the point. When the material in the fresh condition was sectioned and tests made with Melzer's reagent the spores appeared inamyloid under the microscope, yet when a drop of Melzer's was placed on the gleba, a greenish to blue or violet spot was noted. In the dried condition all of the collections had distinctly amyloid spores. None of the spores were dark violet as seen individually; instead they were pale blue, but in masses along the hymenium the blue reaction was very obvious and applied equally to both mature and immature spores. No difference in color was noted between the spores in the cavities near the peridium and those from the center of the basidiocarp. It now appears that the amyloid spores are the major distinction separating *R. idahoensis* from *R. subcaerulescens*. Thus distinguishing between fresh material of the two species may at times be difficult. In general, however, the basidiocarp in *R. subcaerulescens* var. *subcaerulescens* is more cartilaginous than that in *R. idahoensis*. The $FeSO_4$ reaction of *R. idahoensis* was consistently olive to olive-black or inky black, whereas in *R. subcaerulescens* var. *subcaerulescens* it is blue. The study of the Cusick specimens makes clear that the KOH reaction does not separate the two species. In coll. 74131, the glass tip of the dropper bottle was pressed lightly on the peridium and a streak of KOH made. The streak stained both blue and vinaceous—in different places. This agrees with the staining of the cut peridium—both colors were also evident but in different areas. In about forty of the collections, tested, the KOH reaction varied greatly between olive, blue, lilac, vinaceous, or (finally) fuscous. In one collection (Sm 73743), Melzer's gave no reaction on the white gleba; sectioning the gleba revealed that it was practically sterile.

A curious variation was observed in nos. 73926, 74121, 74123, 74125, and 74126. In these the hyphae of the tramal plates, when revived in Melzer's, showed an ochraceous to orange-ochraceous to orange-red homogeneous content. This type of reaction with Melzer's has been observed in variants of *Macowanites americanus* and in a number of species *Leccinum* (Boletaceae). It appears to be rather constant in *Leccinum* but not reliable in *Macowanites*; in *Rhizopogon* at present it can only be reported. The fact that all of the collections but one were found in a small area on a single day could easily mean that we are dealing with a recognizable variant, but the great range in the intensity of the color would suggest caution in using color as a taxonomic character, especially when it was apparently absent in fresh material.

Rhizopogon occidentalis Zeller & Dodge

Before the 1966 season, I had not collected or studied this species in the fresh condition with all stages of development of the basidiocarp present. During 1966 in a mixed conifer wood containing Douglas fir among other conifers, it was collected frequently and in quantity. As a result of the study then made, certain aspects of the species and its relationships to other species were ascertained.

First, it has been clearly demonstrated that the color and rhizomorph pattern of the dried basidiocarps are characteristic (ground color pale to dull lemon or greenish yellow and with a conspicuous over-lay of brownish appressed rhizomorphs). When the specimen is young and fresh, the color is white, and the rhizomorphs, if already developed, are a dingy honey color to dingy brown. The color changes and chemical tests were noted or made on basidiocarps in various stages of development. Collections 73647, 73653, 64037, 74036, 74033, 74040, 74046, 74045, and 74143 did not stain or change color when handled or bruised, and when sectioned the peridium typically remained white to yellow, depending on the color of the particular specimen. However, 74036 and 73146 stained reddish slowly, and in section the peridium also changed to red to some extent. In these collections KOH caused a slow change to scarlet and $FeSO_4$ a slight change to pale olive. In the collections that did not stain, the KOH reaction was typically red on the white stage and rarely red but usually reddish brown on the yellow stage. No brick red or reddish areas were observed to develop on any of the dried material.

The spore features were very constant for all collections and were as published by Smith and Zeller (1966).

As revived in KOH, the peridium is clean or with a slight amount of debris; in mature specimens the layer is lemon yellow in KOH. Inflated cells are present singly or in groups in the peridium throughout and the only amyloid hyphae seen were fragmentary, had clamps at the cross walls, and are considered as foreign mycelium, very likely from a species of *Chroogomphus.*

A study of the specimens cited demonstates that there are stages in the development of the basidiocarp when *R. ochraceorubens, R. occidentalis* and *R. sublateritius* are difficult to distinguish in the field. The following observations will aid in solving this problem: First, *R. sublateritius* is white at first and does not become bright yellow; instead, it becomes unevenly dingy yellow, at which point patches or areas of brick reddish also show. It is definitely positive with $FeSO_4$ with addition of C_2H_5OH. Although in the type the ground color of *R. sublateritius* is not bright yellow, in some of the 1966 collections the dried specimens have a bright yellow ground color about like that of *R. occidentalis,* but the brick reddish areas are clearly defined. Harrison 6355 appears to be an intermediate between the two species: in it the cuticle as revived is orange-brown in KOH in the surface area, the basidiocarps stain olive with $FeSO_4$ and as dried show only slight areas of brick reddish. It is filed as *R. sublateritius.*

In the dried condition *R. ochraceorubens* can be distinguished at once from both *R. sublateritius* and *R. occidentalis* because it becomes bay red after exposure to naphthalene flakes, and because the pattern of rhizomorphs is very indistinct in dried material. In sections mounted in KOH, the red to magenta reaction of the debris in the peridium and the fact that numerous rhizomorphs have visibly been incorporated in the peridium are diagnostic characters. In immature stages *R. occidentalis* and *R. ochraceorubens* are often easily confused in the field. To date, however, the evidence indicates that *R. occidentalis* is associated with Douglas fir, *R. ochraceorubens* with *Pinus contorta* and *R. sublateritius* with *Pinus ponderosa.* Since in Idaho these species of trees grow in mixed stands, ample opportunity exists for gene exchange between the populations, and the observations made in 1966 indicate clearly that this occurs. Benedict, Tyler, and Brady (1966) have postulated on the basis of the presence or absence of isoxazole compounds such as ibotenic acid that hybridization exists in the *Amanita pantherina* group. The results of the study presented here indicate a similar pattern of the distribution of certain pigments and thus lend support to the results of the above-mentioned authors.

Rhizopogon semitectus **sp. nov.**

Fructificationes 2–2.5 cm. latae, globosae, pallidae tactu vinaceo-brunneae fibrillulae demum cinnamomeo-brunneae; sporae 9–12 x 3–4.5 μ, inamyloideae, versiformes. Specimen typicum in Herb. Univ. Mich. conservandum est; legit prope Binarck Creek, Priest Lake, Idaho, 17 Sept. 1966, Smith 73443.

Basidiocarps 2–2.5 cm. diam., globose or nearly so, ground color white, covered with fine rhizomorphs, slowly developing an epicutis of colored hyphae causing it to be cinnamon brown and matted fibrillose, slowly staining dark vinaceous brown from handling; $FeSO_4$ on cutis olive fuscous; KOH greenish fuscous on surface, reddish next to the cavities. Gleba white becoming gray and in age cinnamon brown, chambers minute, consistency cartilaginous and rubbery, hard when dry.

Spores 9–12 x 3–4.5 μ, smooth, thin-walled, suboblong, some irregular or with one or two protuberances, dingy cinnamon buff in KOH and in Melzer's as revived and when fresh (not amyloid at any stage). Basidia 20–30 x 5–7.5 μ, 4-spored, spores sessile, thin-walled and hyaline but not gelatinous, occasional basidia with thickened walls hyaline to brownish; basidioles remaining thin-walled. Subhymenium of hyphae 2–4 μ diam. and in a palisade ending in the hymenium, not gelatinous and only somewhat refractive. Tramal plates of gelatinous and only somewhat refractive to highly refractive hyphae 2–5 μ diam. Peridium when mature with an epicutis of hyphae 4–15 μ diam.; the cells short to long, many inflated to 15–20 μ finally, the short cells keg-shaped to fusoid or clavate, the walls yellow ochre in KOH fresh and greenish as revived, with copious granules and debris adhering; walls 0.5–0.8 μ thick and hyaline in KOH as revived in the thick-walled cells; pigment balls dull brownish orange in Melzer's; inner layer of hyphae reddish in KOH and with some inflated cells. Clamp connections none.

Under a cedar shake, Binarck Creek, Priest Lake, Idaho, Sept. 17, 1966, Smith 73443.

This is a most unusual species. In some respects it recalls *R. subcaerulescens* var. *subpannosus,* especially in the manner in which the colored epicutis develops and in the fact that the hyphae of this layer are not colored as they are in section Villosulus. However, the spores are larger than those in *R. subcaerulescens* var. *subpannosus* and the details of the peridial epicutis are different. *R. semitectus* is closest to *R. subgelatinosus,* but it shows many more rhizomorphs than that species, and the gleba in the latter was never observed to be cinnamon brown when fresh and mature. Also, *R. semitectus* stains dark brown from handling. Finally, the basidia are gelatinous in *R. subgelatinosus* but not in *R. semitectus,* and in the latter spores were not observed to be amyloid at any stage of development. *R. semitectus* is somewhat similar to *R. clavitisporus,* but the latter has larger spores, a cellular subhymenium, and flagellate cells in the epicuticular layer of the peridium.

LITERATURE CITED

SMITH, ALEXANDER H., AND S. M. ZELLER. 1966. A preliminary account of the North American species of *Rhizopogon.* Mem. New York Bot. Gardens **14**(2) : 1–178.

SOEHNER, ERT. 1965. Süddeutsche *Rhizopogon.* Arten Zeitschr. Pilzk. **22**(3) : 65–80.

www.ingramcontent.com/pod-product-compliance
Lightning Source LLC
Chambersburg PA
CBHW081150290426
44108CB00018B/2505
9780807868775